TABLE OF ATOMIC WEIGHTS

Name	Symbol	Atomic number		
Mendelevium	Md	101	*	257
Mercury	Hg	80	200.6	201
Molybdenum	Mo	42	95.94	96
Neodymium	Nd	60	144.2	144
Neon	Ne	10	20.18	20
Neptunium	Np	93	*	237
Nickel	Ni	28	58.71	59
Niobium	Nb	41	92.91	93
Nitrogen	N	7	14.01	14
Nobelium	No	102	*	254
Osmium	Os	76	190.2	190
Oxygen	O	8	16.00	16
Palladium	Pd	46	106.4	106
Phosphorus	P	15	30.97	31
Platinum	Pt	78	195.1	195
Plutonium	Pu	94	*	244
Polonium	Po	84	*	210
Potassium	K	19	39.10	39
Praseodymium	Pr	59	140.9	141
Promethium	Pm	61	*	147
Protactinium	Pa	91	231.0	231
Radium	Ra	88	226.0	226
Radon	Rn	86	*	222
Rhenium	Re	75	186.2	186
Rhodium	Rh	45	102.9	103
Rubidium	Rb	37	85.47	85
Ruthenium	Ru	44	101.1	101
Samarium	Sm	62	150.4	150
Scandium	Sc	21	44.96	45
Selenium	Se	34	78.96	79
Silicon	Si	14	28.09	28
Silver	Ag	47	107.9	108
Sodium	Na	11	22.99	23
Strontium	Sr	38	87.62	88
Sulfur	S	16	32.06	32
Tantalum	Ta	73	180.9	181
Technetium	Tc	43	98.91	99
Tellurium	Te	52	127.6	128
Terbium	Tb	65	158.9	159
Thallium	Tl	81	204.4	204
Thorium	Th	90	232.0	232
Thulium	Tm	69	168.9	169
Tin	Sn	50	118.7	119
Titanium	Ti	22	47.90	· 48
Tungsten	W	74	183.9	184
Uranium	U	92	238.0	238
Vanadium	V	23	50.94	51
Xenon	Xe	54	131.3	131
Ytterbium	Yb	70	173.0	173
Yttrium	Y	39	88.91	89
Zinc	Zn	30	65.37	65
Zirconium	Zr	40	91.22	91

* The isotopic composition of natural or artificial radioactive elements usually varies in specific samples, depending upon their origin. For accurate weights of isotopes that have the longest half-life or are best known see the *Comptes Rendus* of the Montreal IUPAC conference.

Inorganic
Medicinal and Pharmaceutical
Chemistry

Inorganic
Medicinal and Pharmaceutical
Chemistry

JOHN H. BLOCK, Ph.D.

Associate Professor of
Pharmaceutical Chemistry
Oregon State University
School of Pharmacy
Corvallis, Oregon

EDWARD B. ROCHE, Ph.D.

Associate Professor of
Medicinal Chemistry
The University of Nebraska
College of Pharmacy
Lincoln, Nebraska

TAITO O. SOINE, Ph.D.

Professor of Medicinal Chemistry and
Chairman, Department of Medicinal Chemistry
University of Minnesota
Minneapolis, Minnesota

CHARLES O. WILSON, Ph.D.

Dean, School of Pharmacy and
Professor of Pharmaceutical Chemistry
Oregon State University
School of Pharmacy
Corvallis, Oregon

Lea & Febiger · 1974 · *Philadelphia*

Library of Congress Cataloging in Publication Data
Main entry under title:

Inorganic medicinal and pharmaceutical chemistry.

1. Chemistry, Medical and pharmaceutical.
I. Block, John H. [DNLM: 1. Chemistry, Pharmaceutical. QV744 I58 1974]
RS403.I48   615'.2   73–19604
ISBN 0–8121–0443–9

Preface

Inorganic Medicinal and Pharmaceutical Chemistry has been designed as a classroom textbook written with two purposes in mind. The first is to present a review of those principles of inorganic chemistry that apply to medicinal and/or pharmaceutical chemistry. In that regard, the first two chapters are devoted to explanations of atomic structure as it relates to bonding forces and complexation, and a summary of the important physical properties of each element group from the periodic table. The second purpose is to present detailed discussions of those inorganic agents used as pharmaceutical aids and necessities or as therapeutic and diagnostic agents. Those products used as pharmaceutical aids and necessities include acids and bases, buffers, antioxidants, water, and selected tableting aids. Inorganic compounds used therapeutically include products containing fluid electrolytes, biochemically important ions, and therapeutically important ions. Other inorganic products described are antacids, cathartics, topical agents, dental products, inhalants, antidotes, etc. Radiopharmaceuticals are discussed both as diagnostic and as therapeutic agents. The toxicity problems associated with some of the inorganic cations are reviewed.

The general format is to define the class of products under discussion, to describe the rationale for their use, and then to discuss the specific agents. The latter usually includes the official description of the product, contraindications, therapeutic and pharmaceutical incompatibilities where appropriate, the official use, and, in many cases, alternate uses. Pertinent references have been provided.

Those who have taught inorganic pharmaceutical chemistry will note the occasional use of an illustration and some of the text from the eighth edition of *Rogers' Inorganic Pharmaceutical Chemistry*. However, the clinical emphasis in pharmacy education requires that topics be regrouped away from a chemical classification and classified according to their use. Selected chapters can be used as needed depending on where material is presented in a school's curriculum. Those schools using courses in introductory principles followed by a team teaching approach involving pharmaceutical chemists and pharmacologists should have no trouble using sections of this text in many parts of their courses.

We wish to thank Dr. Robert F. Doerge, head of the Department of Pharmaceutical Chemistry, Oregon State University, who read the entire manuscript. Appreciation is expressed to Mr. Martin Dallago of Lea & Febiger who continually gave us guidance and encouragement.

CORVALLIS, OREGON   *John H. Block*
LINCOLN, NEBRASKA   *Edward B. Roche*
MINNEAPOLIS, MINNESOTA   *Taito O. Soine*
CORVALLIS, OREGON   *Charles O. Wilson*

Contents

Contents

1

Atomic and Molecular Structure/Complexation

Electronic Structure of Atoms

The fundamental unit of all matter is the atom. The various chemical and physical properties of matter are determined by its elemental composition, and elements are composed of like atoms and their isotopes. In order to be able to predict the properties of matter, molecules, or elements, it is important to understand the structure of atoms. This chapter will discuss primarily electronic structure, since this is what largely controls chemical properties and the combination of various elements. Nuclear properties of atoms will be the subject of the chapter on radiopharmaceuticals (Chapter 11).

Subatomic Particles. Atoms are composed of a central *nucleus* surrounded by *electrons* which occupy discrete regions of space. The nucleus is considered to contain two types of stable particles which comprise most of the mass of the atom. These particles are "held" within the nucleus by various "nuclear forces." The discussion of these forces and the theories involved is beyond the scope of this book.

One of these particles is called a *neutron*. It is an uncharged species with a mass of 1.675×10^{-24} g or approximately 1.009 mass units on the atomic scale. The other particle is termed a *proton*. This particle has a positive charge of essentially *one* electrostatic unit (e.s.u.). Its mass is close to that of the neutron at 1.672×10^{-24} g or approximately 1.008 atomic mass units (a.m.u.).

Every stable nucleus contains a certain number of protons (equal to the number of electrons in the neutral atom) and a particular number of neutrons. The sum of the masses of the protons and neutrons accounts for most of the *atomic mass* (or weight) of the element, and the number of protons is equal to the *atomic number*. Isotopic forms of a particular element differ in the number of neutrons, and, therefore, in the atomic mass. More will be said about these properties later in this discussion.

A third subatomic particle is the *electron*, which has a *negative* charge of *one* e.s.u. and a mass of 9.107×10^{-28} g or approximately 0.0006 a.m.u. Its

1

charge is opposite in sign and equal in magnitude to that of a proton, so a neutral atom will have the same number of electrons as protons. The mass of the electron is about 1/1840 of that of the proton, thus providing only a small contribution to the atomic mass. Electrons occupy regions of extranuclear space at various distances from the nucleus according to the laws of quantum mechanics.

In addition to these three stable particles, atoms also contain certain unstable species which are observed when atomic nuclei are bombarded with various types of particles. Over the last 20 years nuclear physicists have discovered numerous subnuclear particles, and the list of such entities has grown rapidly. Notable among these particles is the *positron*. This might be thought of as the direct counterpart of the electron in that it is similar in mass and opposite in charge. The *betatron* (negatron) or beta particle is an electron emitted from the nucleus. Its site of origin is the major factor distinguishing it from the electron. There are some indications that the positron and beta particle are components of the neutron. The *neutrino* may also be part of the neutron. This is an uncharged species of zero mass. Its presence can be observed in tracking chambers when nuclei are bombarded with energy or other subatomic particles. Aspects of unstable nuclei or subnuclear particles will be discussed in Chapter 11.

Atomic Orbitals. The remainder of this chapter will be devoted to discussion of electronic structure in atoms and molecules. A knowledge of these areas allows a better understanding of the physical and chemical properties of compounds, and the reader is urged to attempt to master the concepts developed here. These ideas and "pictures" will aid in learning the material to be discussed in later chapters.

The early quantitative description of electronic structure came from Niels Bohr[1] in 1913, and involved a planetary picture of the atom. Electrons were considered as particles which revolved around the nucleus in stationary planar orbits and which had definite energies. Bohr's model involved the use of classical Newtonian mechanics, and suitably predicted the electronic spectrum of hydrogen. However, atoms with more than one electron did not yield satisfactory results.

In the 1920s the theory of quantum mechanics for the description of ultrasmall particles was developed, as was the quantum theory of atomic structure. Although a number of prominent scientists were involved in the historical aspects of the theory, De Broglie's[2] work in 1924, which related velocity and momentum to the dual particle-wave nature of the electron, and Schrödinger's[3] development of the wave equation in 1926 are particularly noteworthy. It is beyond the scope of this text to discuss details of these developments, but the interested reader may pursue this topic in any of several books treating the subject of quantum mechanics or quantum chemistry.[4-6] Relevant to the present discussion is the description of electrons, placing them in discrete volumes of space about the nucleus.

These volumes of space are referred to by the term *atomic orbitals*, and the electrons contained within their boundaries are described by a set of four numbers called *quantum numbers*.

The uncertainty principle of Heisenberg[7] states that it is not possible to fix simultaneously the momentum and the position of an electron. The very act of observing an electron involves sufficient energy to cause it to move. Thus, it is necessary to discuss the "location" of electrons in atoms and molecules in terms of probability. The four quantum numbers set the probability limits within which an electron can be found. The first three quantum numbers refer to some property of the space or orbital, while the fourth quantum number describes the spin of the electron.

(1) *The Principal Quantum Number*. This number is given the symbol n. Quantum theory states that electrons in atoms exist in discrete energy levels. The energy associated with the electron increases as it locates farther from the nucleus. The principal quantum number describes the relative positions of these energy levels, their distance from the nucleus, and the possibility of discontinuities or points of zero probability in the levels. The values this number can assume are integers from $n = 1, 2, 3, \ldots, \infty$. When $n = 1$ the electron is found in the energy level closest to the nucleus. In the older literature, the shells referred to as K, L, M, etc., correspond to $n = 1, 2, 3$, etc.

(2) *The Suborbital Quantum Number*. This number is given the symbol l. The volume of space which represents the region of greatest probability of finding an electron varies in shape and size, depending upon the energy level. The suborbital quantum number may be said to describe the shape of the orbital, or the "electron cloud." This number can assume integer values limited by the corresponding value of n such that $l = 0, 1, 2, \ldots,$ $(n - 1)$. Thus, when $n = 1$, the only permissible value is $l = 0$. When $n = 2$, l can take two values, $l = 0$ and $l = 1$.

A second designation for the orbitals described by various values of l is in common use. The use of the letters s, p, d, and f for orbitals having the following suborbital quantum numbers is more meaningful to chemists:

$$l = 0 \quad s \text{ orbital}$$
$$l = 1 \quad p \text{ orbital}$$
$$l = 2 \quad d \text{ orbital}$$
$$l = 3 \quad f \text{ orbital}$$

Representative shapes for s, p, and d orbitals are shown in Fig. 1–1. Note that s orbitals are spherical, with the nucleus at the center of the sphere. The other orbitals have points with zero probability of finding an electron. These are called *nodes*, and the nodal planes pass through the nucleus of the atom. The signs on the lobes are of mathematical significance in determining overlap or bonding between orbitals. It should also be noted

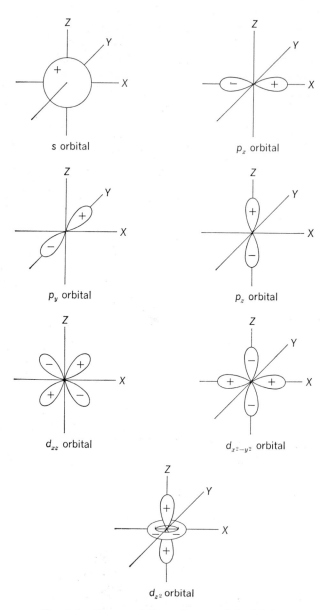

FIG. 1–1.   Representations of s, p, and d orbitals.

that the p and d orbitals have directional properties by virtue of their orientation along axes in three-dimensional space.

(3) *The Magnetic Quantum Number.* The symbol for this quantum number is m_l. Basically, this number describes the spatial orientation of the orbital. The allowed values are restricted by the value of l and can be positive or negative integer values according to: $m_l = -1, \ldots, 0, \ldots, +1$. Obviously, when $l = 0$ the orbital is s (spherical) and can only have one orientation, $m_l = 0$. When $l = 1$, there are three possible orientations of the associated p orbital, $m_l = -1, 0, +1$. These correspond to the three p orbitals shown in Fig. 1–1 along the x, y, and z axes. Likewise, when $l = 2$, there are five possible orientations for the d orbital, $m_l = -2, -1, 0, +1, +2$. Three of these are shown in Fig. 1–1. The other two are similar to the d_{xz} orbital except that they are in the xy and yz planes, and are called d_{xy} and d_{yz}, respectively.

Table 1–1 summarizes the relationships between these three quantum numbers. No attempt has been made to assign values of m_l to particular orientations. The choices in Table 1–1 are purely arbitrary.

(4) *The Spin Quantum Number.* This number is represented by the symbol m_s. The electron can be envisioned in its particle state as a spinning mass. Since it is charged, it will have a magnetic moment which is directionally oriented. Depending upon the direction of spin, there are two orientations of the magnetic moment, $+\frac{1}{2}$ or $-\frac{1}{2}$. These are the only two allowed values of m_s. The significance of this is that for two electrons to occupy the same orbital they must have opposing spins. If one has $m_s = +\frac{1}{2}$, the other must have $m_s = -\frac{1}{2}$.

TABLE 1–1. VALUES OF QUANTUM NUMBERS AND ORBITAL STATES

Quantum Numbers			Orbital
n	l	m_l	
1	0	0	$1s$
2	0	0	$2s$
2	1	+1	$2p_x$
2	1	0	$2p_y$
2	1	−1	$2p_z$
3	0	0	$3s$
3	1	+1	$3p_x$
3	1	0	$3p_y$
3	1	−1	$3p_z$
3	2	+2	$3d_{xy}$
3	2	+1	$3d_{xz}$
3	2	0	$3d_{yz}$
3	2	−1	$3d_{x^2-y^2}$
3	2	−2	$3d_{z^2}$

The Aufbau Process. With a knowledge of the quantum description of electrons in orbitals, it is now possible to approach the electronic structures of atoms in the periodic table of elements. For the most part, electronic structures follow a regular pattern from one atom to the next. Rules can be set down as a guide to determining the structure of any particular atom.

The procedure used has no relationship to physical reality because atoms are not actually formed in a stepwise process of adding electrons to empty orbitals about a nucleus. This process of atom "buildup" is called the *Aufbau process,* and is used only for its convenience.

Before the process can be properly employed, there are some fundamental rules that must be followed. The first of these is known as the *Pauli Exclusion Principle,* and is an outgrowth of quantum theory. A very loose statement of this principle is: In any atom, no two electrons may be described by the same set of values for the four quantum numbers. This is tantamount to saying that a maximum of two electrons may occupy a single orbital, and these two electrons must be of opposed spin.

Other rules which apply to the descriptive process are known as *Hund's Rules.* There are two of these which may be paraphrased:

(1) In the ground state of any atom, an electron may enter only the vacant orbital of lowest energy. In other words, lower energy orbitals must be filled before higher energy orbitals.

(2) Electrons must enter degenerate orbitals, that is orbitals having the same energy, i.e., $2p_x$, $2p_y$, and $2p_z$, singly and with parallel spins. Stated another way, electrons should remain unpaired in degenerate orbitals as long as possible.

When these rules are followed, it will be noticed that the electronic configuration of most atoms is the same as the atom having the next lowest atomic number, with the exception of the added electron. Of course, in moving from atom to atom, the correct number of protons and neutrons must be present in the nucleus. The number of protons is equal to the atomic number, and the number of neutrons can be obtained by subtracting that number from the atomic mass.

A schematic representation of the order in which orbitals are filled is shown in Fig. 1–2. Certain peculiarities associated with this diagram can be noted. Orbitals in the first and second principal quantum levels ($n = 1$, 2) are filled in order, as expected. Starting with the third principal quantum level, the ns orbital must be filled before electrons can be added to an $(n - 1)d$ orbital. Starting with the third principal quantum level, the $(n - 1)d$ orbitals must be filled before electrons can enter the np orbital. In elements where f orbitals are being filled, it is probably not practical to state a particular rule. The f orbitals are low-lying orbitals, and some of the atoms will have an electron or two in the next highest d orbital instead of in the f orbital.

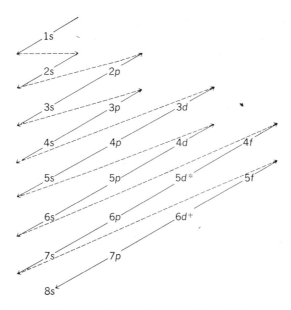

FIG. 1–2.   Order of filling of orbitals. *A single 5*d* is added before the 4*f* orbitals can be filled. ⁺One or more 6*d* electrons must be added before the 5*f* orbitals can be filled.   From *Modern Inorganic Pharmaceutical Chemistry*, by C. A. Discher, John Wiley & Sons, New York, 1964.   By permission.

Table 1–2 shows the electronic configurations of the known elements in the periodic table. One of the details seen in this table concerns the concept of electron correlation. In certain elements in the transition series where *d* orbitals are being filled, an *ns* level will be only half-filled, and the $(n - 1)d$ orbital will be either half-filled or full. These represent more stable configurations than may be achieved by simply filling orbitals according to the diagram in Fig. 1–2. For example, chromium (At. No. 24) has an outer structure of $3d^5 4s^1$ and copper (At. No. 29) has $3d^{10}4s^1$.

As a shorthand means of writing electronic configurations, it is possible to use the inert gas "core" that precedes the element being considered. For example, sodium (At. No. 11), which has the electronic configuration Na: $1s^2 2s^2 2p^6 3s^1$, can be written using the neon core for the first ten electrons: Na: [Ne] $3s^1$. Similarly, manganese (At. No. 25) can be written using the argon core for the first 18 electrons: Mn: [Ar] $3d^5 4s^2$.

Ionization.   The foregoing discussion dealt with a descriptive process of atom "buildup." The process of losing one or more electrons by chemical or physical means is known as *ionization*, and the positive ion produced is termed a *cation*. This process is distinctly different from the Aufbau process in that it is based in physical reality, and *should not* be taken as the exact opposite of the process of atom buildup.

TABLE 1–2. ELECTRONIC DISTRIBUTIONS*

Element	Atomic No.	1s	2s	2p	3s	3p	3d	4s	4p	4d	5s
H	1	1									
He	2	2									
Li	3	2	1								
Be	4	2	2								
B	5	2	2	1							
C	6	2	2	2							
N	7	2	2	3							
O	8	2	2	4							
F	9	2	2	5							
Ne	10	2	2	6							
Na	11	2	2	6	1						
Mg	12	2	2	6	2						
Al	13	2	2	6	2	1					
Si	14	2	2	6	2	2					
P	15	2	2	6	2	3					
S	16	2	2	6	2	4					
Cl	17	2	2	6	2	5					
Ar	18	2	2	6	2	6					
K	19	2	2	6	2	6		1			
Ca	20	2	2	6	2	6		2			
Sc	21	2	2	6	2	6	1	2			
Ti	22	2	2	6	2	6	2	2			
V	23	2	2	6	2	6	3	2			
Cr	24	2	2	6	2	6	5	1			
Mn	25	2	2	6	2	6	5	2			
Fe	26	2	2	6	2	6	6	2			
Co	27	2	2	6	2	6	7	2			
Ni	28	2	2	6	2	6	8	2			
Cu	29	2	2	6	2	6	10	1			
Zn	30	2	2	6	2	6	10	2			
Ga	31	2	2	6	2	6	10	2	1		
Ge	32	2	2	6	2	6	10	2	2		
As	33	2	2	6	2	6	10	2	3		
Se	34	2	2	6	2	6	10	2	4		
Br	35	2	2	6	2	6	10	2	5		
Kr	36	2	2	6	2	6	10	2	6		
Rb	37	2	2	6	2	6	10	2	6		1
Sr	38	2	2	6	2	6	10	2	6		2
Y	39	2	2	6	2	6	10	2	6	1	2
Zr	40	2	2	6	2	6	10	2	6	2	2
Cb	41	2	2	6	2	6	10	2	6	4	1
Mo	42	2	2	6	2	6	10	2	6	5	1
Tc	43	2	2	6	2	6	10	2	6	6	1
Ru	44	2	2	6	2	6	10	2	6	7	1
Rh	45	2	2	6	2	6	10	2	6	8	1
Pd	46	2	2	6	2	6	10	2	6	10	
Ag	47	2	2	6	2	6	10	2	6	10	1
Cd	48	2	2	6	2	6	10	2	6	10	2

TABLE 1–2.   ELECTRONIC DISTRIBUTIONS (continued)

Element	Atomic No.	1s	2s	2p	3s	3p	3d	4s	4p	4d	4f	5s	5p	5d	5f	6s	6p	6d	7s
In	49	2	2	6	2	6	10	2	6	10		2	1						
Sn	50	2	2	6	2	6	10	2	6	10		2	2						
Sb	51	2	2	6	2	6	10	2	6	10		2	3						
Te	52	2	2	6	2	6	10	2	6	10		2	4						
I	53	2	2	6	2	6	10	2	6	10		2	5						
Xe	54	2	2	6	2	6	10	2	6	10		2	6						
Cs	55	2	2	6	2	6	10	2	6	10		2	6			1			
Ba	56	2	2	6	2	6	10	2	6	10		2	6			2			
La	57	2	2	6	2	6	10	2	6	10		2	6	1		2			
Ce	58	2	2	6	2	6	10	2	6	10	2	2	6			2			
Pr	59	2	2	6	2	6	10	2	6	10	3	2	6			2			
Nd	60	2	2	6	2	6	10	2	6	10	4	2	6			2			
Pm	61	2	2	6	2	6	10	2	6	10	5	2	6			2			
Sm	62	2	2	6	2	6	10	2	6	10	6	2	6			2			
Eu	63	2	2	6	2	6	10	2	6	10	7	2	6			2			
Gd	64	2	2	6	2	6	10	2	6	10	7	2	6	1		2			
Tb	65	2	2	6	2	6	10	2	6	10	9	2	6			2			
Dy	66	2	2	6	2	6	10	2	6	10	10	2	6			2			
Ho	67	2	2	6	2	6	10	2	6	10	11	2	6			2			
Er	68	2	2	6	2	6	10	2	6	10	12	2	6			2			
Tm	69	2	2	6	2	6	10	2	6	10	13	2	6			2			
Yb	70	2	2	6	2	6	10	2	6	10	14	2	6			2			
Lu	71	2	2	6	2	6	10	2	6	10	14	2	6	1		2			
Hf	72	2	2	6	2	6	10	2	6	10	14	2	6	2		2			
Ta	73	2	2	6	2	6	10	2	6	10	14	2	6	3		2			
W	74	2	2	6	2	6	10	2	6	10	14	2	6	4		2			
Re	75	2	2	6	2	6	10	2	6	10	14	2	6	5		2			
Os	76	2	2	6	2	6	10	2	6	10	14	2	6	6		2			
Ir	77	2	2	6	2	6	10	2	6	10	14	2	6	9					
Pt	78	2	2	6	2	6	10	2	6	10	14	2	6	9		1			
Au	79	2	2	6	2	6	10	2	6	10	14	2	6	10		1			
Hg	80	2	2	6	2	6	10	2	6	10	14	2	6	10		2			
Tl	81	2	2	6	2	6	10	2	6	10	14	2	6	10		2	1		
Pb	82	2	2	6	2	6	10	2	6	10	14	2	6	10		2	2		
Bi	83	2	2	6	2	6	10	2	6	10	14	2	6	10		2	3		
Po	84	2	2	6	2	6	10	2	6	10	14	2	6	10		2	4		
At	85	2	2	6	2	6	10	2	6	10	14	2	6	10		2	5		
Rn	86	2	2	6	2	6	10	2	6	10	14	2	6	10		2	6		
Fr	87	2	2	6	2	6	10	2	6	10	14	2	6	10		2	6		1
Ra	88	2	2	6	2	6	10	2	6	10	14	2	6	10		2	6		2
Ac	89	2	2	6	2	6	10	2	6	10	14	2	6	10		2	6	1	2
Th†	90	2	2	6	2	6	10	2	6	10	14	2	6	10		2	6	2	2
Pa†	91	2	2	6	2	6	10	2	6	10	14	2	6	10	2	2	6	1	2
U†	92	2	2	6	2	6	10	2	6	10	14	2	6	10	3	2	6	1	2
Np†	93	2	2	6	2	6	10	2	6	10	14	2	6	10	4	2	6	1	2
Pu†	94	2	2	6	2	6	10	2	6	10	14	2	6	10	5	2	6	1	2
Am†	95	2	2	6	2	6	10	2	6	10	14	2	6	10	6	2	6	1	2
Cm†	96	2	2	6	2	6	10	2	6	10	14	2	6	10	7	2	6	1	2
Bk†	97	2	2	6	2	6	10	2	6	10	14	2	6	10	8	2	6	1	2
Cf†	98	2	2	6	2	6	10	2	6	10	14	2	6	10	9	2	6	1	2

* From Inorganic Chemistry by T. Moeller, John Wiley & Sons, New York, pp. 98–101, 1952. By permission.
† Probable structures.

It is always the most loosely "held" electrons which are lost first when an atom ionizes. However, the electronic structure of the ion may not reveal the level from which the electron was lost. This is particularly true for transition elements.

There are several reasons for this. Relative orbital energies are subject to change as electrons are "placed" in them. This means that a high energy orbital in one atom may be of lower energy in a neighboring atom where it might be completely filled. Another phenomenon to be noted is the possibility of rearrangement of the remaining electrons in an ion to a more stable configuration.

Usually atoms in the transition series with incompletely filled d orbitals will ionize to leave d ions, that is, ions in which the outer "shells" are d orbitals which may contain from one to ten electrons, depending upon the atom in question. For example, cobalt (At. No. 27) would ionize:

$$\text{Co}^0 \xrightarrow{\quad -2e^-\quad} \text{Co}^{+2} \qquad\qquad \text{(i)}$$
$$[\text{Ar}]\ 3d^7 4s^2 \qquad\qquad\qquad [\text{Ar}]\ 3d^7$$

This does not necessarily mean that both electrons were lost from the $4s$ orbital, even though the structure of the ion would seem to indicate that such was the case. However, one or both electrons could have been removed from the $3d$ orbital followed by rearrangement of all the valence electrons into this orbital.

Atoms in which s or p orbitals are being filled will usually ionize to form ions with either *inert gas* or *expanded* outer shells. Those which form cations with inert gas shell structures include elements like sodium (At. No. 11) and magnesium (At. No. 12):

$$\text{Na}^0 \xrightarrow{\quad -1e^-\quad} \text{Na}^+ \qquad\qquad \text{(ii)}$$
$$[\text{Ne}]\ 3s^1 \qquad\qquad\qquad [\text{Ne}]$$

$$\text{Mg}^0 \xrightarrow{\quad -2e^-\quad} \text{Mg}^{+2} \qquad\qquad \text{(iii)}$$
$$[\text{Ne}]\ 3s^2 \qquad\qquad\qquad [\text{Ne}]$$

In the periodic table (Table 1–3), elements with ions having inert gas configurations correspond to the elements in Groups IA and IIA, and the inert gas shell of the ion is in Group VIIIA of the preceding period.

Similarly, elements in Periods 2 and 3 of Groups IIIA, IVA, and VA tend to form cations with the preceding inert gas structure. For example, consider aluminum (At. No. 13):

$$\text{Al}^0 \xrightarrow{\quad -3e^-\quad} \text{Al}^{+3} \qquad\qquad \text{(iv)}$$
$$[\text{Ne}]\ 3s^2 3p^1 \qquad\qquad\qquad [\text{Ne}]$$

Differences are found in the structures of cations from these same groups when we reach Periods 4, 5, and 6. The cations formed from these atoms tend to have *expanded* or *18-electron* valence shells. Consider the case of gallium (At. No. 31) which is in the same group as aluminum:

$$Ga \xrightarrow{\quad -3e^- \quad} Ga^{+3} \qquad (v)$$
$$[Ar]\,3d^{10}4s^24p^1 \qquad\qquad [Ne]\,3s^23p^63d^{10}\ \text{or}\ [Ar]\,3d^{10}$$

The 18 electrons are those contained in the $3s$, $3p$, and $3d$ orbitals.

Elements in Groups VIA and VIIA which have larger numbers of electrons in their p orbitals tend to ionize by accepting electrons to form *anions*. These ions have completely filled p orbitals so that the valence shell structure is the same as the inert gas in the same period as the neutral element. Examples of this can be seen in oxygen (At. No. 8) and bromine (At. No. 35):

$$O \xrightarrow{\quad +2e^- \quad} O^{-2} \qquad (vi)$$
$$[He]\,2s^22p^4 \qquad\qquad [He]\,2s^22p^6\ \text{or}\ [Ne]$$

$$Br \xrightarrow{\quad +1e^- \quad} Br^- \qquad (vii)$$
$$[Ar]\,3d^{10}4s^24p^5 \qquad\qquad [Ar]\,3d^{10}4s^24p^6\ \text{or}\ [Kr]$$

Other elements which have incompletely filled f orbitals (lanthanides and actinides) tend to form cations with electrons remaining in the f orbitals.

Periodic Law. In the previous discussion, elements in various groups and periods of the periodic table were mentioned. A table of this type was suggested independently around 1868–1870 by the Russian chemist Mendeléeff and the German chemist Lothar Meyer. They had recorded data that indicated a periodic relationship based on the atomic weights of the elements.

Today it is known that the properties of elements are based primarily upon the outer or valence shell electronic structure. Thus, it can be seen that the periodic table, shown in Table 1–3, lists elements in groups having the same valence orbital structures. However, the table was formulated before much was known about electronic structures. A general statement of the *periodic law* is that "the properties of elements are a periodic function of their atomic numbers." Because of the relationship between atomic numbers and the number of electrons, this is equivalent to the original statement concerning valence orbital structures.

Table 1–3 is generally referred to as a "long form" of the periodic table because the A and B subgroups are spread across the chart. There are eight groups which correspond to the filling of s and p orbitals having principal quantum numbers equal to the number of the period or row of

TABLE 1-3. PERIODIC TABLE OF THE ELEMENTS

Metals

Nonmetals

KEY

ATOMIC WEIGHT (1)

ATOMIC NUMBER

OXIDATION STATES

BOILING POINT, °C

MELTING POINT, °C

DENSITY (g/ml) (2)

SYMBOL

30 65.37 2
906 419.5 7,14
419.5 7.14
Zn
name
Zinc

NOTES:
(1) Atomic weights are 1971 values. Parentheses indicate most stable or best known isotope.
(2) Density values for gaseous elements are for liquids at the boiling point.
(3) Names and symbols for elements 104 and 105 are proposed but not yet officially accepted.

Transition Metals

GROUP																	
IA																	VIIIA

Period 1

1 1.0079 1
−252.7
−259.2 0.071
H
Hydrogen

2 4.0026
−268.9
−269.7 0.126
He
Helium

Period 2

3 6.939 1
1330
180.5 0.53
Li
Lithium

4 9.0122 2
2770
1277 1.85
Be
Beryllium

5 10.811 3
(2030)
2.34
B
Boron

6 12.01115 ±2,4
4830
3727b 3.52
C
Carbon

7 14.0067 ±3,5,4,2
−195.8
−210 1
N
Nitrogen

8 15.9994 −2
−183.0
−218.8 1.14
O
Oxygen

9 18.9984 −1
−188.2
−219.6 1.505
F
Fluorine

10 20.183
−248.6
−248.6 1.20
Ne
Neon

Period 3

11 22.9898 1
892
97.8 0.97
Na
Sodium

12 24.312 2
1107
650 1.74
Mg
Magnesium

13 26.9815 3
2450
660 2.70
Al
Aluminum

14 28.086 ±4
2680
1410 2.33
Si
Silicon

15 30.9738 ±3,5,4
280w
44.2w 2.07
P
Phosphorus

16 32.064 ±2,4,6
444.6
119 2.07
S
Sulfur

17 35.453 ±1,3,5,7
−34.7
−101.0 1.56
Cl
Chlorine

18 39.948
−185.8
−189.4 1.40
Ar
Argon

Period 4

19 39.102 1
760
63.7 0.86
K
Potassium

20 40.08 2
1440
838 1.55
Ca
Calcium

21 44.956 3
2730
1539 3.0
Sc
Scandium

22 47.90 4,3
3260
1668 4.51
Ti
Titanium

23 50.942 5,4,3,2
3450
1900 6.1
V
Vanadium

24 51.996 6,3,2
2665
1875 7.19
Cr
Chromium

25 54.938 7,6,4,2,3
2150
1245 7.43
Mn
Manganese

26 55.847 2,3
3000
1536 7.86
Fe
Iron

27 58.933 2,3
2900
1495 8.9
Co
Cobalt

28 58.71 2,3
2730
1453 8.9
Ni
Nickel

29 63.54 2,1
2595
1083 8.96
Cu
Copper

30 65.37 2
906
419.5 7.14
Zn
Zinc

31 69.72 3
2237
29.8 5.91
Ga
Gallium

32 72.59 4
2830
937.4 5.32
Ge
Germanium

33 74.922 ±3,5
613*
817 5.72
As
Arsenic

34 78.96 −2,4,6
685
217 4.79
Se
Selenium

35 79.909 ±1,5
58
−7.2 3.12
Br
Bromine

36 83.80
−152
−157.3 2.6
Kr
Krypton

Period 5

37 85.47 1
688
38.9 1.53
Rb
Rubidium

38 87.62 2
1380
768 2.6
Sr
Strontium

39 88.905 3
2927
1509 4.47
Y
Yttrium

40 91.22 4
3580
1852 6.49
Zr
Zirconium

41 92.906 5
3300
2468 8.4
Nb
Niobium

42 95.94 6,5,4,3,2
5560
2610 10.2
Mo
Molybdenum

43 (99) 7
5030
2140 11.5
Tc
Technetium

44 101.07 2,3,4,6,8
4900
2500 12.2
Ru
Ruthenium

45 102.905 2,3,4,6
4500
1966 12.4
Rh
Rhodium

46 106.4 2,4
3980
1552 12.0
Pd
Palladium

47 107.870 1
2210
960.8 10.5
Ag
Silver

48 112.40 2
765
320.9 8.65
Cd
Cadmium

49 114.82 3
2000
156.2 7.30
In
Indium

50 118.69 2,4
2270
231.9 7.30
Sn
Tin

51 121.75 ±3,5
1380
630.5 6.24
Sb
Antimony

52 127.60 −2,4,6
989.8
449.5 6.24
Te
Tellurium

53 126.904 ±1,5,7
183
113.7 4.94
I
Iodine

54 131.30
−108.0
−111.9 3.06
Xe
Xenon

Period 6

55 132.9054 1
690
28.7 1.90
Cs
Cesium

56 137.34 2
1640
714 3.5
Ba
Barium

57 138.91 3
3470
920 6.17
La ★★
Lanthanum

72 178.49 4
5425
2222 13.1
Hf
Hafnium

73 180.948 5
5930
2996 16.6
Ta
Tantalum

74 183.85 6,5,4,3,2
5930
3410 19.3
W
Wolfram (Tungsten)

75 186.2 7,6,4,2,−1
5900
3180 21.0
Re
Rhenium

76 190.2 2,3,4,6,8
5500
3000 22.6
Os
Osmium

77 192.2 2,3,4,6
5300
2454 22.5
Ir
Iridium

78 195.09 2,4
4530
1769 21.4
Pt
Platinum

79 196.967 3,1
2970
1063 19.3
Au
Gold

80 200.59 2,1
357
−38.4 13.6
Hg
Mercury

81 204.37 3,1
1457
303.5 11.85
Tl
Thallium

82 207.19 2,4
1725
327.4 11.4
Pb
Lead

83 208.980 3,5
1560
271.3 9.8
Bi
Bismuth

84 (210) 2,4
(962)
9.2
Po
Polonium

85 (210) ±1,3,5,7
(302)
At
Astatine

86 (222)
(−61.8)
(−71)
Rn
Radon

Period 7

87 (223) 1
(27)
Fr
Francium

88 (226) 2
1140
700 5.0
Ra
Radium

89 (227) 3
1050
Ac ★★
Actinium

104 [Rf]
Ku
(Rutherfordium) (Kurchatovium)

105 Ha
Hahnium

★
58 140.12 3,4	59 140.907 3,4	60 144.24 3	61 (147) 3	62 (147) 3,2	63 150.35 3,2	64 151.96 3	65 157.25 3	66 158.924 3	67 162.50 3	68 164.930 3	69 167.26 3	70 168.934 3,2	71 173.04 3,2
3468 795 6.67 Ce Cerium	3127 935 6.77 Pr Praseodymium	3027 1024 7.00 Nd Neodymium	(1027) Pm Promethium	1900 1072 7.54 Sm Samarium	1439 826 5.26 Eu Europium	3000 1312 7.89 Gd Gadolinium	2800 1356 8.27 Tb Terbium	2600 1407 8.54 Dy Dysprosium	2600 1461 8.80 Ho Holmium	2900 1497 9.05 Er Erbium	1727 1545 9.33 Tm Thulium	1427 824 6.98 Yb Ytterbium	3327 1652 9.84 Lu Lutetium

★★
90 232.038 4	91 (231) 5,4	92 238.03 6,5,4,3	93 (237) 6,5,4,3	94 (242) 6,5,4,3	95 (243) 6,5,4,3	96 (247)	97 (247)	98 (249)	99 (254)	100 (253)	101 (256)	102 (254)	103 (257)
3850 1750 11.4 Th Thorium	(1230) 15.4 Pa Protactinium	3818 1132 19.07 U Uranium	637 19.5 Np Neptunium	3235 640 19.8 Pu Plutonium	11.7 Am Americium	Cm Curium	Bk Berkelium	Cf Californium	Es Einsteinium	Fm Fermium	Md Mendelevium	No Nobelium	Lw Lawrencium

(Courtesy of Sargent-Welsh Scientific Company.)

12

the table. These are designated "A groups" in this version of the table. Group VIIIA contains the so-called inert gases which end each period. These elements have filled outer p orbitals and thus are relatively unreactive. Compounds of many of the inert gases are known, but rather vigorous conditions are required to cause reaction. *Inert* gases may be a questionable term considering what is now known about these elements.

Through Periods 2 and 3 s and p orbitals are filled in normal fashion. These are sometimes referred to as the "typical elements." In the fourth and successive periods, d orbitals are filled "between" the s and p orbitals, giving rise to the "B groups" intervening between Groups IIA and IIIA. The filling of these "low-lying" orbitals with one or two electrons in the "overlying" s orbitals causes the properties of these elements to differ somewhat from their A-group counterparts.

In Periods 4, 5, and 6, toward the center of the table, there is a triad of elements designated Group VIII, with no associated subgroup. These are the "real" *transition elements* which occur between the group with elements having half-filled d orbitals (VIIB) and the group with elements having full d orbitals (IB). Most frequently, all the elements in the B groups, including Group VIII, are thought of as the *transition series* of elements.

The reason for Group IB and IIB following the other B groups is based on the filled d orbital structure in these elements, resulting in the situation of "refilling" the same s orbitals that were filled in Groups IA and IIA.

The separated long periods below the main portion of the table contain the elements known as the lanthanides (atomic numbers 57 to 71) and the actinides (atomic numbers 89 to 103). These elements are formed by the "filling" of very low-lying f orbitals giving all of them quite similar properties. Thus, both groups of elements are considered to be members of the same group that contains their parent member. The lanthanides are in Group IIIB in the sixth period, and the actinides are in Group IIIB in the seventh period.

Some general properties of the elements in the periodic table may be seen by referring to both the periods and groups. Properties within a group of elements are generally the same. The major differences, for the most part, are quantitative and are based upon the large increase in atomic radii (or principal quantum number) as one descends in a particular group. For example, in Group IA, electrons are "lost" (or the elements are oxidized) from each of these elements under certain conditions. The ease with which these elements undergo oxidation increases as the group is descended. In the larger elements, the outer electrons are farther away from the nucleus, and the underlying electrons provide a better shield for these electrons from nuclear influences. These factors produce a gradation in the observed ease of oxidation.

This property may be viewed in another way. An examination of the property of *electronegativity*, the affinity an element has for its electrons and

TABLE 1–4. ELECTRONEGATIVITIES OF THE ELEMENTS, ACCORDING TO PAULING[8]

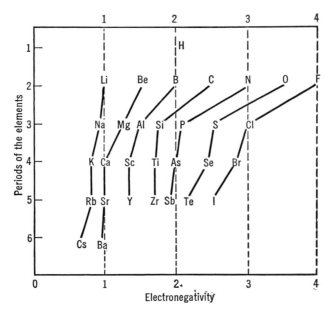

From Roger's Inorganic Pharmaceutical Chemistry, 8th ed., by T. O. Soine and C. O. Wilson, Lea & Febiger, Philadelphia, p. 8, 1967. By permission.

the ability to "take on" additional electrons, will show that this property increases from left to right across any period and from bottom to top in any group (except VIIIA). The most electronegative element is fluorine, and the least electronegative element is francium. Table 1–4 shows the Pauling electronegativity scale[8] for many of the elements. The opposite concept is that of *electropositivity* which varies in directions opposite to those of electronegativity.

Metallic properties of elements generally decrease in a given period as the atomic number increases. However, this same property seems to increase with increasing atomic number within any particular group. The heavy line (Table 1–3) passing between elements from boron to polonium represents an amphoteric area. Elements on both sides show both metallic and nonmetallic character.

A discussion of some of the group properties of elements appears in Chapter 2.

Electronic Structure of Molecules

Moving from the subject of atomic structure, it is now logical to examine some fundamental aspects of chemical bonding and molecular structure.

As a foundation for this discussion, it will be convenient to have some sort of image of a molecule. One such "picture" may be visualized as the nuclei of the constituent atoms at their most stable internuclear distances embedded in a matrix of electrons. Most of the electrons are in atomic orbitals surrounding the individual nuclei, and the remainder (valence electrons) are in more generalized multinuclear *molecular orbitals*.

When atoms are incorporated into molecules, there are three major forces that are involved in the overall combination. *Coulombic attraction* occurs between the negatively charged electrons in the valence orbitals on one atom and the positively charged nucleus of another atom. As the atoms approach each other, there are two repulsive forces that tend to "push" the atoms away: (1) *electron-electron repulsion* between valence electrons on neighbor atoms and (2) *nuclear repulsion* between neighboring nuclei. A stable molecule is possible when the proper balance between these forces exists, and the energy of the resulting system of atoms is less than the sum of the energies of the "isolated" atoms. The equilibrium distances that are evident between atoms in molecules (bond distances) are established largely by the interaction of these forces.

However, simply having an overriding attractive force between two atoms does not assure the formation of a bond between them. Depending upon the type of bonding interaction which is likely to occur (e.g., ionic, covalent, etc.), there are other criteria based on the differences in electronegativity, availability of electrons, and the nature of the valence state atomic orbitals. Bond formation is also affected by the number of electrons in the valence shell orbitals, and by their orbital distribution.

The bonding types that are possible vary with the amount of "sharing" of electrons between the two atoms participating in the bond. *Covalent bonding* ranges from an equal sharing of a pair of electrons in homonuclear diatomic molecules (e.g., H_2, Cl_2, I_2, etc.) to a polar or unequal sharing of the electron pair in heteronuclear diatomic molecules (e.g., HCl). *Ionic bonding* is more of an electrostatic interaction resulting from the transfer of an electron from an electropositive atom to an electronegative atom (e.g., $Na^+ Cl^-$). More will be said about the types of bonds later in this chapter.

Orbital Hybridization. There are many bonding situations in chemistry which cannot be explained in terms of the ground state valence shell orbitals available in most atoms. In this section, we will briefly discuss the concept of *hybrid orbitals* which is necessary to explain covalent bonding in molecules properly.

Covalent bonding is presumed to occur through the overlapping of atomic orbitals between two atoms. In the usual case, overlapping each orbital will contribute one electron to the electron pair in the resulting molecular orbital or bond. Therefore, bonds formed in this manner must involve atomic orbitals which are singly occupied (contain one unpaired electron).

Referring to the electronic configurations given in Table 1–2, it should be apparent that for a majority of atoms the number of singly occupied orbitals does not correspond to their covalent bonding capability (or valence) in molecules. For example, the electronic structures of the second row elements (Table 1–3), beryllium, boron, and carbon, illustrate discrepancies common to their groups in the periodic table. From the configurations shown below (structures I, II, and III), Be would be expected to form no covalent bonds, B appears to be monovalent, and C should bond with a maximum of two other elements. In reality, these atoms and the other members of their groups are di-, tri-, and tetravalent, respectively.

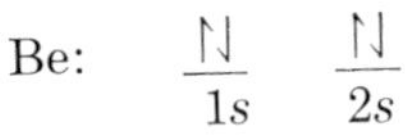

Be: ⇅ ⇅   B: ⇅ ⇅ ↑___
    1s 2s   1s 2s   2p

         I 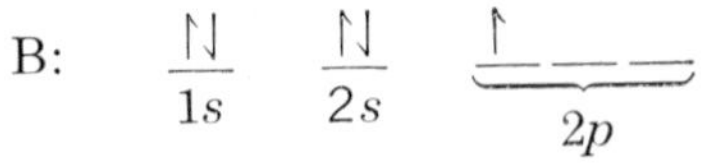   II

C: ⇅ ⇅ ↑ ↑___
   1s 2s   2p

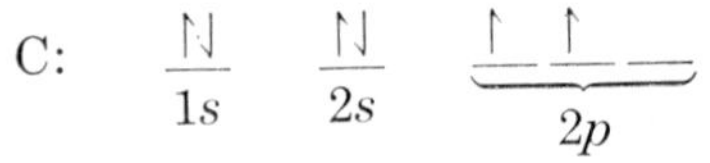

         III

The presumed mechanism allowing these and other elements to increase covalent bonding capacity involves *promotion* to the *valence state*, a situation requiring energy. This is a nonobservable hypothetical state of the atom which is justifiable on certain theoretical grounds and on the basis of the stability of the resulting molecules.[8,9] Figure 1–3a shows a portion of an energy diagram for the promotion of gaseous beryllium atoms ($Be_{(g)}$) to the divalent valence state. This state is usually viewed as hybrid orbital state and is labeled sp in this particular case. Similar diagrams are shown for boron (Fig. 1–3b) and carbon (Fig. 1–3c) indicating that their tri- and tetravalent states also involve hybrid orbitals designated as sp^2 and sp^3, respectively. The tetravalent state for carbon is limited to saturated molecules.

The process of orbital hybridization may be envisioned as a "mixing" of the atomic orbitals to provide a new set of degenerate (energetically equivalent) orbitals having different spatial orientations and directional properties than the original atomic orbitals. The number of hybrid orbitals produced is equal to the number of atomic orbitals involved in the hybridization, and the electrons contained in the original orbitals occupy the hybrids according to Hund's rules.

Directing attention to the construction of hybrid orbitals, the first case to be considered is that of sp orbitals. Figure 1–4 illustrates the result of combining an s orbital with a p orbital having the same principal quantum number. In this case, the p orbital was chosen, necessarily forcing

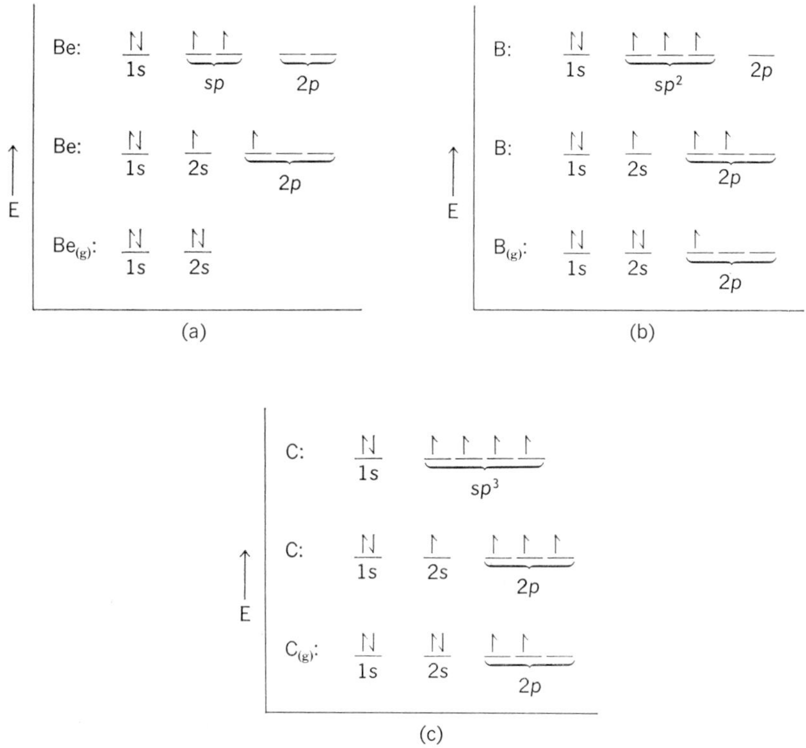

FIG. 1–3. Partial energy diagrams showing the promotion of atoms in their gas phases to their valence state configurations. The distances between states are not meant to indicate relative energy barriers.

the hybrids to be oriented along the z axis. The s orbital has no influence on the orientation due to its spherical symmetry. The two sp orbitals are equivalent (degenerate), symmetrical about the bonding axis, and oriented $180°$ away from each other. The improved directional characteristics of the hybrid should also be noted. The largest proportion of the electron density is oriented in a single direction, thus improving overlap with orbitals from other atoms and increasing the strength of the resulting bond. This pair of hybrids is often referred to by the term *digonal* to describe the fact that they are two opposing orbitals.

This type of hybridization (sp) is evident in the covalent compounds of Group II elements and the unsaturated "acetylenic" compounds of carbon. Linear covalent molecules of gaseous halides of Be, Mg, and Ca, such as $MgCl_2$, and the solid divalent compounds of Cd and Hg are indicative that the bonds are formed through sp hybrids on the Group II element. The energy required to promote the element to its digonal valence state (Fig. 1–3a) is released when the bonds are formed, resulting in stable

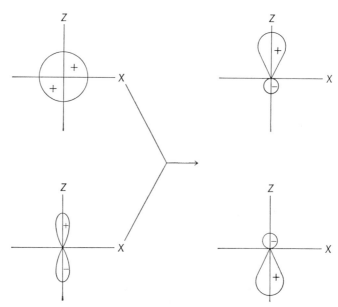

Fɪɢ. 1–4.   Illustration of the formation of *sp* hybrid orbitals
from one *s* and one *p* orbital.

molecules. A diagram more complete than those shown in Fig. 1–3 would include the other atoms which are reacting with the elements pictured and the ultimate energy level of the final molecule, which, for stable compounds, would be somewhat below the starting energy level for the composite elements in their ground states.

In a manner similar to that described above, elements of Group III may be promoted to a valence state in which singly occupied *s* and two *p* orbitals combine to form three equivalent sp^2 hybrid orbitals (Fig. 1–3b). The overall appearance of one of these hybrids is similar to the *sp* orbital except that the inclusion of an additional *p* orbital causes the hybrid to be somewhat more elongated. The three hybrids are located in the same plane, and are oriented toward the points of an equilateral triangle, 120° apart. Figure 1–5 shows the three *trigonal* hybrids on the same set of axes.

The monomeric covalent compounds of boron, aluminum, and other Group III elements, as well as unsaturated "ethylenic" compounds of carbon, show sp^2 hybridization. The empty *p* orbital remaining in the valence state (Fig. 1–3b) of the Group III elements leaves their compounds electron-deficient. These molecules, therefore, react as Lewis acids, and will be discussed in Chapter 4. It should be noted in Figure 1–5 that these orbitals also provide improved overlap due to a concentration of the electron density of each hybrid in a single direction. As was mentioned above, this results in increased stabilization in the molecular energy of the compounds.

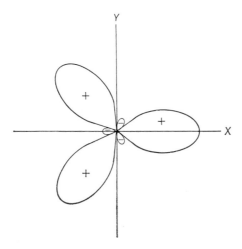

FIG. 1–5.   Illustration of sp^2 hybrid orbitals showing the orientations
to the three corners of an equilateral triangle.

The final extension of hybridization between s and p orbitals is involved
in the tetravalent state of Group IVA elements (Fig. 1–3c).   When one
s and three p orbitals combine, the result is a set of four equivalent sp^3
hybrid orbitals pointing to the four corners of a tetrahedron.   Therefore,
the geometry of a molecule formed through bonding with these orbitals is
tetrahedral, and the bond angles are approximately 109°.   It is difficult to
show all four hybrid orbitals in their proper orientation, but the two representations of the structure of methane below (structure IV) is illustrative
of the tetrahedral geometry.   The shape of each sp^3 hybrid is similar to the
sp^2 hybrid shown in Fig. 1–5.

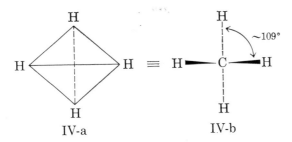

IV-a   IV-b

Hybridization schemes involving d orbitals are involved in transition
metal complexes, and will be discussed more under that topic.   Two of the
more important hybrids in this group include a set of six orbitals with
octahedral geometry termed d^2sp^3 orbitals, and a set of four orbitals with
square planar geometry termed dsp^2 orbitals.

Some general points about hybrid orbitals can now be made.

1. Although the term "equivalent" has been used in referring to sets of orbitals, it should be noted that the equivalency is destroyed when the orbitals on a particular atom form bonds to different elements in the formation of molecules.

2. In the considerations above, the orbitals promoted to the valence state were singly occupied. Promotion to hybridized states is presumed to occur with doubly occupied orbitals as well. In other words, it is possible to have a nonbonded pair of electrons in a hybrid orbital.

3. Bond strengths tend to increase as the amount of s character in a hybrid decreases. This should be understandable since the p character (or d character) would alternately increase, thus providing better directional properties to the hybrid.

4. Finally, bonds formed with hybrid orbitals will not necessarily have the exact geometries presented above. In many molecules, the bond angles will vary from that predicted because repulsive forces will cause changes in the s and p character leading to some contribution intermediate between those discussed above. Table 1–5 illustrates how the angle between orbitals changes with the various s and p orbital contributions. It should be noted that as the p character increases the angle becomes smaller until it reaches $90°$ for two unhybridized p orbitals.

An example which serves to illustrate some of these points can be found in the water molecule H_2O. In the ground state electronic structure of oxygen, there are two p orbitals containing one electron each. If the two hydrogen atoms became bonded to the oxygen through these two orbitals, the water molecule would be expected to have an H—O—H bond angle of $90°$. In fact, the bond angle is closer to $104°$. The larger angle cannot be explained on the basis of repulsion between the two polar O—H bonds. A logical explanation is found in the presumption that the valence shell orbitals on oxygen achieve a hybridized valence state. Assuming sp^3 hybridization, the water molecule would have two lone pairs of electrons

TABLE 1–5. ORBITAL CONTRIBUTION AND HYBRID GEOMETRY

Orbital	Geometry	Contribution		Angle
		s	p	
p^2	(unhybridized p orbitals)	0	1	$90°$
sp^3	tetrahedral	$\frac{1}{4}$	$\frac{3}{4}$	$109°$
sp^2	trigonal	$\frac{1}{3}$	$\frac{2}{3}$	$120°$
sp	digonal	$\frac{1}{2}$	$\frac{1}{2}$	$180°$

Adapted from Advanced Inorganic Chemistry, by F. A. Cotton and G. Wilkinson, Interscience Publishers, New York, p. 72, 1962. By permission.

in hybridized orbitals as shown below (structure V). Repulsive forces between these two orbitals will cause the angle between them to enlarge and the orbitals to lose p character. Such a change would in turn cause the H—O—H angle to become smaller, with a corresponding gain in p character for the bonding orbitals. Thus, the two lone pairs of electrons would occupy orbitals which are hybridized somewhere between sp^2 and sp^3, while the bonding orbitals on the oxygen are between sp^3 and pure p orbitals.

V

Types of Bonding Interactions. In the previous section, a concept was dealt with which is peculiar to covalent bonding in molecules. This is only one of several different types of bonding that can take place between atoms and molecules. The relative occurrence of the various types of interactions depends on a number of factors which have been mentioned previously. At this time, it would be well to examine the types of forces or interactions involved in chemical bonding more closely, including the factors that affect them.

(1) *Ionic Bonding.* Sometimes referred to as *electrovalence,* ionic bonding is the electrostatic force that exists between two chemical entities of opposite charge. The species bearing the positive charge is known as the *cation,* and the negative species is the *anion.* When the two reacting entities are sufficiently far apart in their respective electronegativities (or far apart in the periodic table), the least electronegative entity loses one or more of its valence electrons to the more electronegative entity (or entities) to produce the respective cation(s) and anion(s) [rx (viii)].

$$M\cdot \ + \ \cdot \ddot{X}: \longrightarrow M^+ \ + \ :\ddot{X}:^- \qquad \text{(viii)}$$

In most cases, when the entities are simple atoms, the ionic species take on inert gas shell electronic structures. Since the valence shell of all inert gases except helium contain eight electrons, this kind of structure is associated with stability, and has led to the octet theory of chemical bonding. It will be found that most stable ions have inert gas valence shell structures. Cations will have structures resembling the inert gas in the period above them in the periodic table, and anions resemble the inert gas located in the same period as the neutral element.

Ionic bonding is usually found in associations between metallic, strongly electropositive elements (Groups IA and IIA) and nonmetallic, strongly electronegative elements (Group VIIIA) (see Tables 1–3 and 1–4). It is also found in most salts where the anion is complex, such as SO_4^{-2}, PO_4^{-3}, NO_3^-, and the like. Ionic interactions are also present, in many cases, when polar compounds are dissolved in polar solvents.

The formation of sodium chloride is similar to the general reaction shown above [rx (viii)], and is representative of ionic bonding. Sodium has one electron in its valence shell (Table 1–2) and an electronegativity on the Pauling scale of 0.9 (Table 1–4), while chlorine has seven electrons in its valence shell and a Pauling electronegativity of 3.0. The difference in electronegativity is sufficient to allow the transfer of one electron from sodium to chlorine, thereby forming two oppositely charged ions, each having the inert gas configuration of eight electrons in their valence shells [rx (ix)]. The electrostatic attraction between the oppositely charged species is the force holding the ions together.

$$Na\cdot \ + \ \cdot \ddot{\underset{\cdot\cdot}{Cl}}: \longrightarrow Na^+ \ \ :\ddot{\underset{\cdot\cdot}{Cl}}:^- \tag{ix}$$

0.9   3.0

Electronegativity

Similarly, calcium chloride is formed by calcium transferring its two electrons to each of two chlorine atoms [rx (x)].

$$Ca: \ + \ 2\cdot\ddot{\underset{\cdot\cdot}{Cl}}: \longrightarrow :\ddot{\underset{\cdot\cdot}{Cl}}:^- \ \ Ca^{++} \ \ :\ddot{\underset{\cdot\cdot}{Cl}}:^- \tag{x}$$

The geometry of the chloride ions about the calcium is linear because this is the most stable electrostatic arrangement. Hybridization is not involved in the formation of ionic compounds.

In the transition series, the octet theory is less obvious in ion formation. These metals will form variously charged cations, usually with electrons remaining in the d orbital valence shells. The low-lying d orbitals are responsible for the variable valences and electronic structures of these ions. These factors will be covered more completely when complexation is discussed.

Generally speaking, metals lose electrons to form cations, and nonmetals attract electrons to form anions. Metals that have one valence electron tend to ionize more readily than those having two or more. As was mentioned earlier, the ability to lose electrons increases as we descend in a particular group. This is due to the increased shielding effect of the inner electrons which diminishes the attractive force of the nucleus on the valence electrons. An index to this property is the electronegativity shown in Table 1–4.

(2) *Covalent Bonding.* Unlike ionic bonding, covalent bonding is the attractive force that exists between two chemical entities due to their "sharing" a pair of electrons. The bonding pair of electrons between two atoms is presumed to be in a *molecular orbital* embracing both nuclei. This type of bonding prevails when the electronegativity difference between the atoms is not sufficient to produce ions. Although quite common in inorganic compounds, covalent bonding is most prevalent in organic chemistry. In a simple context, the chemical stability of a covalent compound depends upon each bonded atom sharing the electron pair in a manner producing a valence shell with an inert gaslike structure (octet theory) in the vicinity of the atom.

The most idealistic covalent bonds occur in homonuclear diatomic molecules such as H_2, Cl_2, N_2, etc. [rx (xi)]. In these situations, the electron pair is shared equally by the two bonded atoms. Larger systems of atoms are also covalent with an equal distribution of the bonding electrons, e.g., S_8, P_4 (structure VI), and diamond carbon. The covalent bonds in saturated hydrocarbons (structure VII) approach an ideal sharing of electrons between the carbon atoms, but the carbon-hydrogen bonds depart from this to some extent. For all practical purposes, however, these molecules are considered to be nonpolar.

$$\text{H} \cdot \quad + \quad \cdot \text{H} \longrightarrow \text{H:H} \quad \text{or} \quad \text{H—H} \qquad \qquad \text{(xi)}$$

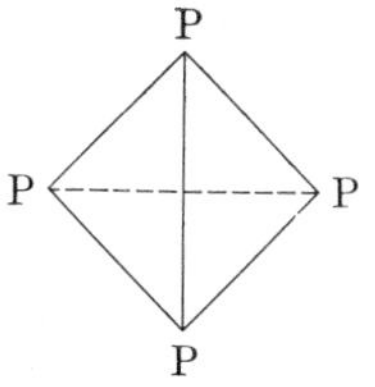

VI 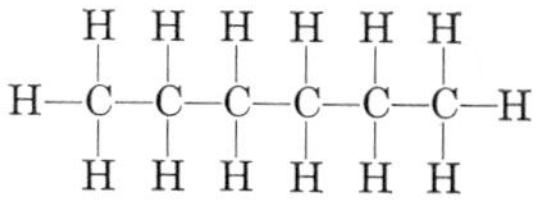   VII

Most covalent molecules, however, cannot be represented by an equal distribution of the bonding electrons. The electron density tends to be shifted toward the more electronegative member of the bond. The bond is still covalent rather than ionic, but the molecule is definitely polar. Water is a good example of a primarily covalent molecule which is strongly polar, with the more electronegative oxygen taking the largest share of the bonding electron density (structure VIII). On the structure below, this polarity is indicated by noting partial charges.

2

$$O^{\delta-}$$

$$\delta+H \qquad H\delta+$$

VIII

Thus far single covalent bonds have been considered which are also termed *sigma* bonds (σ bonds). In this type of bonding, the molecular orbitals (or electron distributions) are symmetrical about the bond axes. Attention is now directed toward double and triple covalent bonds. These are bonds in which two and three pairs of electrons are being shared, respectively, between two atoms. It should be recalled that only one pair of electrons can occupy a single molecular orbital; therefore, atoms bonded in this fashion need to overlap other valence orbitals to form two and three bonding molecular orbitals, respectively. Further, only one of the bonds can be a σ bond. The other pairs of electrons will occupy molecular orbitals which are distributed on both sides of and perpendicular to a plane (or planes) passing through the bond axis (see structures X). This type of covalent bond is termed a *pi* bond (π bond).

Carbon dioxide, CO_2, serves as an illustration of a double covalently bonded molecule. Each C—O bond consists of two pairs of electrons, one in the σ bond and one in the π bond (structure IX).

$$:\overset{..}{O}::C::\overset{..}{O}:$$

IX

The σ bonds are formed by overlapping *sp* orbitals on the carbon with singly occupied *p* orbitals on the oxygens. This leaves two singly occupied *p* orbitals on the carbon and one such orbital on each of the oxygens. The oxygens can be rotated in turn to bring their *p* orbitals parallel to one of the carbon *p* orbitals. Overlapping of parallel *p* orbitals on carbon and oxygen leads to two volumes of electron density separated by a nodal plane at the bond axis. The carbon thus forms a π bond with each oxygen. Structure X shows the overlapping involved in the π bond formation. The σ bonding orbitals and the lone (nonbonded) pairs of electrons on the oxygens are not shown.

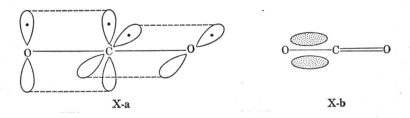

X-a   X-b

Similarly, hydrogen cyanide, HCN, will serve as an illustration of a triple bonded molecule where two π bonds are formed between two atoms (structure XI).

$$H : C \vdots \vdots N$$

XI

The carbon is sp-hybridized as in CO_2, but now both p orbitals overlap with two p orbitals on the nitrogen (structure XII).

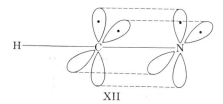

XII

(3) *Coordinate Covalent Bonding.* In the above discussion, each atom participating in a covalent bond contributed one electron to make up the pair. Coordinate covalent bonding is still a covalent interaction but, in this case, both electrons in the bond arise from a single orbital on one of the atoms forming the bond. This type of bonding is found most frequently between complex chemical entities. The entity providing the pair of electrons is generally referred to as the *donor* species. The *acceptor* species is electron-deficient and has an empty orbital which can overlap with the orbital from the donor.

Donor-acceptor complexes, coordination compounds, etc., are molecules or complex ions which involve coordinate covalent bonds as a major aspect of their chemistry. The representation of the bond, in order to distinguish it from other covalent bonds, is an arrow drawn so that it points from the donor atom to the acceptor atom (structure XIII). An example of this type of interaction is the donor-acceptor complex boron-trifluoride etherate (structure XIII).

$$F \blacktriangleright B \leftarrow O \underset{C_2H_5}{\overset{C_2H_5}{<}}$$

XIII

This type of bond formation also occurs in acid-base chemistry, and is frequently the type of bonding one finds between sulfur and oxygen. In particular, it is found in the oxyacids, e.g., sulfuric, nitric, phosphoric, and chloric.

(4) *Hydrogen Bonding.* Unlike the primary bonding interactions previously discussed, hydrogen bonding is a secondary interaction. It may be described as an attractive force that occurs between certain types of molecules. When hydrogen is covalently bonded to the more electronegative elements, e.g., O, N, F, Cl, etc., it becomes somewhat electron-deficient, taking on a partial positive charge. Thus, the hydrogen atom has an increased affinity for the nonbonded electrons on other electronegative atoms in neighboring molecules. This attraction is not usually sufficient to cause the original covalent bond to break. The interaction between water molecules is illustrative of this type of interaction where the actual hydrogen bond is represented by dashed lines in structure XIV.

$$\overset{\delta^+}{}\quad\overset{\delta^-}{}\quad\overset{\delta^+}{}$$

$$\text{---H---}\overset{..}{\underset{..}{O}}:\text{---H---}\overset{..}{\underset{..}{O}}:\text{---H---}\overset{..}{\underset{..}{O}}:\text{---}$$

$$\underset{\text{H}_{\delta^+}}{|}\qquad\underset{\text{H}}{|}\qquad\underset{\text{H}}{|}$$

XIV

Hydrogen bonding is responsible for many of the physical and chemical properties of water and similar molecules. For example, the relatively high boiling point of water is due to the strong association between the molecules through hydrogen bonding. This type of association can also occur between unlike molecules, and plays an important role in solution formation (see Chapter 3) and in water of crystallization.

Hydrogen bonding is also important in interactions between complex molecules, and in the secondary structure of proteins. It is also a secondary binding force in drug-receptor interactions.

(5) *Van der Waals (London) Forces.* These are very weak electrical forces sometimes referred to as induced dipole-induced dipole interactions. The nature of the interaction can be envisioned as the electrons in one atomic or molecular species inducing a repulsive distortion in the electron cloud of a neighboring species. The result is a weak induced dipole. The positive end of the dipole, which is essentially produced by protons in the nuclei, then has an attraction for the oppositely charged electrons in the same or in a neighboring species. Obviously, forces of this nature are active over a very short range, and their strength is dependent upon the polarizabilities of the interacting entities. Van der Waals forces are virtually the only attractive forces between nonpolar molecules. The associations between aromatic hydrocarbon molecules, such as benzene, are due to van der Waals forces. Small disturbances in the electrical balance are present in these molecules because of the motion of the π electrons. The electron distributions do not always coincide with the center of density of protons in the nuclei, thereby producing weak dipoles. These forces also function

in the liquefaction and solidification of the inert gases, a process which requires extremely cold temperatures.

A nuclear repulsion known as van der Waals repulsion occurs when chemical entities approach each other too closely. A balance between the attractive and repulsive forces is responsible for interatomic distances in crystals.

(6) *Other Electrostatic Interactions.* There are several other interactions between atoms, ions, and molecules which will be discussed in more detail in Chapter 3. All of these are, of course, attractive forces between oppositely charged species. Ion-dipole and dipole-dipole interactions take place between polar molecules and ions or other polar species, respectively. Ions and polar molecules are also capable of inducing dipoles in normally non-polar species to set up other types of interactions. These interactions are dependent upon a phenomenon known as polarization which is discussed in the next section.

Polarization. With the exception of homonuclear diatomic molecules and possibly the ionic interactions between Group IA metals and the halogens, pure ionic or covalent bonds are not found in very many inorganic compounds. The common situation is for bonds or molecules to exhibit varying degrees of ionic or covalent character described in terms of *polarity*, which are dependent upon certain factors.

(1) *Polarizability.* The electrons surrounding atomic nuclei in molecules and ions can be influenced by attractive (or repulsive) forces, causing them to shift from a normal distribution to one which is distorted or "lopsided." The more polarizable elements are those which have large atomic radii in which the outer electrons are shielded from the influence of the positively charged nucleus. In general, this property increases from top to bottom in a group of elements. Among ions, as might be expected, anions are more polarizable than cations.

(2) *Polarizing Power.* The influence that an atom, ion, or molecule has to cause polarization in another species is determined by its polarizing power. This is generally viewed as the strength of the electrostatic field; for instance, a small cation can produce extreme polarization in an anion with a large ionic radius or in an easily polarizable molecule. In a given period of the Periodic Table, electron attracting power increases from left to right. This is primarily due to an increase in the number of protons without a significant increase in the atomic radii. This same property decreases from top to bottom for a given group of elements.

(3) *Dipole Moment.* The extent of polarization or polarity in a bond or molecule is frequently expressed as the dipole moment. In polar molecules, one portion is relatively positive and the other relatively negative due to the displacement of electrons in relation to the atomic centers. In diatomic molecules, this situation occurs because of the difference in electronegativities of the two atoms. Such a molecule has a positive pole and a negative

pole separated by some distance, d, and is said to have a dipole moment.
The magnitude of the moment is determined by the product of the magnitude of the charge at one end, and the distance, d, between it and the center
of charge at the other end. If the charge is unity, equal to 4.80×10^{-10}
e.s.u., and the distance is 1 Å, the dipole moment in Debye units, D., is
given by:

$$\frac{(4.80 \times 10^{-10}) \ (1)}{1 \times 10^{-10}} = 4.80 \ D. \qquad \text{(Eq. 1)}$$

Molecules which exhibit polarization by themselves are termed *permanent dipoles*. The term *induced dipole*, used previously, indicates that the
polarity in the molecule was created by the polarizing power of surrounding
atoms, ions, or molecules.

Dipole moments can be experimentally determined and are reflected in
the measurable quantity known as the *dielectric constant*. This property
will be discussed in Chapter 3.

The structural aspects of polarization will now be considered in more
detail. Water constitutes a good example of a polar molecule. The polarization in the O—H bond has already been mentioned in connection with
hydrogen bonding. Since oxygen is more electronegative than hydrogen, it
has a greater attraction for the electrons in the bonds. The polarity in the
bonds gives the water molecule some ionic character, but it is *not* ionized
in the technical sense. However, the dynamic motions of electrons do
allow some ionization at given instants. Other polar molecules that exhibit
limited ionization similar to water include H_2S, H_2O_2, NH_3, HCN, C_2H_5OH,
and $C_6H_5NH_2$. Structure XV below illustrates the polarity in the water
molecule, and the direction of dipole moment is indicated by the arrow
pointing toward the negative end.

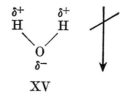

XV

When oppositely charged ions approach each other the cation attracts
the electrons surrounding the anion. Ultimately, this leads to a distortion
of the electron cloud around the anion, termed *polarization*. Ions may be
visualized as spherically symmetrical distributions of charge about the
respective nuclei. A polarized ion has an elongated or lopsided appearance
in which the electron distribution is no longer spherical (structure XVI).

Polarization occurs to a greater extent in anions because of their relatively large size and minimized nuclear attraction. Cations are polarized to a much smaller extent. As ion polarization increases, the electrons begin to be shared to a greater degree, thus increasing the covalent character of the bond.

XVI

The effect of polarization upon the passage from ionic to covalent bonding was described many years ago by K. Fajans. His description aids in the conceptual development of this process, and is based on the ionic charge, ionic radius, and valence shell structure. Briefly, the following points illustrate Fajans' concept.

TABLE 1-6.   THE CRYSTAL RADII OF IONS

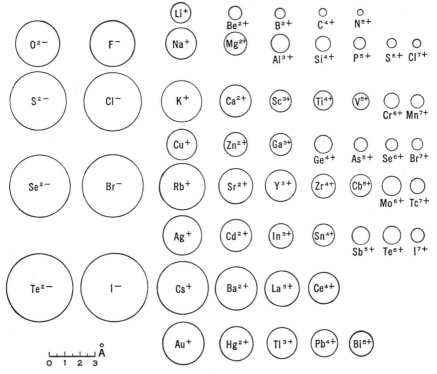

From The Nature of the Chemical Bond, 3rd ed., by L. Pauling, Cornell University Press, Ithaca, N.Y., p. 516, 1960. By permission.

(a) A sizeable charge on a cation, particularly if the radius is small, will have a pronounced polarizing effect on other entities in comparison to less highly charged ions. Thus, one would expect a greater distortion of an anion by Al^{+3} than by Mg^{+2} which in turn would have a greater effect than Na^+ (see Table 1–6).

(b) A small cationic radius coupled with a high charge, in other words, a large charge-to-radius ratio, q/r, improves the polarizing power. Thus, for cations having the same charge, the polarizing effect will increase as the ionic radius decreases ($Be^{+2} > Mg^{+2} > Ca^{+2} > Sr^{+2} > Ba^{+2}$).

(c) Anions with large radii, particularly those having more than one negative charge (e.g., S^{-2}, Se^{-2}), are easily polarized. This is partly due to the fact that the electrons are at a relatively great distance from the positively charged nucleus, and can be influenced more readily by a neighboring cation. Combinations of small cations and large easily distorted anions almost always lead to covalent bonding (e.g., HI).

(d) Cations that have noninert gas electron configurations seem to have a greater polarizing power on anions than do those possessing inert gas configurations. This is believed to be due to the rather incomplete shielding of the nuclear charge in the noninert gas configurations, permitting more of the excess nuclear charge to penetrate the electron shell. The difference is particularly apparent when comparing the polarizing abilities of cations with the same charge-to-radius ratios but with differing electron configurations. Thus, the covalent character of CuCl as compared to the ionic character of NaCl can be rationalized.

Each compound would be expected to be affected differently (for instance, different degrees of ionization). The larger an ion and the less compact its electronic structure, the more easily the electrons are polarized. Similarly, the smaller ions (fewer electrons) will be less easily distorted. A main point is that small ions have the positive charge from the nucleus nearer the "surface," thus exerting a greater polarizing power.

Resonance. In the above discussions, the dynamic motion of electrons has been mentioned frequently. The impact of this motion is illustrated by the fact that molecules can be represented on paper by various electronic "pictures" known as *canonical structures* or *forms*. Each of these structures makes a certain contribution to the actual structure of the molecule, and the actual structure is said to be a *resonance hybrid* of its canonical forms. The term *resonance* indicates a tendency for the electrons to be somewhat *delocalized*, a further reference to their dynamic properties. Simple covalent molecules tend to be more localized than more complex systems. For example, consider the diatomic molecule HCl as represented by the following canonical structures:

$$H^+ \quad :\overset{..}{\underset{..}{Cl}}:^- \longleftrightarrow H:\overset{..}{\underset{..}{Cl}}: \longleftrightarrow H:^- \quad \overset{..}{\underset{..}{Cl}}:^+$$

$$\text{XVII-a} \qquad \text{XVII-b} \qquad \text{XVII-c}$$

The double-tipped arrows indicate that these structures are in resonance. This is not meant to show an equilibrium situation; in fact, equilibrium between these structures does *not* exist. Structures XVII-a and XVII-b represent familiar bonding extremes. Structure XVII-a is ionic with the more electronegative chlorine taking the bonding pair of electrons. Structure XVII-b is the covalent situation with the bonding pair being shared between the two atoms; it should be recalled that the greater electronegativity of the chlorine makes structure XVII-b a *polar* covalent form. Structure XVII-c is not likely to contribute very much to the over-all structure of HCl because of the relatively low electron affinity of the hydrogen for the bonding pair of electrons. For all practical purposes, the structure of HCl can be said to be a resonance hybrid of structures XVII-a and XVII-b. The structures are not weighted equally, but each makes some mathematical contribution to a valence bond or quantum mechanical description of the molecule.

Unsaturated compounds, due to the greater mobility of the electrons, yield more possibilities for canonical forms. Using carbon dioxide (CO_2) as an example, the following structures can be drawn:

$$\overset{+}{:}\text{O}:::\text{C}:\overset{-}{\overset{..}{\underset{..}{\text{O}}}}: \longleftrightarrow \overset{+}{:}\overset{-}{\text{O}}::\text{C}:\overset{..}{\underset{..}{\text{O}}}: \longleftrightarrow :\overset{..}{\underset{..}{\text{O}}}::\text{C}::\overset{..}{\underset{..}{\text{O}}}: \longleftrightarrow :\overset{..}{\underset{..}{\text{O}}}:\text{C}::\overset{+}{\overset{-}{\text{O}}}: \longleftrightarrow :\overset{..}{\underset{..}{\text{O}}}:\text{C}:::\overset{+}{\text{O}}:$$

XVIII-a	XVIII-b	XVIII-c	XVIII-d	XVIII-e

Once again, carbon dioxide is *not* in equilibrium between these structures. Structure XVIII-c is the one usually drawn for CO_2, but the actual structure is a resonance hybrid in which each of the others makes a weighted contribution to the overall electronic picture. Note that structures XVIII-a and XVIII-e and structures XVIII-b and XVIII-d are equivalent on the basis of symmetry. A similar group of structures can be drawn for the sulfate ion, SO_4^{-2}, where the canonical forms would show that both the single and double S—O bonds in the resonance hybrid are identical. Experimental data seem to corroborate these concepts.

Canonical structures may be drawn for any molecule, and their relative contributions to the resonance hybrid may be calculated theoretically using quantum mechanics.

Coordination Compounds and Complexation

It is now desirable to discuss some rather specialized compounds involving metallic elements. This is the area of *coordination chemistry;* it is largely transition metal chemistry, but also involves metals beyond the second period outside the transition series to some extent. The peculiarity of these compounds is that the metallic cation appears to be able to bond with additional anions or molecules after the normal valence requirements have been satisfied. The additional bonding species are usually termed *ligands,* and appear to bond directly to the metal cation in accordance with *maximum coordination numbers* (Table 1–7). As the term implies, this is the maximum number of ligands that can be accommodated by a metal ion, and is a property of the metal and its charge. The ligands occupy space about the metal known as the *coordination sphere,* and, although they may be displaced by other ligands, they do not normally dissociate (ionize) from the metal. Therefore, the metal and its associated ligands constitute what is known as a *complex ion,* if charged; the neutral complex or the complex ion with its counter ions is known as a *coordination compound.* Some complexes are stable in crystalline form and decompose in solution, while others are stable only in solution.

The most widely accepted theory of the formation of complex ions was proposed by Alfred Werner, a Swiss chemist, who devoted some twenty years of his life to studying them, an effort which earned him the Nobel Prize in chemistry. In further recognition of his work, these entities are sometimes referred to as Werner compounds or Werner complexes. An example of the formation of some coordination compounds can be seen with $FeCl_3$, a simple compound of trivalent iron (III) and chlorine. When this compound is dissolved in water and/or hydrochloric acid the following coordination compounds are formed:

$$FeCl_3 + 3H_2O \rightarrow [Fe(H_2O)_3Cl_3] \tag{xii-a}$$

$$FeCl_3 + HCl + 2H_2O \rightarrow H[Fe(H_2O)_2Cl_4]$$
$$= H^+ + [Fe(H_2O)_2Cl_4]^- \tag{xii-b}$$

$$FeCl_3 + NaOH + 2H_2O \rightarrow Na[Fe(H_2O)_2(OH)Cl_3]$$
$$= Na^+ + [Fe(H_2O)_2(OH)Cl_3]^- \tag{xii-c}$$

The brackets enclose the metal and its coordinated groups or ligands, and the quantities on the right-hand side of the equals sign indicate how the compound would dissociate. The main points of Werner's theory are:[9,10]

(1) Two types of valency are observed for the metals. One is the primary (ionizable) or principal valence, and the other is the secondary (nonionizable) valence.

(2) Each metal exhibits a specific, maximum number of secondary valences called the coordination number (see Table 1–7).

(3) The primary valences of a metal are filled by anions, but the secondary valences (in the coordination sphere) may be satisfied by anions and/or neutral molecules known as ligands. Cationic groups sometimes (but rarely) are present.

(4) The ligands are arranged around the metallic ion in certain characteristic geometries. Those compounds having coordination number 2 are either linear or angular, 3 are trigonal-coplanar or trigonal-pyramidal,

TABLE 1-7. METALLIC ION COORDINATION NUMBERS

Cr^{+++} 6		Pt^{++} 4	
Mo^{+++} 8		Cu^{+} 2	
Mn^{+++} 6		Cu^{++} 4	
Fe^{++} 6		Ag^{+} 2	
Fe^{+++} 6		Au^{+} 2	
Co^{+++} 6		Au^{+++} 6	
Ni^{+++} 6		Zn^{++} 4	
Ni^{++} 4		Cd^{++} 4	
Pd^{++++} 6		Hg^{++} 4	
Pd^{++} 4		Al^{+++} 6	
Pt^{++++} 6			

From Roger's Inorganic Pharmaceutical Chemistry, 8th ed., by T. O. Soine and C. O. Wilson, Lea & Febiger, Philadelphia, 1967. By permission.

4 are usually tetrahedral or square-planar, 5 are square-pyramidal or trigonal-bipyramidal, and 6 are ligands arranged octahedrally.

The most stable complexes are formed by cations of the transition series, and particularly the transition elements in Group VIII. The groups immediately preceding and following in Groups VIB, VIIB, IB, IIB, and IIIA also form stable coordination compounds. The major criteria for maximal stability of the metal in a complex involve a high positive charge, a small cationic radius, and unoccupied d orbitals. As these criteria become more difficult to meet, the stability of the complexes diminishes. This would be the expected result of using metals from groups on either side of Group VIII. Table 1–7 shows the coordination numbers for some of the more common complex-forming metal ions.

Properties of Ligands. It was mentioned above that ligand species in complexes are generally anions or neutral molecules. Neutral atoms are not usually found as coordinating agents. The one feature that all ligands have in common is the possession of at least one nonbonded pair of electrons which is used to form a coordinate covalent bond with the metal ion. It should be pointed out that although a large body of evidence supports coordinate covalent bonding as the major bonding type in complexes, it is not yet universally accepted, and some complexes do show other types of ligand bonding (e.g., ionic). However, in this discussion, the assumption will be made that the ligand bonds to the metal through coordinate covalent interactions in which the ligand is the donor species, and the metal cation is the acceptor.

The more stable complexes are formed with anionic or molecular ligands involving the elements of Groups VA, VIA, or VIIA. Generally speaking, the order of stability of a ligand in a complex follows the order of basicity of the ligand. This actually refers to the strength of the electrostatic field emanating from an anion or, in the case of a neutral molecule, the "availability" of the lone pair of electrons. The basicity may be more correctly

TABLE 1-8. COMPLEXING LIGANDS

Neutral Ligands			Anionic Ligands
NH_3	H_2O	R_3As	CN^-
NH_2R	HOR	R_3P	$S_2O_3^{-2}$
NHR_2	ROR	R_2S	F^-
NR_3	RCOR	PX_3	OH^-
	RCHO		Cl^-
			Br^-
			I^-

decreasing stability (Neutral Ligands) — decreasing stability (Anionic Ligands)

R = Alkyl or aryl radical; X = halogen.

From T. Moeller, Inorganic Chemistry, New York, John Wiley & Sons, p. 236, 1952. By permission.

related to the Lewis base concept rather than to that of Brönsted (Chapter 4). The fallacy in considering only the basicity of the ligand in complex stability is that the metal ion also has some effect, and therefore relative stabilities differ in some cases from what would have been predicted on the basis of ligand properties alone. However, for general purposes, Table 1-8 lists some of the more important neutral and anionic ligands in the order of their stability in complexes.

For the most part, the ligands in Table 1-8 effectively have one non-bonded pair of electrons to donate to the metal ion; thus, they are termed *unidentate*, indicating that they each form *one* coordinate covalent bond in a complex. Other ligand species may then be classified according to the number of positions on the molecules capable of coordinating with a metal.

Bidentate ligands have two positions arranged so that they both can act simultaneously as donor sites in a complex. Some of these important to pharmacy and medicine are ethylenediamine (en), glycinate (gly), the dianion of oxalic acid (oxalate), and the anion of 8-hydroxyquinoline (oxinate) (see Table 1-9). Other polydentate ligands are similarly classified as *tridentate, tetradentate, pentadentate,* and *hexadentate*. Representative examples of some of these are shown in Table 1-9.

When polydentate ligands complex a metal ion a ring structure is produced composed of the metal and the ligand molecule. These ring structures have special significance, and are termed *chelates* from the Greek word *chele,* meaning claw. The structure below (XIX) shows a simple chelate between a metal ion and ethylenediamine. In general, the more stable chelates are those where the total number of atoms in the ring

TABLE 1–9.   SOME COMMON POLYDENTATE LIGANDS

Bidentate Ligands

Ethylenediamine (en)

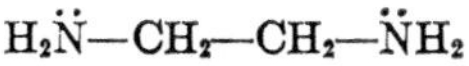

$$H_2\ddot{N}-CH_2-CH_2-\ddot{N}H_2$$

Glycinate (gly)

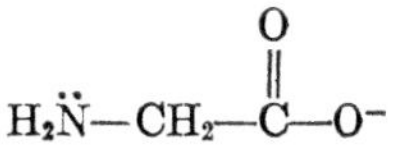

$$H_2\ddot{N}-CH_2-\overset{\displaystyle O}{\overset{\|}{C}}-O^-$$

Oxalate

$$^-O-\overset{\displaystyle O}{\overset{\|}{C}}-\overset{\displaystyle O}{\overset{\|}{C}}-O^-$$

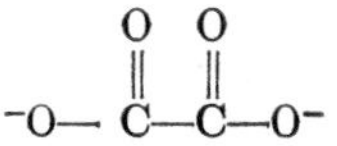

Anion of 8-hydroxy-
quinoline (oxinate)

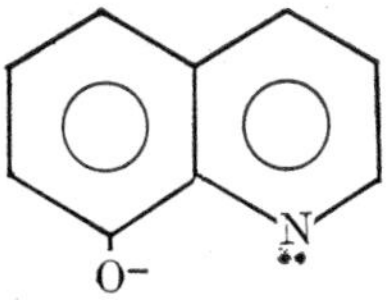

Tridentate Ligand

Diethylenetriamine
(den)

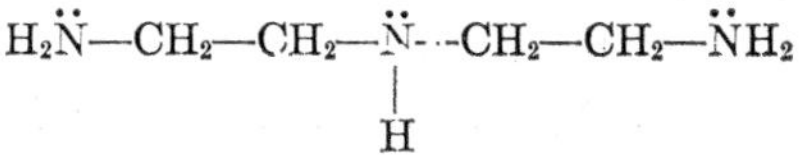

$$H_2\ddot{N}-CH_2-CH_2-\overset{\displaystyle }{\underset{H}{\ddot{N}}}-CH_2-CH_2-\ddot{N}H_2$$

Tetradentate Ligand

Triethylenetetramine
(trien)

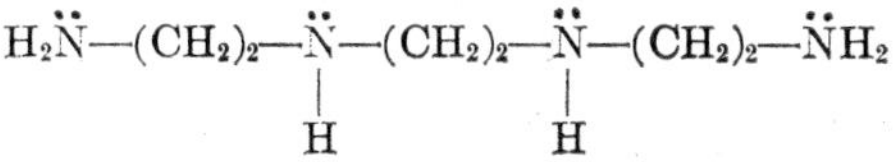

$$H_2\ddot{N}-(CH_2)_2-\overset{\displaystyle }{\underset{H}{\ddot{N}}}-(CH_2)_2-\overset{\displaystyle }{\underset{H}{\ddot{N}}}-(CH_2)_2-\ddot{N}H_2$$

Hexadentate Ligand

Ethylenediamine-
tetraacetate
(EDTA)

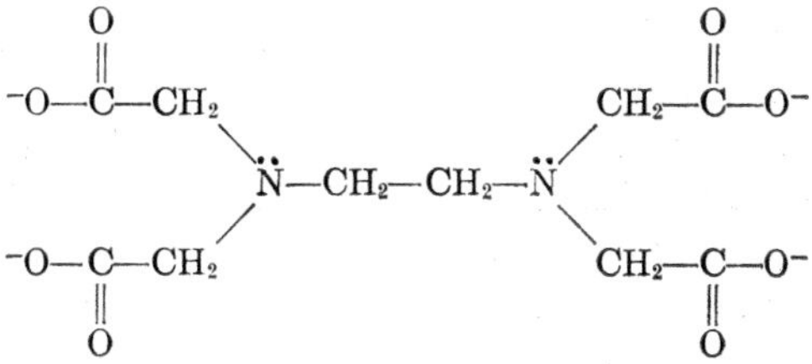

Octadentate Ligand

Diethylenetriamine-
pentaacetate
(DTPA)

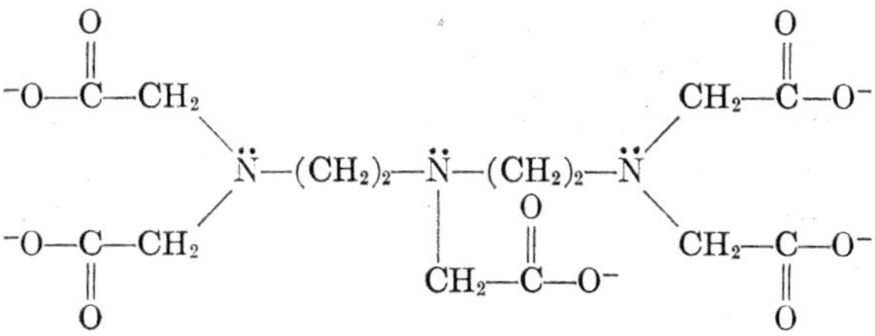

including the metal are *five, six,* or *seven.* Four- and eight-membered rings are usually unstable.

$$H_2N \diagup \overset{M^{+n}}{\diagdown} NH_2$$
$$\diagdown CH_2-CH_2 \diagup$$

XIX

The process of *chelation* is employed in pharmaceuticals and in drug therapy. The polydentate ligands used for chelate formation are generally referred to as *chelating agents.* The term *sequestering agent* is usually applied when a polydentate ligand is used to improve the solubility and/or to stabilize a metal ion by chelation (*sequestration*). Chelating agents will be discussed in more detail later in this chapter.

Bonding in Complexes. There are several approaches to a discussion of bonding in metal complexes, none of which is entirely satisfactory in explaining all the properties of these compounds. Working on a coordinate covalent bond concept, Pauling developed a *valence bond theory* which is a useful qualitative picture of bonding in coordination compounds. Around the same time that this theory was introduced (late 1920s to early 1930s), Bethe and Van Vleck independently introduced an electrostatic approach called *crystal field theory.* This theory completely neglects covalent bond formation, but offers more quantitative information. In the 1950s Van Vleck introduced two more comprehensive approaches. One was a modification of crystal field theory which includes covalent bond formation through orbital overlap and is called *ligand field theory.* The other is *molecular orbital theory* which probably offers the most complete qualitative and quantitative picture of bonding, but has many computational problems. The interested student is directed to reference 9 for a more complete discussion of all of these theories.

Since the primary goal of this section is directed toward obtaining a qualitative picture of bonding in complexes, the valence bond approach will be used, recognizing its rather severe limitations in explaining chemical events. One attractive feature of this theory is that it lends itself to an easily formed picture of hybridization and bonding, although, in certain instances, that picture is not entirely correct.

With the above cautions in mind, a valence bond discussion may begin by first reviewing the orientations of d orbitals in space. Reference to Fig. 1–1 shows that only two of the d orbitals are oriented along the axes of a cartesian coordinate system, the $d_{x^2-y^2}$ and the d_{z^2} orbitals. The other three orbitals are directed between the axes. Now consider a hexacoordinate complex with the six ligands each located on an axis of the cartesian coor-

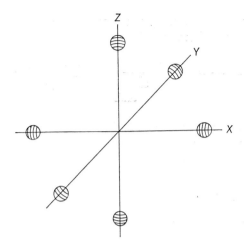

FIG. 1-6.   Six ligands organized on the axes of a three-dimensional cartesian coordinate system.

dinate system as shown in Fig. 1-6.   If overlap of the ligand lone pair orbitals with d orbitals on the metal ion is to take place, the $d_{x^2-y^2}$ and d_{z^2} orbitals are the only ones properly oriented.   The problem, of course, is that according to Hund's rules these two orbitals cannot accept all of the six pairs of electrons being donated by the six ligands.   It is generally accepted that the $(n-1)d$, ns, and np orbitals are close enough in energy to become hybridized into six bonding orbitals which are directed along the same axes occupied by the ligands.   These hybrid orbitals are designated d^2sp^3 hybrids and are equivalent as long as the six ligands are equivalent.

The valence bond problem can now be expressed as one involving an examination of the atomic orbitals on the metal ion to discover which ones are occupied by electrons on the ion and which ones are available (empty) for hybridization to accept ligand electrons.   Of course, this is an ultra-simplification!   Nevertheless, consider the trication of chromium, Cr(III), complexing with six cyanato, CN^-, ions to form $[Cr(CN)_6]^{-3}$.   Chromium (III) is a d^3 ion; that is, it contains three electrons in its $3d$ valence orbital as shown below (XX):

$$\text{Cr(III)}\quad \underline{\uparrow\ \uparrow\ \uparrow\ \_\ \_}\qquad \_\qquad \underline{\_\ \_\ \_}$$
$$\qquad\qquad\quad 3d \qquad\qquad 4s \qquad 4p$$
$$\text{XX}$$

These electrons are unpaired and occupy the three off-axis d orbitals (d_{xy}, d_{yz}, d_{xz}), thus leaving two d, one s, and the three p orbitals empty for bonding with the six cyanato groups.   If these six orbitals hybridize, a valence bond picture of the complex ion can be drawn (XXI) where the x's represent the electrons from the CN^-:

$$[Cr(CN)_6]^{-3} \quad \underbrace{\uparrow\ \uparrow\ \uparrow}_{3d} \quad \underbrace{\times\times\ \times\times\ \times\times\ \times\times\ \times\times\ \times\times}_{d^2sp^3}$$

XXI

Notice that the complex is a trianion because the negative charges from the six cyanato groups are in excess by three of what would be required to neutralize the +3 charge on the chromium.

Similar valence bond reasoning can be applied to other complexes formed with metal ions having one to three electrons in their d orbitals. A problem arises when there are four or more electrons in these orbitals on the metal. For example, complexes formed with iron(III) (a d^5 ion). according to this theory, must alter the normal ground state arrangement of the electrons, use different orbitals for bonding, or bond in a noncovalent manner. If the electronic configuration on the metal was in fact changed from the ground state configuration, it could be measured experimentally by determining the magnetic moment of the complex. This is a measure of the "unpairedness" of electrons in that unpaired electrons are attracted to a magnetic field, *paramagnetic*, and paired electrons are repelled by a magnetic field, *diamagnetic*.

The electronic structure of the valence shell of iron(III) is shown in XXII:

$$Fe(III) \quad \underbrace{\uparrow\ \uparrow\ \uparrow\ \uparrow\ \uparrow}_{3d} \quad \underset{4s}{\text{—}} \quad \underset{4p}{\text{— — —}}$$

XXII

When complexed with water molecules the hexaaquoiron(III) ion is formed, and has the same magnetic properties as would be expected from the electronic structure shown above. The magnetic moment is 5.9 B.M. (Bohr magnetons) which is consistent with five unpaired electrons. This is generally referred to as a *high-spin complex* because the electrons are unpaired as much as possible. Within the framework of valence bond theory, the problem of finding six empty atomic orbitals to overlap with the donor water molecules can be accomplished by assuming that the $4d$ orbitals are of appropriate energy to hybridize with the $4s$ and $4p$ orbitals. Hybridization of this type is termed *outer orbital* hybridization, and still provides six octahedrally arranged hybrids designated sp^3d^2. The electronic configuration of the complex may then be represented by XXIII:

$$[Fe(H_2O)_6]^{+3} \quad \underbrace{\uparrow\ \uparrow\ \uparrow\ \uparrow\ \uparrow}_{3d} \quad \underbrace{\times\times\ \times\times\ \times\times\ \times\times\ \times\times\ \times\times}_{sp^3d^2}$$

XXIII

If the water molecules in complex XXIII are replaced with cyanato groups, the hexacyanoferrate(III) ion results which has distinctly different magnetic properties. The magnetic moment is about 2 B.M., which is indicative of *one* unpaired electron, and the ion is termed a *low-spin complex*. An explanation of this may be found in the fact that the cyanato anion has a negative electrostatic field of sufficient strength to repel the electrons in the two d orbitals which directly oppose the approaching ligands ($d_{x^2-y^2}$ and d_{z^2}). The repulsive interaction forces the electrons to become paired with those in the other d orbitals. The effect of strong field ligands producing low-spin complexes has been noticed with other metals as well. The electronic configuration of this complex involves the $3d$ orbitals in an *inner orbital* hybrid as was seen previously to produce the valence bond picture XXIV:

$$[Fe(CN)_6]^{-3} \quad \underbrace{\text{N} \ \text{N} \ \text{N}}_{3d} \quad \underbrace{\text{XX XX XX XX XX XX}}_{d^2sp^3}$$

XXIV

The octahedral arrangement of hexacoordinate complexes is shown in Fig. 1–7a.

Metal ions with seven, eight, or nine d electrons generally have a coordination number of 4 which leads to either a square planar or a tetrahedral arrangement of the ligands. Here again the field strength of the ligand and the formation of high- and low-spin complexes may be predictive of the type of hybridization and, therefore, the geometry of the complex. For example, a d^8 ion complexing with a ligand having a relatively weak electrostatic field has no d orbitals available for bonding. However, the ligands can bond through sp^3 hybrids formed on the metal to give a tetrahedral complex. If the ligand field strength is sufficient to force the metal into a low-spin diamagnetic state, one d orbital would be vacant and could be used to form a square planar dsp^2 hybrid. Both of these geometries are illustrated in Fig. 1–7b and 1–7c, respectively.

The valence bond approach to bonding and the use of magnetic criteria to predict hybridization and geometry are not successful in many cases. One of the more comprehensive theories (e.g., ligand field or molecular orbital) would actually give a more theoretically satisfying account of bonding.

Complexes and Chelating Agents. Complexes and complexation are important aspects of chemistry and pharmacy. Of course, it is not possible to discuss all applications, and the detailed discussion will be limited to those products used in drug therapy. However, a few other instances where complexation is important can be mentioned.

Complexation plays an important role in analytical chemistry where, for example, concentrations of metals can be determined by titration with

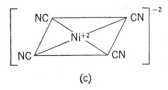

FIG.1–7.   Representative complex ion geometries: (a) octahedral;
(b) tetrahedral; (c) square planar.

complexing agents.   In some analytical solutions containing metal ions, chelating agents are used to solubilize the metal and to stabilize its oxidation state.   Two rather classical examples are found in solutions employed in the identification of reducing substances (e.g., sugars), Benedict's solution and Fehling's solution.   Both of these solutions contain copper(II) ions which are chelated by citric acid in Benedict's and by tartaric acid in Fehling's solution.   Chelating agents are also found as preservatives in preparations subject to decomposition due to trace quantities of metals, such as preparations containing hydrogen peroxide.   Throughout this text, other examples will be found where complexation plays important roles.

Chelating agents occupy a rather unique place in drug therapy.   They are essentially the only compounds which have shown much efficacy in the treatment of heavy metal poisonings from such elements as lead, mercury, iron, etc.   In addition to their usefulness in toxicological problems such as these, they are also being used to treat certain metabolic disorders where metals such as iron and copper are accumulated in abnormal amounts in various tissues. The particular chelating agents discussed in the sections to follow include calcium disodium edetate (EDTA), dimercaprol (BAL), penicillamine, and deferoxamine.

Calcium Disodium Edetate, U.S.P. XVIII (Calcium Disodium Versenate; Calcium Disodium Ethylenediaminetetraacetate; $C_{10}H_{12}CaN_2Na_2O_8 \cdot xH_2O$; Mol. Wt. (anhydrous) 374.28)

XXV

This compound is a mixture of the dihydrate and trihydrate (predominantly the dihydrate) which exists as a white crystalline granule or a white crystalline powder. It is odorless, slightly hygroscopic, and has a faint saline taste. It is stable in air, freely soluble in water, and the pH of an aqueous solution is between 6.5 and 8.0.

The compound is actually the calcium complex of the disodium salt of ethylenediaminetetraacetic acid (EDTA). It is used in the treatment of heavy metal poisoning, primarily that caused by *lead* (plumbism). It may also be employed in poisonings due to copper, nickel, cadmium, zinc, chromium, and manganese, but it is of *no value* in the treatment of toxicities produced by mercury, arsenic, or gold. EDTA preparations have a strong affinity for calcium; therefore, the disodium calcium form is used to avoid inducing hypocalcemic states (low serum calcium). This chelating agent removes lead from the tissues by forming an inactive soluble complex which can be removed from the circulation by the kidneys and excreted in the urine. Reports have indicated that the urinary excretion of lead increases as much as 40 times through the use of calcium disodium edetate. The compound is poorly absorbed from the gastrointestinal tract, and may even aid the absorption of lead which may be in the gut, thereby producing toxic reactions or aggravating an established toxicity.

The usual route of administration is by intravenous (I.V.) injection. The official *Calcium Disodium Edetate Injection,* U.S.P. XVIII, contains not less than 180 mg and not more than 220 mg of the compound in each ml. Intramuscular (I.M.) administration is employed sometimes in diagnosis of metal poisonings. An increase in the excretion of the metal in the urine (500 μg/liter/24 hours or greater for lead) is indicative of toxicity.

Doses. (a) Intravenous infusion of 75 mg/kg of body weight in two divided doses per day, administered in 250 to 500 ml of isotonic sodium chloride or 5% dextrose. One course of treatment would normally last about five days, with up to two-week intervals between courses.

(b) Intramuscular injection of 75 mg/kg administered as a 20% solution in 0.5% to 1.5% procaine which serves to nullify the pain associated with the injection.

Preparations. Calcium Disodium Versenate® (Riker), a solution containing 200 mg/ml for injection.

Disodium Edetate, U.S.P. XVIII (Disodium Ethylenediaminetetraacetate; $C_{10}H_{14}N_2Na_2O_8 \cdot 2H_2O$; Mol. Wt. 372.24)

XXVI

This compound is a white crystalline powder which is soluble in water, providing an aqueous solution of pH between 4.0 and 6.0.

Disodium edetate will chelate the same metals as the disodium calcium form (see above). However, its added affinity for calcium limits its usefulness as an agent for the treatment of toxicities due to these other metals. The chance of hypocalcemia during such therapy definitely exists. The primary use of disodium edetate is in conditions related to *hypercalcemic* states (high serum calcium). The compound may be useful in treating such problems as occlusive vascular disease and cardiac arrhythmias when associated with high blood levels of calcium. Other problems resulting in hypercalcemia may be symptomatically aided by treatment with this chelating agent. It is apparently of no value in aiding dissolution of urinary calculi (calcium-containing stones in the urinary tract).

The usual route of administration is by intravenous injection. The official Disodium Edetate Injection, U.S.P. XVIII, contains varying amounts of the disodium and trisodium salts due to the effects of pH adjustment.

Doses. Intravenous infusion of 50 mg/kg of body weight dissolved in 500 ml of isotonic sodium chloride or 5% dextrose. The infusion is done over a period of three to four hours once a day for a term of five days. Two or three therapeutic courses may be required. There is general disagreement concerning proper dosage and therapeutic value.

Preparations. Endurate® (Abbott), a solution containing 150 mg/ml for injection.

Dimercaprol, U.S.P. XVIII (2,3-Dimercapto-1-propanol; BAL; $C_3H_8OS_2$; Mol. Wt. 124.22)

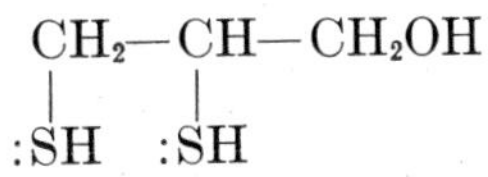

$$CH_2—CH—CH_2OH$$
$$|\qquad |$$
$$:\overset{..}{S}H \quad :\overset{..}{S}H$$

XXVII

This compound is a colorless or almost colorless liquid having a disagreeable, mercaptan-like odor. It is soluble in water, alcohol, and benzyl benzoate.

Certain heavy metals, such as trivalent arsenic, owe their cellular toxicity to the "tying up" of sulfhydryl (—SH) groups present in enzymes which are responsible for oxidation-reduction reactions in tissues. Presumably, this inactivation involves covalent bond formation between the metal and the sulfhydryl groups, as illustrated in XXVIII.

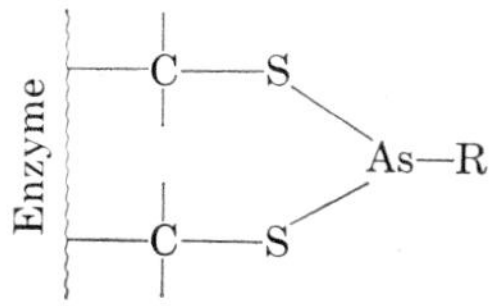

XXVIII

Knowledge of this association led to the idea that the use of simple dithiol compounds (those containing sulfhydryl groups) as competitors with the enzymes for these metals might serve to prevent toxic reactions. The idea proved to be successful and subsequent work resulted in the introduction of dimercaprol or BAL (British anti-lewisite) as an effective neutralizing agent for arsenical war gases, such as lewisite. Following this use, the compound was used with marked success in the treatment of arsenic poisoning from other sources, and has been extended to the treatment of mercury and gold poisoning.

The compound forms stable mercaptides (see structure XXVIII) of the metals which are excreted in the urine. It appears to be of value in the treatment of toxic reactions due to arsenic and gold. Its effectiveness in the treatment of mercury poisoning is dependent upon its use within a few hours following ingestion. The lapse of a longer period of time diminishes its efficacy. Dimercaprol has also been shown to improve the excretion of lead and copper (Wilson's disease), but it is not the agent of choice for these metals. It is contraindicated in poisonings due to iron, cadmium, or selenium because the resulting complexes have greater renal (kidney) toxicities than do the free metals. Dimercaprol-metal chelates tend to dissociate in acid media; therefore in therapy the urine should be alkalinized (e.g., with sodium bicarbonate, see Chapter 5) to prevent the release of free metal, producing renal toxicity.

The usual route of administration is by intramuscular injection. The official *Dimercaprol Injection*, U.S.P. XVIII, is a solution of dimercaprol in a mixture of benzyl benzoate and vegetable oil. The solution contains the equivalent of 100 mg of the compound in each ml.

Doses. The recommended dosage schedule varies with the severity of the toxicity. In severe arsenic or gold poisoning, intramuscular injection of 3 mg/kg is given six times a day for two days, four times on the third day, then twice daily for the next ten days.

Mercury poisoning requires 5 mg/kg initially (within a few hours of ingestion) followed by 2.5 mg/kg twice daily for ten days.

Preparations. BAL in Oil (Hynson, Westcott and Dunning), a solution containing 100 mg/ml in peanut oil for I.M. injection.

Penicillamine, U.S.P. XVIII (D-(−)-3-Mercaptovaline; β,β-Dimethyl-cysteine; $C_5H_{11}NO_2S$; Mol. Wt. 149.21)

$$CH_3\underset{:SH}{\overset{CH_3}{\underset{|}{\overset{|}{C}}}}\underset{:NH_2}{\overset{H}{\underset{|}{\overset{|}{C}}}}\overset{O}{\overset{\|}{C}}OH$$

XXIX

This compound is a white or offwhite crystalline powder, having a slight characteristic odor. It is freely soluble in water, and slightly soluble in alcohol. The pH of an aqueous solution is between 4.5 and 5.5.

Penicillamine is a chelating agent capable of forming soluble complexes with copper, iron, mercury, lead, gold, and other metals. However, its use has been reserved for the improvement of copper excretion in patients with hepatolenticular degeneration (degenerative changes in the brain associated with increased levels of copper in the tissues and degeneration of the liver, also known as Wilson's disease). This is a rare disease resulting from a familial inability to regulate copper balance, with the consequence that toxic amounts of copper are deposited in tissues such as the eye, liver, brain, and kidney. The disease was customarily treated with calcium disodium edetate or dimercaprol. In recent years penicillamine[11,12] has been shown to be more effective in promoting urinary excretion of the excess copper in the chelated form. The effectiveness of penicillamine is related to its resistance to metabolic inactivation by amino acid oxidase since it lacks a hydrogen on the beta-carbon atom. A further aspect has been proposed[13] relating its superiority to the ability of its sulfhydryl group to reduce the copper(II) in the tissues to copper(I). The proposal is that protein-copper(II) complexes have square planar geometries. Copper(I) must be complexed tetrahedrally, which may limit competing reactions between tissue protein and the tetrahedral dipenicillamine-copper(I) complex. A probable structure of the complex between penicillamine and copper is illustrated in structure XXX:

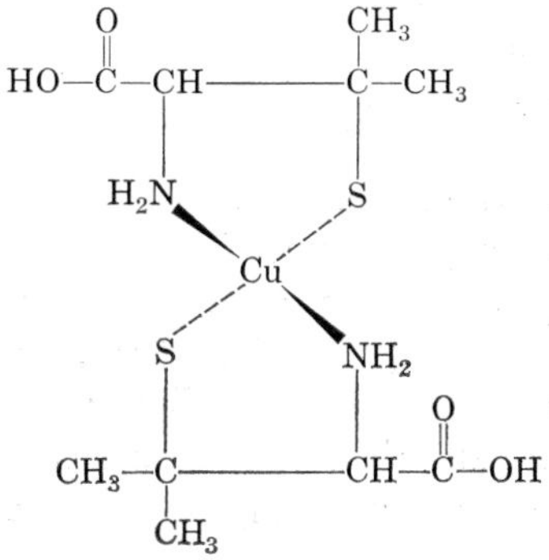

XXX

Another use of penicillamine which is being investigated is the treatment of gold dermatitis in patients on chronic gold therapy.[14] It has also been used in the treatment of cystinuria (the presence of crystals of the amino acid cystine in the urine) which is not related to its metal chelating abilities.

Unlike the other chelating agents discussed heretofore, the usual route of administration of penicillamine is oral. *Penicillamine Capsules* are official in the U.S.P. XVIII.

Doses. The usual oral dose is 250 mg given four times a day. Doses may be gradually increased on an individual basis to a maximum of 5 g daily. Low copper diets or a cation exchange resin is frequently employed during therapy. Sulfurated potash (N.F. XIII) has also been used in doses of 40 mg to minimize the absorption of dietary copper.

Preparations. Cuprimine® (Merck Sharp and Dohme), capsules containing 250 mg of penicillamine for oral administration.

Deferoxamine Mesylate, (Desferrioxamine B; N-(5-[3-[(5-amino-pentyl)-hydroxycarbamoyl]propionamide]pentyl) - 3 - [[5 - (N - hydroxyacetamido)-pentyl]-carbamoyl]propionhydroxamic acid)

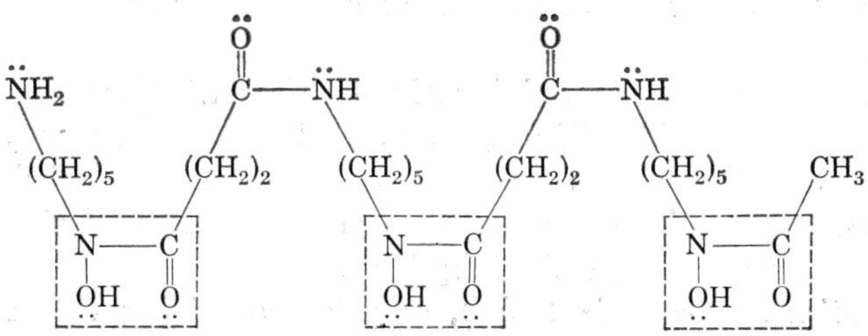

XXXI

This compound is the methylsulfonic acid salt of structure XXXI and is usually available as a white, crystalline, lyophilized powder. It is soluble in water and the aqueous solution is stable at room temperature for two weeks. The outlined groups in structure XXXI are involved in the chelation.

Deferoxamine is produced naturally by *Streptomyces pilosus* as a ferric [Fe(III)] complex. After chemical removal of the iron, the chelating agent is purified as the methylsulfonate (mesylate) salt.

It is a polydentate ligand with a particular affinity for ferric ions with which it forms stable, water soluble, octahedral complexes (XXXII). It does not have a very strong affinity for ferrous or other divalent metal ions. Deferoxamine is used with other indicated drugs and procedures for the treatment of acute iron toxicity (see Chapter 6). It is also under investigation for the treatment of iron storage diseases (e.g., hemochromatosis). The compound is poorly absorbed from the gastrointestinal tract. Administration by this route is not generally recommended.

XXXII

The usual route of administration is by intramuscular or intravenous injection. The former is preferred. Intravenous administration is generally done by slow infusion with isotonic sodium chloride or other electrolyte solution.

Doses. Either route of administration (I.M. or I.V.) has a usual dose of 1 g initially, followed by 500 mg every 4–12 hours which may be given depending on the clinical response of the patient. A total dose of 6 g should not be exceeded in any 24-hour period. The rate of intravenous infusion should not exceed 15 mg/kg/hour.

Preparations. Desferal® (Ciba), ampules containing 500 mg of the lyophilized powder for injection.

A number of other chelating agents are being investigated for various uses. One interesting area of study deals with improving the excretion of some long-lived radioactive isotopes. An agent which has shown some value in this area is trisodium calcium diethylenetriaminepentaacetate (trisodium calcium pentetate), the pentaacid form of which is known as DTPA (see Table 1–9).

References

1. Bohr, Niels. On the constitution of atoms and molecules. Part II. Phil. Mag., **26**:476, 1913.
2. De Broglie, L. A tentative theory of light quanta. Phil. Mag., **47**:446, 1924.
3. Schrödinger, E. Quantisierung als Eigenwertproblem. Ann Phys. (Leipzig), **79**: 361, 1926.
4. Saxson, D. S. Elementary Quantum Mechanics. San Francisco: Holden-Day, 1968.
5. Streitwieser, A., Jr. Molecular Orbital Theory for Organic Chemists. New York: John Wiley & Sons, 1961.
6. Daudel, R., Lefebvre, R., and Moser, C. Quantum Chemistry Methods and Applications. New York: Interscience, 1959.
7. Heisenberg, W. Uber den anschaulichen Inhalt der Quantentheoretischere Kinematik und Mechanik. Z. Physik, **43**:172, 1927.
8. Pauling, Linus. The Nature of the Chemical Bond, 3rd ed. Ithaca, Cornell University Press, p. 88, 1960.
9. Cotton, F. A., and Wilkinson, G. Advanced Inorganic Chemistry. New York, Interscience Publishers, John Wiley & Sons, 1962.
10. Moeller, T. Inorganic Chemistry, New York: John Wiley & Sons, p. 230, 1952.
11. Walsh, J. M. Treatment of Wilson's disease with penicillamine. Lancet, **1**:188, 1960.
12. Scheinberg, I. H., and Sternlieb, I. The long term management of hepatolenticular disease (Wilson's disease). Amer. J. Med., **29**:316, 1960.
13. Peisach, J., and Blumberg, W. E. A mechanism for the action of penicillamine in the treatment of Wilson's disease. Molec. Pharmacol., **5**:200, 1969.
14. Davis, C. M.: D-Penicillamine for the treatment of gold dermatitis. Amer. J. Med., **46**:472, 1969.

2

Group Properties of Elements

Group IA, The Alkali Metals

The principal metals comprising this group are lithium, sodium, potassium, rubidium, cesium, and more recently, francium. The last-named, francium, will not be discussed further.

Table 2–1 indicates a number of regularities in the variation of properties indicative of the fact that Group I metals are among the least complicated in the periodic table. The properties they have in common include: one

TABLE 2–1. PROPERTIES OF THE ALKALI METALS

Properties	Lithium	Sodium	Potassium	Rubidium	Cesium
Atomic Number	3	11	19	37	55
Atomic Weight	6.939	22.9898	39.102	85.47	132.905
Isotopes	6, 7	23	39, 40, 41	85, 87	133
Electrons	2–1	2–8–1	2–8–8–1	2–8–18–8–1	2–8–18–18–8–1
Density	0.534	0.97	0.862	1.525	1.873
Melting Point °C.	180	97.7	63.65	39	28.5
Boiling Point °C.	1336	883	774	696	705
Oxidation States	0, +1	0, +1	0, +1	0, +1	0, +1
Radius (covalent) (Angstroms)	1.23	1.57	2.03	2.16	2.35
Radius (ionic) (Angstroms)	0.60	0.95	1.33	1.48	1.69
Ionization Potential (volts)	5.36	5.12	4.32	4.16	3.87
Oxidation Potential (volts)	+3.02	+2.71	+2.92	+2.93	+3.02
Hydration Energy (Kcal/mole)	136	114	94	—	—
Flame Color	Red	Yellow	Violet	Red	Blue

All tables, unless otherwise indicated, are from Roger's Inorganic Pharmaceutical Chemistry, 8th ed., by T. O. Soine and C. O. Wilson, Lea & Febiger, Philadelphia, 1967. By permission.

valence electron outside of a well-shielded core; and the core representing that part of the atom within the valence shell, i.e., the nucleus and inner electrons. This valence electron is capable of easy removal (ionization), with the relative difficulty of removal decreasing as the atomic radius increases, i.e., as the group is descended. This, of course, is to be expected as a consequence of the increased distance of the electron from the central nuclear charge, and is further reflected by the decreasing ionization potentials when progressing from lithium to cesium. The removal of a second electron, however, presents quite a different picture. For example, the energy required to remove a second electron from the sodium ion, which has an inert gas (neon) configuration, is approximately ten times that required to remove the first electron. A rough approximation of the relative volumes occupied by the valence electron as compared to the core may be found by comparing the atomic radii with the ionic radii. Using sodium as an example, a simple comparison of the relative volumes of the atom versus the ion reveals the fact that the core occupies roughly 20% of the volume of the atom, whereas the valence electron occupies 80% of the volume. It is, therefore, not too surprising that this somewhat voluminous and loosely held electron is so easily removed to form monovalent ions. The above picture also accounts for the low ionization potential of alkali metals, and for the fact that their compounds are predominately found in an ionic form in the solid state. Another way of expressing the "looseness" of the valence electron is to say that the alkali metals have pronounced electropositive character (tendency to go to a positive state). In general, the electropositive character lessens as one proceeds from left to right in the periods of the periodic table and increases in progressing downward in any particular group.

The alkali metals form white solid hydrides quite readily when heated in hydrogen gas. They are among the most stable of the metallic hydrides and exist in an ionic crystal lattice (sodium chloride arrangement) consisting of sodium ions and hydride (H^-) ions. Only lithium reacts readily with nitrogen, even at room temperature, to form the nitride (Li_3N), which decomposes in water to form ammonia and lithium hydroxide.

The alkali metals have a high affinity for oxygen and readily form the ordinary oxides, as well as the higher oxides, with both the affinity and the tendency for higher oxide formation increasing with atomic weight. They also react vigorously with water to form hydrogen gas and the metallic hydroxide. All the alkali metals bring about this facile reduction of water with the activity increasing as the atomic weight increases, and, for this reason, must be stored under kerosene, coated with paraffin, or protected in some other fashion. Although lithium has a low reactivity (as compared to other alkali metals) with water, much of this can be attributed to the fact that it is the only one that does not melt below the boiling point of water. When a metal melts in hot aqueous solution it has continuous opportunity to expose a new surface to the action of the water. The oxidation potential, given in Table 2–1, measures the extent to which each metal forms its hydrated ion in aqueous solution (also a measure of its action as a reducing agent). It may be noted that the oxidation potential should increase with a decreasing ionization potential and, indeed, this does occur, with the exception of lithium. The deviation of lithium is probably due to the strong hydration of the ion on account of its small size. The oxidation potential of lithium is approximately equal to that of cesium; it is a better reducing agent than the latter because of the strong hydration effect.

The alkali metal hydroxides are all alkaline in aqueous solution (see Chapter 4) with the alkalinity naturally increasing as the ionic radius increases. In a similar way, the alkalinity of the alkali metal hydroxides is greater than that of the alkaline earth metal hydroxides (Group IIA) which, in turn, is greater than that of the Group III hydroxides. Thus, a generalization can be made that the alkalinity of metallic hydroxides increases from right to left and from top to bottom of the periodic table stating, in effect, that the greater the radius of the cation the greater will be the tendency for alkalinity.

Virtually all salts of the alkali metals are water soluble (see Chapter 3) and, as a consequence, chemical incompatibilities of a solubility nature due to the metal cation are rare. The general rule is that the salts of alkali metals are more soluble than the salts of any other periodic group. According to Sidgwick,[2] there is a rather simple generalization concerning the order of solubilities of alkali metal salts. Briefly, he notes that when the anions are derived from strong acids, the lower atomic weight alkali metals form the most soluble salts; with anions from weak acids the opposite is true. These solubility generalizations can be rationalized on the basis that with weak acids the hydrogen must be held rather firmly and, because the lithium ion is small in size, it too will be held strongly in the solid state. On the other hand, the anions of strong acids are usually rather large, and it is well known that the stability of crystals is lowered somewhat by large discrepancies in the relative sizes of the cation and anion (see Chapter 3).

Generally speaking, the alkali metals do not form complexes, although a few are known. If one includes the hydrated and ammoniated ions as complexes then these solvated forms represent perhaps the most common types that will be encountered. The degree of solvation of the ions tends to diminish in going from lithium to cesium, as indicated by the increase in mobility of the ions with increasing ionic radius. Although the smaller cations have a stronger tendency to hydrate, the number of dipoles (water molecules) that can be held is limited, so that Li^+ can hold four, Na^+ and K^+ can hold six, and Rb^+ and Cs^+ can hold eight. However, the potential coordinating capacities of K^+, Rb^+, and Cs^+ are rarely, if ever, realized.

The tendency for alkali metal ion hydration to carry over to the solid salts is, for all practical purposes, limited to lithium and sodium. Although potassium salts may possess some water of hydration, this may be ascribed to anion water and, indeed, there is a parallelism between several potassium and ammonium (which cannot hold cation water) salts insofar as comparative hydrates are concerned. The above statements also hold for ammoniates (solid compounds of salts with ammonia or amines).

Reference should be made, finally, to the behavior of the alkali metals with liquid ammonia. If oxidizing impurities are rigidly excluded, all the alkali metals will dissolve in ammonia with the formation of a blue solution. The nature of the process is still obscure. The blue solution can be evaporated to recover the unchanged metal eventually. However, in the presence of catalytic amounts of oxidizing agents, ammonia will rapidly react with the metals to form the alkali amides (e.g., $NaNH_2$). The rapidity of reaction is correlated with increasing atomic weight (i.e., $Cs > Rb > K > Na > Li$).

Group IB, The Coinage Metals

The three members of this family are designated as the "coinage metals" because, from early times, they have been employed for ornamental and coinage purposes. These elements form Division B of Group I in the periodic table, but their properties differ in many respects from those of the alkali metals which comprise Division A of this same group. It is quite evident that Mendeléeff was cognizant of these facts because he gave them an alternate place in Group VIII. This latter classification associated them with nickel, palladium, and platinum, to which they are closely related. The differences between the alkali metals and the coinage metals are tabulated on page 52.

All these elements occupy positions in their respective periods representing the end of a transition sequence. Table 2–3 indicates that each has a core with an outer shell of 18 electrons, together with a single valence electron in the outermost shell. They resemble the alkali metals with respect to the single valence electron, but the resemblance can hardly be construed as going any further. The differences have been well illustrated in the comparative table given (see Table 2–2).

The table of physical properties (Table 2–3) indicates that the monovalent cations of Group IB are smaller in size than those of the alkali metals and that they have a higher ionization potential. Because of their size and because of the somewhat imperfect screening (as compared to inert gas configuration kernels) of nuclear charge by the 18-electron shell, these elements have a tendency toward covalent bond formation with increasing atomic number. This is certainly much stronger and in a reverse direction in the group than the corresponding tendency of the alkali metals, although the ions are approximately in the same size range. In the polyvalent

TABLE 2–2. COMPARISON OF GROUP IA AND IB METALS

Alkali Metals	Copper, Silver, Gold
A. Do not occur free in nature.	A. Occur free in nature and are easily recovered from their compounds by reduction.
B. Very active chemically; displace all other elements from their compounds. The chemical activity increases as the atomic weight increases.	B. Are low in the electromotive series and hence are not very active chemically; they are displaced by most other metals. The chemical activity decreases as the atomic weight increases.
C. Oxides and hydroxides are strongly basic.	C. Oxides and hydroxides are feebly basic (except Ag_2O which is an active basic oxide).
D. Alkali halides are soluble in water, and are not hydrolyzed.	D. Silver, Copper(I) and Gold(I) halides are nearly insoluble in water. With the exception of the silver halides they are readily hydrolyzed and form numerous basic salts.
E. Univalent, forming but one series of compounds.	E. Copper(I) and Copper(II) each form a series of compounds; Silver(I), one series; and Gold(I) and Gold(III), one series each.
F. Form simple cations, never occur in complex anions, and do not form complex cations with ammonia.	F. All of them form complex anions, e.g., $Cu(CN)_2^-$, $Ag(CN)_2^-$, $Au(CN)_2^-$, and complex cations with ammonia, e.g., $Ag(NH_3)_2^+$, $Cu(NH_3)_4^{++}$ and $Au(NH_3)_2^+$.
G. All are rapidly oxidized in air.	G. Copper is only slowly oxidized in air, but is rapidly oxidized when finely divided and heated in oxygen.

TABLE 2–3. PROPERTIES OF THE COINAGE METALS

Properties	Copper	Silver	Gold
Atomic Number	29	47	79
Atomic Weight	63.54	107.870	196.967
Isotopes	63, 65	107, 109	197
Electrons	2–8–18–1	2–8–18–18–1	2–8–18–32–18–1
Density	8.94	10.5	19.32
Melting Point °C.	1083	960.5	1063
Boiling Point °C.	2595	2000	2600
Common Oxidation States	0, +1, +2	0, +1	0, +1, +3
Radius (cov.) Å	1.17	1.34	1.34
Radius (ionic) Å	0.69 (+2) 0.95 +1)	1.13	1.2 (probable)
Ionization Potential, first (volts)	7.724	7.574	9.223
Oxidation Potential, $M \rightarrow M^+$ (volts)	−0.522	−0.7995	∼1.68

states, the increased ionic charge results in greater covalency in the compounds, notably in the case of trivalent gold where the covalency has been increased by coordination to 4 (e.g., $HAuCl_4$). In fact, there is good reason to believe that polyvalent gold does not exist as such at all but is always present as a complex with a coordination number of 4. It is curious to note that, judging from the crystal structures of the monovalent halides of copper and silver, the copper compounds are all covalent in nature whereas the silver compounds (with the exception of the iodide) are all ionic. It has been shown, however, that in the gaseous state the silver halides are all covalent, a not too surprising circumstance if one considers that in the gaseous state the ion pair is freed from other ion attractions and can lead to greater deformation of the anion. The potentiality of this happening has already been indicated by the covalent nature of solid AgI, inasmuch as iodide is one of the most deformable ions.

Two different valence states exist, at least for copper and gold and very probably for silver, and give rise to compounds which can almost be considered as having originated from two different elements. Although the cores have a complete shell of 18 electrons, this shell can contribute one or more of these electrons to form bonds in addition to using the normal valence electron. When the element is in the monovalent state, that is, when the cation is the normally expected one, the ion is colorless and diamagnetic. However, in the polyvalent form where the 18-electron outer shell has become deficient in electrons, the ions correspond in structure to the transition elements and, accordingly, are colored and paramagnetic. The transition elements with partially filled d shells are usually colored, and other ions that possess this incomplete shell are no exception.

There is a definite tendency toward disproportionation of the monovalent ions, particularly of copper and gold, to the free element and to the higher oxidation state (autooxidation). Thus, the following reactions can be written:

$$2Cu^+ \rightleftharpoons Cu^{+2} + Cu^0 \tag{i}$$
$$2Ag^+ \rightleftharpoons Ag^{+2} + Ag^0 \tag{ii}$$
$$3Au^+ \rightleftharpoons Au^{+3} + 2Au^0 \tag{iii}$$

The point of equilibrium in the above is far to the right in the case of copper and gold, but to the left for silver. Thus, salts of monovalent copper and gold are unstable in a medium that permits ionization, whereas salts of monovalent silver are stable. It is because of these instabilities that the commonly occurring valence states are 2 for Cu, 1 for Ag and 3 for Au.

Group IIA, The Alkaline Earth Metals

Consultation of the accompanying table of properties (Table 2–4) illustrates the fact that the typical elements of Group IIA are all bivalent. They

Group Properties of Elements

TABLE 2-4. PROPERTIES OF THE ALKALINE EARTH METALS OF GROUP IIA

Properties	Beryllium	Magnesium	Calcium	Strontium	Barium	Radium
Atomic Number	4	12	20	38	56	88
Atomic Weight	9.0122	24.312	40.08	87.62	137.34	226.05
Isotopes	9	24, 25, 26	40, 42, 43, 44, 46, 48	84, 86, 87, 88	130, 132, 134-8	223, 224, 226, 228
Electrons	2-2	2-8-2	2-8-8-2	2-8-18-8-2	2-8-18-18-8-2	2-8-18-32-18-8-2
Density	1.84	1.74	1.54	2.6	3.5	ca. 6
M.P. °C.	1284-1300	651	850	757	850	700
B.P. °C.	1500	1100	1440	1366	1140	<1737
Oxidation States	0, +2	0, +2	0, +2	0, +2	0, +2	0, +2
Radius (cov.), Å	0.89	1.36	1.74	1.91	1.98	—
Radius (ionic), Å	0.30	0.65	0.94	1.10	1.29	1.52
Ionization Potential, first (volts)	9.28	7.61	6.09	5.67	5.19	—
Oxidation Potential, M→M++ (volts)	+1.70	+2.34	+2.87	+2.89	+2.90	—
Hydration Energy (Kcal/mole)	—	490	410	376	346	—
Flame Color	—	—	Brick-red	Crimson	Yellow-green	Carmine-red

all possess the core of the preceding inert gas and strongly resist the removal of any more than the two valence electrons. The ionization potentials in the table are for the removal of the first electron, and it may be noted that as the atomic radius increases the ease of removal of the electron increases. The ionization potential indicates how well the core electrons are screening the central nuclear charge from exerting its attractive effect on the valence electrons. Actually, the ions formed are divalent and two electrons must be removed to form the ion. The second electron requires approximately twice as much energy to be removed as does the first. This suggests the possibility of a monovalent ion, but the possibility is dispelled when the heat of hydration is taken into account. The heat of hydration of a divalent ion is so much greater than that of a monovalent ion that the energy released on hydration is sufficient to satisfy the energy requirements for removal of the second electron.

Although the compounds of these metals are largely ionic in their salts and oxides, there is, nevertheless, a definite trend toward a covalent type of linkage as the ion size becomes smaller. This, naturally, is a result of greater deformation of the electron cloud on the anion by the greater density of positive charge on the smaller-sized cations. This trend is not peculiar to the alkaline earth metals but is a general property of other families of elements as well.

The chemical activities of these metals increases as the atomic radius increases, indicating a greater availability of the electrons for bond formation. This activity, however, is not as great as that of the alkali metals, as evidenced by their activity in reducing water, because beryllium and magnesium do not reduce water although the other alkaline earth metals do so with the liberation of hydrogen and the formation of an alkaline solution. All the metals, with the exception of beryllium, are attacked by atmospheric oxygen but in the case of magnesium its oxide coating protects it from further attack. The tendency to form and stability of the peroxides of these metals increases markedly with a rise in atomic number. The stability of the carbonates with respect to their dissociation to the oxide and to carbon dioxide varies similarly.

It should be noted that the greatest differences in these elements occur in passing from beryllium to magnesium and the next greatest difference is in passing from magnesium to calcium. From then on the alkaline earth group behaves in a very regular fashion. Beryllium has many similarities to aluminum, and magnesium to zinc, which illustrates the general principle that the first element in a group has a diagonal relationship with the second element of the next group, and that the second element of a group has a close relationship with its own B subgroup. In fact, magnesium also has some similarities to lithium, again illustrating the diagonal relationship. These relationships are due in some cases to similarities in ion sizes (Li^+ and Mg^{+2}) and in other cases to similarities in the charge-to-radius ratios (especially if the ion sizes vary considerably). Where these criteria are important considerations, the properties of the two ions will also be markedly similar.

The salts of these metals are not as soluble as are those of the alkali metals. Although there is a tendency toward covalency among the smaller cations, the chemistry of Group IIA metals is that of an ionic species, namely the divalent cations. Many of the salts in this group are hydrated, with the smaller cations having a high hydration affinity because of the greater density of positive charge. There is no simple distinction between the solubilities of salts derived from weak and strong acids as exists with the alkali metals. In general, the relative solubilities of any one anion combined with the cations of the alkaline earth metals will be Ca > Sr > Ba > Ra (see Chapter 3). It is only with the hydroxides that there is a reverse order of solubility than the above (e.g., Ca < Ba). In general, one can say that the most soluble salts of these metals are derived from monovalent anions and that the least soluble ones are derived from the divalent or polyvalent anions. This probably reflects the increased interionic forces holding the polyvalent ions together against the dipole-ion attractions of the solvent molecules. However, water of hydration within the crystal may alter this picture somewhat (e.g., $MgSO_4 \cdot 7H_2O$).

Group IIB Metals

The core of these metals does not have the configuration of the previous inert gas, but has, instead, an additional ten electrons which gives it a "pseudo-inert gas" or 18-electron structure. They utilize both of the electrons in the outer shell for bonding purposes but do not use any others. All these metals form the normal divalent ions, but mercury, in addition, has the unique property of having a monovalent ion. They are smaller in size than the Group IIA elements and are therefore less active. Consider, for example, their inertness to the action of water and oxygen as compared to the A-group. These elements tend to form covalent compounds more frequently than the A-group elements, probably due to the more incomplete screening of the nuclear charge by the 18-electron core. Although magnesium and zinc are approximately the same size the tendency toward covalency is greater with zinc. Similarly, cadmium and mercury tend toward formation of covalent compounds to a greater extent than calcium, and cadmium shows a greater tendency than zinc although the latter has a smaller radius. Divalent mercury has an even greater tendency toward covalent bond formation than its congeners in the subgroup. Although there are some compounds where mercury is definitely in an ionic state the covalent type of compound predominates.

The B-group elements show much more individual character than do the A-group elements (alkaline earths). This is a general characteristic of

TABLE 2–5. PROPERTIES OF GROUP IIB

Properties	Zinc	Cadmium	Mercury
Atomic Number	30	48	80
Atomic Weight	65.37	112.40	200.59
Isotopes	64, 66, 67, 68, 70	106, 108, 110– 114, 116	106, 198–202, 204
Electrons	2–8–18–2	2–8–18–18–2	2–8–18–32–18–2
Density	7.14	8.65	13.5939
M.P. °C.	419.4	321	−39
B.P. °C.	907	767	356.9
Oxidation States	0, +2	0, +2	0, +1, +2
Radius (cov.), Å	1.25	1.41	1.44
Radius (ionic), Å	0.70	0.92	1.05
Ionization Potential, first (volts)	6.92	8.99	10.42
Oxidation Potential, $M \rightarrow M^{++}$ (volts)	+0.762	+0.4020	−0.854

B groups in comparison to A groups, and was also evident in the IB group (Cu, Ag, Au).

Basicity increases in the group with increase in ionic radius, with mercuric oxide the most basic and even zinc oxide exhibiting the property of dissolving in alkali. This property of dissolving in basic solution always decreases in groups with increasing atomic number.

The oxides of these metals become less stable to heat as the atomic number increases. The chlorides are hydrolyzed with decreasing ease in the order Zn > Cd > Hg.

The elements cadmium and zinc form complex ions with considerable ease, although cadmium has a stronger tendency toward coordination than zinc, the increase with increasing atomic number being a common characteristic of B-group elements in contrast to A-group elements, where the trend is normal. The common complexes are the ammine, cyano-, and halo-, and the coordination number of zinc and cadmium is 6 in the divalent state. Mercury, on the other hand, shows a peculiarity in not entering into complex formation with any degree of readiness.

The solubilities of these metals are quite similar to those of Group IIA, particuarly those of magnesium. The halides of zinc, where covalent linkage is somewhat evident, are soluble in both water and organic solvents. Cadmium salts, in general, are less soluble than the corresponding zinc salts, indicating a greater degree of covalency. Indeed, the cadmium and zinc salts which are predominantly covalent do not exhibit this disparity in solubilities and are approximately equal.

Group IIIA Elements

This group of elements is composed of boron, aluminum, gallium, indium, and thallium, of which the first two are most important to pharmacy. Only one of these elements, aluminum, can be considered to be abundant, and, in fact, it is the most common metal and the third most common element in the earth's crust. Only oxygen and silicon are more abundant.

The first element in the group, boron, is a nonmetal, but is sometimes termed a *metalloid* in reference to its somewhat hybrid behavior as a borderline element possessing both metallic and nonmetallic character. Comparison of the electronic structures of the elements reveals that boron is the only element with less than four electrons in its valence shell that is not a metal. Once again we find the diagonal relationship of an element with the element to its right in the next period when we consider the many similarities of boron to silicon.

In general, the tendency to form covalent bonds in Group III is greater than in the preceding groups as a result of the small ion size (see Table 2–6) and the greater charge, both of which result in a greater ability to deform anions (a measure of covalency). Boron almost invariably forms covalent linkages, whereas aluminum is on the borderline but has a strong tendency toward covalency. The tendency toward covalent bond formation is even greater in aluminum than it is in beryllium with which it has a diagonal relationship in the periodic table. Beryllium is also a borderline case with respect to covalent and ionic bonding. The strong covalent tendency of boron compared to aluminum is not too surprising in view of the fact that the ionic radius ratio of boron to aluminum is approximately 2:5, and the ionic volumes have a ratio of approximately 1:16.

TABLE 2–6. PROPERTIES OF GROUP IIIA ELEMENTS

Properties	Boron	Aluminum	Gallium	Indium	Thallium
Atomic Number	5	13	31	49	81
Atomic Weight	10.811	26.9815	69.72	114.82	204.37
Isotopes	10, 11	27	69, 71	113, 115	203, 205
Electrons	2–3	2–8–3	2–8–18–3	2–8–18–18–3	2–8–18–32–18–3
Density	2.34 (amor.) 3.33 (cryst.)	2.7	5.91	7.31	11.85
M.P. °C.	2300	660	29.8	156.4	300 ± 3
B.P. °C.	>2550	1800	2070	>1450	1460
Common Oxidation States	0, +3	0, +3	0, +2, +3	0, +1, +3	0, +1, +3
Radius (cov.), Å	0.80	1.25	1.26	1.44	1.47
Radius (ionic), Å (trivalent)	0.20	0.50	0.62	0.81	0.95
Ionization Potential, first (volts)	8.28	5.96	6.00	5.785	6.106
Oxidation Potential $M \rightarrow M^{+++}$ (volts)	—	—	+0.52	+0.38	+0.336*

* For $Tl \rightarrow Tl^+$.

The remaining Group IIIA elements (Ga, In, Tl) are not of importance in pharmacy. They exhibit a valence of 3, but they also show a marked tendency to deviate from this group valence. For example, gallium can be divalent and indium and thallium monovalent. The latter deviations may be explainable in terms of some sort of stability associated with the outermost s orbitals of the elements contrasted with the evident lability of the single electron in the associated p orbitals. The stability of the monovalent cation is much greater in thallium than in indium. The divalent character of gallium has no ready explanation.

The metals of this group are readily oxidized when heated in air, although they are stable at ordinary temperatures. The oxides are readily reduced back to the free metal. The metals react readily with sulfur and the halogens. The hydroxides tend to show amphoteric properties, with the exceptions of the top and bottom elements of the group. Indeed, boron trihydroxide (boric acid) is a weak acid. Aluminum hydroxide is a weak amphoteric base, as are the hydroxides of gallium and indium (forming gallates and indates), but not thallium. The amphoteric character of gallium and indium hydroxides may be related to a decreased ability of the core electrons to shield the nuclear positive charge.

Group IIIB Elements

This is a very long group of elements which includes scandium, yttrium, lanthanum, actinium, and two large groups of elements known as the lanthanides (*rare earth elements*, At. Nos. 58–71) and the actinides (At. Nos. 90–103). Little will be said about the rare earths in this chapter, and the actinides will not be considered at all.

In general, the Group IIIB elements have an increasing metallic character as the atomic number increases, with less tendency toward covalency. An illustration of this is the increasing basicity and decreasing degree of hydrolysis of the hydroxides. Scandium hydroxide is a weak base, yttrium hydroxide is a stronger base, and so forth down the group. On the other hand, the basicity of the rare earth metals decreases as the atomic number increases. This may be ascribed to the peculiarity in this series of elements known as the "lanthanide contraction."[3] It is described as the small but consistent decrease in the trivalent ion size of each succeeding rare earth element (La = 1.22 Å to Lu = 0.99 Å). Naturally, as the ion size decreases it has greater polarizing power (tendency to hold the oxygen of the OH^- more firmly), and thus would decrease the ionization.

The Group IIIB elements are the first elements large enough to permit the addition of electrons to the d orbitals, beginning with the third principal quantum number. This single electron and the electron pair in the low-lying s orbital are involved in the trivalency of these metals. Lanthanum in the fifth period presents the first opportunity to add electrons to the $4f$ orbital

TABLE 2–7. PROPERTIES OF GROUP IIIB ELEMENTS

Properties	Scandium	Yttrium	Rare Earth Metals*
Atomic Number	21	39	57→71
Atomic Weight	44.956	88.905	138.91→174.97
Isotopes	45	89	(138–9)→(175)
Electrons	2–8–(8 + 1)–2	2–8–18–(8 + 1)–2	2–8–18–18–(8 + 1)–2
			to
			2–8–18–32–(8 + 1)–2
Density	2.5	5.51	6.15→9.74
M.P. °C.	1200	1490	860 →?
B.P. °C.	2400	2500	1800 →?
Common Oxidation			
States	0, +3	0, +3	0, +2, +3, +4
Radius (cov.), Å	1.51	1.8	1.87→1.74
Radius (ionic), Å	0.81	0.93	1.22→0.99
Ionization Potential,			
first (volts)	6.7	6.5	5.6→?

* The elements here included represent the atomic numbers 57 (lanthanum) to 71 (lutetium) inclusive. The outside limits of this large group are given. Other information concerning these elements will be found in various reference texts such as the Handbook of Chemistry and Physics published by the Chemical Rubber Co., Cleveland, Ohio.

to complete the fourth quantum level. The rare earth elements are involved in filling this level, thus making them somewhat different from the normal groups in the periodic table. Since the low-lying s and d orbitals do not change significantly during this process, the entire group of elements is placed in Group IIIB. Similar arguments can be made for the actinides. The normal valence for the rare earth elements is 3, but some of them have abnormal valencies of 2 or 4. Of particular interest to pharmacy is cerium, which exhibits valence states of 3 (cerous) and 4 (ceric). Praseodymium and terbium can also be tetravalent. Among the elements forming divalent ions are samarium, europium, and ytterbium. The chemical behavior of these elements lies between that of aluminum and the alkaline earth metals. The solubilities of many of the comparable salts are similar.

Group IVA Elements

Carbon and silicon are the first two "short period" members of this group, and, unlike the other members, are nonmetallic in their chemical behavior. But, like all the other members of this group (germanium, tin, and lead), they have a maximum valence of 4. Carbon, but not silicon, can exhibit a valence of 2. Some other members of this group also show bivalency.

In general, the properties of Group IVA have a different chemical character than those of Groups I, II, and III. These elements now are able to make up the valence octet without the use of coordination. Because of the small size and high charge on the tetravalent ions, the bonding in this group is predominantly covalent. However, bonding with the divalent ions shows considerable ionic character which increases with the size of the ion. The metallic character of the elements beyond silicon is more noticeable in the lower valence state.

The importance of carbon both in the elemental state and in its combined forms can scarcely be overestimated. By far the greatest importance attaches itself to its function as the basic building unit of organic compounds. These compounds will not be discussed in any detail in this text. In comparison, silicon seems to form compounds similar to the saturated molecules of carbon, but to date it has not been possible to achieve the long alkane-like chains with silicon.

The other three elements of this group exhibit valences of either 2 or 4. The divalent form implies the presence of an "inert" pair of electrons. The stability of the divalent character increases with the atomic number. Divalent germanium is a good reducing agent, whereas tetravalent lead (as in PbO_2) is a good oxidizing agent. Indeed, the stability of the divalent form of lead is such that frequently it is assumed to be its characteristic oxidation state. Although germanium is classed as a metal its metallic character only slightly exceeds its nonmetallic character, so that it can properly be called a metalloid. Nonmetallic character even extends to tin inasmuch as one of its allotropic forms (low temperature form) is definitely nonmetallic. The other form is the familiar metallic tin. Lead, on the other hand, is definitely metallic. The tetrahalides of these elements are volatile (with the exception of SnF_4) and are mainly of the covalent type.

TABLE 2–8. PROPERTIES OF GROUP IVA ELEMENTS

Properties	Carbon	Silicon	Germanium	Tin	Lead
Atomic Number	6	14	32	50	82
Atomic Weight	12.01115	28.086	72.59	118.69	207.19
Isotopes	12, 13	28–30	70, 72–3, 76	112, 114–20, 122, 124	204, 206–8
Electrons	2–4	2–8–4	2–8–18–4	2–8–18–18–4	2–8–18–32–18–4
Density	2.22	2.33	5.323	7.31	11.34
M.P. °C.	3700 ± 100	1420	936	231.9	327.4
B.P. °C.	4830	2300	2700	2260	1740
Common Oxidation States	−4 to +4	−4 to +4	0, +2, +4	0, +2, +4	0, +2, +4
Radius (cov.), Å	0.771	1.17	1.22	1.40	1.74
Radius (ionic), Å, tetravalent	0.15	0.41	0.53	0.71	0.84 1.21 (+2)
Ionization Potential, first (volts)	11.264	8.149	8.09	7.30	7.38
Oxidation Potential, $M \rightarrow M^{++}$ (volts)	—	—	—	+0.13	+0.12

In the tetravalent state, the only element in the series that forms stable compounds with oxyacids is lead, with one of the most important ones being lead tetraacetate, a good oxidizing agent. Here again the behavior is that of a covalent compound rather than that of a salt. The hydroxides in the tetravalent state of germanium, tin, and lead are all very weak acids.

Group IVB Elements

These elements are transitional in character. Hafnium is so similar to zirconium in all respects that whatever applies to zirconium in a chemical way applies as well to hafnium. Because of the electronic configuration, these elements can exhibit valences of 2, 3, and 4; however, the lower valence forms are less stable than the tetravalent forms. Trivalent titanium is important, but the divalent form is mostly a curiosity. The trivalent states of zirconium and hafnium exist but are of little importance. In general, compounds containing the tetravalent form of these elements do not exhibit color, whereas compounds containing the lower valence states are colored. This is frequently the case among ions of transition elements and their complexes where the d orbital is only partially filled. The color is a result of the promotion of d electrons to higher energy levels through the absorption of visible light.

The cations in this group are not the normal type but instead are oxo-ions of the type TiO^{+2} and ZrO^{+2}. The halides of titanium and zirconium both behave as Lewis acids, forming adducts with nucleophilic organic oxygen compounds (e.g., alcohols, ethers, esters, etc.). Although they already possess complete octets, the central atoms are capable of holding more electrons than the normal complement of eight. Tetravalent titanium and

TABLE 2–9.　PROPERTIES OF GROUP IVB ELEMENTS

Properties	Titanium	Zirconium	Hafnium
Atomic Number	22	40	72
Atomic Weight	47.90	91.22	178.49
Isotopes	46–50	90–2, 94, 96	174, 176–80
Electrons	2–8–(8 + 2)–2	2–8–18–(8 + 2)–2	2–8–18–32–(8 + 2)–2
Density	4.51	6.53	11.4
M.P. °C.	1725	1857	1700
B.P. °C.	5100	3577	5390
Common Oxidation States	0, +2, +3, +4	0, (+2), (+3), +4	0, (+2), (+3), +4
Radius (cov.), Å	1.32	1.45	1.48
Radius (ionic), Å	0.68	0.80	0.87
Ionization Potential, first (volts)	6.82	6.84	~5.5
Oxidation Potential M→MO⁺⁺ (volts)	+0.95	+1.53	+1.68

zirconium also form stable complexes with halogens. The hexafluorides (TiF_6^{-2} or ZrF_6^{-2}) are the most important of these. The general properties of this group are illustrated in Table 2–9.

Group VA, The Nitrogen Family

Nitrogen and phosphorus are both nonmetallic elements, arsenic is a metalloid, and antimony and bismuth are usually classed as metals. However, antimony has some nonmetallic character and could be classed as a metalloid. Nitrogen is a somewhat atypical member of this family. In combination with carbon, oxygen, and hydrogen nitrogen is usually covalent; in the nitride form (N^{-3}) its bonding may be considered ionic. Nitrogen compounds usually have three covalent bonds and an unshared electron pair. It is also quite able to act as a donor of its unshared electron pair to cations or to neutral atoms, forming tetravalent compounds or radicals (e.g., NH_4^+, $R_3N{\rightarrow}O$). This type of bonding also occurs with other elements in this family. The covalency of 5 is not found with nitrogen, which might be expected because of its 5-valence electrons, but it is found with other elements of Group VA. However, the 3 and 4 covalent states, common with nitrogen, are found among some of the other members of the group as a typical type of bonding. It is believed that two of the electrons often behave as an "inert pair" analogous to that previously outlined for the IVA group. Nitrogen also seems to form triple bonds ($N{\equiv}N$) quite readily, which accounts for its great stability to oxidation or reduction and its relatively nonpolar character. This triple link is not found in the later elements of the group, where the P_4 and As_4 molecules are found instead in a single-bonded tetrahedral arrangement.

Although N, P, and As in the oxidation state of $+3$ act almost exclusively as the oxyacids or their derivatives, Sb and Bi may exist as positive ions of the type SbO^+ and BiO^+. The typical oxidation states for this family are -3, 0, $+3$, and $+5$, which are normal for elements with five electrons in the valence shell. The -3 state is found in such compounds as NH_3, PH_3, AsH_3, Li_3N, etc., where the bonding occurs because of combination with three other electrons to form the nitride anion or to form three covalent bonds with less electronegative elements. In the $+3$ state, the "inert pair" is apparent and bonding occurs with three other more electronegative elements as in H_3PO_3, PCl_3, $AsCl_3$, etc. The highest oxidation state of $+5$ utilizes all five electrons to form covalent bonds, as in H_3PO_4, PCl_5, H_3AsO_4, etc.

Although nitrogen does not enter into the 5-covalent state, the other elements can expand their shells beyond the usual octet (i.e., to ten) of electrons. The oxyanions from acids such as nitric, phosphoric, and arsenic all have equivalent oxygen bonds, although the convention is to show them as possessing one coordinate covalent bond which might indicate that, as

TABLE 2-10. PROPERTIES OF GROUP VA ELEMENTS

Properties	Nitrogen	Phosphorus	Arsenic	Antimony	Bismuth
Atomic Number	7	15	33	51	83
Atomic Weight	14.0067	30.9738	74.9216	121.75	208.980
Isotopes	14, 15	31	75	121, 123	209
Electrons	2-5	2-8-5	2-8-18-5	2-8-18-18-5	2-8-18-32-18-5
Density	0.81 (−195°)	1.83 (wh.) 2.34 (red)	2.0 (yellow) 3.70 (amor.) 5.727 (met.)	6.68	9.78
M.P. °C.	−209.9	44.1	818	631	271
B.P. °C.	−195.8	280	Sublimes	1380	1438
Common Oxidation States	0, −3, +3, +5	0, −3, +3, +5	0, −3, +3, +5	0, −3, +3, +5	0, −3, +3, +5
Radius (cov.), Å	0.70	1.10	1.12	1.41	1.52
Radius (ionic), Å, M^{3+} (Pauling)	0.11	0.34	0.47	0.62	0.74
Ionization Potential, first (volts)	14.54	11.10	10.5	8.5	8.0
Electronegativity (after Pauling)	3.0	2.1	2.0	1.8	—
Physical State	Gas	Solid	Solid	Solid	Solid
Color	Colorless	White	Gray	Shiny gray	Shiny gray

an anion, there is something different about one oxygen as contrasted to the others. The explanation for this is that in the anion form they must have equivalent bonds that are intermediate between single and multiple covalent bonds, a phenomenon called *resonance* (see Chapter 1). As we proceed down the group to higher atomic numbers the negative oxidation state becomes less stable. Likewise, as the electronegativity decreases in going down the group and the electropositivity increases, the tendency to form cations, already cited, is evident. It is interesting to note the increase in the number of oxygens associated with the central element in this group. These are $N = 3$, P and $As = 4$, Sb and $Bi = 6$. Thus, it follows that as the central atom gets larger it is possible for more oxygens to get close enough to form stable bonds.

All of the elements in this family form hydrides analogous to ammonia. The tendency to form hydrides from the binary combinations with metals (e.g., Mg_3P_2) by the action of water or dilute acids decreases as the atomic number increases. All of them, except NH_3, tend to be inflammable and are poisonous. These elements react readily with the halogens to form covalent tri- and pentahalides; they tend to hydrolyze readily with water to the hydrogen halides and oxyacids with P and As, and to hydroxy and oxychlorides (e.g., BiOCl) with Sb and Bi. All of them form oxides with oxygen and sulfides with sulfur to give a number of different combinations. As the atomic number increases, the oxides in this group are less acidic and, in fact, become somewhat basic (or at least amphoteric) in the last members of the group. The sulfides are highly colored and have, historically, been used as pigments. With active metals, these elements readily form binary combinations known as phosphides, arsenides, etc.

Group VB Elements

The elements of Group VB do not show much similarity to the other Group V elements except as related to their electronic structure. These elements have their valence electrons in both the outermost and the next outermost shells. The difference of niobium from vanadium and tantalum regarding the distribution of the valence electrons does not appear to bring about any significant differences. As the accompanying table shows (Table 2–11), they exhibit variable valences, but with a principal oxidation state of $+5$. In some respects the resemblance to Group IVB is noticeable, especially in the differences between the first member and the second and the third members. For all practical purposes, the resemblance of niobium and tantalum is identical to that of zirconium and hafnium. As has been previously noted, the lower valence states become less stable as the atomic number increases. The acidity of the oxides decreases with progressing atomic number, with vanadium being amphoteric to a notable degree in its highest oxidation state. The basicity of the hydroxides decreases with

TABLE 2–11.   PROPERTIES OF GROUP VB ELEMENTS

Properties	Vanadium	Niobium (Columbium)	Tantalum
Atomic Number	23	41	73
Atomic Weight	50.942	92.906	180.948
Isotopes	51	93	181
Electrons	2–8–(8 + 3)–2	2–8–18–(8 + 4)–1	2–8–18–32–(8 + 3)–2
Density	6.11	8.57	16.69
M.P. °C.	1717	2415	2996
B.P. °C.	3000	>3300	6100
Common Oxidation States	0, +2, +3, +4, +5	0, +3, +4, +5	0, +4, +5
Radius (cov.), Å	1.22	1.34	1.34
Radius (ionic), Å, M^{5+}	0.40	0.70	0.73
Ionization Potential, first (volts)	6.71	6.77	$\sim$6
Oxidation Potential, $M \rightarrow M^{++}$ (volts)	+1.5	$\sim$ +1.1 ($M \rightarrow M^{+++}$)	—

increasing oxidation state. In the +5 state vanadium has some resemblance to phosphorus but there are numerous differences. The oxides are more acidic than the comparable IVB elements, as expected on the basis of increased nuclear charge. Hydrolysis of the covalent halides takes place readily in aqueous solution. The compounds of these elements in the lower oxidation states are usually colored, which is in keeping with their character as transition metals. In the lower oxidation states vanadium is basic and forms ionic bonds. However, in the higher oxidation states, the bonding becomes more covalent along with increasing volatility of the compounds.

These metals show a passivity unless finely divided. That is, although they show a good reducing action on the basis of oxidation potentials, for all practical purposes they are inert to chemical action. It is this property of tantalum which permits it to be used for corrosion-resistant applications in medicine (e.g., screens, plates, and wires to be left in the body).

Group VIA Elements

Oxygen and sulfur are the typical elements in Group VIA. Sulfur, and the elements selenium, tellurium, and polonium, constitute a family with a gradation of properties that merits the name of "sulfur family" as well as consideration as a unit. Polonium, being the product of radioactive decay, will not be discussed here. Group VIB, composed of chromium, molybdenum, tungsten (wolfram), and uranium, differs greatly from this group. This difference is characteristic of groups occurring late in the periodic table. The principal similarity between the two groups stems from the fact that they all have six valence electrons and it is only when these are all involved that similarities are found (e.g., sulfates and chromates).

TABLE 2-12. PROPERTIES OF GROUP VIA ELEMENTS

Properties	Oxygen	Sulfur	Selenium	Tellurium	Polonium
Atomic Number	8	16	34	52	84
Atomic Weight	16.000	32.064	78.96	127.60	210
Isotopes	16-8	32-4, 36	74, 76-8, 80, 82	120, 122-6, 128, 130	—
Electrons	2-6	2-8-6	2-8-18-6	2-8-18-18-6	2-8-18-32-18-6
Density	1.14 (−184°)	1.96-2.06	4.3-4.8	6.25	—
M.P. °C.	−218.8	214.5	217	450	—
B.P. °C.	−183	444.6	685	1390	—
Common Oxidation States	−2, −1, 0	−2, 0, +2, +4, +6	−2, 0, +4, +6	−2, 0, +4, +6	—
Radius (cov.), Å	0.74	1.04	1.17	1.45	—
Radius (ionic), N−, Å	1.45	1.95	2.02	2.21	—
Ionization Potential, first (volts)	13.61	10.36	9.75	8.96	—
Electronegativity (after Pauling)	3.5	2.5	2.4	2.1	—
Physical State	Gas	Solid	Solid	Solid	Solid
Color	Colorless	Yellow	Gray	Gray	—
Oxidizing Ability	→→→→→→→→→→→→→→→→→→→→→→→→→→→ Decreasing →→→→→→→→→→→ - - - - - - - - - →

Table 2–12 shows that the melting and boiling points, densities, and atomic volumes increase with atomic number. As indicated above, the principal difference in this particular family is between oxygen and sulfur, the relationship of the rest of the elements being rather close. Sulfur and the higher elements exist in different allotropic forms and have oxidation states of -2, 0, $+2$, $+4$, and $+6$, whereas oxygen has only the oxidation states -2, -1, and 0. It should be noted that, in common with Group VA elements, the negative oxidation states become less important as the atomic number increases. These elements are among the most electronegative in the periodic table, with oxygen being the second most electronegative of all elements (fluorine is first). The electronegativity decreases down the group, as expected. This decrease in electronegativity, concomitant with a decrease in oxidizing activity, is attributed to the fact that the nuclear attraction for an incoming electron is decreased as the distance to the outer shell is increased. Although the nucleus has an increasing charge as the atomic number increases, the additional electrons in the larger atoms have an additional repelling effect that reinforces the distance factor.

The elements are nonmetallic, with oxygen being a gas and sulfur a solid. The solid character of the later members of the group is attributed to the fact that they exist in chain structures whereas oxygen exists as a diatomic molecule. Sulfur, for example, has the S_8 molecule which permits a greater amount of intermolecular attraction than can be achieved in the smaller molecules; thus, the melting and boiling points tend to be higher. Selenium and tellurium are best characterized as metalloids although tellurium has considerably more metallic character than selenium. The general rule is that the metallic character increases in progressing down the group.

These elements are the first to be mentioned that can form monatomic anions (oxide, sulfide, etc.) and for this reason can form both ionic as well as covalent compounds. The formation of covalent bonds is limited to two for oxygen, but may rise to three, four, and six with the other elements. In general, it may be stated that, in binary combinations of these elements, they tend to be in the -2 oxidation state except for the oxides which, like ternary combinations, usually assume the higher oxidation states. Wherever the binary combinations involve nonmetals (e.g., SeS) the negative oxidation state is assigned to the most electronegative element (in this case, S).

This group forms hydrides of the general type RH_2 (e.g., H_2O, H_2S, etc.) which decrease in stability and increase in acidity as the molecular weight increases. The hydrides are all gaseous with the exception of H_2O which, of course, is liquid and reflects the fact that strong association exists between the dipole molecules. This association is not evident with the hydrides of the other elements. With the exception of water the hydrides are all highly odoriferous and poisonous. The sulfides, selenides, and tellurides of alkali and alkaline earth metals are water soluble but the corresponding salts of the other metals are highly water insoluble.

Although several oxides of these elements are known, the most common are those characterized by the general formulae RO_2 and RO_3 which are the anhydrides of the acids represented by H_2RO_3 and H_2RO_4. Sulfurous acid (H_2SO_3) and sulfuric acid (H_2SO_4) may be cited as examples. Selenium and tellurium form similar acids. The salts of the "ous" acids have the familar "ite" ending and the salts of the "ic" acids have the "ate" ending. The acidity of the oxyacids of a given element in this group is always greatest when the oxidation number is highest. Thus, sulfuric acid (oxidation number = 6) will be more acidic than sulfurous acid (oxidation number = 4). This is related to the relative covalent characters of the oxides; the most covalent oxides are the ones with the highest oxidation numbers. The size of the central ion influences the relative acidity of the corresponding oxyacids in this group (as well as other groups); thus, sulfurous acid is a slightly stronger acid than selenous acid. This is ascribed to the fact that sulfur is smaller than selenium. Therefore, we find that a combination of small size and high charge on the central ion contributes to high acidity.

Group VIB Elements

The members of Group VIB are chromium, molybdenum, tungsten (wolfram), and uranium. They are all distinctly metallic and form oxides of which those of higher molecular weight are acidic. These oxides form a series of compounds, such as the chromates, the molybdates, etc. In this respect they resemble Group VIA. The metals of this group have a tendency to unite with oxygen and in such combination (CrO^{+2}, MoO_2^{+2}, UO_2^{+2}) replace the hydrogen of acids to form salts. This property increases with atomic number. Chromium forms two basic hydroxides, $Cr(OH)_2$ and $Cr(OH)_3$, which are the parent substances of a great many chromous

TABLE 2–13. PROPERTIES OF GROUP VIB ELEMENTS

Properties	Chromium	Molybdenum	Tungsten
Atomic Number	24	42	74
Atomic Weight	51.996	95.94	183.85
Isotopes	50, 52–4	92, 94–8, 100	180, 182–4, 186
Electrons	2-8-(8 + 5)-1	2-8-18-(8 + 5)-1	2-8-18-32-(8 + 4)-2
Density	7.14	10.2	19.35
M.P. °C.	1900	2622	3410
B.P. °C.	2480	4510	5900
Common Oxidation States	2, 3, 6	2, 3, 4, 5, 6	2, 4, 5, 6
Radius (cov.), Å	1.17	1.29	1.30
Radius (ionic), Å	0.52	0.62	0.62
Ionization Potential, first (volts)	6.77	7.38	7.98

and chromic salts. Molybdenum and tungsten are distinctly acidic, whereas uranium is both acid- and base-forming. The similarity in the properties of successive horizontal elements in the periodic table is particularly in evidence here. For example, vanadium, chromium, and manganese are closely related to one another by virtue of very similar physical and chemical properties. Molybdenum and columbium, and also tungsten and tantalum, exhibit close relationships.

Group VIIA, The Halogen Family

Berzelius suggested the word "halogen," derived from the two Greek words meaning "sea salt" and "to produce," which thus means "the producer of sea salt." The term is applied to the four elements—fluorine, chlorine, bromine, and iodine— because the sodium salts of their respective hydroacids are very similar to ordinary sea salt. These four elements and their compounds show a great resemblance to one another in general chemical properties. The physical properties of the elements exhibit a gradual transition that is evident upon consulting the accompanying table (Table 2–14). Thus, as the atomic weight increases: the physical state changes from that of a gas (F and Cl) to that of a liquid (Br) and then to that of a solid (I); the melting and boiling points rise; the colors deepen; the densities increase; and so forth. There appears to be a relationship between the ease of removal of an electron (ionization potential) and the energy necessary for excitation of the halogen molecule as far as color formation is concerned. Iodine requires the least energy for excitation and absorption of visible light, whereas fluorine requires high energy radiation. Thus, iodine absorbs low energy yellow and green radiation to give a violet color, and fluorine absorbs high energy violet radiation to give a yellow color.

The gradual transition from gaseous fluorine to solid iodine may be explained on the basis of increasing atomic radius. As the radius increases the outermost electrons get further away from the influence of the nucleus and thus have a greater opportunity to exert an influence on neighboring nuclei. As the attraction for neighboring nuclei increases it is to be expected that liquefaction and finally solidification will take place. This attraction may be ascribed to van der Waals forces (weak attractive forces between uncharged bodies; see Chapter 1), which should be greater between heavier molecules than between lighter ones, and also to the greater deformability of the electron clouds of the heavier elements.

The important chemical properties of the halogens are those in which they are reduced and promote oxidation of some other substance. Fluorine is the best oxidizing agent and consideration of the oxidation potentials shows that the oxidizing property decreases with increasing atomic number. On the other hand, the tendency to be oxidized is, of course, greatest with iodide and least with fluoride. In fact, the two heaviest halides may be used as reducing agents, whereas the two lightest halogens are commonly used as oxidizing agents. A halogen of lower atomic weight will always displace one of higher atomic weight from its binary hydrogen compounds or from the salts thereof [rx (iv)]. Again, this illustrates that the force with which the respective halogens hold electrons diminishes from fluorine to iodine.

$$2HI + Cl_2 \leftrightharpoons 2HCl + I_2 \tag{iv}$$

$$2I + Cl_2 \leftrightharpoons I_2 + 2Cl^- \tag{v}$$

All four halogens unite with hydrogen but the affinity toward this element decreases as the atomic weights increase. Thus it is found that, although hydrogen and fluorine unite explosively, hydrogen and iodine need a catalyst to promote the reaction. The hydrides of the halogens are all colorless gases and, as anhydrous gases, have considerable covalent character. This character increases with the size of the halogen. However, when dissolved in water, it is found that the hydrogen halide with the greatest ionic character, HF, is the weakest acid (and, in fact, is classed as a weak acid; see Chapter 4). The order of decreasing acidity in aqueous solution is: HI > HBr > HCl > HF. With the exception of HF, the hydrogen halide molecules ionize almost completely in water and are classed as strong acids. In general, series of halides such as the above, as well as others (e.g., stannic salts), will have the greatest ionic character with the smallest halogen.

The affinity of the halogens for oxygen increases as the atomic number increases. Thus, iodine pentoxide is a well-defined crystalline solid while chlorine monoxide, peroxide, and heptoxide are very unstable even at ordinary temperatures. The oxides of fluorine and bromine are exceedingly difficult to make and are correspondingly unstable. When treated with water, the oxides of the halogens yield acids, thereby indicating that the simple halogens are nonmetals.

In combination with hydrogen or the metals, the halogens are univalent and negative. When combined with oxygen or in the form of their oxygen salts, they have valences that are often greater than one and are positive. Of the halogens, fluorine exhibits the most marked tendency to become negative and the greatest resistance toward becoming positive, whereas iodine shows the greatest inclination to become positive and the least to become negative. The simple derivatives of the halogens (binary combinations) can form either ionic or covalent compounds. With the ionic types derived from combinations with metals early in the periodic table it is noted that the greatest ionic character is associated with the smallest halide. This is apparent from the trends in melting and boiling points, which are high with lithium fluoride but decrease with increasing atomic weight of the halogen.

On the other hand, the covalent halides (e.g., HX, CX_4, SiX_4, GeX_4) show low melting and boiling points with the lower halogens, increasing with the heavier ones. These results may be rationalized on the basis that, with the ionic compounds, the electrostatic attraction diminishes as the ions are further apart and therefore less work is required to separate them as the ionic radius increases. The covalent molecules do not have a strong external field, and it is to be expected that as the atomic weight increases the work necessary to fuse or to boil them should increase. Of course, the covalent compounds have lower boiling and melting points than the ionic compounds to begin with. The covalent types of combinations are found with the metals in the middle groups of the periodic table, where the combination of small size and high charge lends itself to this type of bonding.

There are four types of oxyacids of the halogens (not including F): (1) HOX (hypohalous acid); (2) HOXO (halous acid); (3) $HOXO_2$ (halic acid); and (4) $HOXO_3$ (perhalic acid). Accordingly, their salts are the hypohalites, halites, halates, and perhalates. A number of them will be encountered among the salts used in pharmaceutical practice. Insofar as the acids are concerned, their acidity and thermal stability increase for any particular halogen with the increasing number of oxygens coordinated with the central atom. In general, the hypohalous and halous acids are classed as weak acids while the halic and perhalic acids are strong acids.

TABLE 2–14. PROPERTIES OF THE GROUP VIIA ELEMENTS
(THE HALOGENS)

Properties	Fluorine	Chlorine	Bromine	Iodine
Atomic Number	9	17	35	53
Atomic Weight	18.998	35.453	79.909	126.904
Isotopes	19	35, 37	79, 81	127
Electrons	2–7	2–8–7	2–8–18–7	2–8–18–18–7
Density	1.14	1.56	3.12	4.94
	$(-200°)$	$(-33.6°)$		
M.P. °C.	-223	-102	-7.3	114
B.P. °C.	-187	-34.5	58.8	183
Common Oxidation States	$0, -1$	$0, -1, +1,$ $+3, +5, +7$	$0, -1, +1,$ $+3, +5$	$0, -1, +1,$ $+3, +5, +7,$
Radius (X$^-$), Å	1.36	1.81	1.95	2.16
Radius (cov.), Å	0.64	0.99	1.14	1.33
Electronegativity, Pauling's scale	4.0	3.0	2.8	2.4
Oxidation Potential, $2X^-\rightarrow X_2+2e$ (volts)	-2.8	-1.359	-1.065	-0.535
Ionization Potential, first (volts)	17.3	13.0	11.8	10.6
Usual Physical State	gas	gas	liquid	solid
Color	Pale yellow	Greenish yellow	Reddish brown	Black (solid) Violet (gas)

This may be considered to be a result of each additional oxygen atom (each of which has a strong electron affinity) exerting its attraction to help pull the bonding electrons away from the hydrogen in the acid (which is itself bonded to oxygen). The progressive "loosening" of the hydrogen as a proton leads to stronger acids.

Although the thermal stability increases with increasing oxygen attached to the central atom, the stability of all of these acids is low. The oxidizing action of these acids is usually believed to be greatest with the greater number of oxygens. Actually, this measures the capacity for oxidation rather than the readiness to oxidize and, indeed, the lower the thermal stability, the greater the ability to oxidize. As an illustration of this, consider the well known activity of sodium hypochlorite as an oxidizing antiseptic (see Chapter 9) and the recognized deficiency of potassium chlorate as an oxidizing antiseptic.

Fluorine exhibits some peculiar differences from the other members of this family. For example, silver chloride, bromide, and iodide are nearly insoluble in water, whereas silver fluoride is appreciably soluble. Sodium chloride, bromide, and iodide are readily soluble in water, whereas sodium fluoride is much less soluble and interesting because it is the only one having antiseptic properties.

Concerning the solubility of halides, a general rule is that they are soluble except in the case of salts of Ag^+, Hg_2^{+2}, and Pb^{+2} which are insoluble. Obviously this pertains to the halides formed from metallic elements and not to those from nonmetallic ones, which present a more complicated picture.

Group VIIB Elements

The only metal of pharmaceutical importance in this group is manganese. Technetium is a product of radioactive decay, and will be discussed in the

TABLE 2–15. PROPERTIES OF GROUP VIIB ELEMENTS

Properties	Manganese	Technetium	Rhenium
Atomic Number	25	43	75
Atomic Weight	54.9380	99	186.2
Isotopes	55	99	185, 187
Electrons	2-8-(8 + 5)-2	2-8-18-(8 + 6)-1	2-8-18-32-(8 + 5)-2
Density	7.23	—	20.5
M.P. °C.	1247	—	3180
B.P. °C.	2032	—	5900
Common Oxidation States	0, +2, +3, +4, +6, +7	—	+3, +4, +5, +6, +7
Ionization Potential, first (volts)	7.43	7.23	7.87
Radius (cov.), Å	1.17	—	1.36
Radius (ionic), Å	0.80 (+2)	—	—

chapter on radioisotopes (Chapter 11). Rhenium is extremely rare although a fair amount of knowledge is available concerning its properties.

In the oxidation state of $+7$, manganese and rhenium form MnO_4^- and ReO_4^- ions, respectively. It is only in this oxidation state that there are any real resemblances to other Group VII elements. The higher valence states of rhenium are more stable than those of manganese, whereas the opposite is true of the lower valence states. In common with most transition metals possessing an incomplete inner subshell these elements form colored ions.

Group VIII Elements

This is a rather unusual group of metals and is composed of nine elements arranged as groups of three in Periods 4, 5, and 6 following Group VIIB and preceding Group IB. Mendeléeff originally termed these triads the "transition metals" because they enabled the bridging necessary for his "short form" of the periodic table. Since then the concept of transition metals has been expanded concomitant with increased knowledge of electron arrangement.

Quite generally, these elements are considered as horizontal triads rather than as vertical groups although there is probably as much reason to consider them vertically as horizontally. Nevertheless, convention dictates the consideration of iron, cobalt, and nickel as a triad. The rest of the metals are termed the "platinum metals," with ruthenium, rhodium, and palladium being the "light" platinum metals and osmium, iridium, and platinum the "heavy" platinum metals.

The electronic configurations of the iron triad show two $4s$ electrons in the outermost shell, together with a partially filled $3d$ orbital in the next outermost shell. It is the $4s$ electrons that are lost in the formation of divalent ions, whereas a $3d$ electron must be lost to form the trivalent ion. Observation of the relative stabilities of the divalent and trivalent states of the three elements leads to the conclusion that the higher valence state becomes less stable as the atomic number increases. This is tantamount to saying that the $3d$ electron becomes more difficult to utilize in bonding as the $3d$ shell becomes more saturated. For all practical purposes, the higher valence states of the iron triad are very unstable. Much of the chemistry of these elements is concerned with complexes which were discussed in Chapter 1.

One of the principal differences between the iron triad and the platinum metals is the marked tendency of the iron group to form simple cations in contrast to the definite reluctance to do so by the platinum group. The platinum metals seem to prefer the higher oxidation states. Whereas there is an increase in atom size when comparing the iron triad with the platinum metals, the latter do not differ greatly in size. This accounts for the much

TABLE 2–16. PROPERTIES OF GROUP VIII ELEMENTS

IRON TRIAD

Properties	Iron	Cobalt	Nickel
Atomic Number	26	27	28
Atomic Weight	55.847	58.9332	58.71
Isotopes	54, 56–8	59	58, 60–2, 64
Electrons	2–8–(8 + 6)–2	2–8–(8 + 7)–2	2–8–(8 + 8)–2
Common Oxidation States	0, +2, +3, +4, +6	0, +2, +3, +4	0, +2, +3, +4
Density	7.86	8.92	8.9
M.P. °C.	1535	1493	1455
B.P. °C.	3000	3550	3075
Ionization Potential, first (volts)	7.83	∼8.5	7.6
Oxidation Potential, $M \rightarrow M^{++}$ (volts)	+0.44	+0.277	+0.250

LIGHT PLATINUM METALS

Properties	Ruthenium	Rhodium	Palladium
Atomic Number	44	45	46
Atomic Weight	101.07	102.905	106.4
Isotopes	96, 98–102, 104	103	102, 104–6, 108, 110
Electrons	2–8–18–(8 + 7)–1	2–8–18–(8 + 8)–1	2–8–18–(8 + 9) 1
Common Oxidation States	0, 2, 3, 4, 5, 6, 7, 8	0, 2, 3, 4, 6	0, 2, 3, 4, 6
Density	12.2	12.42	12.0
M.P. °C.	2450	1966	1555
B.P. °C.	4150	4500	3980
Ionization Potential, first (volts)	7.37	7.7	8.3
Oxidation Potential, $M \rightarrow M^{++}$ (volts)	−0.45	∼ −0.6	−0.98

HEAVY PLATINUM METALS

Properties	Osmium	Iridium	Platinum
Atomic Number	76	77	78
Atomic Weight	190.2	192.2	195.09
Isotopes	184, 186–90, 192	191, 193	192, 194–6, 198
Electrons	2–8–18–32–(8 + 6)–2	2–8–18–32–(8 + 7)–2	2–8–18–32–(8 + 8)–2
Common Oxidation States	0, 2, 3, 4, 6, 8	0, 2, 3, 4, 6	0, 2, 3, 4, 6
Density	22.5	22.4	21.45
M.P. °C.	2700	2450	1773
B.P. °C.	5500	5300	4530
Ionization Potential, first (volts)	8.7	9.2	8.96
Oxidation Potential, $M \rightarrow M^{++}$ (volts)	−0.7	< −1.0	∼ −1.2

greater density of the heavy platinum metals as compared to the light ones. Indeed, osmium is the heaviest known substance.

All of the Group VIII elements are grayish-white metals with high melting and boiling points. They all have some absorptive ability for hydrogen, with palladium exhibiting unusual activity in this regard. Reactivity with oxygen is noted particularly in the iron triad, with the affinity decreasing rapidly as the atomic number increases. This same trend of high oxygen affinity in the left-hand member of the series is noted in each succeeding triad of the platinum metals as well. The platinum metals are classed as "noble" metals because they show low oxidation potentials and low reactivity. Because of this lack of reactivity they are not easily tarnished. However, the light platinum metals are more susceptible to oxidizing agents than the heavier metals.

The magnetic properties of the iron triad are unique in that these are the only elements possessing this property at room temperature. This property is thought to be due in some way to the incomplete d orbitals.

Most of the metals in Group VIII have the unique property of forming "carbonyls" with carbon monoxide. These are low melting liquids which decompose rather readily to the metal and carbon monoxide on heating.

Group VIIIA Elements, The Inert Gases

As the name indicates, elements of this group are all gases, and by virtue of their closed outer orbitals (a total of eight electrons in the outer shell except helium, of course) they are very stable to any kind of chemical reaction. During the early 1960s some inert gas compounds were reported, notably compounds of xenon (tetrafluoride and hexafluoride) which is one of the more reactive elements of the group. These elements have very high ionization potentials, and for our purposes can be considered as inert. Argon is the most widely distributed and most abundant of the inert gases. It is present in air (0.94%, v/v), in natural gas, occluded in minerals, and dissolved in the oceans and all fresh waters. Radon is an inert gas given off by radium salts. It is itself radioactive, being a short half-life alpha

TABLE 2–17. PROPERTIES OF GROUP VIIIA ELEMENTS
(THE INERT GASES)

Properties	Helium	Neon	Argon	Krypton	Xenon
Atomic Number	2	10	18	36	54
Atomic Weight	4.0026	20.183	39.948	83.80	131.30
Density	0.126	1.20	1.40	2.6	3.06
M.P. °C.	−269.7	−248.6	−189.4	−157.3	−111.9
B.P. °C.	−268.9	−246	−185.8	−152	−108.0
Ionization Potential, first (volts)	24.6	21.6	15.8	14.0	12.1

emitter, and this property has a limited utility in the treatment of cancer (see Chapter 11). Krypton and xenon have been investigated as general anesthetics and found to be quite active in this respect. Table 2–17 lists some of the properties of the inert gases.

References

1. A large portion of this chapter, including many of the tables, was compiled from Roger's Inorganic Pharmaceutical Chemistry, 8th ed., by T. O. Soine and C. O. Wilson, Lea & Febiger, Philadelphia, 1967. Used by permission.
2. Sidgwick, N. V. The Chemical Elements and Their Compounds. Oxford: Clarendon Press, p. 95, 1950.
3. Douglas, B. E. The lanthanide contraction. J. Chem. Ed., **31**:598, 1954.

3
Solutions and Solubility

In this chapter certain chemical and physical properties of homogeneous systems known as solutions will be discussed. In contrast to heterogeneous systems, which are composed of matter in different phases, homogeneous systems have uniform compositions throughout.

Solutions, then, may be described as single phase systems composed of two or more chemical substances representing homogeneous molecular dispersions. In general, the components of a solution retain their individual identities, and, to a degree, their properties. Thus, a solution is properly termed a homogeneous mixture on the basis of this variability of composition. The properties of a solution are uniform throughout the mixture because the dispersion of the solute molecules in the solvent is on a molecular scale, making the molecules indistinguishable by usual observation procedures. Colloidal solutions in contrast to true solutions contain very small particles, but these are not of molecular dimensions and may be observed by various techniques.

The components of a solution are the *solute(s)* and the *solvent*. The solvent is usually the component that establishes the phase of a solution and is present in the largest concentration. For example, a solution in which liquid water is the solvent exists as a single liquid phase established by the solvent regardless of whether solids or gases are dissolved in it. In situations where all the components are in roughly equal concentrations, any one of them could be considered as the solvent.

Either gases, liquids, or solids may act as solvents for one or more of these three states of matter. The one exception to this is gases, which can only dissolve other gases. Liquid solutions of gases, other liquids, or solids are by far the most important to pharmacists, and, therefore, the other types of solutions will not be considered in this chapter.

Concentration Expressions

The concentration of solute in a solution may be expressed in many ways, depending upon the convenience to those concerned with its use. Chemists more frequently prefer to work with the number of moles or equivalents of a particular solute. These quantities are also of importance to pharmacists, as will be seen in the use of milliequivalents per liter (mEq/l) for electrolyte solutions. Pharmacists will also encounter percentage concentrations or some other expression of the constituents by parts. The commonly employed concentration expressions are reviewed in the following paragraphs.

(1) **Molarity.** The molarity of a solution expresses the number of moles (gram-molecular weights) of solute contained in 1000 ml (1 liter) of solution. A solution containing 1 mole of solute in each liter of the total solution is said to be a one molar (M) solution. Equation 1 illustrates the relationship between the weight of solute in grams, the number of moles, and the molarity of a solution.

$$\frac{\text{g (solute)/Mol. Wt. (solute)}}{\text{1 liter (solution)}} = \frac{x \text{ moles}}{1 \text{ l}} = x \text{ M } (x \text{ molar}) \qquad \text{(Eq. 1)}$$

Solutions containing very small amounts of solute may be expressed in millimolar (mM) concentrations, defined as the number of millimoles/ml of solution (1 mM = 1×10^{-3} M).

(2) **Normality.** The normality of a solution expresses the number of equivalents (gram-equivalent weights) of the solute in one liter of solution. This is generally a much more useful expression since it is directly related to reactive concentrations of various species in solution. The number of equivalents of a substance is determined by its valence in the molecule or in the reaction in which it is taking part. The equivalent weight of a substance is its gram-molecular weight divided by the total valence of the particular element or complex ion under consideration. Sample calculations of milliequivalents are given in Chapter 5. Consider here that a solution containing one gram-molecular weight of sodium chloride (NaCl) also contains one equivalent weight or one equivalent of sodium chloride, since the valence of either sodium ion or chloride ion is *unity*. In the case of sodium sulfate (Na_2SO_4), the equivalent weight is one half the molecular weight because the two sodium ions contribute a total valence of two and sulfate is a bivalent ion. Dividing the molecular weight by two for this and similar compounds provides the equivalent weight. In compounds like sodium bisulfate ($NaHSO_4$), the equivalent weight depends upon its use. If it is going to be used as a source of sodium ion or as an acid (for its hydrogen ion), its equivalent weight is equal to its molecular weight, but if it is to be used for the sulfate ion, the equivalent weight is one half the molecular weight. Similarly, in the case of acids and bases, the equivalent weight is that amount which will provide or accept one gram-atomic weight of hydrogen ion (1.008 g), respectively. For example, one equivalent of a monoprotic acid, e.g., HCl, is the same as one mole of the acid, but

one equivalent of a diprotic acid, e.g., H_2SO_4, is one half mole, etc. (see Eq. 2). After the number of equivalents are calculated, the normality of the solution is found in a manner similar to Eq. 1 above, where the number of equivalents replaces the number of moles, and a solution containing one equivalent per liter is said to be a one normal (1 N) solution.

$$\text{Equiv. Wt. } H_2SO_4 = \frac{\text{Mol. Wt.}}{\text{No. Eq./mole}} = \frac{98 \text{ g/mole}}{2 \text{ Eq./mole}} = 49 \text{ g/Eq.} \quad \text{(Eq. 2)}$$

Normality may also be applied to solutions of oxidizing and reducing agents by using the "redox" valence or oxidation number of the compound in question. For example, when nitric acid is used as an oxidizing agent, HNO_3 is reduced to NO, a process involving a change from positive penta-valent nitrogen to positive divalent nitrogen (change in oxidation number from $+5$ to $+2$). The change in valence as an oxidizing agent is, therefore, three $(+5 - (+2) = +3)$, and a one normal solution is then prepared by dissolving $\frac{1}{3}$ mole of nitric acid in enough water to make one liter of solution.

The concentrations of various cations and anions in solutions used for electrolyte therapy are expressed in milliequivalents (mEq) per liter. Once again the valence of the species involved determines the milliequivalent weight. This will be illustrated further in the chapter on *Fluid Electrolytes* (Chapter 5).

(3) **Molality.** The molality of a solution expresses the number of moles of a solute contained in 1000 g of a *solvent*. Note that the quantity of total solution is not mentioned. This method of denoting concentration is used in many equations to express thermodynamic properties of solutions. An example of molal concentration calculation is shown in Appendix A.

(4) **Percent.** Percent is an expression of "parts of solute per one hundred parts of solution." Aside from the analytical aspects of pharmacy, the percent concentration designation is one of the most important to pharmacists. Pharmacists use three official interpretations of this expression. The U.S.P. XVIII (p. 12) and the N.F. XIII (p. 15) state the following concerning percentage concentrations:

Percentage Measurements—Percentage concentrations of solutions are expressed as follows:

Percent weight in weight—(w/w) expresses the number of g of a constituent in 100 g of solution.

Percent weight in volume—(w/v) expresses the number of g of a constituent in 100 ml of solution, and is used regardless of whether water or another liquid is the solvent.

Percent volume in volume—(v/v) expresses the number of ml of a constituent in 100 ml of solution.

The term percent used without qualification means, for mixtures of solids, percent weight in weight; for solutions or suspensions of solids in liquids, percent weight in volume; for solutions of liquids in liquids, percent volume in volume; and for solutions of gases in liquids, percent weight in volume. For example, a 1 percent solution is prepared by dissolving 1 g of a solid or 1 ml of a liquid in sufficient of the solvent to make 100 ml of the solution.

In the dispensing of prescription medications, slight changes in volume owing to variations in room temperatures may be disregarded.

(5) **Saturated Solution.** *A saturated solution* is one which has dissolved all the solute it is capable of holding at a given temperature. The temperature is a very crucial aspect of saturated solutions, and, unless otherwise specified, the temperature is assumed to be 25° C. As implied by the fact that a solution can be saturated with a particular solute, the constituents of a solution are not always miscible in all proportions to form a homogeneous mixture. Thus, if potassium iodide (KI) is dissolved in water to the limit of its solubility, any excess of the salt forms a separate solid phase distinct from the solution phase. At the point of maximum solubility, the salt crystallizes back onto the surface of this solid phase at the same rate as salt is being dissolved into the solution phase. Increasing the amount of solid phase increases the surface for dissolution, but also increases the surface area for crystallization of the salt. Therefore, the total amount of potassium iodide in solution is not increased. This is an equilibrium situation and the solution is said to be saturated with respect to the solute (potassium iodide). The concentration of the salt in this solution is termed the *solubility* of the salt in that solvent and at that temperature.

An example of a saturated solution is boric acid solution used as an eye wash (see Chapters 4 and 9). The concentration of this solution is near enough to saturation (4.5–5%) that a drop to below usual room temperature will cause the boric acid to crystallize from the solution. This represents a caution to its use in the eye. If crystals are present, the solution should be warmed to dissolve them.

Solubility Expressions

The solubility of a compound may be expressed in many ways. The official compendia have adopted a system of stating the amount of a particular solvent necessary to dissolve 1 g of the substance in question at 25° C. When special quantitative solubility tests are given in the compendia, these solubilities can be used as a criterion for assessing the purity of the compound. Whenever the exact solubility of a pharmaceutically important compound is not known or designated, the following descriptive terms (U.S.P. XVIII, p. 8, and N.F. XIII, p. 11) can be used:

Descriptive Term	Parts of Solvent for 1 Part of Solute
Very soluble	Less than 1
Freely soluble	From 1 to 10
Soluble	From 10 to 30
Sparingly soluble	From 30 to 100
Slightly soluble	From 100 to 1000
Very slightly soluble	From 1000 to 10,000
Practically insoluble, or insoluble	More than 10,000

The chemical literature frequently expresses solubility in terms of the number of grams of solute that can be dissolved in 100 g of solvent. An alternate method reports the volume of solvent required to dissolve 1 g of the compound. No matter which of the various methods is employed the solvent must be specified, as well as the temperature. The effect of temperature on solubility can be illustrated by constructing a solubility curve which is a graphic representation of the concentration of a saturated solution in equilibrium with the solid phase at different temperatures. Figure 3–1 is a composite of such solubility curves for a number of inorganic salts. This graph illustrates some important solubility/temperature concepts. For example, if a curve rises steeply on the graph, increasing the temperature has the pronounced effect of increasing the solubility. On the other hand, if a curve has little or no rise, the temperature has a minimal effect on the solubility. In some cases, a curve will decline rather than rise,

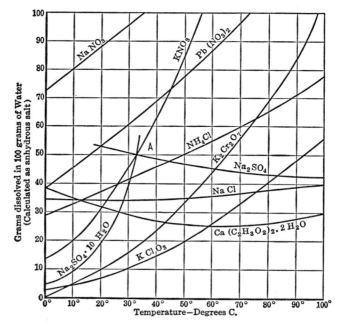

FIG. 3–1. Typical solubility curves. (From *General Chemistry* by H. G. Deming, courtesy of John W. Wiley & Sons Inc.)

indicating a decreasing solubility with increasing temperature as is seen with sodium sulfate, point A in Fig. 3–1. This is an indication that a hydrated salt has reverted to its anhydrous form or to a lower hydrate. Depending upon whether the solubility of a compound increases or decreases with increasing temperature, the substance may be said to have a positive or negative temperature coefficient of solubility, respectively. The variation of solubility with temperature can sometimes be used to advantage in the preparation of various compounds or altered crystalline forms.

At the end of the previous section, the situation was mentioned where an undissolved compound was in equilibrium with its saturated solution. Since the chemical entities that are in equilibrium with each other are found in two distinct phases, i.e., solid and liquid, this is an example of a heterogeneous equilibrium. This is to be distinguished from a homogeneous equilibrium where the equilibrium is established within a single phase system. Heterogeneous equilibria involving slightly soluble or insoluble salts give rise to another solubility expression known as the solubility product.

Consider the mixing of equal volumes of 0.1 M solutions of the two water soluble compounds barium chloride and sodium sulfate. This mixture results in the immediate precipitation of insoluble barium sulfate [rx (i)]:

$$BaCl_2 + Na_2SO_4 \leftrightarrows BaSO_4 \downarrow\; + 2\,NaCl \qquad\qquad (i)$$

In a short period of time, an equilibrium is established between the solid barium sulfate and the barium and sulfate ions in solution [rx (ii)]:

$$\underset{\text{Solid}}{BaSO_4} \leftrightarrows \underset{\text{Solution}}{Ba^{+2} + SO_4^{-2}} \qquad\qquad (ii)$$

An equilibrium constant expression for this reaction (K) can initially be written according to Eq. 3:

$$K = \frac{[Ba^{+2}]\,[SO_4^{-2}]}{[BaSO_4]} \qquad\qquad (Eq.\ 3)$$

Assuming that virtually all the barium sulfate in solution is ionized, and that the concentration of solid $BaSO_4$ remains constant, due to its low solubility, the above expression can be rewritten to give Eq. 4 where the product of the equilibrium constant and the concentration of $BaSO_4$ solid provides a new constant.

$$K\,[BaSO_4] = K_{sp} = [Ba^{+2}] \cdot [SO_4^{-2}] \qquad\qquad (Eq.\ 4)$$

The product of the ion concentrations is termed the *solubility product*, and K_{sp} is the *solubility product constant*.

A general statement of the solubility product principle says that, at a given temperature, the product of the molar concentrations of the ions in a saturated solution of a slightly soluble salt is a constant, and that the concentrations of the ions in the solubility product expression (e.g., Eq. 4) are raised to a power corresponding to the occurrence of the ion in the ionization reaction [e.g., rx (ii)]. The principle is of value in determining whether precipitation will occur or whether precipitates will dissolve under certain conditions.

The solubility product behaves in a predictable manner under most imposed conditions; that is, increasing the concentration of either of the ions to the point that the K_{sp} is exceeded results in additional precipitation of the salt. This is generally known as the *common ion effect*, and the solubility product provides a quantitative means of assessing this phenomenon.

In a manner opposing the common ion effect, removal of one of the ions in the solubility product results in an ion product which is less than the K_{sp}, and more solid will dissolve in the solution. An example of this is the dissolution of silver chloride by the addition of ammonia. A soluble diammine-silver(I) complex ion is formed in the presence of excess ammonia,

$$Ag^+ + 2NH_3 \leftrightharpoons [Ag(NH_3)_2]^+ \tag{iii}$$

and effectively removes Ag^+ from the silver chloride ion product. More silver ions accompanied by chloride ions can enter the solution phase resulting, ultimately, in the total dissolution of the solid if sufficient ammonia is added. The removal of silver ion causes a reestablishment of the equilibrium between the ions in solution and the silver chloride solid. A similar situation occurs when a large excess of *chloride ion* is added to silver chloride solutions. Unlike the situation described as the common ion effect, large amounts of chloride ion can result in the formation of a soluble complex [rx (iv)],

$$Ag^+ + 2Cl^- \leftrightharpoons [AgCl_2]^- \tag{iv}$$

removing silver ion and effectively increasing the solubility of the silver chloride solid. This is sometimes described as the *complex ion effect*.

The Solution Process

The general statement that "like dissolves like" serves as a good empirical guide to the process of solution formation. This process is intimately involved with the mutual electrostatic forces between solvent molecules and solute molecules or ions. To effect solution of a given solute, the solvent molecules must possess sufficient energy to overcome the electrostatic forces between the chemical entities comprising the solute. Likewise, the solute must be able to disrupt the forces existing between solvent molecules. When both of these can be accomplished, the individual solute molecules or ions then become dispersed throughout the solvent, and are maintained in their separated molecular or ionic state by solute-solvent interactions. It is reasonable, then, to examine the properties of solvents and solutes as they effect the solution process.

Properties of Solvents. Solvents can be roughly divided into two major categories, *polar* and *nonpolar*. The molecules of a polar solvent, e.g., water, are attracted to each other through relatively strong electrostatic forces known as *dipole-dipole interactions*. These forces were discussed briefly in Chapter 1 where the hydrogen bonding of water molecules was illustrated. Hydrogen bonding is one very prevalent type of dipole-dipole interaction. Permanent dipoles are produced in unsymmetrical molecules due to differences in electronegativity between bonded atoms (causing an unequal sharing of the bonding electrons) and the presence of pairs of nonbonded electrons.

The magnitude of the dipole is expressed by the dipole moment (see Chapter 1) which is the product of the charge on one of the atoms in the bond and the distance separating the two average centers of positive and negative charge. The experimental measurement of the dipole moment of a particular molecule involves the determination of its dielectric constant, ϵ. This is a measure of the tendency of molecules to orient themselves under the influence of an applied electric field, and, thereby, partially neutralize the electrostatic field between two oppositely charged plates. A capacitance cell is usually used to measure the dielectric constant (see Fig. 3–2) according to the relationship in Eq. 5 where the force, F, between the two plates in the cell is a function of the product of the total charge, Q, on each plate,

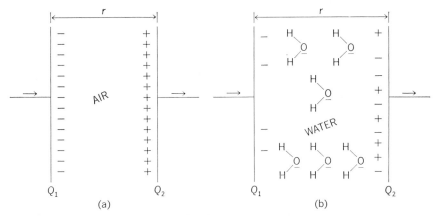

FIG. 3–2. Illustration of the charged plates of a capacitance cell in air (a) and water (b). The arrows indicate the direction of current flow through the cell.

the square of the distance separating them, r, and the dielectric constant of the medium between them, ϵ.

$$F = Q_1 Q_2 / r^2 \epsilon \qquad \text{(Eq. 5)}$$

The dielectric constant is further defined as the ratio of the capacitance of the cell in the specified medium (or solvent), C_m, to the capacitance of the cell in air (or vacuum), C_a (Eq. 6).

$$\epsilon = C_m / C_a \qquad \text{(Eq. 6)}$$

A medium with a high dielectric constant (composed of molecules with high dipole moments) reduces the force between the two plates, relative to air, by facilitating the leakage (transfer or movement) of charges from the negative to the positive plate (see Fig. 3–2b). Thus, if the force between the two plates in air is established at 1, then it will be only 1/78.5 for water, which has a dielectric constant at 25° C of 78.5. Table 3–1 lists the dielectric constants of some common polar and nonpolar solvents.

The solvent power of polar solvents on polar solutes is dependent upon more than the dielectric constant. Molecular size is also important in that solvent molecules must be able to enter the crystal lattice of the solute in order to disrupt its crystal structure. Large polar molecules will dissolve polar compounds with greater difficulty than small polar molecules.

Nonpolar solvents have very low dielectric constants (see Table 3–1). The primary intermolecular forces in these substances are van der Waals forces or induced dipole-induced dipole interactions, as discussed in Chapter 1. The weak electrostatic character of these interactions leaves solvents such as benzene and cyclohexane capable of dissolving only those solutes having similar weak forces between their molecular constituents.

Properties of Solutes. Chemical substances which are usually found as solutes in solutions may generally be divided into the same polar and

TABLE 3-1. DIELECTRIC CONSTANTS OF SOME COMMON
SOLVENTS AT 25° C*

Solvent	ϵ
Water	78.5
Glycerol	42.5
Ethyl alcohol	24.3
Acetone	20.7
Isopropyl alcohol	18.3
Chloroform	4.8
Toluene	2.39†
Benzene	2.27
Cyclohexane	2.05†

* Data taken from various sources.
† Value determined at 20° C.

nonpolar categories used above for solvents. The primary electrostatic forces found in polar solutes are of the ion-ion or dipole-dipole type. Ion-ion interactions are the strongest type of electrostatic attraction and are found in all salts, metal oxides, and metal hydroxides. Salts are considered to be ionic even in the solid state, and exist in the solid form as ionic crystal lattices. The molecular state is not definable in the arrangement of ions in the crystal lattice, and solubility is the result of separating the ionic entities from their lattice neighbors through solute-solvent interactions. The strength of ionic forces is reflected in the hardness and high melting points of these compounds.

The next strongest forces are dipole-dipole forces and are found in covalent inorganic compounds, e.g., boric acid, and most organic compounds. Solid compounds in which the crystal structures are held by these attractive forces are "softer" than the hard crystals mentioned above, and have significantly lower melting points than ionic crystals. Liquid states of dipole-dipole interacting molecules are also found, e.g., ethyl alcohol, whereas ionic substances are not found in liquid state when they are pure. Even some molecules which have dipole moments of zero due to symmetry, e.g., CCl_4 (structure I), have individual bond moments which allow weak dipolar associations of molecules to produce the liquid state.

$$\overset{\displaystyle Cl^{\delta-}}{\underset{\displaystyle Cl^{\delta-}}{\overset{\displaystyle |}{\underset{\displaystyle |}{{}^{\delta-}Cl\!\!=\!\!\!\!-C^{\delta+}\!\!-\!\!\!\!=\!Cl^{\delta-}}}}}$$

I

Nonpolar solutes include those substances with very weak molecular associations and include condensed gases such as H_2, Ne, N_2, etc., and organic compounds such as hydrocarbons. As was the case with the nonpolar solvents mentioned above, the primary associations in these compounds are van der Waals forces, which are weak attractive forces resulting from temporary polarizations of the electron cloud which induce similar polarizations in neighboring molecules (induced dipole-induced dipole). These forces diminish rapidly with increasing distance between molecules, accounting for the solubility of these compounds in nonpolar solvents.

Solute-Solvent Interactions. Having discussed the attractive forces in pure solvents and pure solutes, attention can now be focused on the forces between solutes and solvents responsible for the solution process. Since primary interest is centered on aqueous solutions, the discussion will proceed from the most polar to the least polar types of interactions.

(1) *Ion-Dipole Interactions.* Insofar as inorganic chemicals are concerned, ion-dipole interactions are probably one of the more important contributors to the solvation process. These are the solute-solvent interactions responsible for the dissolution of salts in polar solvents, such as electrolytes in water. As indicated earlier, the polarity of the solvent molecules aids in breaking the crystal lattice of the solute. In turn, the charges on the individual ions "released" from their crystalline ionic interactions attract oppositely charged ends of the polar solvent molecules. This process is generally termed *solvation*, and may be envisioned as a clustering of the solvent molecules around the solute ions due to electrostatic forces. When water is the solvent the process is termed hydration and may be illustrated for cations (II) and anions (III) as shown below.

In addition to the effect of the dielectric constant and the size of the polar molecule on the solvating power of the solvent, the charge and size of the ionic solute are also important factors. In particular, the effect of the charge on the ion-ion interactions in the crystalline structure and the effect of the ionic radius on the electrostatic force should be noted. The ionic radius also affects the ease with which solvent molecules can enter the crystal structure between the constituent ions to initiate the solution process. Of course, both the radius and the charge are also important aspects of the energy released during the solvation process (energy of solvation).

With respect to the effect of charge on the ion-ion electrostatic force or crystal lattice energy, it can be stated generally that ions with low charges are more weakly held in the lattice structure of the crystal, thus causing it to be disrupted more easily by outside forces (solvents). Some general rules regarding the effect of charges in ionic crystals on solubility in polar solvents are:

(a). If both the cation and anion of an ionic compound are monovalent, the ion-ion attractive forces are generally overcome easily by polar solvents and, as a rule, these compounds are quite soluble. As the interaction between elements of the crystal becomes more covalent, solubility becomes more difficult. This factor contributes to the low water solubility of such monovalent compounds as the silver halides. On the other hand, alkali metal halides (e.g., NaCl) which are primarily ionic crystals are generally quite soluble in water.

(*b*). If the compound is composed of two or three monovalent ions with a divalent or trivalent ion of opposite charge (e.g., $CaCl_2$, Na_2SO_4, Na_3PO_4, etc.), the interactions in the crystal become stronger, and the solubility decreases with respect to those compounds described in (*a*) above. However, these compounds are still water soluble, particularly if the cation is an alkali metal or an ammonium ion, or if the anion is nitrate, acetate, or an oxy-halogen ion such as ClO_4^-.

(*c*). If both the cation and anion are divalent or trivalent, the ion-ion interactions are generally strong enough to resist disruption by polar solvent molecules, and these compounds are generally insoluble. There are numerous exceptions to this rule if one considers the high water solubilities of aluminum sulfate, zinc sulfate, ferric sulfate, etc. However, these apparent anomalies have hydrated cations in their crystalline structures (e.g., an octahedral arrangement of water molecules may be found around the aluminum ions in aluminum sulfate crystals) which aid the solution process in aqueous solvents. These compounds may be considered to be partially hydrated (solvated) in their solid forms.

In general, if charge is held constant, the electrostatic field emanating from ions is inversely proportional to the ionic radius; that is, greater attractive forces are found in smaller ions. This fact is evident in the lattice energies of crystals and in the hydration energies of the ions when dissolved in water. For most groups of elements in the periodic table, the hydration energy (the energy released during the hydration process) decreases as one descends in a particular group of ions. For example, the hydration energies of the alkali metal ions *decrease* in the order $Li^+ >$ $Na^+ > K^+ > Rb^+ > Cs^+$ while the ionic radii *increase* in the order $Li^+ < Na^+ < K^+ < Rb^+ < Cs^+$.

To a degree, the hydration energy seems to parallel the solubilities of metal salts, indicating that the strong attraction of the smaller ions for the polar solvent molecules is a major influence in the solution process. However, the diminished ionic attraction of the larger ions in the crystal lattice appears to play a role in the solubility of some salts of rubidium and cesium. Some of the halides of these two metals have greater solubilities than those of potassium.

Parallelism to hydration energies is seen, though, in salts of the alkaline earth metals (Group IIA). These ions have smaller radii than the Group IA ions and their higher charge provides stronger interactions in the crystal state which have to be overcome by the solvent, but their hydration energies are several orders of magnitude greater than those of the alkali metal ions. Therefore, solvation is a major contributor to the solution process involving salts of these metals.

While this discussion of ion-dipole interactions has been directed toward solutes that exist mostly as ions in the solid state, it is well to point out that so-called "nonionic" compounds may dissolve in polar solvents through the intervention of ion-dipole forces. Many polar covalent compounds can be acted upon by polar solvents (particularly water) so as to produce solvated ions in the solution phase. A notable case in point is the dissolution of HCl gas in water to form hydrochloric acid. As will be seen in Chapter 4, this is actually an acid-base reaction where the covalent hydrogen chloride molecule becomes ionized under the influence of the "basic" water molecule to provide hydrated hydrogen (i.e., H_3O^+, hydronium) and chloride ions. The point being made here is that the amount of ionic (or covalent) character of a compound is a function of its surroundings, and can be altered from the theoretical value by placing the molecule in an interacting (in this case polar) medium.

(2) *Dipole-Dipole Interactions.* This type of interaction was discussed earlier under *Properties of Solvents* and in Chapter 1. The concept to be considered is that this type of force can also function in the solution process as a means of taking molecules of the solute into the solution phase. As has been pointed out, hydrogen bonding is probably the most prevalent dipole-dipole interaction and is of considerable energy (5–10 kcal/mole) when compared to most of the other solute-solvent interactions. Hydrogen bonding forces, as illustrated below, are responsible for the solubility of organic carbonyl compounds (IV) (e.g., carboxylic acids, esters, aldehydes, ketones), amines (V), organic hydroxy compounds (VI) (e.g., alcohols, sugars), some inorganic hydroxy compounds (e.g., boric acid), and other dipolar molecules. The inclusion of the various organic compounds requires the *a priori* understanding that these molecules must be of relatively low molecular weight, and that the nonpolar organic portion of these compounds does not overshadow their polar characteristics.

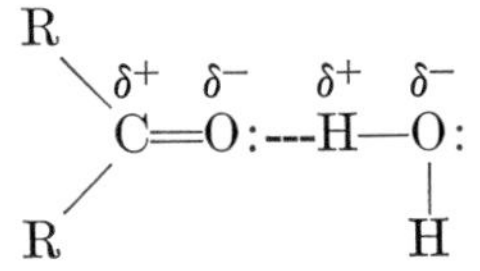

$$
\begin{array}{c}
\text{R} \\
\quad\diagdown \ \ {\scriptstyle \delta^+ \ \ \delta^- \quad \ \delta^+ \ \ \delta^-} \\
\qquad \text{C=O:---H—O:} \\
\quad\diagup \qquad\qquad\quad | \\
\text{R} \qquad\qquad\qquad \text{H}
\end{array}
$$

IV

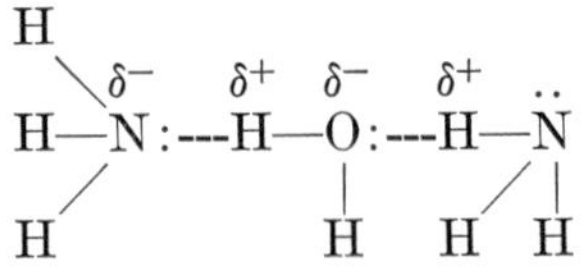

$$
\begin{array}{c}
\text{H} \\
\quad\diagdown \ \ {\scriptstyle \delta^- \quad \ \delta^+ \ \ \delta^- \quad \ \delta^+} \\
\text{H—N:---H—O:---H—N} \\
\quad\diagup \qquad\qquad | \quad\ \ \diagup | \\
\text{H} \qquad\qquad \text{H} \quad \text{H} \ \text{H}
\end{array}
$$

V

$$
\begin{array}{c}
{\scriptstyle \delta^- \quad \ \delta^+ \ \ \delta^- \quad \ \delta^+} \\
\text{R—O:---H—O:---H—O:} \\
\quad | \qquad\quad | \qquad\quad | \\
\text{H} \qquad\quad \text{H} \qquad\quad \text{R}
\end{array}
$$

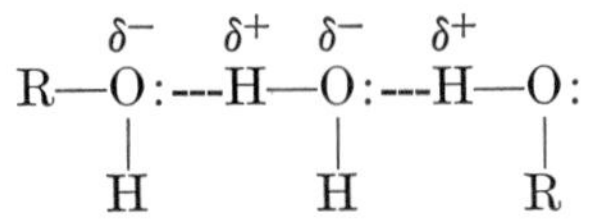

VI

The type of interaction shown in V is responsible for the basic reactions of amines in aqueous media (see Chapter 4). The dipole-dipole interaction promotes the ionization of water to, ultimately, ammonium hydroxide, $NH_4^+OH^-$.

(3) *Ion-Induced Dipole Interactions*. The inductive effect that a charged species has on a normally nonpolar species, whereby it induces some polarity into the molecule, represents an ion-induced dipole interaction. This is not a very common interaction, but any solubility that an ionic compound might have in a nonpolar solvent is most likely due to this type of force. The most common example of this effect is also one of pharmaceutical importance, the interaction between iodide ion and iodine. This interaction cannot be classified technically as a solute-solvent interaction because it occurs between two solutes, but it is ultimately responsible for the solubility of one of the solutes in water. Molecular iodine is not soluble in water, but in the presence of iodide ion (usually from the dissolution of potassium iodide), the iodine will dissolve. Reaction (v) depicts the simple chemical result of this interaction.

$$K^+ + I^- + I_2 \rightleftharpoons K^+ + I_3^- \tag{v}$$

The product in rx (v) indicates that the solubility is due to ion-dipole forces between the triiodide ion and water. A physical picture of the interaction may be envisioned as a perturbation of the normally symmetrical electron distribution in the iodine molecule produced by the electrostatic field from the iodide ion [rx (vi)]:

$$I^- \text{-----} \overset{\delta^+}{I}\text{---}\overset{\delta^-}{I} \equiv I_3^- \tag{vi}$$

An electronic picture can also be represented in rx (vii):

$$:\!\overset{..}{\underset{..}{I}}\!:^- + :\!\overset{..}{\underset{..}{I}}\!:\!\overset{..}{\underset{..}{I}}\!: \; \underset{\longleftarrow}{\overset{\longrightarrow}{\rightleftharpoons}} \; :\!\overset{..}{\underset{..}{I}}\!:\!:\!\overset{..}{\underset{..}{I}}\!:\!\overset{..}{\underset{..}{I}}\!:^- \tag{vii}$$

This reaction illustrates that I_3^- can exist in different resonance hybrids based on shifting the double bond and the negative charge. This type of force is involved in the formulation of aqueous iodine solutions used in the treatment of goiter or as antibacterials (e.g., Strong Iodine Solution, U.S.P. XVIII).

(4) *Dipole-Induced Dipole Interactions*. This is a relatively weak interaction that functions in the dissolution of inert gases in polar solvents. The dipole of water presents a sufficiently strong electrostatic field to induce a distortion in the electron distribution around the otherwise stable elements such as helium and neon. Some low molecular weight hydrocarbons such as propane, butane, pentane, etc., are known to form stable hydrates, undoubtedly through this type of interaction.

(5) *Induced Dipole-Induced Dipole Interactions*. This type of interaction, also known as van der Waals forces, was discussed under *Properties of*

Solvents and in Chapter 1. As a solute-solvent interaction, it does not hold much relevance to the area of inorganic chemistry. It is responsible for the solubility of hydrocarbons in other hydrocarbons. It also functions in the interactions between π orbitals in aromatic rings which are generally considered to be secondary binding forces between organic drug molecules and their biological "receptors."

Heat of Solution

Several times in the previous discussions, the importance of the disruption of the crystal structure of the solute and the solvation of solute entities has been described in terms of the overall solution process. The energetics involved in these two processes invariably lead to temperature changes as the solute dissolves. Of course, disruption of the crystal lattice of a solute is an energy-requiring reaction in which the energy consumed is equal to the mathematical negative of the lattice energy (the energy released when the crystal was formed). Unless energy is supplied from outside the system, the energy needed to break down the crystal lattice is taken from the kinetic energy available in the solvent molecules (solution). On the other hand, solvation of solute ions (or molecules) is an energy-releasing process and therefore represents a stabilizing interaction between the solute and the solvent. The net temperature change during solution formation depends upon the relative magnitudes of energy involved in these two processes. If the lattice energy exceeds the amount of energy available from solvation of the solute, heat energy will be absorbed from the solvent, the solution will cool, and the solution process is described as being *endothermic*. When the solvation energy exceeds the energy required to disrupt the crystal lattice (lattice energy), heat will be released, the solution will warm, and the solution process is said to be *exothermic*. This temperature or heat change, either heating or cooling, is termed the *heat of solution* and is dependent upon the nature of the solute and solvent and the concentration of the solution. The overall energy relationships are presented in the following scheme, rx (viii), which is a representation of an energy cycle for the dissolution of sodium chloride.

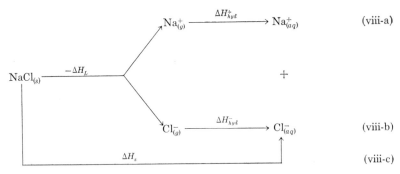

                                                 (viii-a)

                                               (viii-b)

                                               (viii-c)

The subscripts (s), (g), and (aq) in rx (viii) refer to the solid, gas, and solvated (aquated) forms, respectively. The heat of solution, ΔH_S, covers the entire process from the solid state of the solute to the solvated or hydrated state of the respective ions, as shown in rx (viii-c). The lattice energy is represented by $(-)\Delta H_L$ where the minus sign indicates that the process of going from the solid to the separate ions is an energy-consuming reaction, as opposed to the reverse process of forming the crystalline state. The solvation energy when the solvent is water is represented by ΔH_{hyd}, which is the sum of the hydration energies for the positive and negative ions (Eq. 7).

$$\Delta H_{hyd} = \Delta H^+_{hyd} + \Delta H^-_{hyd} \qquad \text{(Eq. 7)}$$

Using standard thermodynamic sign conventions, a minus sign for an energy change indicates that energy is released and the system becomes stabilized (moves to lower energy), while a positive sign indicates that energy is consumed or taken in, and the system becomes destabilized (moves to higher energy).

Aside from the solution process, both the lattice and hydration energies are usually negative, as would be expected in the formation of stable crystalline solids and stable ion hydrates. The relationship in Eq. 8 shows that a *negative* heat of solution (exothermic) should result when $\Delta H_{hyd} > \Delta H_L$ since the ΔH_L term is negative and multiplication by the minus sign makes the entire $-\Delta H_L$ term positive.

$$\Delta H_S = \Delta H_L + \Delta H_{hyd} \qquad \text{(Eq. 8)}$$

Similarly, a *positive* heat of solution (endothermic) results when $\Delta H_L > \Delta H_{hyd}$ due to the large positive value of the $-\Delta H_L$ term associated with a very stable crystalline structure. Salts with negative heats of solution tend to be more soluble than those with positive heats of solution, because the heat energy released through the solvation of ions is available to aid in breaking up the crystal lattice of other solid particles of solute. The solution process can be thought of as a "melting" of the solid into the solvent, a process requiring energy. In fact, in the absence of chemical interactions between the solute and the solvent, the heat of solution is approximately equal to the heat of fusion of the solid. It may be difficult to rationalize a *negative* heat of solution with an *increase* in temperature of the solution, but it should be recalled that negative energy values indicate a release of energy. It is this release of energy which causes the solution to become warm.

The response of a saturated solution of a compound in equilibrium with undissolved solute is in accord with *Le Chatelier's Principle*, which states that *any equilibrium tends to shift in the direction that will neutralize the effect of any stress applied to it.* Referring back to the solubility curves in

Fig. 3–1, it may be noted that those compounds showing an increase in solubility with an increase in temperature have positive heats of solution—the solution process takes place endothermically, e.g., KNO_3. Therefore, the application of external heat will "drive" the reaction to the right, resulting in the dissolution of more solute. Similarly, those compounds showing less solubility at higher temperatures have negative heats of solution. Since dissolution takes place exothermically, the application of heat will cause solute to precipitate from the solution—e.g., Na_2SO_4. Salts like NaCl have heats of solution near zero, and their solubilities are not greatly influenced by heat.

The heat of solution of certain salts can be utilized in providing controlled amounts of heating or cooling for various purposes. For example, hot and cold packs for application to swollen or painful areas produced by injury, infection, and the like can be prepared by mixing a specific volume of solvent with a weighed quantity of a solute having either a negative or positive heat of solution, respectively. Two products known as Redi-Temp®-C and RediTemp®-H contain ammonium nitrate and calcium chloride, respectively, in a special package containing a solvent. At the time of use the package is ruptured, bringing the solute and the solvent together in a pliable plastic bag. The ammonium nitrate dissolves endothermically to provide a controlled temperature cold pack, and the calcium chloride dissolves exothermically to provide a hot pack. The quantities of the ingredients are controlled to produce useful hot and cold temperatures without damage to tissues.

This discussion has dealt primarily with the dissolution of inorganic salts in water. A few statements are in order concerning other kinds of solutes. In the case of liquid solutes, the heats of solution are not as noticeable. This is due to several factors; most important are the smaller amounts of energy involved in the solute and the solute-solvent interactions in comparison to crystalline solutes. Furthermore, dissolution of a liquid solute does not involve a change of phase as is the case with solids.

In general, an increase in temperature will cause an increase in solubility or an increase in the rate of dissolution of most compounds. The major exception to this is solutions of gases. The solute-solvent interactions involved in solutions of most gases are easily overcome by the application of heat, resulting in the expulsion of the gas. The least polar gases are the ones most readily expelled, followed by those which interact more strongly with the solvent (water). This is in accord with *Henry's Law*, which states that *the solubility of a gas in a liquid is directly proportional to its partial pressure above the liquid*. As a solution of a gas is heated, the partial pressure is diminished as the molecules of the gas are carried away from the surface by the escaping solvent vapors. In many instances, it is not possible to expel a gas from a solution completely. Hydrochloric acid in particular is typical of several *constant boiling acids*. When concentrated hydrochloric acid (saturated) is heated to boiling the solution will lose HCl faster than water molecules until it reaches a concentration of approximately 20% w/w. At this point, water and HCl molecules escape at the same rate, and the concentration of the boiling liquid remains unchanged. This same phenomenon occurs when a very dilute solution of hydrochloric acid is boiled; water molecules escape faster than HCl molecules until the concentration is reached where both molecules escape at the same rate. The concentration at which this occurs depends on the barometric pressure, and the mixture is termed a *constant boiling solution*.

Finally, there is another heat effect which is observed when concentrated solutions are diluted, called the *heat of dilution*. This may be either exothermic or endothermic and, in some cases, may be the opposite of the heat of solution. More quantitative treatments of the solution process may be found in References 1 and 2.

Prediction of Solubility

It is not often possible to afford the time required to assess the solubility of a compound from the standpoint of ionic radii, charge-to-radius ratios, hydration energies, and other important structural and energetic properties. The above discussion of ion-dipole interactions has presented a brief outline of relative solubilities based on ionic charge. The following paragraphs summarize general solubility expectations based on cationic elemental groups. There are some notable exceptions to these so-called solubility rules, but for the most part they lend themselves to a rapid consideration and prediction of water solubility. Further data are presented in Table 3–2.

In general, all salts of the alkali metals are water soluble. This includes salts formed with di- and trivalent anions. The same rule applies to salts of the ammonium ion and its derivatives. In fact, the general chemistry of the ammonium ion very closely resembles that of the alkali metal cations.

Salts of alkaline earth metal cations (Group IIA) with monovalent anions (halides and nitrates) are also water soluble. Most other salts of these metals are slightly soluble to insoluble, with variations within the group. For example, a general trend is illustrated by the water solubility of magnesium sulfate (Epsom Salts) and the insolubility of the radiopaque substance barium sulfate (see Chapter 11). Solubilities vary accordingly between these two extremes. It is important to note that lithium salts resemble the alkaline earth metal salts more closely than the more closely related alkali metal salts. The solubility of most of the insoluble salts of Group IIA metals can be improved through the formation of acid salts in acidic media (hydrogen sulfates, mono- and dihydrogen phosphates, etc.).

The remaining metals are difficult to discuss in terms of group solubilities. Most lead salts are insoluble or slightly soluble. The carbonates, phosphates, borates, and similar anions of most of the heavier metals (iron,

TABLE 3–2.   WATER SOLUBILITY CHARACTERISTICS OF
INORGANIC SALTS

Soluble Salts   a) All salts of Na^+, K^+, and NH_4^+.

b) All NO_3^-, ClO_3^-, and OAc^- of any metal.

c) All Cl^- except $AgCl$, $PbCl_2$, and Hg_2Cl_2 (Mercurous).

d) All SO_4^{-2} except $CaSO_4$, $SrSO_4$, $BaSO_4$, and $PbSO_4$.

Insoluble Salts a) All CO_3^{-2} and PO_4^{-3} except those of the cations listed in *Soluble Salts* a).

b) All S^{-2} except those of Groups IA and IIA, and ammonium.

---

aluminum, etc.) are not soluble in water (neutral) or basic solution. Salts of low molecular weight organic acids and hydroxy acids, such as acetates, lactates, and gluconates, as well as sulfonic acid salts of most metals, are soluble in water. Some of the acetates, like aluminum, are subject to hydrolysis to less soluble basic acetates (e.g., aluminum subacetate; see Chapter 9). Oxides and hydroxides of heavy metals are usually insoluble; hence, basic iron salts or iron hydroxides generally precipitate from alkaline solution (see Chapter 6). The influence of pH on the apparent solubility or insolubility of salts of transition metals and aluminum is often quite critical. Neutral, alkaline, or acidic hydrolysis can alter the form of the salt to a more or less soluble one. Complexation can also aid solubility of transition metal salts which would normally be insoluble.

The Colligative Properties of Solutions

The *colligative properties* of a solution are those which are independent of the *kinds* of solute entities, but which are totally dependent upon the *number* of such entities in the solution. However, in the usual (nonideal) case the kinds of particles may influence the effective number of "independent" species in solution. For example, the electrostatic fields associated with ions may cause two or more ions to behave as a single species in solutions of significant concentration, due to association. On the other hand, a nonelectrolyte solute (no ionization) should be theoretically equal to any other nonelectrolyte with respect to its ability to impart colligative properties to a solution without considering its size, structure, or component atoms.

The colligative properties of solutions involve vapor pressure, freezing point, boiling point, and osmotic pressure. The concentration expressions utilized in the quantitative aspects of these properties are the molal concentration, m (see *Molality*) and the mole fraction, X. The *mole fraction*, X, is defined as the ratio of the number of moles (M) of any constituent in a system to the total number of moles of all constituents (Eq. 9). The sum of the mole fractions of all components will, therefore, equal unity (Eq. 10):

$$X_1 = \frac{M_1}{M_1 + M_2 + \cdots + M_n} \qquad \text{(Eq. 9)}$$

$$\sum_n X_i = (X_1 + X_2 + \cdots + X_i + \cdots + X_n) = 1 \qquad \text{(Eq. 10)}$$

Sample calculations of these expressions are presented in Appendix A. Essentially all these properties are related to the vapor pressure-lowering effect of solutes or Raoult's Law; therefore, this property will be discussed first. Reference 3 provides a more mathematical development of this topic.

(1) **Vapor Pressure.** The vapor pressure above a solution is lowered relative to that of the pure solvent. This property follows *Raoult's Law*, which states that *the partial pressure of a component of a solution is directly proportional to its mole fraction in the solution.* The proportionality constant in this relationship is the vapor pressure of the pure substance, p^0, and, using the subscript 1 to designate the solvent, the law can be represented by Eq. 11:

$$p_1 = X_1 p_1^0 \qquad \text{(Eq. 11)}$$

In any solution, the mole fraction (X_1) of the solvent (or any other component) will be less than one and, particularly if the solute is non-volatile, the vapor pressure of the solution is equal to the partial pressure of the solvent. Thus, the solute lowers the vapor pressure as its mole fraction (X_2) increases. This law applies directly to *ideal solutions*, those where there is little or no interaction between the components that would not be present in the pure substances. This is not the usual case, and most pharmaceutically important solutions are *nonideal*. These solutions present both positive and negative deviations from Raoult's law. For example, solutions of electrolytes will show a concentration effect in which ions will interact more strongly with each other as their concentration increases in the solution. The effect of these ions on the vapor pressure is less than would be expected on the basis of their mole fraction, since there are fewer "independent" species present in the solution. These solutions will generally behave in an ideal manner as they approach very low concentrations or infinite dilution.

(2) **Freezing Point.** Dissolved compounds depress the freezing point of a solution below that of the pure solvent. In dilute solution, non-electrolytes depress the freezing point in direct proportion to their molal concentration, while electrolytes depress the freezing point to a greater extent than nonelectrolytes of similar molecular weight. This latter effect is due to ionization, which produces a larger number of independent species.

Depression of freezing point can be envisioned in terms of the solute-solvent interactions that must be overcome to allow solvent, upon cooling, to crystallize from the solution. A more accurate explanation may be made on the basis of depression of vapor pressure. At the freezing point of pure solvent (e.g., water), liquid water and ice are in equilibrium. This state can exist only if the vapor pressures above the liquid and solid water (ice) are the same. In aqueous solutions, the vapor pressure is lower than pure water at all temperatures and, therefore, a lower temperature is required before solvent water and ice can exist in equilibrium, i.e., before their vapor pressures can be equal.

The extent of freezing point depression is expressed in terms of the *molal freezing point depression constant*, K_f, which is derived from the change in freezing point imparted to a solvent by dissolving one gram-molecular weight of the solute in 1000 g of the solvent. The general expression for K_f of a solvent is given in Eq. 12,

$$K_f = \frac{RT_0^2 M_1}{1000 \, \Delta H_f} \qquad \text{(Eq. 12)}$$

where R is the gas constant (1.9872 cal/deg·mole), T_0 is the absolute (degrees Kelvin) freezing point of the pure solvent, M_1 is the molecular weight of the pure solvent, and ΔH_f is the heat of fusion of the pure solvent. The change in freezing point is obtained from Eq. 13,

$$\Delta T_f = K_f m \qquad \text{(Eq. 13)}$$

where m is the molality of the solution. In the case of water, $K_f = 1.86$ deg/molal, which means that a one molal solution of a nonelectrolyte in water will depress the freezing point 1.86° C. This is quantitatively true only for dilute solutions. Freezing point depression is employed as a method of determining molecular weights of nonelectrolytes.

(3) **Boiling Point.** The boiling point of a solution is greater than that of pure solvent due to the presence of solute. Once again, the vapor pressure lowering of the solution versus that of the pure solvent indicates that a higher temperature is required to force the more volatile solvent molecules to exert sufficient pressure to escape from the surface of the solution, thus allowing the solution to boil. The quantitative expression of this property is similar to the freezing point expression (Eq. 12), and is termed the *molal boiling point elevation constant*, K_b, shown in Eq. 14:

$$K_b = \frac{RT_0^2 M_1}{1000 \, \Delta H_v} \qquad \text{(Eq. 14)}$$

In Eq. 14, the terms have the same relationship as in the freezing point depression constant (Eq. 12) except that T_0 is now the absolute boiling temperature of the pure solvent and ΔH_v is the heat of vaporization. The change in boiling point is obtained by multiplying K_b by the molality of the solution as shown in Eq. 15:

$$\Delta T_b = K_b m \tag{Eq. 15}$$

For water, $K_b = 0.52$ deg/molal, indicating that a one molal aqueous solution of a nonelectrolyte will boil at $0.52°$ C above the normal boiling point of pure water.

(4) **Osmotic Pressure.** When two solutions of different concentration are separated from each other by a membrane (e.g., parchment) which is permeable to solvent but not solute (semipermeable), the solvent will pass from the solution of greater concentration (solute more dilute) to the solution of lower concentration (solute more concentrated). This spontaneous process will continue until the solvent concentration on both sides of the membrane is equal and the system is said to be in equilibrium. The passage of solvent is known as *osmosis* and may exist between a solution and pure solvent. The *osmotic pressure*, π, is the pressure that must be applied to stop the spontaneous flow of solvent in either direction. Osmotic pressure is dependent upon the *solute concentration* and the *absolute temperature* and, for dilute solutions, can be calculated using one of several variations of the *Van't Hoff equation* as shown in Eq. 16:

$$\pi V = n_2 R T \tag{Eq. 16}$$

In this equation, V is the volume of the solution, n_2 is the number of moles of solute, R is the gas constant (0.08205 liter-atm/deg·mole), and T is the absolute temperature. The magnitude of osmotic pressure can be considerable, and pressures of several atmospheres are common. Osmotic properties are important to the mechanisms of action of drugs such as saline cathartics and osmotic diuretics. They are also important in the maintenance of fluid balance across biological membranes (see Chapter 5).

(5) **Tonicity.** As was indicated above, the osmotic pressure is dependent upon the number of particles (ions, molecules, colloidal particles) in each unit volume of the solution. This is analogous to the pressure of a gas being proportional to the number of molecules per unit volume in the containing vessel.

Physiological cell membranes act as semipermeable membranes with respect to the passage of water by osmosis, while retaining solutes. It should then be expected that solutions coming into contact with tissue cells and having osmotic pressures different from physiological fluids could cause fluid imbalances and cell damage. This becomes a consideration for injectable or parenteral preparations, ophthalmic solutions (collyria), and other solutions applied to mucous membranes or sensitive tissues. Ideally, these solutions are prepared in a manner which would allow minimal interference with the normal tone of cell membranes and fluid balance.

Such solutions are termed *isotonic*, indicating that their effect on *cellular tone, tonicity,* is the same as that of normal physiological fluids. In other words, isotonic solutions have osmotic pressures equal to the osmotic pressure of intracellular fluid ($\pi_{soln} = \pi_{cell}$). These solutions can be applied to tissues or injected without causing damage to cells through osmotic effects.

The effect on cells of nonisotonic solutions follows the physical description of osmotic pressure imbalance mentioned above. If the osmotic pressure of the applied solution is greater than that of the intracellular fluid, the solution is termed *hypertonic* ($\pi_{soln} > \pi_{cell}$). This type of solution will cause water to leave the intracellular compartment with consequent cell shrinkage, a phenomenon known as *plasmolysis* (the term *crenation* is applied to this occurrence in red blood cells).

The opposite situation, in which the osmotic pressure of the solution is less than that of the intracellular fluid, results in a *hypotonic* solution ($\pi_{soln} < \pi_{cell}$). When a solution of this type comes into contact with tissue cells, the cell will imbibe water, which produces swelling, distention, and finally rupture. This course of events is referred to as *plasmoptysis*, or *hemolysis* in the case of red blood cells.

Hypotonic or hypertonic solutions are sometimes used to advantage in electrolyte therapy (see Chapter 5), and the production of hypertonic conditions in kidney tubules and the intestinal tract is responsible for the action of osmotic diuretics and saline cathartics, respectively (see Chapter 8). However, isotonic conditions are required for ophthalmic, nasal, most electrolyte, and other preparations.

Experimental evidence (e.g., freezing point data) shows that a 0.9% w/v aqueous solution of sodium chloride is isotonic with all body fluids (including lachrymal fluid). Since sodium chloride is normally found in extracellular fluid, it follows that this salt can be used as the compound of choice for the adjustment of tonicity. Comparisons of the freezing point depression of various drugs with that of sodium chloride have resulted in the development of *sodium chloride equivalents*. These are factors which, when multiplied by the weight of a corresponding compound, provide a number equivalent to the weight of sodium chloride necessary to produce a solution having the same tonicity, provided that the weight of the compound and the calculated weight of sodium chloride are dissolved in equal volumes of water. This procedure allows the quantity of sodium chloride being replaced in a particular solution by another compound to be determined as well as the amount of sodium chloride to be added to the preparation to make it isotonic on the basis of a 0.9% solution. Of course, hypotonic and hypertonic solutions having a particular tonicity relative to sodium chloride can be prepared using the same factors. A table of sodium chloride equivalents for some commonly used drugs and sample calculations are given in Appendix B.

References

1. Daniels, F., and Alberty, R. A.   Physical Chemistry, 3rd ed. New York: John Wiley and Sons, 1966.
2. Johnson, D. A.   The standard free energies of solutions of anhydrous salts in water. J. Chem. Ed., **45**:236, 1968.
3. Martin, A. N., Swarbrick, J., and Cammarata, A.   Physical Pharmacy, 2nd ed. Philadelphia: Lea & Febiger, Chapter 7, 1970.

4
Pharmaceutical Aids and Necessities

This chapter presents discussions of several topics important to the preparation, preservation, and storage of pharmaceutical products. The major subject areas include: *acids and bases*, frequently employed in the conversion of drugs to chemical forms convenient to their product formulations; *buffers*, used to maintain the pH of various formulations within prescribed limits; *antioxidants*, used to prevent oxidative decomposition of pharmaceutically active components; *water*, the primary solvent or liquid phase in most liquid pharmaceutical preparations; and *glass*, used in storage and dispensing containers for most drug products. Where appropriate, these topics will be introduced from a theoretical point of view with final emphasis on their pharmaceutical and/or therapeutic applications.

Acids and Bases

Acid-Base Theories. Acids and bases may be defined within the framework of several different concepts. The variation in these concepts, aside from historical aspects, is a function of the respective breadth of application intended and is not a matter of their overall correctness. It should be noted, however, that one interpretation may be better suited than another, depending upon the particular acid-base reaction or system involved.

Arrhenius Concept. All present concepts may be viewed as extensions of the classical Arrhenius concept, which defines acids and bases in aqueous systems and, while limited in scope, is a correct and useful interpretation. Arrhenius defined an acid as any substance which is capable of providing hydrogen ions (or protons, H^+) in aqueous solution. A base was defined as a substance containing hydroxy groups and/or capable of providing hydroxide ion (OH^-) in aqueous solution. According to these definitions, a neutralization reaction combines these two ions to form the solvent, i.e., water, and a salt. This can be illustrated in the following general reaction (i) between an acid (HA) and a base (MOH):

$$HA + MOH \rightleftarrows M^+A^- + H_2O \qquad \text{(i)}$$

The distinction accorded to hydrogen and hydroxide ions in this approach is merely a reflection of their being the component ions of the most common solvent, water.

This concept is limited by the fact that nonhydroxide-containing compounds which function as bases or produce hydroxide ion in aqueous solution are not covered by the definition. Secondly, the Arrhenius concept does not lend itself to extension to nonaqueous systems. It is for this reason that broader concepts of acid-base reactions have been sought, resulting in those of Brönsted-Lowry and Lewis, as well as others.

Brönsted-Lowry Concept. This is sometimes termed the protonic concept of acid-base character. The concept was developed independently by J. N. Brönsted and J. M. Lowry in 1923, and has been one of the most popular descriptions of acid-base reactions. According to definition, an acid is any substance capable of *donating* a proton (H^+) in a chemical reaction. This definition is not substantially different from that provided by Arrhenius, with the apparent exception that neither water nor any other solvent is mentioned. However, the definition of a base is significantly different: a base is any substance capable of *accepting* a proton in a chemical reaction. This definition neither excludes hydroxide ion as a base nor limits bases to hydroxide-producing substances. A summary definition can be stated: *an acid is proton donor and a base is a proton acceptor.* The Brönsted-Lowry concept has come to be referred to as simply the Brönsted concept, and this terminology will be used throughout the remainder of this text.

Within the framework of the above definitions, a Brönsted acid ionizes to produce a proton and the *conjugate base* of the acid. This can be illustrated for the general case by the following half-reaction for the acid HA [rx (ii)]:

$$HA \rightleftarrows H^+ + A^-$$

Acid	Conjugate
	Base

(ii)

Some representative acid-conjugate base pairs are illustrated below:

Acid	*Conjugate Base*
HCl	$\rightleftarrows H^+ + Cl^-$
H_2SO_4	$\rightleftarrows H^+ + HSO_4^-$
HSO_4^-	$\rightleftarrows H^+ + SO_4^{-2}$
H_2CO_3	$\rightleftarrows H^+ + HCO_3^-$
H_3PO_4	$\rightleftarrows H^+ + H_2PO_4^-$
$H_2PO_4^-$	$\rightleftarrows H^+ + HPO_4^{-2}$
NH_4^+	$\rightleftarrows H^+ + NH_3$
H_3O^+	$\rightleftarrows H^+ + H_2O$

(iii)

Brönsted bases are usually neutral molecules or anions, although a few cases of cations acting as proton acceptors are known. The chemical species formed when a base accepts a proton is known as the *conjugate acid* of that base, and is illustrated for the general case by the following half-reaction for a neutral base, B: [rx (iv)]:

$$\text{B:} + \text{H}^+ \rightleftarrows \text{BH}^+ \qquad\qquad \text{(iv)}$$

Base   Conjugate Acid

Some representative base-conjugate acid pairs are illustrated below:

Base			Conjugate Acid
OH^-	$+ H^+$	$\rightleftarrows$	H_2O
H_2O	$+ H^+$	$\rightleftarrows$	H_3O^+
SO_4^{-2}	$+ H^+$	$\rightleftarrows$	HSO_4^-
HCO_3^-	$+ H^+$	$\rightleftarrows$	H_2CO_3
HPO_4^{-2}	$+ H^+$	$\rightleftarrows$	$H_2PO_4^-$
$H_2PO_4^-$	$+ H^+$	$\rightleftarrows$	H_3PO_4
NH_3	$+ H^+$	$\rightleftarrows$	NH_4^+

(v)

It should be noted in some of the above reactions that certain species (e.g., HSO_4^- and $H_2PO_4^-$) may react as either an acid or a base. Chemical conditions determine which type of reaction predominates. Water is a very interesting solvent in this respect; in the presence of a strong base it reacts as a proton donor, and in the presence of an acid it reacts as a proton acceptor.

A complete acid-base reaction in the Brönsted sense requires the combination of half-reactions similar to those given in rx's (iii) and (v) to provide a reaction in which an acid donates a proton to a base, producing the conjugate base and conjugate acid, respectively. The following is an illustration of the general case of an acid-base reaction [rx (vi)]:

$$\text{HA} + \text{B:} \rightleftarrows \text{BH}^+ + \text{A}^- \qquad\qquad \text{(vi)}$$

Acid   Base   Conjugate Acid   Conjugate Base

Some typical reactions include:

Acid		Base		Conjugate Acid		Conjugate Base
HCl	$+$	H_2O	$\rightleftarrows$	H_3O^+	$+$	Cl^-
H_2O	$+$	NH_3	$\rightleftarrows$	NH_4^+	$+$	OH^-
H_3PO_4	$+$	OH^-	$\rightleftarrows$	H_2O	$+$	$H_2PO_4^-$
$H_2PO_4^-$	$+$	OH^-	$\rightleftarrows$	H_2O	$+$	HPO_4^{-2}
H_2SO_4	$+$	OH^-	$\rightleftarrows$	H_2O	$+$	HSO_4^-
H_3O^+	$+$	OH^-	$\rightleftarrows$	H_2O	$+$	H_2O

(vii)

The preceding reactions illustrate the acidic and basic character of water, a property which is usually termed *amphoteric*. As indicated previously, this property may be found in a number of other chemicals in addition to water.

A final point which will be expanded later in this chapter deals with the relative strengths of acids and bases and their conjugate species. By definition, a strong acid is one which readily donates its proton to a base; therefore, the conjugate base of such an acid must necessarily be weak. If this was not the case, the conjugate base would have a strong attraction for the proton, thereby weakening the strength of the acid. The general rule to follow, then, is that *strong acids have weak conjugate bases* and, similarly, *strong bases have weak conjugate acids*.

Lewis Acid-Base Concept. This concept represents the ultimate extension of the Brönsted-Lowry approach, and was introduced at about the same time by G. N. Lewis. The Lewis concept is very important to the general understanding of all acid-base reactions in chemistry. It draws upon the idea that a proton is an electron-deficient or *electrophilic* species, and that in reality this is the case with other nonprotonic entities that react as acids in a manner similar to that of the proton. In general, the acidic property of chemical entities is involved with electron "seeking" in a reaction, and the basic property is exhibited by those species which donate electrons in a reaction (*nucleophilic* species).

Thus, the Lewis concept defines an *acid as any substance which can accept a share in a pair of electrons in a reaction*, and a *base is any substance which can donate a pair of electrons to share with an acid in a reaction*. In other words, an acid is an *electron acceptor*, and a base is an *electron donor*.

In this definition, the Brönsted acid, i.e., the proton, is also a Lewis acid, but quite obviously all Lewis acids are not Brönsted acids. By similar reasoning, it can be seen that Brönsted bases are also bases in the Lewis sense because of their electron donor properties. However, since all Lewis bases may not react with a proton, they cannot all be classified as Brönsted bases.

The Lewis concept really widens the field of acid-base reactions, but space does not allow a complete consideration of all the possibilities (see Ref. 1). Some common examples of Lewis acids, other than the previously described proton, include such electron-deficient compounds as BF_3, $AlCl_3$, $FeCl_3$, and such simple metal cations as Ag^+, Fe^{+2}, and Zn^{+2}. These compounds and ions have the obvious feature of possessing at least one empty orbital capable of accepting a pair of electrons. Transition metal complexes, discussed in Chapter 1, are examples of the combination of a Lewis acid (the metal cation) with a Lewis base (the ligand). Other compounds which can be classified as Lewis acids are those which possess a central species capable of expanding the valence shell to accept a pair of electrons; some examples are SiF_4, $SnCl_4$, and SF_4. While the compounds have a complete octet of electrons around the central element, these elements are able to coordinate more ligands by expanding their valence shells to include empty d orbitals. Molecular iodine is also capable of expanding its outer molecular orbitals to accept a pair of electrons from an iodide ion to form the tri-iodide ion (see below). Thus, the ion-induced dipole interaction responsible for the solubilization of iodine in water (see Chapter 3) may be viewed as an acid-base reaction in the Lewis sense. Practically any anion and any neutral molecule having a pair of nonbonded electrons can react as a Lewis base. Water, ammonia, and halide ions are common Lewis bases.

The reaction between a Lewis acid and Lewis base provides a product that is described variously as an *adduct*, acid-base complex, or coordinated complex. The term adduct will be used in this text. Some representative Lewis acid-base reactions are shown below (viii). The first few reactions are shown employing the characteristic Lewis "dot" structures to illustrate the "open sextets" of some of the acids.

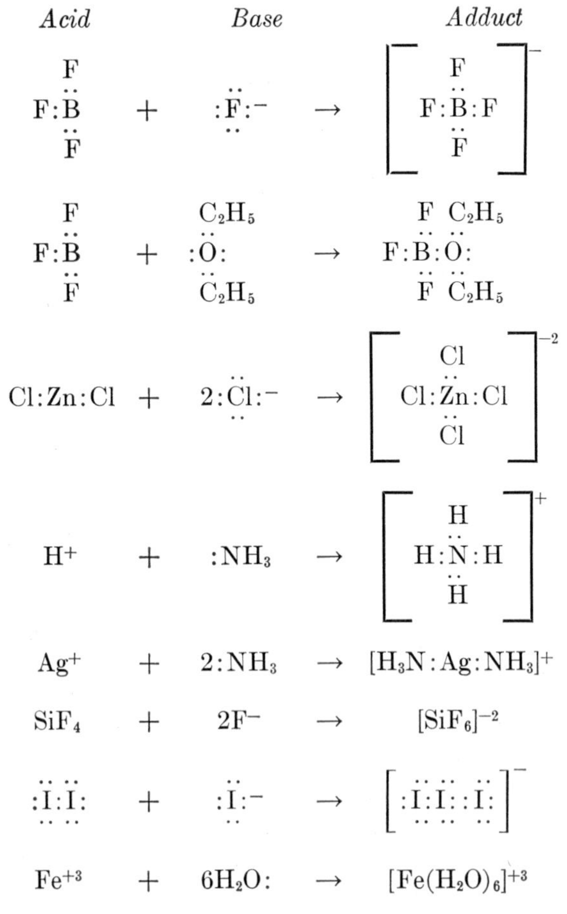

Acid		*Base*		*Adduct*

$$\begin{matrix} F \\ F\!:\!B \\ F \end{matrix} \quad + \quad :\!\ddot{F}\!:^- \quad \rightarrow \quad \left[\begin{matrix} F \\ F\!:\!B\!:\!F \\ F \end{matrix}\right]^-$$

$$\begin{matrix} F \\ F\!:\!B \\ F \end{matrix} \quad + \quad \begin{matrix} C_2H_5 \\ :\!\ddot{O}\!: \\ C_2H_5 \end{matrix} \quad \rightarrow \quad \begin{matrix} F \quad C_2H_5 \\ F\!:\!B\!:\!\ddot{O}\!: \\ F \quad C_2H_5 \end{matrix}$$

$$Cl\!:\!Zn\!:\!Cl \quad + \quad 2:\!\ddot{Cl}\!:^- \quad \rightarrow \quad \left[\begin{matrix} Cl \\ Cl\!:\!Zn\!:\!Cl \\ Cl \end{matrix}\right]^{-2}$$

(viii)

$$H^+ \quad + \quad :\!NH_3 \quad \rightarrow \quad \left[\begin{matrix} H \\ H\!:\!N\!:\!H \\ H \end{matrix}\right]^+$$

$$Ag^+ \quad + \quad 2:\!NH_3 \quad \rightarrow \quad [H_3N\!:\!Ag\!:\!NH_3]^+$$

$$SiF_4 \quad + \quad 2F^- \quad \rightarrow \quad [SiF_6]^{-2}$$

$$:\!\ddot{I}\!:\!\ddot{I}\!: \quad + \quad :\!\ddot{I}\!:^- \quad \rightarrow \quad \left[:\!\ddot{I}\!:\!\ddot{I}\!:\!:\!\ddot{I}\!:\right]^-$$

$$Fe^{+3} \quad + \quad 6H_2O\!: \quad \rightarrow \quad [Fe(H_2O)_6]^{+3}$$

Many other Lewis acid-base reactions could be cited, including most solute-solvent interactions, hydrogen bonding, and most organic chemical reactions. The general utility of this concept to both organic and inorganic chemistry should be evident.

Hard and Soft Acid-Base Concept. In actuality, this concept does not offer another definition of an acid or base, but provides a qualitative view of various factors affecting the stability of acid-base complexes. The concept was developed by R. G. Pearson[2,3] and is sometimes referred to as *Pearson's principle of hard and soft acids and bases*, or *Pearson's HSAB principle.* As in the Lewis concept, acids and bases are still defined as electron acceptor and donor species, respectively. However, the *HSAB principle further categorizes acids and bases according to the properties of charge, size (e.g., ionic radius), polarizability, etc.*

According to the fundamental concepts of the HSAB principle, *hard acids* are electron acceptors having high positive charges, relatively small sizes, and unfilled valence shell orbitals. These are properties which impart high electronegativities (electron-seeking) and low polarizabilities to these species. *Soft acids* have the opposite characteristics—low positive charges, relatively large sizes, and filled valence shell orbitals—all of which impart low electronegativities and high polarizabilities. All hard or soft acids need not have all of the above characteristics simultaneously, and in fact there are acids classified in a borderline category. Table 4–1 lists some of the acids important to pharmacy according to their hard and soft properties.

Bases are donor species which may be described in terms similar to those used for acids. *Hard bases* have high electronegativities, are easily reduced, have stable valence shell structures, and low polarizabilities. It then follows that *soft bases* have low electronegativities, are easily oxidized, have empty low-lying orbitals, and high polarizabilities. As in the case of acids, borderline entities may also be found in this classification. Table 4–2 lists some representative hard, soft, and borderline bases.

This classification of acids and bases arose from observations concerning the types of acids and bases that form stable complexes or adducts with each other. A general rule may be stated from these observations: *the most stable adducts are formed between hard acids and hard bases and/or*

TABLE 4–1. HARD, SOFT, AND BORDERLINE CATEGORIES
OF COMMON LEWIS ACIDS

Hard Acids: H^+, Li^+, Na^+, K^+, Mg^{+2}, Ca^{+2}, Sr^{+2}, Al^{+3}, Fe^{+3}, As^{+3}, Sn^{+4}, I^{+7}, I^{+5}, Cl^{+7}, BF_3, $B(OR)_3$, HX (hydrogen bonding molecules)

Soft Acids: Cu^+, Ag^+, Au^+, Hg^+, Hg^{+2}, Cd^{+2}, I^+, Br^+, I_2, Br_2, HO^+, O, Cl, Br, I, M^0 (metal atoms)

Borderline: Fe^{+2}, Cu^{+2}, Zn^{+2}, Pb^{+2}, Sn^{+2}, Sb^{+3}, Bi^{+3}

TABLE 4–2. HARD, SOFT, AND BORDERLINE CATEGORIES
OF COMMON LEWIS BASES

Hard Bases:	H_2O, OH^-, F^-, $CH_3CO_2^-$, PO_4^{-3}, SO_4^{-2}, CO_3^{-2}, ClO_4^-, NO_3^-, ROH, R_2O, NH_3, RNH_2, Cl^{-*}
Soft Bases:	I^-, CN^-, H^-, $S_2O_3^{-2}$, R_2S, RSH, R_3As
Borderline:	Br^-, SO_3^{-2}, aromatic amines, pyridine

* Some reactions of Cl^- indicate that it might be more correctly classed as a borderline base.

between soft acids and soft bases. The relative stabilities of various adducts is dependent upon the relative softness or hardness of the acids and bases involved. Tables 4–1 and 4–2 are not intended to illustrate any trends in these characteristics, but it should be noted that the harder acids have the higher oxidation states. Also, hardness and softness can be influenced by the attached groups in a complex ion. Softness would be expected to increase as one descends in any group of similarly substituted elements in the periodic table. Similarly, softness should decrease from left to right in any period of comparably substituted elements.

The HSAB principle is applicable to many aspects of chemistry. For example, an explanation of relative Brönsted acidities and basicities may be found in this concept. The combination of a proton (hard acid) with water (hard base) should yield a fairly stable or relatively undissociated complex. The hydronium ion, H_3O^+, should be a much weaker acid (proton donor) than the combination of proton with iodide ion (soft base).

Although the Lewis and HSAB concepts are of tremendous general utility, it is usually more convenient to consider acids and bases from the Brönsted point of view. This does not mean that one should lose sight of the former concepts. While the proton donor-acceptor model is most easily adaptable to the chemical events to be discussed, the HSAB principle will be of value in determining aspects of relative acid-base strength.

The Specification of Acidity and Basicity. The reactivity of acids and bases is a function of their strength. Within the framework of the Brönsted definition, the strength of an acid is a measure of the ease of proton donation, and in the case of a base, the ease of protonation. These properties are dependent upon the conditions present in the total system containing the acid or base. For example, an acid which is weak in a water solution may be a good proton donor in a solution containing a base stronger than water. The conditions may suppress or promote the donor-acceptor properties of the acidic or basic entities. This discussion will center on the relative strength of acids and bases in aqueous solution. Water has been chosen because it is the common vehicle for liquid pharmaceuticals, and biological fluids are similarly aqueous. Proton donor-acceptor properties are also well demonstrated in water.

Dissociation Constants. The strength of a Brönsted acid is reflected by the degree of dissociation of the proton from the conjugate base when the acid is dissolved in water. This reaction is illustrated by the following equilibrium [rx(ix)]:

$$HA + H_2O \rightleftharpoons H_3O^+ + A^- \tag{ix}$$

The degree of dissociation or ionization is expressed in the equilibrium constant which may be written for the above reaction (Eqs. 1a and 1b):

$$K_{eq} = \frac{[H_3O^+] \, [A^-]}{[HA] \, [H_2O]} \tag{Eq. 1a}$$

$$K_a = K_{eq} \, [H_2O] = \frac{[H_3O^+] \, [A^-]}{[HA]} \tag{Eq. 1b}$$

The term K_a (Eq. 1b) is the ionization constant for the acid, and is usually expressed as the product of the equilibrium constant and the concentration of water. Therefore, the K_a is written omitting the water concentration term (Eq. 1b). The second expression given in Eq. 1b will suffice for most of the further work to be done with acid ionization constants. The K_a varies directly with the strength of the acid. Strong acids (e.g., HCl, H_2SO_4, HNO_3) have large ionization constants ($K_a > 1$), while weaker acids have successively smaller values.

Another expression of the strength of an acid is the hydrogen ion concentration. As the strength of an acid increases it would be expected to furnish more hydrogen ion according to the equilibrium shown in rx (ix). In most polar solvents, the hydrogen ion will react with solvent to form a stable complex, and the concentration of this complex becomes equated with the hydrogen ion concentration. In aqueous solution, then, the concentration of proton is usually considered to be the concentration of hydronium ion H_3O^+. Pure water ionizes to a small degree as illustrated in the equilibrium [rx (x)]:

$$2H_2O \rightleftharpoons H_3O^+ + OH^- \tag{x}$$

An ionization constant expression can be written similar to Eq. 1b for the acid ionization equilibrium. However, the most frequently employed relationship involves only the concentrations of the ionic species on the right-hand side of rx (x). This is termed the *ion product* constant, K_w, and is represented by Eq. 2 below:

$$K_w = [H_3O^+] \, [OH^-] = 1 \times 10^{-14} \tag{Eq. 2}$$

As indicated in Eq. 2, the ion product constant for water has a constant value at $25°$ C of 1×10^{-14}. In pure water, then, the concentrations of both ions are equal, each having the value of 1×10^{-7} M (Eq. 3):

$$[H_3O^+] = [OH^-] = 1 \times 10^{-7} \qquad \text{(Eq. 3)}$$

If, through the addition of acid or base, the concentration of one of the above species is increased, the concentration of the other species will be decreased to a quantity such that the constant value of K_w is maintained. For example, acid may be added to an aqueous solution to the extent that $[H_3O^+] = 1 \times 10^{-3}$ M; the corresponding hydroxide ion concentration then becomes $[OH^-] = 1 \times 10^{-11}$ M.

Hydronium ion concentration in the above expressions is usually expressed in units of moles/liter or as gram-equivalents/liter, and is written in complex form as an exponential power of ten. This is not a very convenient numerical form to use in making comparisons of the acidities of solutions containing varying amounts of H_3O^+. Therefore, Sorenson introduced a more practical concept of expressing *acidity as a negative logarithm to the base 10 of the hydrogen (or hydronium) ion concentration*. This serves as *the definition of pH* ("p" from the German word for "power" [Potenz] and "H" for hydrogen), and is represented by the expression given in Eq. 4:

$$pH = -\log [H^+] = -\log [H_3O^+] = \log \frac{1}{[H_3O^+]} \qquad \text{(Eq. 4)}$$

TABLE 4–3.　pH AND pOH RELATIONSHIPS†

$[H^+]$ $\quad$ or moles/l	$[H_3O^+]$ g-equiv/l	pH*			pOH
1.0	1×10^0	0 ↑		↑	14
0.1	1×10^{-1}	1			13
0.01	1×10^{-2}	2	Increasing		12
0.001	1×10^{-3}	3	acidity		11
0.0001	1×10^{-4}	4			10
0.00001	1×10^{-5}	5			9
0.000001	1×10^{-6}	6			8
0.0000001	1×10^{-7}	7	neutral		7
0.00000001	1×10^{-8}	8			6
0.000000001	1×10^{-9}	9			5
0.0000000001	1×10^{-10}	10	Increasing		4
0.00000000001	1×10^{-11}	11	alkalinity		3
0.000000000001	1×10^{-12}	12			2
0.0000000000001	1×10^{-13}	13			1
0.00000000000001	1×10^{-14}	14 ↓		↓	0

* A tenfold change in $[H^+]$ takes place from one pH unit to another.
† From Roger's Inorganic Pharmaceutical Chemistry, 8th ed., by T. O. Soine and C. O. Wilson, Lea & Febiger, Philadelphia, p. 108, 1967. By permission.

Table 4–3 illustrates the relationship between hydrogen ion concentration and pH. Note that as the concentration of hydrogen ion increases the acidity increases and the pH decreases.

While it is relatively easy to arrive at the pH of a solution from the hydrogen ion concentrations given in Table 4–3, it becomes more difficult when these concentrations are not simple powers of 10. Examples illustrating the mathematical manipulations follow.

Example 1. What is the pH of a 3.43×10^{-5} N solution of hydrochloric acid? Since hydrogen chloride is virtually totally ionized in water, the above solution contains 3.43×10^{-5} gram-equivalents of H_3O^+.

$$\log 3.43 = 0.54$$
$$-\log [H_3O^+] = -\log 3.43 \times 10^{-5} = -(0.54 - 5) = 4.46$$
$$pH = 4.46$$

Another approach:

$$[H_3O^+] = 3.43 \times 10^{-5} = 1 \times 10^{(-5 + 0.54)} = 1 \times 10^{-4.46}$$
$$pH = -\log 1 \times 10^{-4.46} = 4.46$$

The process is simply reversed to calculate the hydrogen ion concentration of a solution of known pH. If, however, the pH is not a simple whole number, it must be converted to a combination of a negative exponent and a log having a zero characteristic. It is necessary, then, to take the antilog of the latter to arrive at the multiplier of 10 raised to the negative power.

Example 2. What is the hydronium ion concentration of a solution of hydrochloric acid having a pH of 4.52?

Since the pH $= 4.52$

$$[H_3O^+] = 1 \times 10^{-4.52} = 1 \times 10^{(-5 + 0.48)} = \text{antilog } 0.48 \times 10^{-5}$$
$$[H_3O^+] = 3.2 \times 10^{-5} \text{ N}$$

Another approach may be used:

$$[H_3O^+] = -\text{antilog } (5 - 0.48) = 3.2 \times 10^{-5} \text{ N}$$

It should also be noted in Table 4–3 that a change of one whole pH unit represents a tenfold change in hydronium ion concentration. However, due to the exponential relationship between hydronium ion concentration and pH, these two quantities do not vary linearly between unit changes in pH. Table 4–4 illustrates this relationship for changes in hydrogen ion concentration between a pH of 4 and 5.

TABLE 4–4. THE RELATIONSHIP BETWEEN pH AND
HYDROGEN ION CONCENTRATION*

$[H^+]$ moles/l	pH
0.00001	5.0
0.00002	4.69
0.00003	4.52
0.00004	4.39
0.00005	4.30
0.00006	4.22
0.00007	4.15
0.00008	4.09
0.00009	4.04
0.0001	4.0

* From Roger's Inorganic Pharmaceutical Chemistry, 8th ed., by T. O. Soine and
C. O. Wilson, Lea & Febiger, Philadelphia, p. 112, 1967. By permission.

The measurement of pH normally involves the potentiometric method
using a pH meter. The U.S.P. and N.F. describe a method of accurate
measurement utilizing two standard buffer solutions which bracket the
expected pH range (see U.S.P. XVIII, p. 938, or N.F. XIII, p. 824). In
situations where potentiometric measurements cannot be performed or
only an approximate pH value is required, indicator solutions or pH papers
(indicator or test papers) may be employed. Indicators are themselves
acids or bases which exist as molecules having a different color in their
acid form than they do in their conjugate base form, or vice versa. As
with any other acid-base reaction, an equilibrium is involved; e.g., the
indicator represented by HInd reacts in water according to rx (xi-a) or (xi-b):

$$\underset{\text{Acid}}{\text{HInd}} + H_2O \rightleftarrows H_3O^+ + \underset{\text{Conj. Base}}{\text{Ind}^-} \tag{xi-a}$$

or

$$\text{Yellow molecules} + H_2O \rightleftarrows H_3O^+ + \text{Blue anions}^- \tag{xi-b}$$

Therefore, the point of color change does not provide a definite pH value;
rather, a narrow pH range is observed. Solutions of indicators at particular
pH values can be compared to give an approximate pH for an unknown
solution. Table 4–5 lists some common indicators and their respective
pH ranges. Information concerning the choice of an appropriate indicator
may be found in most analytical chemistry textbooks (see Refs. 4 and 5).
Indicator papers accomplish a purpose similar to the indicator solutions.
The paper is dipped into the solution to be measured and the color compared to a standard chart. These papers can be used very easily by
chemists and pharmacists to assess the approximate pH of a preparation

TABLE 4-5. SOME COMMON INDICATORS AND THEIR pH RANGES

Indicator	Range of pH and Color Changes
Metacresol-purple	1.2 (red) to 2.8 (yellow)
Hellige orange	2.6 (red) to 4.2 (yellow)
Bromophenol blue	3.0 (yellow) to 4.6 (blue)
Methyl orange	3.1 (red) to 4.4 (yellow)
Bromocresol green	4.0 (yellow) to 5.6 (blue)
Methyl red	4.2 (red) to 6.3 (yellow)
Chlorophenol red and bromocresol purple	5.2 (green) to 6.8 (purple)
Bromothymol blue	6.0 (yellow) to 7.6 (blue)
Phenol red	6.8 (yellow) to 8.4 (red)
Cresol red	7.2 (yellow) to 8.8 (red)
Thymol blue	8.0 (yellow) to 9.6 (blue)
Phenolphthalein	8.3 (colorless) to 10.0 (red)
Thymolphthalein	9.4 (colorless) to 10.6 (blue)
Nitro yellow	10.0 (faint yellow) to 11.6 (deep yellow)
Violet	12.0 (purple) to 13.6 (blue)

quickly. They are also of value to diabetic patients for testing the acidity of urine. Papers of this type are marketed under various brand names—for example, pHydrion® paper and Nitrazine® paper.

The Relationship Between Dissociation Constants and pH. The two previously mentioned expressions of acidity, K_a and pH, can be related to each other through a very useful equation. Referring back to Eq. 1, this equation can be rewritten in the form:

$$K_a = [\text{H}_3\text{O}^+] \frac{[\text{A}^-]}{[\text{HA}]} \qquad \text{(Eq. 5)}$$

Taking the negative logarithm of both sides of this equation:

$$-\log K_a = -\log [\text{H}_3\text{O}^+] -\log \frac{[\text{A}^-]}{[\text{HA}]} \qquad \text{(Eq. 6)}$$

Applying the definition of pH from Eq. 4 and making the additional definition that:

$$-\log K_a = \text{p}K_a \qquad \text{(Eq. 7)}$$

Equation 6 may be rewritten to give:

$$\text{p}K_a = \text{pH} -\log \frac{[\text{A}^-]}{[\text{HA}]} \qquad \text{(Eq. 8a)}$$

or

$$\text{p}K_a = \text{pH} + \log \frac{[\text{HA}]}{[\text{A}^-]} \qquad \text{(Eq. 8b)}$$

Equations 8a and 8b are forms of the well known Henderson-Hasselbalch equation which will be utilized in the discussion of buffer solutions (see below).

There are several points to be considered with respect to Eqs. 7 and 8 and the concept of pK_a. For example, as with the concept of pH, the lower the value of the pK_a the stronger the acid (i.e., the more ionized the acid is in water). Very strong acids (100% ionized) will actually have negative pK_a values. Weak acids will have positive pK_a values, usually somewhere between 5 and 10. Compounds with pK_a's much greater than 10 cannot be considered for all practical purposes to be acids, at least in aqueous solution. In the case of weak acids, it is possible to reach a point where the concentrations of acid and conjugate base are equal, $[HA] = [A^-]$, i.e., the acid is 50% ionized. Equation 8 shows that this point of equal concentrations occurs when the pH of the medium is equal to the pK_a of the acid. A discussion of percent ionization will be taken up later in this section.

Dissocation Constants of Bases and pOH. The emphasis thus far has been placed on the development of the concepts of pH and pK_a. Although it will be seen that these two values will suffice in most discussions of either acids or bases, relationships similar to those stated above for acids can be developed for bases—i.e., K_b, pK_b, and pOH.

Consider the following equilibrium for a general base B: (according to the Brönsted definition any substance which can act as a proton acceptor):

$$B: + H_2O \rightleftarrows B:H^+ + OH^- \qquad \text{(xii)}$$

In a manner similar to that shown in Eq. 1, an ionization constant can be written for rx (xii) and termed K_b (Eqs. 9):

$$K_{eq} = \frac{[B:H^+][OH^-]}{[B:][H_2O]} \qquad \text{(Eq. 9a)}$$

$$K_b = K_{eq}[H_2O] = \frac{[B:H^+][OH^-]}{[B:]} \qquad \text{(Eq. 9b)}$$

The K_b bears the same relationship to the ionization of neutral (uncharged) bases [rx (xii)] as the K_a to the ionization of neutral acids [rx (ix)]. That is, the stronger the base the more the equilibrium in rx (xii) lies toward the right and the larger is the K_b.

As the hydrogen ion concentration may be considered a measure of acidity, the hydroxide ion concentration is a measure of the basicity of a compound in water. For all but very strong bases, the numerical values that must be compared in relating the basic strength of solutions are unwieldy exponential numbers of the type mentioned earlier. These numbers can be treated like those involved in hydrogen ion concentrations according to the definition:

$$-\log [OH^-] = pOH \qquad \text{(Eq. 10)}$$

The pOH values obtained by using Eq. 10 express the basicity of a solution with the smallest numbers corresponding to the greatest alkalinity (see Table 4–3).

The relationship between K_a and K_b and pH and pOH can be easily illustrated by the ion product constant for water, K_w. Just as rx (xii) illustrates the ionization of the base, B:, a similar reaction can be written to show the ionization of its conjugate acid in water [rx (xiii)]:

$$B:H^+ + H_2O \rightleftarrows B: + H_3O^+ \qquad \text{(xiii)}$$

Since this represents the ionization of an acid, the corresponding ionization constant is the K_a for the reaction:

$$K_a = \frac{[B:] [H_3O^+]}{[B:H^+]} \qquad \text{(Eq. 11)}$$

Assuming that the bases and conjugate acids in rxs (xii) and (xiii) are the same, multiplying Eqs. 9b and 11 together provides the results shown in Eqs. 12a and 12b:

$$K_a \cdot K_b = \frac{[B:] [H_3O^+]}{[B:H^+]} \cdot \frac{[B:H^+] [OH^-]}{[B:]} = [H_3O^+] [OH^-] = K_w \qquad \text{(Eq. 12a)}$$

$$K_a \cdot K_b = K_w = 1 \times 10^{-14} \qquad \text{(Eq. 12b)}$$

The relationships shown in Eqs. 12a and 12b can now be easily extended to show that, for water, the following are true:

$$pK_a + pK_b = pK_w = 14 = pH + pOH \qquad \text{(Eq. 13)}$$

As a further illustration of the relationships shown in Eqs. 12a, 12b, and 13, consider the following:

(a) If an acid has a $K_a = 1 \times 10^{-5}$, its conjugate base, when dissolved in water, would be expected to have a $K_b = 1 \times 10^{-9}$ (see Eq. 12b). This is a further indication that a strong acid has a weak conjugate base.

(b) A base having a $pK_b = 4.5$ would be expected to have a conjugate acid with a $pK_a = 9.5$ (see Eq. 13). This would be true in the case of a strong base with a weak conjugate acid.

(c) A strongly acidic solution might have pH $= 3$; this same solution would thus have pOH $= 11$ (see Eq. 12b). On the other hand, a strongly basic solution might have pOH $= 2$ and pH $= 12$. Both of these situations illustrate the point that the product of the hydrogen (hydronium) ion and hydroxide ion concentrations is equal to the constant value of 1×10^{-14} (see Table 4–3).

In an attempt to keep the relationships between acid and base strength on the same scale, the concepts of pH and pK_a will be used from this point on without regard to whether the discussion is involved with acids or bases. This should cause no problems in the case of acids, but some rearrangement of thinking must be done in the case of bases. The designation of a pK_a value for a base refers to the conjugate acid of the base instead of the base itself. For example, the base ammonia, NH_3, would be expected to react in water in the following manner:

$$NH_3 + H_2O \rightleftarrows NH_4^+ + OH^- \tag{xiv}$$

The K_b for this reaction is given by Eq. 14:

$$K_b = \frac{[NH_4^+][OH^-]}{[NH_3]} \tag{Eq. 14}$$

From Eqs. 12a and b, the following relationship can be derived:

$$K_a = \frac{K_w}{K_b} + \frac{[H_3O^+][\cancel{OH^-}]}{1} \cdot \frac{[NH_3]}{[NH_4^+][\cancel{OH^-}]} \tag{Eq. 15a}$$

$$K_a = \frac{[NH_3][H_3O^+]}{[NH_4^+]} \tag{Eq. 15b}$$

Proceeding from Eq. 15b to the ionic equilibrium reaction from which it was derived demonstrates that the pK_a of the base ammonia is actually the pK_a of the conjugate acid or the ammonium cation [rx (xv)]:

$$NH_4^+ + H_2O \rightleftarrows NH_3 + H_3O^+ \tag{xv}$$

Extending this concept to its final conclusion, the pK_a of a strong base will be a relatively large number—12 or 13 (weak conjugate acid), whereas the pK_a of a weak base will be a relatively small number—2 or 3 (strong conjugate acid). Of course, the relationship between acid strength and pK_a is opposite to this, with stronger acids having smaller pK_a values (weak conjugate bases), and weaker acids having larger pK_a values (strong conjugate bases).

Percent Ionization. As was previously indicated, acids and bases can exist in various degrees of ionization depending upon the strength of the acid and the pH of the solution. This is particularly true of so-called weak acids or bases which are not 100% ionized in aqueous solution. The concept of the degree or percent ionization of a Brönsted acid or base is important to pharmacists when considering the biopharmaceutical parameters of absorption, distribution, and excretion of a drug molecule. Since most drugs fall into the weak acid or weak base category, and since ionized molecules do not easily cross biological membranes, the absorption and excretion rate of drugs can frequently be modified by adjusting pH factors and thus affecting the percent ionization. These types of pH adjustments can be utilized in certain drug toxicities where excretion rates of the drug may be improved by acidifying or alkalinizing the patient's urine. For example, a barbiturate, an acidic drug, is excreted more rapidly in cases of toxicity if the pH of the urine is made alkaline through the administration of sodium bicarbonate (see Chapter 5).

The calculation of percent ionization is accomplished through use of Eq. 8b transposed to the forms shown below for acids and bases, Eqs. 16 and 17, respectively:

$$pK_a - pH = \log \frac{[HA]}{[A^-]} \qquad \text{(Eq. 16)}$$

$$pK_a - pH = \log \frac{[B:H^+]}{[B:]} \qquad \text{(Eq. 17)}$$

From Eqs. 16 and 17, it can be shown that when the pH of a solution of an acid or a base is the same as the pK_a of the acid or the conjugate acid of the base, the particular molecule is 50% ionized:

$$pK_a - pH = 0 = \log \frac{1}{1} = \log 1$$

$$\% \text{ ionization} = \frac{1}{1+1} \times 100 = 50\%$$

The calculation of percent ionization for situations where the pH and the pK_a are not equal is shown in the following examples.

Example 3. An acid having $pK_a = 4$ was present in a solution having pH = 3. What is the percent ionization of the acid?

$$pK_a - pH = 4 - 3 = 1$$

Therefore, from Eq. 16:

$$\log \frac{[HA]}{[A^-]} = 1,$$

and, since the number having a log equal to 1 is 10, the log ratio of acid to conjugate base is

$$\log \frac{[HA]}{[A^-]} = \log \frac{10}{1} = 1.$$

Therefore, the ratio of the concentration of nonionized acid to conjugate base is 10 to 1, and the percentage of total acid in the conjugate base form, or percent ionization, is

$$\% \text{ ionization} = \frac{1}{(10 + 1)} \times 100 = 9.09\%.$$

The above result would be expected for this particular weak acid because the relatively low pH would keep most of the acid in the nonionized form. If the situation described in Example 3 was altered so that the conditions applied to a base (i.e., the value of the pK_a would be for the conjugate acid), according to Eq. 17 the ionized form, or the conjugate acid, would be present in the largest quantity, and the percent ionization would be 90.91%.

Example 4. A base having $pK_a = 5$ for its protonated form was present in a solution having pH = 7. What is the percent ionization of the base?

$$pK_a - pH = 5 - 7 = -2$$

Since the number having a log equal to -2 is 0.01, use of Eq. 17 shows that

$$\log \frac{[B:H^+]}{[B:]} = \log \frac{1}{100} = -2$$

and

$$\% \text{ ionization} = \frac{1}{(100 + 1)} \times 100 = 0.99\%.$$

The pH of the above solution is high enough to convert most of the conjugate acid to the nonionized base. According to Eq. 16, an acid having the same pK_a and in solution at the same pH would be ionized to the extent of 99.01%.

Other situations where the difference between pK_a and pH are not whole numbers can be calculated in the same manner by looking up the antilog of the difference in a log or antilog table, and constructing the appropriate ratio. The practical need for such a calculation does not usually extend beyond a $pK_a - pH$ value of ± 4. At this point, 99.99% of an acid or base will be in one form or the other.

The above examples illustrate the procedures involved in the calculation of percent ionization. These procedures can be expressed more concisely in two separate formulas given in Eqs. 18 and 19. As indicated below, Eq. 18 applies to the ionization of a neutral (uncharged) acid represented by HA, and Eq. 19 applies to the ionization of a protonated base, e.g., NH_4^+, as represented by BH^+.

For HA $\rightleftarrows$ H$^+$ + A$^-$

$$\% \text{ ionization} = \frac{100}{1 + \text{Antilog } (pK_a - pH)} \qquad \text{(Eq. 18)}$$

For B:H$^+$ $\rightleftarrows$ H$^+$ + B:

$$\% \text{ ionization} = \frac{100}{1 + \text{Antilog } (pH - pK_a)} \qquad \text{(Eq. 19)}$$

Salts of Acids and Bases. The concept of conjugate acids and bases and their strength in relation to the parent bases or acids has been mentioned in a previous section. The importance of this concept in the prediction of acidity, basicity, or neutrality of solutions of salts of acids and bases will now be discussed.

As was mentioned earlier in this chapter, conjugate bases and conjugate acids are related inversely in their strength with respect to the parent acid or base, respectively. That is, a weak acid will have a strong conjugate base; a weak base will have a strong conjugate acid, and vice versa. This information is quite useful in considering the properties of salts formed between acids and bases. For example, the salt formed in the neutralization reaction occurring between hydrochloric acid and sodium hydroxide, **rx (xvi)**, is sodium chloride:

$$\text{HCl} + \text{NaOH} \rightarrow \text{H}_2\text{O} + \text{NaCl} \qquad \text{(xvi)}$$

Since HCl is a strong acid and NaOH is a strong base, their conjugate species are necessarily weak. Sodium chloride, then, being a salt formed from a strong acid and a strong base, when dissolved in water would not be expected to produce much of a change in the ionic state of the water such as would be necessary to cause a change in the pH of the medium. In other words, sodium chloride is a neutral salt unable to impose either acidic or basic properties on an aqueous solution. This would be the situation expected with every salt that can be traced to a neutralization reaction between a strong acid and a strong base, e.g., Na_2SO_4, NaNO_3, etc. This does not mean that there are no associations or chemical reactions between these salts and water, but that the net effect of such reactions does not significantly alter the concentrations of acidic and basic species in the water (i.e., H_3O^+ and OH^-).

Unlike the situation described above, salts formed from strong acids and weak bases or from weak acids and strong bases can impart acidic or basic properties, respectively, to a solution. For example, consider the salt, sodium acetate, formed from sodium hydroxide and the relatively weak acetic acid ($pK_a = 4.7$). Because this salt ionizes in aqueous solution to provide acetate ion, and because acetate ion is the conjugate base of acetic acid and, therefore, relatively strong, the solution will have a decidedly alkaline reaction due to hydrolysis, as shown in rx (xvii):

$$Na^+ + CH_3COO^- + H_2O \rightleftharpoons CH_3COOH + Na^+ + OH^- \qquad (xvii)$$

This reaction illustrates that dissolving sodium acetate in water will cause an increase in the hydroxide ion concentration and a decrease in the hydrogen ion concentration through the formation of weakly dissociated acetic acid. The net result is an increase in the pH of the solution.

An illustration of the acidic properties of a salt formed from a strong acid and a weak base may be found in the case of ammonium chloride (from hydrochloric acid and ammonia). When this salt is dissolved in water an acidic solution is formed through the following hydrolysis [rx (xviii)]:

$$NH_4^+ + Cl^- + H_2O \rightleftharpoons NH_4OH + H^+ + Cl^- \qquad (xviii)$$

In rx (xviii), the ammonium ion, as the conjugate acid of the relatively weak base ammonia (or ammonium hydroxide), may be considered to be functioning as a proton donor through its hydrolysis to the weakly ionized ammonium hydroxide and proton. The proton formed will not associate significantly with the chloride ion, which is a weak base. The net effect is a lowering of the pH of the solution. It should also be mentioned here that ammonium salts will react with strong bases to cause the liberation of ammonia gas, rx (xix):

$$NH_4Cl + NaOH \rightleftharpoons NH_3 \uparrow + H_2O + NaCl \qquad (xix)$$

This reaction further illustrates the acid reaction of ammonium salts, and can be utilized in testing for the presence of ammonium compounds.

A final point concerns those salts which are formed from weak acids and weak bases. These salts will undergo hydrolysis, like the salts illustrated above, to form slightly ionized compounds, but the acidic or basic character of the resulting solutions is dependent upon the relative strength of the acidic or basic species formed. If the weak acid and weak base produced in solution are approximately equal in strength, a neutral solution will be formed, as in the case of ammonium acetate shown in rx (xx). Hydrolysis of ammonium acetate provides acetic acid and ammonium hydroxide, both of which are slightly ionized in water and comparable in their respective acid and base strength.

$$CH_3COO^- + NH_4^+ + H_2O \rightleftharpoons NH_4OH + CH_3COOH \qquad (xx)$$

The effect on relative hydrogen and hydroxide ion concentrations in the solvent water would for all practical purposes be negligible.

Acid and Base Products

This section will discuss those compounds used primarily for their acid-base properties in various chemical reactions and pharmaceutical products. There are numerous acids and bases with specific therapeutic attributes which will be discussed in succeeding chapters and which will not be covered now.

Official Inorganic Acids

Boric Acid, N.F. XIII (H_3BO_3; Mol. Wt. 61.83)

$$\begin{array}{c} OH \\ | \\ B \\ \diagup \; \diagdown \\ HO \qquad OH \end{array}$$

Boric acid is a solid available in three forms: (1) colorless, odorless, pearly scales; (2) six-sided triclinic crystals; and (3) white, odorless powder which is unctuous to the touch (having a soapy feeling). It is stable in air, and has a density of 1.46. Boric acid is soluble in water and alcohol, and freely soluble in glycerin, boiling water, and boiling alcohol. Clear solutions of boric acid are obtained when 1 g is dissolved in 25 ml of water or 10 ml of boiling alcohol. The scale and crystalline forms of the compound are more suitable for preparing aqueous solutions since the powder tends to float on top of the water.

Boric acid is a weak acid having $pK_a = 9.19$ for the ionization of the first proton at 25° C. This level of acidity is too low to permit accurate titration with standard base in aqueous solution. The official assay requires titration with 1 N sodium hydroxide in a solution of equal parts glycerin and water. The presence of glycerin causes the acid to behave as a strong monoprotic acid which can be titrated to a phenolphthalein end point. The monoprotic character of boric acid in glycerin is thought to occur through esterification to a tetravalent boron ester known as glyceroboric acid, with the net release of one hydronium ion into the solution [rx (xxi)]:

$$2 \begin{array}{c} CH_2OH \\ | \\ CHOH \\ | \\ CH_2OH \end{array} + \begin{array}{c} OH \\ | \\ B \\ \diagup \; \diagdown \\ HO \qquad OH \end{array} \longrightarrow \left[\begin{array}{c} H_2C-OH \quad HO-CH_2 \\ | \qquad\qquad\qquad | \\ HC-O \diagdown \quad \diagup O-CH \\ \qquad\;\; B \\ | \diagup \quad \diagdown | \\ H_2C-O \qquad O-CH_2 \end{array} \right]^{-} + H_3O^+ + 2H_2O$$

$$(xxi)$$

Similarly, boric acid can be esterified with other polyhydroxy compounds such as glycol, mannitol, and catechol.

Due to the weak acid nature of boric acid, only salts produced by the replacement of one proton per molecule (primary salts) can be formed in aqueous solution. Those salts formed with alkali metals are the only ones that are soluble in water, and these produce a very alkaline solution. Salts formed with other metals are hydrolyzed in water to produce insoluble metal hydroxides. The triprotic nature of the acid, however, can be shown by esterification reactions similar to rx (xxi) above and by the formation, with ethyl alcohol, of ethyl orthoborate having the formula $B(OC_2H_5)_3$.

The name boric acid refers to what is technically known as orthoboric acid. Heating orthoboric acid to certain temperatures produces various dehydration products important to the overall chemistry of boric acid and borates. For example, heating orthoboric acid to $100°$ C causes the loss of one molecule of water to produce metaboric acid, HBO_2 [rx (xxii)]:

$$H_3BO_3 \xrightarrow[\Delta]{100°\ C} HBO_2 + H_2O \qquad (xxii)$$

Heating to approximately $160°$ C causes a further loss of water to produce tetraboric (or pyroboric) acid, $H_2B_4O_7$ [rx (xxiii)]:

$$4HBO_2 \xrightarrow[\Delta]{160°\ C} H_2B_4O_7 + H_2O \qquad (xxiii)$$

Heating to still higher temperatures will produce the anhydride of boric acid, boron trioxide, B_2O_3, a glassy-appearing solid [rx (xxiv)]:

$$H_2B_4O_7 \xrightarrow[\Delta]{>160°\ C} 2B_2O_3 + H_2O \qquad (xxiv)$$

Most solutions of boric acid contain only minute amounts of tetraboric acid, but when neutralized with base, e.g., NaOH, and concentrated to cause precipitation of solid, the stable crystalline salts are tetraborates, e.g., $Na_2B_4O_7$. Sodium metaborate can be crystallized with one molecule of H_2O_2 and $3H_2O$ to yield the well known sodium perborate (see Chapter 9).

Uses. The therapeutic aspects involving the antimicrobial action of boric acid and borates will be discussed in more detail in Chapter 9.

The N.F. XIII specifies that boric acid must contain not less than 99.5% and not more than 100.5% of H_3BO_3 calculated as an anhydrous (dried) form. It further categorizes it as a buffer. Since it is a weak acid, it constitutes a portion of what would be required for a buffer solution, but used alone it does not have much buffer capacity (see *Buffers*). However, solutions of boric acid are utilized in various topical medications to maintain an acidic pH in the medium. The N.F. XIII (p. 833) describes a boric acid vehicle for ophthalmic solutions containing 1.9% boric acid and having a pH slightly below 5.0. A few ophthalmic decongestant products (used for their ability to constrict blood vessels and to dilate pupils in the eye) use boric acid as a buffer to maintain the pH around 6. It also contributes to the isotonicity of these solutions (see Chapter 3). Boric acid is also used as a buffer in *Aluminum Acetate Solution* U.S.P. XVIII and *Aluminum Subacetate Solution* U.S.P. XVIII where it may be present in a concentration of not more than 0.6% and 0.9%, respectively. The toxicity of boric acid precludes its use in products that are to be taken internally (see Chapter 9).

The reaction of boric acid with equimolar quantities of glycerin at 140° to 150° C produces a compound known as *Boroglycerin Glycerite* (N.F. XI) ($C_3H_5BO_3$) which has found some use as a suppository base. This product is formed in preference to the glyceroboric acid shown in rx (xxi), which requires a 1:2 ratio of boric acid and glycerin.

Hydrochloric Acid, U.S.P. XVIII (HCl; Mol. Wt. 36.46)

Hydrochloric acid is an aqueous solution of hydrogen chloride gas containing not less than 35.0% and not more than 38.0% by weight of HCl. It is described as a colorless, fuming liquid having a pungent odor and a specific gravity of about 1.18. A nonfuming solution can be prepared by diluting hydrochloric acid with two volumes of water.

Hydrogen chloride is a colorless gas having an acrid irritating odor and an acid taste. It is about 25% heavier than air, as reflected by its density of 1.2681. The gas can be liquefied by pressure alone, and both the liquid and vapor states are nonconductors of electricity. Hydrogen chloride is very soluble in water, as illustrated by the fact that 503 volumes at standard pressure will dissolve in one volume of water at 0° C, and 460 volumes at standard pressure will dissolve in one volume of water at 20° C. A 0.1 N solution of hydrogen chloride in water is 92% ionized at 18° C and conducts electricity readily. As was indicated in Chapter 3, distillation of a saturated solution (43.4% HCl) at standard pressure will cause a loss of HCl until a concentration of 20.24% HCl is obtained. This concentration is a constant boiling mixture of water and HCl (b.p. 110° C) and will continue to distill with no further change in concentration; i.e., the concentrations of water and HCl in the vapor are the same as in the liquid.

Muriatic acid (from the Latin *muria*, meaning brine) is a technical grade of hydrochloric acid containing 35 to 38% HCl and a number of impurities including chlorine, arsenous and sulfurous acids, and iron. The impurities give it a characteristic yellow color.

Hydrochloric acid is a strong monoprotic acid which can be assayed conveniently by titrating a weighed sample against standard 1 N sodium hydroxide [rx (xxv)] using methyl red as an indicator:

$$HCl + NaOH \rightarrow NaCl + H_2O \qquad (xxv)$$

It reacts as well with other metal oxides and hydroxides to produce the chloride of the metal and water. The hydrogen in hydrochloric acid is displaced by metals preceding hydrogen in the electromotive series to cause the evolution of hydrogen gas [rx (xxvi)]:

$$Zn + 2HCl \rightarrow ZnCl_2 + H_2 \uparrow \qquad\qquad\text{(xxvi)}$$

The presence of chloride ion will cause the precipitation of insoluble chlorides of metals such as Ag, Pb, and Hg(I) [rx (xxvii)]:

$$AgNO_3 + HCl \rightarrow AgCl \downarrow + HNO_3 \qquad\qquad\text{(xxvii)}$$

Hydrochloric acid can be oxidized by strong oxidizing agents, resulting in the evolution of chlorine gas [rx (xxviii)]:

$$2KMnO_4 + 16HCl \rightarrow 2MnCl_2 + 2KCl + 5Cl_2 \uparrow + 8H_2O \quad\text{(xxviii)}$$

Uses. The biochemical and pharmacological aspects of the chloride ion present in hydrochloric acid and various chloride salts will be discussed in Chapter 5. The primary chemical uses of *Hydrochloric Acid* are categorized by the U.S.P. XVIII as a pharmaceutical aid or, specifically, as an acidifying agent. By virtue of its strong acid character it is capable of reacting with organic molecules which are weakly basic to form usually water soluble hydrochloride salts. A general reaction with a primary organic amine is shown in rx (xxix), where R is any organic group:

$$R\text{—}NH_2 + HCl \rightarrow R\text{—}NH_3^+ \; Cl^- \qquad\qquad\text{(xxix)}$$

This type of reaction is utilized to convert normally water insoluble organic bases into a water soluble form for extraction or other separation purposes. Since salts of most organic amines are solids, treatment of liquid organic bases with hydrochloric acid can render the compound suitable for incorporation into a solid dosage form. However, one disadvantage of using hydrochloric acid for this purpose is that some hydrochlorides are hygroscopic. Nevertheless, most drugs are available as hydrochloride salts. The reasons for the popularity of hydrochloride salts are related to their ease of preparation and to the very low toxicity of the chloride ion.

Diluted Hydrochloric Acid, N.F. XIII, is a solution of hydrogen chloride containing not less than 9.5 g and not more than 10.5 g of HCl in each 100 ml of solution. It is prepared from the more concentrated *Hydrochloric Acid*, U.S.P. XVIII. When *Diluted Hydrochloric Acid* is further diluted with 25 to 50 volumes of water it may be used as a gastric acidifier when the level of hydrochloric acid in the gastric juice is low (achlorhydria; see Chapter 8).

Diluted hydrochloric acid can be used as an acidifying agent in much the same manner as the more concentrated solution. That is, organic bases will form water soluble hydrochloride salts in dilute solutions of HCl. Dilute solutions may actually be employed more frequently than *Hydrochloric Acid*, U.S.P., in the chemical treatment of compounds to form salts, to extract basic drugs, and to test for alkaline properties.

Nitric Acid, N.F. XIII (HNO_3, Mol. Wt. 63.01)

$$HO-N\begin{array}{c} O \\ O \end{array}$$

Nitric acid is an aqueous solution containing not less than 69.0% and not more than 71.0% by weight of HNO_3. It is described as a highly corrosive fuming liquid having a characteristic, highly irritating odor. It has a boiling point of about 120° C and a specific gravity of about 1.41. Nitric acid will produce a yellow stain on animal tissues due to nitration of the aromatic amino acids, phenylalanine, tyrosine, and tryptophan, found in the protein of the skin. This is known as the *xanthoproteic test.*

Nitric acid is a strong monoprotic acid; a 0.1 N solution is about 93% ionized at 25° C. It is assayed by titrating a weighed sample against 1 N sodium hydroxide [rx (xxx)], using methyl red as an indicator:

$$HNO_3 + NaOH \rightarrow NaNO_3 + H_2O \hspace{2cm} (xxx)$$

In addition to its ability to react as a strong acid, nitric acid is an oxidizing and a nitrating agent. Its oxidizing properties, apparent even in very dilute solutions, can be observed on both metals and nonmetals. In many instances, the oxidizing power is enhanced by the presence of small amounts of nitrous acid, HNO_2, produced by photochemical decomposition of nitric acid. It oxidizes all common metals except gold and platinum to produce the nitrate salt of the metal and, depending upon the concentration and position of the metal with respect to hydrogen in the electromotive series, various oxides of nitrogen or ammonia. With metals above hydrogen in the electromotive series, such as Zn, concentrated nitric acid produces ammonia in the form of ammonium nitrate, while dilute acid yields principally nitrous oxide, N_2O. Metals below hydrogen in the electromotive series, such as Cu, when treated with concentrated nitric oxide yield primarily nitrogen dioxide, NO_2. On the other hand, dilute acid produces the more reduced nitric oxide, NO. It should be noted that the metal is oxidized to the same extent with either concentrated or dilute nitric acid. Oxidation of nonmetals is also possible with nitric acid in that sulfuric and phosphoric acids can be produced by treating elemental sulfur and phosphorus, respectively, with the acid,

The nitrating properties of nitric acid are used extensively in organic chemistry. The reactions are usually carried out in a "mixed acid" medium containing nitric and sulfuric acids mixed in various proportions. The reaction involves the substitution of the nitro group, $-NO_2$, for some other group, usually hydrogen, on the organic molecule. For example, this reaction is used to nitrate toluene to produce trinitrotoluene (TNT). Nitration also occurs when nitric acid comes into contact with protein-containing materials or animal tissues (see above).

Uses. Although the acidic properties of nitric acid are similar to those of hydrochloric acid, making it useful as an acidifying agent with bases, its oxidizing and nitrating properties add a unique dimension to its use as compared with most other acids. It is used in the manufacture of sulfuric acid, coal tar dyes, and explosives. It is also used as a nitrating agent in *Pyroxylin*, U.S.P. XVIII, and as a source of nitrate ion in the preparation of *Milk of Bismuth*, N.F. XIII (see Chapter 8). It has been used externally to destroy chancres and warts, but is rarely, if ever, prescribed for internal use.

Nitric acid is highly corrosive and is capable of causing damage to tissues in contact either with the solution or with vapors of the acid. However, the corrosive properties of the acid are not the only dangers, since the gases evolved from spillage on heavy metals or on organic materials are also exceedingly toxic. These gaseous oxides of nitrogen may produce little or no discomfort at the time of inhalation, so damage to lung tissue can occur until the individual experiences difficulty in breathing. Untreated cases frequently terminate fatally due to suffocation from pulmonary edema (congested lungs).

Phosphoric Acid, N.F. XIII (H_3PO_4; Mol. Wt. 98.00)

$$\begin{array}{c} O \\ \uparrow \\ HO-P-OH \\ | \\ OH \end{array}$$

Phosphoric acid is an aqueous solution containing not less than 85.0% and not more than 88.0%, by weight, of H_3PO_4. It is a colorless, odorless, syrupy liquid having a specific gravity of about 1.71 at 25° C. *Diluted Phosphoric Acid*, N.F. XIII, contains not less than 9.5 g and not more than 10.5 g of H_3PO_4 in 100 ml of solution. It has a specific gravity of about 1.057 at 25° C.

Orthophosphoric acid is a relatively strong triprotic acid ionizing in three steps [rx (xxxi-a, -b, and -c)]:

$$H_3PO_4 + H_2O \rightleftarrows H_3O^+ + H_2PO_4^- \quad pK_{a1} = 2.12 \qquad \text{(xxxi-a)}$$
$$H_2PO_4^- + H_2O \rightleftarrows H_3O^+ + HPO_4^{-2} \quad pK_{a2} = 7.21 \ (7.12) \quad \text{(xxxi-b)}$$
$$HPO_4^{-2} + H_2O \rightleftarrows H_3O^+ + PO_4^{-3} \quad pK_{a3} = 12.32 \qquad \text{(xxxi-c)}$$

The pK_a's of the above reactions illustrate the decrease in acid strength with each successive ionization, which is characteristic of polyprotic acids. In fact, very little of the phosphate trianion, PO_4^{-3}, is produced in the simple ionization process. The official assay procedure involves titration of only the first two protons. The assay is accomplished by titrating a weighed sample of approximately 1 g against 1 N sodium hydroxide to a thymolphthalein end point. Corrections are made through a blank titration, and each ml of 1 N sodium hydroxide is equivalent to 49.00 mg of H_3PO_4.

The acid-base properties of the various phosphate anions are more complicated than rxs (xxxi-b) and (xxxi-c) appear to indicate. Considering first the dihydrogen phosphate anion which ionizes according to rx (xxxi-b), it is also possible for this ion to react as a base (a property termed either amphiprotic or amphoteric) to provide the hydrolysis reaction shown in rx (xxxii):

$$H_2PO_4^- + H_2O \rightleftharpoons H_3PO_4 + OH^- \qquad \text{(xxxii)}$$

Although this reaction does occur to some extent, the strength of phosphoric acid relative to the dihydrogen phosphate anion dictates that the equilibrium will be directed strongly toward the left. Therefore, aqueous solutions of salts of this anion will be definitely acidic, due to the predominance of rx (xxxi-b). In fact, this anion effectively serves as the weak acid in phosphate buffer systems (see *Buffers* and Chapter 5).

The monohydrogen phosphate anion which ionizes according to rx (xxxi-c) can also hydrolyze as shown in rx (xxxiii):

$$HPO_4^{-2} + H_2O \rightleftharpoons H_2PO_4^- + OH^- \qquad \text{(xxxiii)}$$

The product of the ionization reaction [rx (xxxi-c)] is the very basic PO_4^{-3}, which would tend to react strongly with the hydronium ion to force the equilibrium in rx (xxxi-c) toward the left. However, the product of the hydrolysis reaction [rx (xxxiii)] is a weak acid, and the equilibrium would be expected to lie more toward the right. The predominance of the hydrolysis reaction indicates that the monohydrogen phosphate anion is a base, and solutions of salts of this anion will have an alkaline pH. This anion also serves as the basic member of the buffer pair in phosphate buffer systems (see *Buffers* and Chapter 5).

A final point should be made concerning the phosphate trianion, PO_4^{-3}, mentioned above as being a strong base. Salts of this ion, e.g., Na_3PO_4, produce a strongly alkaline solution when dissolved in water. It is unlikely that this ion exists in solution at all, and the hydrolysis reaction shown in rx (xxxiv) is directed almost totally toward the production of hydroxide ion:

$$PO_4^{-3} + H_2O \rightleftharpoons HPO_4^{-2} + OH^- \qquad \text{(xxxiv)}$$

The only water soluble phosphates are those of ammonium and the alkali metals and the dihydrogen phosphates of the alkaline earth metals. The insoluble monohydrogen phosphates of calcium and magnesium can be converted to the more soluble dihydrogen phosphates in acidic media.

When orthophosphoric acid is heated to 200° C it loses a molecule of water for every two molecules of phosphoric acid to form pyrophosphoric acid [rx (xxxv)]:

$$
2 \; \underset{\displaystyle HO}{\overset{\displaystyle HO}{HO-P\rightarrow O}} \xrightarrow[\Delta]{200°\,C} \quad O\leftarrow \underset{\displaystyle OH}{\overset{\displaystyle OH}{P}}-O-\underset{\displaystyle OH}{\overset{\displaystyle OH}{P}}\rightarrow O + H_2O \qquad (xxxv)
$$

When either orthophosphoric or pyrophosphoric acid is heated to 300° C, metaphosphoric acid is produced. This reaction can be more easily represented for orthophosphoric acid as an intramolecular dehydration with the loss of one molecule of water for every molecule of acid [rx (xxxvi)]:

$$
\underset{\displaystyle HO}{\overset{\displaystyle HO}{HO-P\rightarrow O}} \xrightarrow[\Delta]{300°\,C} HO-\overset{\displaystyle O}{\underset{}{P}}\rightarrow O + H_2O \qquad (xxxvi)
$$

Metaphosphoric acid constitutes the "glacial phosphoric acid" of industry. It is usually sold in the form of sticks or pellets formed by the addition of small amounts of sodium metaphosphate ($NaPO_3$) to harden the metaphosphoric acid. Solutions of this acid react slowly with water to produce orthophosphoric acid; therefore, metaphosphoric acid solution is virtually impossible to keep, and is no longer recognized officially.

Orthophosphoric acid is nonvolatile and has no oxidizing properties, thus enabling its use wherever a nonoxidizing acid is required—e.g., to prepare HBr from NaBr where sulfuric acid is unsatisfactory because of its oxidizing action.

Uses. The pharmacological and biochemical aspects of phosphates are discussed in Chapter 5. Phosphoric acid can be used as an acidifying agent where its limitations will be primarily in the solubilities of the various phosphate salts produced. It can be treated with sodium hydroxide at particular concentrations to produce mixtures of HPO_4^{-2} and $H_2PO_4^-$ which serve as the basic and acidic species, respectively, of the phosphate buffer system. The N.F. XIII categorizes phosphoric acid as a solvent useful in the preparation of *Anileridine Injection*, N.F. XIII.

Nonofficial Inorganic Acid

Sulfuric Acid, $(H_2SO_4$; Mol. Wt. 98.06)

$$\begin{array}{c} O \\ \uparrow \\ HO-\!\!S\!\!-OH \\ \downarrow \\ O \end{array}$$

Sulfuric acid, as described in the N.F. X, is a colorless, odorless liquid of oily consistency containing not less than 94% and not more than 98% of H_2SO_4. It has a specific gravity of not less than 1.84. When heated strongly, sulfuric acid is vaporized and gives off dense, white fumes of sulfur trioxide, SO_3. It dissolves in water and alcohol with the evolution of large amounts of heat. Although it can be heated sufficiently to drive off SO_3, it does not volatilize at lower levels of heat, e.g., the boiling point of water. Therefore, it is usually classified as a nonvolatile acid.

Sulfuric acid is a strong diprotic acid. The ionization shown in rx (xxxvii-a) is essentially complete in dilute solutions. The second ionization, rx (xxxvii-b), has a pK_a of approximately 2.

$$H_2SO_4 + H_2O \rightarrow H_3O^+ + HSO_4^- \qquad \text{(xxxvii-a)}$$

$$HSO_4^- + H_2O \rightarrow H_3O^+ + SO_4^{-2} \qquad \text{(xxxvii-b)}$$

The acid may be assayed by titrating a weighed sample against 1 N sodium hydroxide, using methyl red as an indicator.

In addition to the acidic properties of sulfuric acid, its other chemical properties make it useful as a dehydrating agent, oxidizing agent, and sulfonating or sulfating agent. The ability of sulfuric acid to take up water is evident when the concentrated acid is added to water, with large amounts of heat being liberated. There is also a volume contraction noticeable in the final mixture. The production of heat is believed to be due to the formation of hydrates of sulfuric acid according to rx (xxxviii):

$$H_2SO_4 + nH_2O \rightarrow H_2SO_4 \cdot nH_2O \qquad \text{(xxxviii)}$$

Concentrated sulfuric acid can absorb enough water to overflow the container if left exposed to the atmosphere for long periods of time.

The oxidizing properties of sulfuric acid are observed primarily with hot acid on metals, nonmetals, and many other compounds. Metals above hydrogen in the electromotive series are oxidized to form sulfate salts of the metal with the evolution of hydrogen gas. Although this reaction oxidizes the metal from zero to a positive valence, it is characteristic of all acids, oxidizing or nonoxidizing, and results from the action of hydronium ion on those metals above hydrogen in the electromotive series. The true oxidizing properties are illustrated in those metals below hydrogen in the electromotive series where the oxidation occurs with a corresponding reduction in the sulfur valence, as shown by the evolution of SO_2. This can be illustrated in the case of mercury [rx (xxxix)]:

$$Hg + 2H_2SO_4 \rightarrow HgSO_4 + SO_2 \uparrow + 2H_2O \qquad \text{(xxxix)}$$

Sulfuric acid will also oxidize bromides and iodides (but not chlorides) to produce molecular bromine and iodine [rx (xl)]:

$$H_2SO_4 + 2HBr \rightarrow Br_2 \uparrow + SO_2 \uparrow + 2H_2O \qquad \text{(xl)}$$

The sulfonating action of sulfuric acid occurs with organic compounds to produce sulfonic acids. A general reaction of this type is illustrated in rx (xli), and usually requires fuming sulfuric acid as the sulfonating agent:

$$R—H + HO—SO_3H \rightarrow R—SO_3H + H_2O \qquad \text{(xli)}$$

$$R = \text{aliphatic or aromatic group}$$

Sulfation is an esterification reaction taking place between sulfuric acid and alcohols. Monoalkyl sulfates can usually be prepared at room temperature [rx (xlii)] while dialkyl sulfates usually require a catalyst and heat:

$$R—OH + \begin{matrix} HO \\ \diagdown \\ \\ \diagup \\ HO \end{matrix} SO_2 \rightarrow \begin{matrix} RO \\ \diagdown \\ \\ \diagup \\ HO \end{matrix} SO_2 + H_2O \qquad \text{(xlii)}$$

The sulfate salts of most metals are soluble (see Chapter 3) with the notable exceptions of Ba, Sr, and Pb. The sulfates of Ca, Ag, and Hg(I) are only slightly soluble. Hydrolysis to basic salts occurs with sulfates of Bi, Sb, and Fe, but acidic solutions will stabilize the normal sulfate against the hydrolytic reaction. Sulfate salts in solution have no oxidizing ability. Bisulfates (hydrogen sulfates or acid sulfates) are known for the alkali metals; however, this type of salt is most unusual for other metals.

Uses. The pharmacological aspects of sulfates are discussed in Chapter 8. Sulfuric acid finds chemical use as a strong diprotic acid which can be utilized to form salts of basic organic drug molecules. It is also utilized as a dehydrating agent—for example, in the preparation of *Pyroxylin*, U.S.P. XVIII.

Diluted sulfuric acid is also known, and was official in the N.F. X. This acid contains approximately 10% H_2SO_4.

Official Inorganic Bases

Strong Ammonia Solution, N.F. XIII (NH_3; Mol. Wt. 17.03; Ammonium Hydroxide, Stronger Ammonia Water)

Strong Ammonia Solution contains not less than 27.0% and not more than 30.0% by weight of NH_3. Upon exposure to air, it loses ammonia rapidly. The solution is a clear, colorless liquid, having an exceedingly pungent, characteristic odor, and a specific gravity of about 0.90. The N.F. XIII issues the following caution concerning *Strong Ammonia Solution: Caution: Use care in handling Strong Ammonia Solution because of the caustic nature of the Solution and the irritating properties of its vapor. Cool the container well before opening, and cover the closure with a cloth or similar material while opening. Do not taste Strong Ammonia Solution, and avoid inhalation of its vapor.*

The above description of *Strong Ammonia Solution* is due to the presence of ammonia itself, which is a colorless gas having a strong, pungent, characteristic odor. The gas is lighter than air, having a specific gravity of 0.5967. It may be liquefied at atmospheric pressure by cooling to $-60°$ C, or by cooling to $10°$ C at pressures of 6.5 or 7 atmospheres. Liquid ammonia is a good solvent and ionizing medium. It will dissolve salts to form ionic solutions. It will also dissolve alkali and alkaline earth metals to form blue solutions which decompose slowly in the presence of impurities, liberating hydrogen and leaving the amide of the metal, e.g., $NaNH_2$.

One milliliter of water dissolves 1298.9 volumes of ammonia at $0°$ C, and 727 volumes at $15°$ C. The pH of a 1 N solution is 11.6 and that a 0.1 N solution is 11.1. Ammonia is also soluble in alcohol and ether. All the gas may be expelled from solutions by boiling. In aqueous solution, ammonia exists primarily in the form of NH_3, with only a small amount reacting with water to form ammonium hydroxide, NH_4OH. However, the reactions of ammonia solutions are such that for all practical purposes they may be considered as containing ammonium hydroxide.

Ammonia is a relatively weak base, having a pK_a in aqueous solution of approximately 9.2. Strong ammonia solution is assayed via residual titration. The precise procedure is described in the N.F. XIII, and essentially involves the addition of approximately 2 ml of the solution to a weighed flask containing 35.0 ml of 1 N sulfuric acid. The increase in weight is the weight of the ammonia solution sample. The excess acid in the flask is titrated against 1 N sodium hydroxide, using methyl red as the indicator. Each ml of 1 N sulfuric acid neutralized by the ammonia solution is equivalent to 17.03 mg of NH_3. The proton acceptor properties of ammonia can be further illustrated by its reaction with very weak acids, e.g., HCO_3^-, which are usually neutral in solution [rx (xliii)]:

$$Ca(HCO_3)_2 + 2NH_3 \rightarrow CaCO_3 \downarrow + (NH_4)_2CO_3 \qquad \text{(xliii)}$$

Since the ammonia molecule possesses an unshared pair of electrons, it can serve as a ligand in forming soluble complex ions with many metal cations, notably Cu, Ag, Au, Zn, Cd, Cr, Ni, Co, Mn, and Pt. The hydroxides or insoluble salts of these metals become soluble in ammonia solution. A common example mentioned in Chapter 3 is the solubilization of silver chloride in excess ammonia solution.

Similar to the manner in which water hydrolyzes certain salts, ammonia can form ammonia-basic salts through a reaction designated as *ammonolysis*. This is illustrated with mercuric chloride, which undergoes both processes [rx (xliv-a) and (xliv-b)]:

$$HgCl_2 + H_2O \rightarrow Hg(OH)Cl + HCl \text{ (hydrolysis)} \qquad \text{(xliv-a)}$$

$$HgCl_2 + NH_3 \rightarrow Hg(NH_2)Cl + HCl \text{ (ammonolysis)} \quad \text{(xliv-b)}$$

Salts of ammonia (ammonium compounds) will generally react as acids in the presence of bases. This is due to the previously mentioned weak acid nature of the ammonium ion, $pK_a = 4.9$. A representative example is shown in rx (xlv). Depending upon the particular salt, the pH of aqueous solutions of ammonium compounds will range from neutral to acidic. Further discussion of ammonium chloride appears in Chapter 5.

$$NH_4Cl + NaOH \rightarrow NH_3 \uparrow + NaCl + H_2O \qquad \text{(xlv)}$$

Uses. Ammonia solution is used as a Brönsted base in many applications to form ammonium salts of acids. It is also used in the manufacture of nitric acid and sodium bicarbonate. *Strong Ammonia Solution* is used in the preparation of *Aromatic Ammonia Spirit*, N.F. XIII, where it serves as a source of ammonia to stabilize the ammonium carbonate against hydrolysis. It is also used in the preparation of ammoniacal silver nitrate solution (N.F. XII).

A *Diluted Ammonia Solution* was official in the U.S.P. XVI and contained not less than 9 g and not more than 10 g of NH_3 in each 100 ml. Diluted ammonia solution can be prepared by diluting strong ammonia solution with the appropriate quantity of *Purified Water*, U.S.P. XVIII. This product is also known as "ammonia water" or "household ammonia."

Calcium Hydroxide, U.S.P. XVIII ($Ca(OH)_2$; Mol. Wt. 74.09; Slaked Lime)

Calcium hydroxide is official as a white powder having an alkaline and slightly bitter taste. It is slightly soluble in water (1 g in 630 ml) and very slightly soluble in boiling water (1 g in 1300 ml). It is soluble in glycerin and in syrup, but is insoluble in alcohol.

The official solution is *Calcium Hydroxide Solution,* U.S.P. XVIII (*Lime Water*), and is described as a solution containing, in each 100 ml, not less than 140 mg of $Ca(OH)_2$. It is a clear, colorless liquid having an alkaline taste, and is alkaline to litmus. It is prepared by adding 3 g of calcium hydroxide to 1000 ml of *Purified Water*. The mixture is agitated repeatedly over a period of one hour, and the excess calcium hydroxide is allowed to settle. The clear supernatant liquid is used; the undissolved portion is not suitable for preparing additional quantities of the solution. The solubility of calcium hydroxide diminishes with increasing temperature; thus, at 15° C the concentration is about 170 mg per 100 ml. The official concentration is based on the solubility at 25° C.

Calcium hydroxide is manufactured from lime or calcium oxide, CaO, through the addition of water in limited amounts. The process is known as "slaking" (hence the name, slaked lime) and is characterized by the avid absorption of water by the oxide, accompanied by the evolution of much heat, swelling of the CaO lumps, and a final disintegration into a fine powder. The difference between lime and slaked lime should be noted.

Calcium hydroxide solutions are basic, having pH = 12.3, and are capable of neutralizing acids. The solution is assayed by titrating 50 ml against 0.1 N hydrochloric acid, using phenolphthalein as an indicator. Each ml of acid used is equivalent to 3.705 mg of $Ca(OH)_2$. The reaction is illustrated in rx (xlvi):

$$Ca(OH)_2 + 2HCl \rightleftarrows CaCl_2 + 2H_2O \qquad \text{(xlvi)}$$

A special type of neutralizing action may be illustrated by the ability of calcium hydroxide to absorb carbon dioxide from the air, with the formation of calcium carbonate [rx (xlvii-a) and (xlvii-b)]:

$$CO_2 + H_2O \rightleftarrows H_2CO_3 \qquad \text{(xlvii-a)}$$
$$Ca(OH)_2 + H_2CO_3 \rightarrow CaCO_3 \downarrow + 2H_2O \qquad \text{(xlvii-b)}$$

The precipitate of calcium carbonate is the source of the cloudy appearance of calcium hydroxide solutions.

Uses. Calcium is used medicinally as a fluid electrolyte (see Chapter 5) and a topical astringent (see Chapter 9). Calcium hydroxide is used in pharmaceutical preparations for its potentially high hydroxide ion concentration. This alkalinity allows it to react with the free fatty acids in various oils, e.g., oleic acid, to form calcium soaps (calcium salts of the fatty acids) which have emulsifying properties that aid in the suspension or mixing of other ingredients. The carbon dioxide-absorbing properties of calcium hydroxide are useful in certain types of gas traps. Calcium hydroxide is combined with sodium hydroxide in a mixture known as *Soda Lime* which is used for its ability to absorb CO_2 from expired air in metabolic function tests (see *Sodium Hydroxide*).

Potassium Hydroxide, U.S.P. XVIII (KOH; Mol. Wt. 56.11; Caustic Potash)

Potassium hydroxide is official as white or nearly white fused masses, small pellets, flakes, sticks, or other dry forms that have a crystalline fracture. The various forms must all contain not less than 85.0% total alkali calculated as KOH, including not more than 3.5% K_2CO_3. It is very deliquescent and rapidly absorbs both moisture and carbon dioxide from the air.

One gram of potassium hydroxide dissolves in 1 ml of water, in about 3 ml of alcohol, and in about 2.5 ml of glycerin at 25° C. One gram is soluble in 0.6 ml of boiling water, and it is very soluble in boiling alcohol. The base will fuse at 360° C, and is appreciably volatilized at higher temperatures.

Potassium hydroxide is a strong base which can be assayed by titrimetry. The procedure involves titrating a solution containing a weighed sample of KOH against 1 N sulfuric acid. Two end points are determined. The first, using phenolphthalein as the indicator, is due to the neutralization of KOH. A second titration is performed in the same solution, using methyl orange as the indicator, to determine the amount of K_2CO_3 in the sample. Each ml of 1 N sulfuric acid consumed in the total titration is equivalent to 56.11 mg of total alkali, calculated as KOH. Each ml of acid consumed in the titration with methyl orange is equivalent to 138.2 mg of K_2CO_3. This type of assay is required because of the ease with which potassium hydroxide absorbs carbon dioxide from the air. The chemical properties of potassium hydroxide are similar to those of sodium hydroxide and will be discussed in that section (see below).

Uses. Potassium hydroxide is a very strong base having a caustic or corrosive effect on tissue. The U.S.P. states: *Caution—Exercise great care in handling Potassium Hydroxide, as it rapidly destroys tissues.* It enjoys its widest use as a base or alkaline reagent. It is used in official preparations as a saponifying agent to hydrolyze esters of fatty acids (fats and/or oils) into their constituent alcohols and the potassium salt. The saponification value, U.S.P. XVIII, p. 906, of fatty substances is defined in terms of this base. Aqueous and alcoholic volumetric solutions for titrating acids are widely used (U.S.P. XVIII, p. 1036). Test solutions in water and alcohol are also available (U.S.P. XVIII, p. 1030). Potassium hydroxide can also be used in place of sodium hydroxide in soda lime (see *Sodium Hydroxide*).

Sodium Carbonate, U.S.P. XVIII ($Na_2CO_3 \cdot H_2O$; Mol. Wt. 124.00; Monohydrated Sodium Carbonate)

Sodium carbonate is official as the monohydrate, and unless specified otherwise this will be the form implied throughout this discussion. Sodium carbonate is a colorless crystal or white, crystalline powder. It is odorless and has a strong alkaline taste. It will absorb small amounts of moisture from the air, but in warm, dry air at 50° C or above, it effloresces, becoming anhydrous at 100° C.

One gram of sodium carbonate dissolves in 3 ml of water and in 7 ml of glycerin at 25° C, and in 1.8 ml of boiling water. It is insoluble in alcohol.

Sodium carbonate is a reasonably strong base, as may be evidenced by the fact that a 1 M solution has a pH of 11.6. This high alkalinity is the result of hydrolysis of the carbonate anion, which is a strong base [rx (xlviii)]:

$$CO_3^{-2} + H_2O \rightleftarrows HCO_3^- + OH^- \qquad \text{(xlviii)}$$

The bicarbonate ion formed will also hydrolyze to a lesser extent [rx (xlix)]:

$$HCO_3^- + H_2O \rightleftarrows H_2CO_3 + OH^- \qquad \text{(xlix)}$$

These two reactions account for the basic character of the compound in solution, and allow for its method of assay. The assay procedure involves dissolving a weighed sample of anhydrous sodium carbonate in water and titrating against 1 N sulfuric acid to a methyl orange end point. Each ml of sulfuric acid is equivalent to 52.99 mg of Na_2CO_3. The reaction taking place in the assay is typical of the reaction of this compound with all acids [rx (l)]:

$$Na_2CO_3 + H_2SO_4 \rightarrow Na_2SO_4 + H_2O + CO_2 \uparrow \qquad \text{(l)}$$

Solutions of sodium carbonate, when treated with carbon dioxide in the cold, will form sodium bicarbonate. This reaction may be considered as a half neutralization with carbonic acid, which is formed to a slight extent in the solution [rxs (li)]:

$$CO_2 + H_2O \rightleftarrows H_2CO_3 \qquad \text{(li-a)}$$

$$Na_2CO_3 + H_2CO_3 \rightleftarrows 2NaHCO_3 \qquad \text{(li-b)}$$

Sodium carbonate exists in three well characterized hydrates in addition to the anhydrous form (soda ash). These are:

Monohydrate $= Na_2CO_3 \cdot H_2O$
Heptahydrate $= Na_2CO_3 \cdot 7H_2O$
Decahydrate $= Na_2CO_3 \cdot 10H_2O$ (sal soda, washing soda)

At temperatures above 35° C, a saturated solution of sodium carbonate will deposit only the monohydrated form. If the hot saturated solution is allowed to cool undisturbed in an atmosphere free of dust, it is often possible to obtain crystals of the heptahydrate. However, if the solution cools below 33° C, the decahydrate will crystallize because the heptahydrate is more soluble and remains in solution. The maximum solubility of sodium carbonate occurs at approximately 35° C. The decahydrate is prone to effloresce under storage conditions.

Uses. Medicinal uses of bicarbonates are discussed under electrolytes (Chapter 5) and antacids (Chapter 8). Essentially, carbonates have no therapeutic uses.

Sodium carbonate is used for its basicity in pharmaceutical preparations, where it will form sodium salts of acidic drugs. An example of this is the preparation of *Nitromersol Solution,* N.F. XIII, where together with sodium hydroxide it is used to form the water soluble sodium salt of the water insoluble nitromersol. It may also be used as a source of carbonate in the preparation of carbonates of various metals (e.g., ferrous carbonate).

Sodium Hydroxide, U.S.P. XVIII (NaOH; Mol. Wt. 40.00; Caustic Soda, Soda Lye)

The description, properties, and solubilities of sodium hydroxide are practically identical to those of potassium hydroxide. The method of assay involves the same double titration discussed under potassium hydroxide because sodium hydroxide will also absorb CO_2 to form sodium carbonate. The primary difference is that the total alkali calculated for sodium hydroxide must be not less than 95.0% and not more than 100.5%, including not more than 3.0% Na_2CO_3.

Sodium hydroxide is used as a base in pharmaceutical preparations more frequently than potassium hydroxide due to the pharmacological activity of potassium and its possible toxicity (see Chapter 5). As is potassium hydroxide, sodium hydroxide is highly ionized in solution making it one of the strongest bases available. The alkalinity of its solutions dictates certain precautions concerning filtration and storage. Solutions should be filtered through glass wool or asbestos, and the solution should be stored in hard glass bottles, using *rubber* stoppers. If glass-stoppered bottles are used, the solution will cause the stoppers to "freeze" in the neck of the bottle due to the solubilization of the glass by the NaOH solution. Glass-stoppered bottles can be used if a little petrolatum or other laboratory lubricant is spread around the stopper.

Sodium hydroxide reacts with the salts of all metals in solution precipitating almost all of them, except those of the alkali metals and ammonium [see rx (xlv)], as the insoluble metal hydroxides.

Sodium hydroxide will catalyze the hydrolysis of esters, and therefore is often employed as a saponifying agent. The base reacts with fats (glyceryl esters of fatty acids) to produce glycerin and the sodium salt of the fatty acid [rx (lii)]:

$$
\begin{array}{ccc}
C_{17}H_{35}COO-CH_2 & & HO-CH_2 \\
| & & | \\
C_{17}H_{35}COO\ \ CH\ +\ 3NaOH\ \rightarrow\ 3C_{17}H_{35}COO^-Na^+\ +\ HO-CH & & (lii) \\
| & & | \\
C_{17}H_{35}COO-CH_2 & & HO-CH_2 \\
\text{Stearin} & \text{Sodium stearate} & \text{Glycerin} \\
\text{(a fat)} & \text{(a soap)} &
\end{array}
$$

Uses. Sodium hydroxide is a very strong base capable of damaging tissue. In fact, the U.S.P. cautions: *Exercise great care in handling Sodium Hydroxide, as it rapidly destroys tissues.* However, solutions of sodium hydroxide can be employed in pharmaceutical preparations as an alkalinizing agent to form soluble sodium salts of various drugs. For example, *Nitromersol Solution,* N.F. XIII, *Nitromersol Tincture,* N.F. XIII, and *Phenolsulfonphthalein Injection,* U.S.P. XVIII, all utilize sodium hydroxide to solubilize the active component. Sodium hydroxide is also used in the soap industry for its saponification properties.

Soda Lime, U.S.P. XVIII (p. 842)

Soda lime is a mixture of calcium hydroxide and sodium or potassium hydroxide or both, intended for use in metabolism tests, anesthesia, and oxygen therapy. The U.S.P. states that it may contain an indicator which will not react with the common anesthetic gases, and which will change color when the absorption capacity of the soda lime for CO_2 is exhausted. The function of the soda lime in a closed system is to absorb the CO_2 which would otherwise accumulate in the system. The sodium hydroxide is capable of picking up carbon dioxide, but its combining power is soon exhausted. The calcium hydroxide present in the mixture will react with the accumulated carbon dioxide (in the form of Na_2CO_3) to form calcium carbonate, thereby "regenerating" the sodium hydroxide [rx (liii)]:

$$Na_2CO_3 + Ca(OH)_2 \rightarrow 2NaOH + CaCO_3 \downarrow \qquad \text{(liii)}$$

This process can continue until the calcium hydroxide and sodium hydroxide are exhausted. If an indicator is present in the mixture, the color will indicate when the soda lime is exhausted. When no indicator is present, a record of the length of time of usage must be kept to determine when the soda lime must be changed.

Buffers

The control of pH in solutions is a very important aspect of pharmaceutical chemistry and practice. The necessity for controlling pH within certain specified limits is found in the broad areas of chemical stability and solubility of the drug, and in patient comfort. Even though the pH of a solution can be carefully adjusted with acid or base to a prescribed level, it will not remain at that point indefinitely. Some factors which can produce alterations in pH include alkali in the glass of certain inexpensive containers, and gases in the air, such as CO_2 and NH_3, which can dissolve in the solution accompanied by acidic or basic reactions. The effect that these factors have on the established pH of a medium can be controlled through the use of chemical systems known as *buffers*.

Theory and Mechanism. *Buffer systems are pairs of related chemical compounds capable of resisting large changes in the pH of a solution caused by* *the addition of small amounts of acid or base.* More specifically, a buffer system is composed of a weak acid and its salt (conjugate base) or a weak base and its salt (conjugate acid; see Brönsted-Lowry concept). The latter are not widely employed because of the stability problems encountered with most of the available weak bases; therefore, this discussion will be devoted to systems composed of weak acids and their salts.

The operating theory behind these mixtures of compounds is that the two components of the system, the buffer pair, will complement each other. When small amounts of hydrogen ion are introduced into the medium they will react with the conjugate base or basic member of the buffer pair to form the weak acid which, by definition, will be only slightly ionized. Similarly, when small amounts of hydroxide ion (base) are introduced into the medium, they will react with the weak acid or acidic member of the buffer pair to form water and the conjugate base. Hence, each component of the buffer pair will react with either acid or base to form the other component, thereby repressing large changes in hydrogen ion concentration.

The above mechanism is illustrated further by the following equilibria. Consider a phosphate buffer system composed of sodium dihydrogen phosphate, $H_2PO_4^-$, as the acid member and disodium hydrogen phosphate, HPO_4^{-2}, as the basic member of the buffer pair. If a small amount of hydrochloric acid is added to a nonbuffered solution, it will essentially be ionized completely to chloride and hydronium ions, resulting in a significant lowering of the pH [rx (liv)]:

$$HCl + H_2O \rightarrow H_3O^+ + Cl^- \qquad \text{(liv)}$$

In a solution containing the phosphate buffer pair, the hydronium ion produced in the above reaction will react further with the basic member of the buffer according to rx (lv):

$$HPO_4^{-2} + H_3O^+ \rightleftarrows H_2PO_4^- + H_2O \qquad \text{(lv)}$$

Since the dihydrogen phosphate ion is a weak acid, the equilibrium in this reaction will lie strongly toward the right, resulting in very little of the added hydronium ion remaining in the solution.

Similarly, small amounts of sodium hydroxide in a nonbuffered solution will be largely ionized to hydroxide ion [rx (lvi-a)], which will ultimately reduce the hydronium ion concentration [rx (lvi-b)]:

$$NaOH \rightarrow Na^+ + OH^- \qquad \text{(lvi-a)}$$

$$OH^- + H_3O^+ \rightleftarrows 2H_2O \qquad \text{(lvi-b)}$$

A solution buffered with the same phosphate system used above would react with the added hydroxide ion according to rx (lvii):

$$H_2PO_4^- + OH^- \rightleftarrows HPO_4^{-2} + H_2O \qquad \text{(lvii)}$$

Essentially, the presence of base in the solution promotes the ionization of the dihydrogen phosphate ion to produce additional hydronium ions, which will in turn react with the hydroxide ion to form slightly ionized water.

The above reactions will take place whenever the acid added is stronger than the buffer acid, or if the base added is stronger than the buffer base. If the acid or base added is weaker than the buffering species, usually no reaction will take place. In any case, these would necessarily be quite weak in their acid-base properties, and would be expected to produce only slight alterations in the pH.

Buffer solutions, whether the phosphate system or others to be mentioned later, can be formulated to produce specific pH's within particular ranges. The pH of a buffer solution is related to the pK_a of the buffer acid and the log of the ratio of buffering species. This relationship is described by the *Henderson-Hasselbalch equation*, which can be derived from the ionization constant expression for the weak acid (see Eq. 5). A form of this equation was derived earlier and is shown in Eq. 8. Rearranging Eq. 8a provides the Henderson-Hasselbalch equation for the pH of a buffer solution containing the general weak acid, HA, and its conjugate base (salt), A⁻ (Eq. 20):

$$pH = pK_a + \log \frac{[A^-]}{[HA]} = pK_a + \log \frac{[Conj\ Base]}{[Acid]} \qquad (Eq.\ 20)$$

From Eq. 20 it should be noted that, when the buffer pair is present in equal concentrations, the pH of the buffer solution is equal to the pK_a of the buffer acid because the logarithmic term becomes zero. This further indicates that the pK_a of any buffer acid should be in the center of the desired pH range for the system, and serves as a basis for choosing a buffer for the maintenance of a particular pH. Usually, if a system must be kept within certain pH limits, a buffer having a pK_a close to the center of this range would provide the most efficient pH control.

It should also be apparent from the preceding discussion that the more nearly equal the concentrations of the buffering species, the greater will be the ability of the buffer to resist pH changes due to both acid and base. As the ratio of the concentrations of conjugate base to acid, $[A^-]/[HA]$, increases or decreases from the optimal value of *one*, the change in pH (ΔpH) produced by adding a constant amount of either acid or base will increase, as shown in Fig. 4–1. The numerical proof of this fact is presented in the data shown in Table 4–6.

In addition to the effect of the relative concentrations of buffering species, the actual concentrations of the acid and conjugate base also play an important role in the efficiency and capacity of buffer solutions. A comparison of the third and last buffer solutions in Table 4–6 illustrates by the

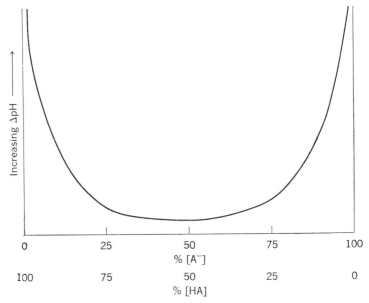

Fig. 4–1. Buffer efficiency illustrated in terms of pH shown as a function of the ratio of buffering species. This type of graph is obtained when a constant amount of strong acid or strong base is added to the buffer solutions indicated and the resulting changes in pH are plotted against the various ratios.

TABLE 4–6. [A⁻]/[HA] RATIOS AND CALCULATED
BUFFERING EFFICIENCY

	+1 mEq *Strong Acid*	*[A⁻]/[HA], mEq* *and Initial pH*	*+1 mEq* *Strong Base*
$[A^-]/[HA]$	9/191	10/190	11/189
pH	5.881	5.929	5.973
ΔpH	0.048	—	0.044
$[A^-]/[HA]$	49/151	50/150	51/149
pH	6.719	6.731	6.742
ΔpH	0.012	—	0.012
$[A^-]/[HA]$	99/101	100/100	101/99
pH	7.199	7.208	7.217
ΔpH	0.009	—	0.009
$[A^-]/[HA]$	149/51	150/50	151/49
pH	7.674	7.685	7.697
ΔpH	0.012	—	0.012
$[A^-]/[HA]$	189/11	190/10	191/9
pH	8.443	8.487	8.535
ΔpH	0.044	—	0.048
$[A^-]/[HA]$	999/1001	1000/1000	1001/999
pH	7.207	7.208	7.209
ΔpH	0.0008	—	0.0008

Adapted from Modern Inorganic Pharmaceutical Chemistry by C. A. Discher, John Wiley and Sons, Inc., N.Y., p. 175, 1964. By permission.

smaller ΔpH, the improved capacity of the solution composed of 1000 mEq of each species over that composed of 100 mEq each of acid and conjugate base. On the basis of the calculated data shown in Table 4–6, a tenfold increase in the concentrations of buffering species produces a greater than tenfold increase in buffer capacity.

Pharmaceutical Buffer Selection.   The selection of a buffer system for pharmaceutical purposes is based on both chemical and pharmacological factors. The chemical aspects should be obvious, in that the buffer should not react with other chemicals in the preparation. That is, the buffer pair should not: participate in oxidation-reduction reactions; alter the solubility of other components; form complexes with active ingredients; or participate in acid-base reactions other than those required as part of the buffer function. The buffer system itself must also exhibit reasonable chemical stability. Volatile species, e.g., NH_3 or CO_2, should be avoided since their loss will alter the pH and the buffer capacity of the system. Similarly, alkaline buffers should be protected against the absorption of CO_2 from the air, which would ultimately produce a drop in the pH. Other chemical factors involved in buffer selection are concerned with the aspects of buffer range and capacity discussed in the previous section. As much as possible, buffer acids should be chosen so that they have a pK_a near the middle of the desired pH range. This allows the buffer pair to have a ratio as close to unity as practicable. Buffer capacity is related to the absolute concentrations of the buffering species. These concentrations must also be tempered with practical considerations related to possible chemical reactivity and stability of the buffer in the pharmaceutical preparation to be protected.

The pharmacological factors of buffer selection are concerned with how the particular product is going to be used. In most instances, the buffer should neither contribute to nor detract from the pharmacological properties of the active ingredients. Certain buffer systems, such as borates, are toxic when administered systemically, and their use must be reserved for those preparations which are used topically. Borate buffers can, however, be used in certain ophthalmic preparations. Those weak acids and conjugate bases lacking in toxicity, at least in the small quantities employed as buffers, are those which are also found as common biochemicals in the body. One such buffer system frequently used in preparations intended for internal use is the phosphate system. This system has an added advantage in injectable preparations because the pK_a of the dihydrogen phosphate anion is reasonably close to the pH of body fluids. Other relatively nontoxic buffers may be found among certain weak organic acids and their conjugate bases, e.g., citrates and acetates.

A final point concerning those buffers composed of common biochemical acids and conjugate bases is based on their ability to support microbial growth. These buffers can serve as nutrient media for certain microorganisms, particularly when the pH is close to neutrality—neither too acidic nor too basic. Solutions of these compounds can be preserved with low concentrations of quaternary ammonium antimicrobial agents, e.g., 0.002% benzalkonium chloride (Zephiran®).

Pharmaceutical Buffer Systems. Buffer systems used in pharmacy can roughly be placed in two categories: (1) standard buffer systems which are designed to provide a solution having a specific pH for analytical purposes; and (2) actual pharmaceutical buffers designed to maintain pH limits in drug preparations.

The U.S.P. XVIII (pp. 858 and 938) and the N.F. XIII (pp. 783, 824, and 905) describe various buffer solutions which reproduce a given pH for standardizing pH meters and for other analytical procedures. For the most part, there are five primary solutions which cover the pH range from 1.2 to 10.0: *Hydrochloric Acid Buffer* (pH 1.2–2.2 in 0.1-unit intervals); *Acid Phthalate Buffer* (pH 2.2–4.0 in 0.2-unit intervals); *Neutralized Phthalate Buffer* (pH 4.2–5.8 in 0.2-unit intervals); *Phosphate Buffer* (pH 5.8–8.0 in 0.2-unit intervals); and *Alkaline Borate Buffer* (pH 8.0–10.0 in 0.2-unit intervals). Each of these buffer solutions is considered to be capable of reproducing the pH indicated within ±0.02 unit at 25° C. Of course, the hydrochloric acid buffer cannot be considered as a buffer at all since HCl is a strong acid. The other buffers contain biologically undesirable potassium salts, i.e., potassium chloride, potassium biphthalate, and potassium phosphates.

Among the inorganic buffers of importance which have particular advantages in pharmaceutical preparations are the phosphate and borate systems. The phosphate buffer system enjoys the advantage of containing the dihydrogen and monohydrogen phosphate anions, one of the physiological buffer pairs normally found in the body (see Chapter 5). Also, the pK_a of the dihydrogen phosphate anion is 7.2, making this a very efficient buffer system at the pH of physiological fluids, including lachrymal fluid (tears, pH = 7.4). The disadvantages of phosphate buffer systems include the insolubility of phosphate salts of such metals as silver, zinc, and aluminum. Phosphates will also support microbial growth, but may be preserved with 0.002% benzalkonium chloride (Zephiran®) or benzethonium chloride (Phemerol®).

A phosphate buffer system that has been modified to include the addition of sodium chloride to make it isotonic with physiological fluid is known as the *Sørensen phosphate buffer*. This system covers a range of pH from 5.9 to 8.0, and the solutions are prepared by adding various quantities of M/15 sodium dihydrogen phosphate and M/15 disodium phosphate according to the data given in Table 4–7. The mechanism involved in phosphate buffers has been illustrated previously [see rxs (lv) and (lvii)].

Borate buffer systems have been used in many pharmaceutical preparations containing metals that would otherwise precipitate in the presence of phosphates. However, there are some disadvantages accompanying their

TABLE 4–7. MODIFIED PHOSPHATE BUFFER SYSTEM (SØRENSEN)
ADJUSTED TO BE ISOTONIC WITH PHYSIOLOGICAL SALINE

Acid Buffer,	Sodium Biphosphate, N.F.	9.2 g
$M/15$ NaH_2PO_4	Purified Water, q.s.	1000.0 ml
Alkaline Buffer,	Disodium Phosphate	17.86 g
$M/15$ Na_2HPO_4	Purified Water, q.s.	1000.00 ml

pH	Acid Buffer, ml	Alkaline Buffer, ml	Sodium Chloride g/100 ml, to Render Isotonic
5.91	90.0	10.0	0.52
6.24	80.0	20.0	0.51
6.47	70.0	30.0	0.50
6.64	60.0	40.0	0.49
6.81	50.0	50.0	0.48
6.98	40.0	60.0	0.46
7.17	30.0	70.0	0.45
7.38	20.0	80.0	0.44
7.73	10.0	90.0	0.43
8.04	5.0	95.0	0.42

use. As indicated previously, borates are toxic, and their use in preparations intended for internal administration (particularly injectables) is
contraindicated. The pK_a of boric acid is 9.2, making borate buffers quite
inefficient at physiological pH, but they can be used in alkaline pH ranges.
It should be remembered that the alkaline pH ranges of this system offer
a further incompatibility for many of the metals with which it might be
used. Borates are weak bacteriostatics (see Chapter 9), but the solutions
will support mold growth when stored for a month or two at room temperature.

There are three primary borate buffer systems presently recognized.
These systems and their pH ranges are: Feldman's (pH 7–8.2); Atkins and
Pantin (pH 7.6–11.0); and Gifford's (pH 6–7.8). The buffer pair in each
of these is the tetraborate anion and boric acid ($B_4O_7^{-2}/H_3BO_3$), but in
the latter two, the sodium tetraborate is formed *in situ* rather than being
added directly to the solution.

The Feldman buffer system was introduced in 1937, but the pH values
for the various mixtures were soon found to be in error. Table 4–8 lists
the revised values in current use. The buffer is prepared by mixing the
indicated volumes of the acid solution containing boric acid and the alkaline
solution containing sodium borate (sodium tetraborate or sodium pyroborate) to make 100 ml of the buffer solution at the required pH.

The borate buffer mechanism is somewhat complicated by the chemistry
relating boric acid, metaboric acid, sodium metaborate, and sodium pyroborate. Boric acid is a weak acid ionizing by the simple reaction shown in
rx (lviii-a), or alternatively as shown in rx (lviii-b), indicating that the

TABLE 4–8. REVISED FELDMAN'S BUFFER MIXTURES

Acid Buffer Solution:	Boric Acid, N.F.	12.4 g
	Sodium Chloride, U.S.P.	2.9 g
	Purified Water, q.s.	1000.0 ml
Alkaline Buffer Solution:	Sodium Borate, U.S.P.	19.07 g
	Purified Water, q.s.	1000.00 ml

pH of Mixture	ml Acid Buffer	ml Alkaline Buffer
7.0	95.0	5.0
7.1	94.0	6.0
7.2	93.0	7.0
7.3	91.0	9.0
7.4	89.0	11.0
7.5	87.0	13.0
7.6	85.0	15.0
7.7	82.0	18.0
7.8	80.0	20.0
7.9	76.0	24.0
8.0	73.0	27.0
8.1	69.0	31.0
8.2	65.0	35.0

intermediate conjugate base is a tetrahedral arrangement of hydroxy groups around the boron.[6] The product of the latter reaction [rx (lviii-b)] is actually a Lewis acid-base adduct.

$$H_3BO_3 + H_2O \rightleftharpoons H_3O^+ + H_2BO_3^- \qquad \text{(lviii-a)}$$

$$B(OH)_3 + 2H_2O \rightleftharpoons H_3O^+ + B(OH)_4^- \qquad \text{(lviii-b)}$$

In any case, the reaction of boric acid in the buffer solution with base, (e.g., sodium hydroxide) results in the formation of a salt [rx (lix)] which will decompose to metaborate (NaBO$_2$) according to rx (lx). Metaborate, in the presence of excess boric acid, will form the pyroborate salt [tetra-borate, Na$_2$B$_4$O$_2$; rx (lxi)]:

$$B(OH)_3 + NaOH \rightarrow Na^+ + B(OH)_4^- \qquad \text{(lix)}$$

$$NaB(OH)_4 \rightarrow NaBO_2 + 2H_2O \qquad \text{(lx)}$$

$$2NaBO_2 + 2B(OH)_3 \rightarrow Na_2B_4O_7 + 3H_2O \qquad \text{(lxi)}$$

The reaction of the sodium borate in the buffer with acid involves the formation of the slightly ionized pyroboric acid (H$_2$B$_4$O$_7$) from the highly ionized sodium pyroborate present, according to rx (lxii). The pyroboric acid, in turn, hydrolyzes to form metaboric acid and boric acid [rx (lxiii)]:

$$2HCl + Na_2B_4O_7 \rightarrow H_2B_4O_7 + 2NaCl \qquad \text{(lxii)}$$

$$H_2B_4O_7 + 3H_2O \rightleftharpoons 2HBO_2 + 2B(OH)_3 \qquad \text{(lxiii)}$$

A significant modification of this system was introduced by Atkins and Pantin. Recognizing that the alkaline buffer solution employed in the Feldman buffer system containing sodium borate, U.S.P., was unstable, they formulated a buffer system using sodium carbonate, U.S.P. ($Na_2CO_3 \cdot H_2O$) in the alkaline solution in place of the sodium borate. When the sodium carbonate solution is mixed with the boric acid solution, the same buffer pair as in Feldman's buffer is available (i.e., $B_4O_7^{-2}/H_3BO_3$). Since sodium carbonate hydrolyzes to sodium hydroxide in water [rx (lxiv)], the chemistry involved in the formation of the tetraborate anion is essentially the same as shown above [rxs (lix)–(lxi)] for the reaction of sodium hydroxide with boric acid:

$$Na_2CO_3 + 2H_2O \leftrightarrows 2NaOH + H_2CO_3 \qquad \text{(lxiv)}$$

The Atkins and Pantin buffer system is useful at alkaline pH although the most efficient range is too alkaline for many drug purposes. The system, however, can be used in ophthalmic solutions, contact lens solutions, and as a solvent for soluble fluorescein. Table 4–9 illustrates the pH values for the various mixtures of the acid and alkaline solutions. It should be noted that additional amounts of sodium chloride have been

TABLE 4–9. ATKINS AND PANTIN ALKALINE BUFFER

Acid Buffer Solution:	Boric Acid, N.F.	12.4 g
	Sodium Chloride, U.S.P.	7.5 g
	Purified Water, q.s.	1000.0 ml
Alkaline Buffer Solution:	Sodium Carbonate, U.S.P.	24.8 g.
	Purified Water, q.s.	1000.0 ml.

pH of Mixture	ml Acid Buffer	ml Alkaline Buffer
7.6	93.8	6.2
7.8	91.7	8.3
8.0	88.8	11.2
8.2	85.0	15.0
8.4	80.7	19.3
8.6	75.7	24.3
8.8	69.5	30.5
9.0	63.0	37.0
9.2	56.4	43.6
9.4	49.7	50.3
9.6	42.9	57.1
9.8	36.0	64.0
10.0	29.1	70.9
10.2	22.1	77.9
10.4	15.4	84.6
10.6	9.8	90.2
10.8	5.7	94.3
11.0	3.5	96.5

added to the acid solution of this system in order to approximate an isotonic product more closely. Since these solutions are used as vehicles for drugs, it is doubtful that isotonicity can be achieved in this manner and, in fact, many products may actually be hypertonic in their final form.

Gifford's buffer is a borate system very similar to that of the Atkins and Pantin system. The major modification is that potassium chloride is used in place of sodium chloride, together with adjustment of the volumes of the acid and alkaline solutions to provide a final volume of 30 ml. The quantity of potassium chloride in Gifford's buffer actually makes the final solution hypertonic. The Gifford's and Atkins and Pantin buffer systems are prone to slow decomposition and a slight increase in alkalinity due to a loss of carbon dioxide from the sodium carbonate [see rx (lxiv)].

To summarize, borate buffers operate best from pH 7 to 11, being most efficient at a pH of 9. Although the buffer efficiency is reduced at acidic pH, borates are used at a pH of 6 with zinc and silver salts, since Sørensen's buffer is incompatible with these metals. Borates are suitable for ophthalmic and nasal solutions and external preparations, but are contraindicated in parenteral solutions.

Other buffer systems employed in pharmaceutical preparations include weak organic acids (e.g., acetic and citric) and their conjugate bases. These are low in toxicity because they are normal body metabolites, and can be used in solutions that are going to be administered internally, including parenteral injection. In common with the phosphates, these buffer solutions should be preserved against bacterial growth. Organic acid salts of various metals frequently exhibit greater water solubility than the corresponding phosphates, e.g., the solubility of silver acetate allows the use of acetate buffers in solutions containing silver ion.

To further illustrate the importance of pH control in pharmaceutical products, the Appendix lists numerous drugs under the title "Effect of pH on Individual Drugs" and should be referred to for aid in assessing potential incompatibilities based on pH changes (see Appendix C).

Antioxidants

Antioxidants may be described as compounds which have the capability of functioning chemically as reducing agents. They are used in pharmaceutical preparations containing easily oxidized substances (e.g., iodide or ferrous ions) in order to maintain these substances in their reduced form. The mechanism of antioxidant action may be seen in two ways, both achieving the same net result: (a) either the antioxidant is oxidized in place of the active constituent; or (b) if the active component is oxidized, the antioxidant reduces it back to its normal oxidation state. For all practical purposes, these two events are indistinguishable.

Theory. The theory involved in antioxidant action is the same as that involved in any oxidation-reduction or redox reaction. Redox reactions may be viewed in a manner analogous to Brönsted acid-base reactions, in that so-called "conjugate pairs" of oxidized and reduced forms of a compound can be separated from the chemical equation. For example, the usual way of examining a redox reaction is to separate it into half-reactions of the oxidizing and reducing reagent. Since redox is essentially the transfer of electrons from one compound to the other, the loss or gain of electrons is used to balance the oxidation states on both sides of the half-reaction. Consider a general reaction between the oxidized form of compound 1, Ox_1, and the reduced form of compound 2, Red_2. Since oxidation is the loss of electrons from a chemical species and reduction is the gain of electrons, the half-reactions may be written:

$$Ox_1 + e \rightleftarrows Red_1 \qquad \text{(lxv-a)}$$

$$Red_2 \rightleftarrows Ox_2 + e \qquad \text{(lxv-b)}$$

The total redox reaction is obtained by adding rxs (lxv-a) and (lxv-b):

$$Ox_1 + Red_2 \rightleftarrows Red_1 + Ox_2 \qquad \text{(lxvi)}$$

The tendency for chemical substances to undergo oxidation-reduction reactions can be determined in electrochemical cells where the electron transfer takes place through a system of electrodes. The electrical potential developed in a cell can be measured with a potentiometer or a voltmeter. Through such measurements it has been found that spontaneous redox reactions generally develop positive potentials, as opposed to those reactions requiring an external driving force (nonspontaneous) which develop negative potentials.

The properties of electrochemical cells are expressed through a relationship known as the *Nernst equation,* shown in Eq. 21:

$$E_{cell} = E^0_{cell} - \frac{RT}{nF} \ln Q \qquad \text{(Eq. 21)}$$

where E_{cell} is the potential of the cell in volts, E^0_{cell} is the standard potential, i.e., when all reactants and products are at unit activity, R is the gas constant, T is the temperature in °K (absolute temperature), F is the Faraday constant, n is the number of electrons transferred in the reaction, and Q is the activity quotient of products over reactants. For a cell at 298° K (25° C), and converting to logarithms to the base 10, Eq. 21 may be rewritten:

$$E_{cell} = E^0_{cell} - \frac{0.059}{n} \log Q \qquad \text{(Eq. 22)}$$

Eq. 22 may be utilized to make decisions concerning the ease with which certain redox reactions will take place. The quantity Q is set up in much the same way as the expression for an equilibrium constant. Using rx (lxvi), Q becomes:

$$Q = \frac{[\text{Red}_1]\,[\text{Ox}_2]}{[\text{Ox}_1]\,[\text{Red}_2]} \qquad \text{(Eq. 23)}$$

The value of E^0_{cell} is determined from a table of standard electrode potentials (see Table 4–10) which lists the potentials for various half-reactions. E^0_{cell} is the sum of the $E^0_{\frac{1}{2}}$ for the oxidation and reduction half-reactions:

$$E^0_{cell} = E^0_{\frac{1}{2}oxid} + E^0_{\frac{1}{2}red} \qquad \text{(Eq. 24)}$$

The number of electrons transferred, n, can be found in the equations for the half-reactions.

Table 4–10 lists some standard electrode potentials of reactions of interest in pharmaceutical chemistry. The half-reactions are all written as oxidations and follow the American sign convention of using positive signs for those reactions with relatively strong oxidation tendencies. The European convention is opposite to the American in that reactions are written as reductions, with those reactions showing strong reduction tendencies given positive potentials. The potentials in Table 4–10 may be used for reductions by simply changing the sign. Tables of this type are frequently presented in decreasing numerical order of the electrode potential; this has the advantage of illustrating the relative tendencies of metals to react with acid to release hydrogen. All metals above hydrogen in such a presentation will evolve hydrogen upon reaction with acid, while those below hydrogen will not. Hydrogen is then established as an arbitrary zero point, and the table is called the *electromotive series of metals*. For convenience in finding the necessary reactants, this table is arranged in alphabetical order by element and, within an elemental series, numerically by oxidation state.

As an example of the use of the data in Table 4–10 to evaluate antioxidants, consider *Dilute Hydriodic Acid*, which was official in N.F. XII, and which contained 10% HI stabilized with 0.8% hypophosphorous acid (HPH_2O_2). The hypophosphorous acid prevents the oxidation of iodide ion to molecular iodine through the following initial reaction [rx (lxvii)]; the complete reaction is discussed later under Hypophosphorous Acid]:

$$HPH_2O_2 + I_2 + 3H_2O \rightleftarrows H_2PHO_3 + 2I^- + 2H_3O^+ \qquad \text{(lxvii)}$$

TABLE 4–10. SELECTED STANDARD ELECTRODE POTENTIALS
(AMERICAN SIGN CONVENTION)

Oxidative Half-Reaction	$E_{\frac{1}{2}}^0$ (volts)
$Ag = Ag^+ + e$	-0.80
$Ag + Cl^- = AgCl + e$	-0.22
$2As + 3H_2O = As_2O_3 + 6H^+ + 6e$	-0.23
$HAsO_2 + 2H_2O = H_3AsO_4 + 2H^+ + 2e$	-0.56
$Bi + H_2O = BiO^+ + 2H^+ + 3e$	-0.32
$2Br^- = Br_2 + 2e$	-1.09
$Ce^{+3} = Ce^{+4} + e$	$+1.70$
$2Cl^- = Cl_2 + 2e$	-1.36
$\frac{1}{2}Cl_2 + 2H_2O = HClO + 2H^+ + 2e$	-1.63
$ClO^- + 2OH^- = ClO_2^- + H_2O + 2e$	-0.66
$Cl^- + 2OH^- = ClO^- + H_2O + 2e$	-0.94
$2Cr^{+3} + 7H_2O = Cr_2O_7^{-2} + 14H^+ + 6e$	-1.33
$Cu = Cu^+ + e$	-0.52
$Cu = Cu^{+2} + 2e$	-0.34
$Fe = Fe^{+2} + 2e$	$+0.44$
$Fe = Fe^{+3} + 3e$	$+0.04$
$Fe^{+2} = Fe^{+3} + e$	-0.77
$H_2 = 2H^+ + 2e$	0.00
$2Hg = Hg_2^{+2} + 2e$	-0.79
$Hg_2^{+2} = 2Hg^{+2} + 2e$	-0.91
$2I^- = I_2 + 2e$	-0.54
$3I^- = I_3^- + 2e$	-0.54
$Li = Li^+ + e$	$+3.03$
$Mn = Mn^{+2} + 2e$	$+1.19$
$Mn^{+2} + 2H_2O = MnO_2 + 4H^+ + 2e$	-1.23
$Mn^{+2} + 4H_2O = MnO_4^- + 8H^+ + 5e$	-1.51
$NO + H_2O = HNO_2 + H^+ + e$	-0.99
$HNO_2 + H_2O = NO_3^- + 3H^+ + 2e$	-0.94
$2H_2O = H_2O_2 + 2H^+ + 2e$	-1.77
$H_2O_2 = O_2 + 2H^+ + 2e$	-0.69
$HPH_2O_2 + H_2O = H_2PHO_3 + 2H^+ + 2e$	$+0.59$
$PH_2O_2^- + 3OH^- = PHO_3^{-2} + 2H_2O + 2e$	$+1.60$
$H_2PHO_3 + H_2O = H_3PO_4 + 2H^+ + 2e$	$+0.28$
$2S_2O_3^{-2} = S_4O_6^{-2} + 2e$	-0.17
$H_2SO_3 + H_2O = SO_4^{-2} + 4H^+ + 2e$	-0.17
$Sn = Sn^{+2} + 2e$	$+0.14$
$Sn^{+2} = Sn^{+4} + 2e$	-0.15
$Zn = Zn^{+2} + 2e$	$+0.76$

This reaction can be factored into the corresponding half-reactions with the $E_{\frac{1}{2}}^0$ values obtained from Table 4–10:

$$HPH_2O_2 + 3H_2O = H_2PHO_3 + 2H_3O^+ + 2e \quad E_{\frac{1}{2}}^0 = +0.59$$

$$I_2 + 2e = 2I^- \qquad\qquad\qquad\qquad\quad E_{\frac{1}{2}}^0 = +0.54$$

$$E_{cell}^0 = +1.13$$

The rather large positive value of E_{cell}^0 is indicative of a facilitated spontaneous reaction, and is further supported when the Nernst equation for the redox reaction is solved:

$$E_{cell} = 1.13 - \frac{0.059}{2} \log \frac{[H_2PHO_3] \, [I^-]^2}{[HPH_2O_2] \, [I_2]}$$

The appropriate concentration values for substitution into the above expression are the molar concentrations of iodide and hypophosphorous acid. The average concentrations formerly official in the N.F. XII provided HI = 0.78 M and HPH_2O_2 = 0.12 M. Therefore, the initial E_{cell} is:

$$E_{cell} = 1.13 - 0.03 \log \frac{[H_2PHO_3] \, [0.78]^2}{[0.12] \, [I_2]}$$

$$= 1.13 - 0.03 \, (0.705)$$

$$= 1.13 - 0.02 = 1.11$$

The values substituted into the above equation represent the preequilibrium conditions. Of course, after equilibrium is established, no net oxidation-reduction takes place and the cell potential is effectively zero. However, the large positive value for E_{cell} indicates that the hypophosphorous acid present should effectively reduce any molecular iodine formed in dilute hydriodic acid solution by air oxidation. However, it should be remembered that the antioxidant power of hypophosphorous acid diminishes as the overall redox reaction approaches equilibrium.

The above example tells only part of the story. The hypophosphorous acid transferred two electrons and reduced one molecule of iodine back to two iodide ions while it was oxidized to phosphorous acid. In turn, phosphorous acid (H_2PHO_3) can transfer two more electrons to reduce another molecule of iodine, which subsequently forms phosphoric acid (H_3PO_4). Thus, one molecule of hypophosphorous acid can actually reduce two molecules of iodine, forming four iodide ions and one molecule of phosphoric acid.

The Selection of Antioxidants. A major consideration in selecting a suitable antioxidant for pharmaceutical purposes has already been illustrated. The probability of the desired redox reaction taking place should be assessed through the use of standard electrode potentials and the Nernst equation. Other aspects involve possible physiological and chemical incompatibilities. An antioxidant in a pharmaceutical preparation should be physiologically inert; usually the concentrations employed are sufficiently low to insure this for most of the common antioxidants. In considering a particular antioxidant, however, the possible toxicity of both the reducing agent and its oxidized product must be assessed.

Chemical problems associated with antioxidants are similar to those discussed for buffers. Antioxidant selection should consider possible solubility problems between the reducing agent and the drug. For example, preparations containing calcium should be adjusted to an acidic pH if an antioxidant containing the sulfite ion (SO_3^{-2}) is going to be used. Sulfite compounds will cause the precipitation of calcium from solutions at neutral to alkaline pH. Also, bisulfites will form addition compounds with many unsaturated organic functional groups, e.g., ethylenic bonds and carbonyl groups.

One very serious problem accompanying the handling of antioxidants concerns mixing them with strong oxidizing agents. Very strong reducing agents (e.g., hypophosphorous acid and hypophosphites) will form explosive mixtures when combined in dry form or in concentrated solutions with strong oxidizing agents.

Those preparations where specific antioxidants find the greatest use will be mentioned in the following sections.

Official Antioxidants

Hypophosphorous Acid, N. F. XIII (HPH_2O_2; Mol. Wt. 66.00)

Hypophosphorous acid is a colorless or slightly yellow, odorless liquid containing not less than 30.0% and not more than 32.0% HPH_2O_2. The solution has a specific gravity of 1.13 at 25° C. The pure acid is a syrupy, colorless liquid which becomes a solid at 17.4° C and melts at 26.5° C.

The compound contains only one ionizable hydrogen, i.e., the one bonded to the oxygen atom in the structure shown above. It therefore reacts as a monoprotic acid, and is assayed by titrating against 1 N sodium hydroxide to a phenolphthalein end point.

Although it is an acid, it is practically never used for its acidic properties. The oxidation state of the central phosphorus atom is +1, making the compound a very powerful reducing agent. As mentioned above, it can easily reduce many compounds to form phosphorous acid (H_2PHO_3) having an oxidation state of +3, and finally phosphoric acid (H_3PO_4) having an oxidation state of $+5$, for a net transfer of four electrons. Its reducing properties are readily illustrated by its reaction with molecular iodine to form iodide ions [rx (lxviii)], and its ability to decolorize acidic solutions of potassium permanganate immediately [rx (lxix)]:

$$H_3PO_2 + 2I_2 + 2H_2O \rightarrow 4HI + H_3PO_4 \qquad \text{(lxviii)}$$

$$5H_3PO_2 + 4KMnO_4 + 6H_2SO_4 \rightarrow 2K_2SO_4 + 4MnSO_4 + $$
$$5H_3PO_4 + 6H_2O \qquad \text{(lxix)}$$

The mixture of hypophosphorous acid with any oxidizing agent in concentrated form can produce an incompatibility and possible explosion. However, when its reaction with reducible substances is desirable, hypophosphorous acid can function in dilute solution as a very effective reducing agent or antioxidant.

Uses. Hypophosphorous acid and its salts (hypophosphites) have no important pharmacological actions. The use of the acid is relegated entirely to that of an antioxidant. It serves to prevent the formation of free iodine in *Diluted Hydriodic Acid* (N.F. XII) and *Hydriodic Acid Syrup* (N.F. XII). It is also present in *Ferrous Iodide Syrup* (N.F. XI), where it prevents the formation of both ferric ions and molecular iodine.

Salts of hypophosphorous acid are also used for their antioxidant properties: sodium hypophosphite is present as a preservative in certain foods, and ammonium hypophosphite may also be found as a preservative in many preparations. Hypophosphites should never be triturated with oxidizing agents such as nitrates, chlorates, or permanganates.

The concentration ranges for hypophosphorous acid and its salts when used as antioxidants are never over 1%, and usually between 0.5 and 1%.

Sulfur Dioxide, U.S.P. XVIII (SO_2; Mol. Wt. 64.06)

Sulfur dioxide is a colorless, nonflammable gas possessing a strong suffocating odor characteristic of burning sulfur. This is the same odor noticed from the burning of high sulfur-containing fuels. It contains not less than 97.0% by volume of SO_2. Under pressure, sulfur dioxide condenses to a colorless liquid which boils at $-10°$ C and has a density of 1.5.

Sulfur dioxide is being discussed at this point because of its intimate relationship to sulfites and bisulfites, covered in the next section. These compounds all contain sulfur in the $+4$ oxidation state and function as strong reducing agents. When sulfur dioxide is passed through an aqueous solution containing iodine, hydriodic acid is formed [rx (lxx)]:

$$SO_2 + I_2 + 2H_2O \rightarrow 2HI + H_2SO_4 \qquad \text{(lxx)}$$

A similar reaction occurs with potassium permanganate [rx (lxxi)].

$$5SO_2 + 2KMnO_4 + 2H_2O \rightarrow K_2SO_4 + 2MnSO_4 + 2H_2SO_4 \qquad \text{(lxxi)}$$

Sulfur dioxide is stable as such only at moderate to strongly acidic pH. As the pH approaches neutral to alkaline values, sulfur dioxide is converted almost totally to bisulfite and sulfite. This fact is used to advantage in the official assay for SO_2 which involves its absorption into 0.1 N sodium hydroxide to form a solution of sodium bisulfite. The bisulfite solution is then titrated against 0.1 N iodine solution to a pale blue end point with starch T.S. The reaction is similar to that shown in the discussion on sodium bisulfite. This reaction should also be recognized as a possible incompatibility when sulfur dioxide is used to stabilize solutions in the alkaline pH range. In unbuffered solutions particularly, the SO_2 will reduce hydroxide ion concentrations and cause a shift to more acidic pH [rx (lxxii)]:

$$SO_2 + OH^- \rightarrow HSO_3^- \qquad \text{(lxxii)}$$

The opposite of the above reaction serves as an extemporaneous preparation of sulfur dioxide. According to rx (lxxiii), treatment of bisulfites with dilute acid causes the formation of SO_2:

$$NaHSO_3 + H_2SO_4 \rightarrow NaHSO_4 + SO_2 \uparrow + 2H_2O \qquad \text{(lxxiii)}$$

Uses. Sulfur dioxide is classified as an antioxidant pharmaceutical aid. It will protect many susceptible compounds from oxidation by reducing the oxidized form back again, or, probably even more important, by reacting with oxygen before the susceptible compounds do. Because of the gaseous nature of the compound, sulfur dioxide is usually used in injectable preparations enclosed in single-dose ampules or multiple-dose vials. The concentrations employed vary with the compound to be protected and the type of dispensing container. The usual concentration is about 0.1% and seldom exceeds 1.0%. Multiple-dose containers will require more SO_2 than single-dose containers because each withdrawal introduces air which depletes the antioxidant more rapidly.

Sulfur dioxide also finds extensive use in industry for such processes as bleaching wood pulp, fumigating grains, and arresting fermentation. It is also used to fumigate houses where it is given off in the burning of sulfur candles (see Chapter 9).

Sodium Bisulfite, U.S.P. XVIII (NaHSO₃; Mol. Wt. 104.06; Sodium Hydrogensulfite, Sodium Acid Sulfite)

Although the above name and formula indicate a pure compound, such is not the case. The U.S.P. describes sodium bisulfite as a mixture of sodium bisulfite ($NaHSO_3$) and sodium metabisulfite ($Na_2S_2O_5$) in varying proportions. The compound is white or yellowish white crystals or a granular powder having the odor of sulfur dioxide, and should yield not less than 58.5% and not more than 67.4% SO_2. The solid is unstable in air, giving off SO_2.

Sodium Metabisulfite, N.F. XIII, 4th Supplement ($Na_2S_2O_5$; Mol. Wt. 190.10; Disodium Pyrosulfite)

This compound is also a white crystal or a white to yellowish crystalline powder having the odor of sulfur dioxide. It should contain an amount of $Na_2S_2O_5$ equivalent to not less than 66.0% and not more than 67.4% of SO_2.

Most commercial sodium bisulfite is actually the metabisulfite. This salt can be used when bisulfite is specified because, when dissolved in water, it is immediately converted to bisulfite [rx (lxxiv)]:

$$Na_2S_2O_5 + H_2O \rightarrow 2NaHSO_3 \qquad \text{(lxxiv)}$$

As has been alluded to above, treatment of bisulfites or the normal salts (sulfites) with aqueous acid yields sulfurous acid, which is essentially a solution of SO_2 in water [rxs (lxxv-a, -b, and -c)]:

$$NaHSO_3 + HCl \rightarrow NaCl + H_2SO_3 \qquad \text{(lxxv-a)}$$

$$Na_2SO_3 + 2HCl \rightarrow 2NaCl + H_2SO_3 \qquad \text{(lxxv-b)}$$

$$H_2SO_3 \rightleftharpoons H_2O + SO_2 \uparrow \qquad \text{(lxxv-c)}$$

Sodium bisulfite, sodium metabisulfite, other bisulfites, and sulfites are strong reducing agents. Like sulfur dioxide, they contain sulfur in the +4 oxidation state, and along with SO_2 they constitute a very closely related family of compounds that can be formed into and from one another, depending primarily upon the pH of the medium. Most of these relationships have been pointed out in some of the previous reactions.

Sodium bisulfite is acidic enough to neutralize the stronger bases, e.g., sodium carbonate [rx (lxxvi)], to form the sulfite:

$$2NaHSO_3 + Na_2CO_3 \rightarrow 2Na_2SO_3 + H_2O + CO_2 \uparrow \qquad \text{(lxxvi)}$$

On the other hand, the addition of sulfites to acidic solutions will result in a shift toward alkaline pH by the formation of bisulfite, which reduces the concentration of hydrogen ions [e.g., see rx (lxxv-b)].

One primary source of chemical incompatibilities with bisulfites and sulfites is solubility. The only soluble salts are those of the alkali metals. The alkaline earth metal bisulfites are less soluble, and the sulfites of these metals are insoluble.

In a manner similar to sulfur dioxide, sodium bisulfite and sodium metabisulfite are assayed on the basis of reducing a 0.1 N iodine solution which is added to the compounds in excess. The residual iodine is titrated against sodium thiosulfate to determine the amount that was reduced by the bisulfite or metabisulfite. The bisulfite reaction and the reaction of the excess iodine are illustrated in rx (lxxvii-a) and (lxxvii-b), respectively:

$$NaHSO_3 + I_2 + H_2O \rightarrow NaHSO_4 + 2HI \qquad \text{(lxxvii-a)}$$

$$2Na_2S_2O_3 + I_2 \rightarrow 2NaI + Na_2S_4O_6 \qquad \text{(lxxvii-b)}$$

Uses. Sodium bisulfite or sodium metabisulfite is used almost exclusively as an antioxidant. It is usually found in solutions of drugs that contain the phenol or catechol nucleus (e.g., phenylephrine hydrochloride and epinephrine hydrochloride solutions) to prevent oxidation of these compounds to quinones and like substances. Its use in this way requires that the solution be at acid pH, since phenolate anions inhibit the reducing action of bisulfites. In addition, the acid pH also helps to prevent oxidation of catechol derivatives. Bisulfite may also be found in ascorbic acid injection as a reducing agent.

The concentrations employed for this purpose are usually around 0.1%. Concentrations equivalent to 0.2% SO_2 are allowed.

Sodium bisulfite may also be used to prepare water soluble derivatives of normally insoluble drugs. An example of this is *Menadione Sodium Bisulfite,* N.F. XIII, which is a bisulfite addition compound formed across the ethylenic linkage of *Menadione,* N.F. XIII. The reaction converts the water insoluble menadione into a water soluble form for preparing parenteral products while maintaining the activity of the parent compound through regeneration of menadione in the tissues.

Sodium bisulfite has also been used topically to treat dermatological problems caused by certain parasites. This, however, does not constitute a major use for the compound.

Sodium Thiosulfate, U.S.P. XVIII ($Na_2S_2O_3 \cdot 5H_2O$; Mol. Wt. 248.18)

$$\begin{array}{c} S \\ \uparrow \\ Na^+ \; {}^-O{-}S{-}O^- \; Na^+ \cdot 5H_2O \\ \downarrow \\ O \end{array}$$

Since sodium thiosulfate is official as an antidote, the physical description and general chemical properties will be discussed in Chapter 12.

Sodium thiosulfate contains sulfur in two different oxidation states. According to the structure shown above, the oxidized sulfur atom is in a +6 state resisting further oxidation, while the remaining sulfur atom is in a zero oxidation state. This allows the compound to act as a reducing agent or as an antioxidant. Its reducing properties are illustrated by its application as the titrating reagent in iodine determinations. It readily converts colored molecular iodine solutions to colorless sodium iodide, with the accompanying oxidation to sodium tetrathionate [see rx (lxxvii-b) above]. The reaction is complete, thus allowing the compound to be used to remove iodine stains and traces of chlorine from industrial bleaching operations. Sodium thiosulfate will also reduce ferric ions to the ferrous form.

The only soluble salts of thiosulfate are the alkali metals. Mixing sodium thiosulfate with solutions containing other metal cations is a source of incompatibility due to the precipitation of the metal thiosulfate. In acid solution, these precipitates may darken due to the formation of the corresponding sulfide.

Uses. Sodium thiosulfate, sometimes called "hypo" from the erroneous nomenclature "sodium hyposulfite," is categorized in the U.S.P. as an antidote for cyanide poisoning (see Chapter 12). Sodium thiosulfate is used somewhat less frequently in combination with acids to treat various dermatological problems. The acid causes the precipitation of sulfur and evolution of SO_2.

The use of sodium thiosulfate as an antioxidant is usually limited to solutions containing iodides. The N.F. XIII allows the use of 0.05% sodium thiosulfate in *Potassium Iodide Solution*, N.F. XIII, if the solution is not going to be used immediately.

Sodium Nitrite, U.S.P. XVIII ($NaNO_2$; Mol. Wt. 69.00)

Since sodium nitrite is official as an antidote, the physical description and general chemical properties will be discussed in Chapter 12.

Chemically, nitrites can act as both reducing and oxidizing agents. Reduction of compounds with sodium nitrite (nitrogen oxidation state $+3$) results in the formation of nitrates (nitrogen oxidation state $+5$). An example of the reducing property of nitrite is illustrated in rx (lxxviii) where the chlorate is reduced to the chloride in acid solution:

$$3HNO_2 + KClO_3 \rightarrow 3HNO_3 + KCl \qquad \text{(lxxviii)}$$

When nitrites react as oxidizing agents, the reduced product will be nitric oxide (NO) in acidic solution, or molecular nitrogen in neutral to alkaline solution.

Uses. Sodium nitrite is classified by the U.S.P. as an antidote for cyanide poisoning by virtue of its ability to induce methemoglobin formation on injection. This will be discussed more completely in Chapter 12.

Pharmacologically, the nitrite ion relaxes the smooth muscle of the blood vessels, giving it a vasodilator action. Sodium nitrite is not used for this particular effect, although organic derivatives containing the nitrite and nitrate groups (e.g., amylnitrite, nitroglycerin, etc.) are used as coronary vasodilators in angina pectoris.

Sodium nitrite is not used specifically as an antioxidant in pharmaceutical preparations as are the previous compounds. It is, however, used as a reducing agent when combined with sodium carbonate. This combination is available as Anti-Rust Tablets.

Nitrites are used in brine solutions for the curing of meats and fish. These compounds serve a threefold function in this process: color development, flavor production, and preservation against bacterial growth (antibacterial). Recent observations that the nitrite ions remaining in cured meats react with organic amines that are also present to form potentially carcinogenic N-nitrosamines have resulted in a great deal of concern about the safety of using nitrites, and the levels that should be allowed to remain in meat products. It is to be pointed out that nitrates which are naturally present in certain vegetables (e.g., spinach) and in varying amounts in certain water supplies are also potential sources of nitrite through reductive metabolism by certain microorganisms. The real health hazard posed by the presence of nitrite, at least in terms of carcinogenicity, is not presently known.[7] Studies to date seem to indicate that, in the healthy human being, conditions of gastrointestinal pH, microbial flora, and concentrations of nitrites may not be suitable for the *in vivo* formation of N-nitrosamines. More research in this area, as well as research on the potential hazard of N-nitrosamines produced in meats during curing and storage, and the hazard presented by nitrites produced by the reduction of nitrates in other foods and water, is needed before these compounds can be indicated as definite causes of cancer in man. There may be a need for some changes in standards and regulations, but the situation is not proving as alarming as originally stated.

Nitrogen, U.S.P. XVIII (N_2; Mol. Wt. 28.01)

Nitrogen is an inert gas (see Chapter 12). The reactions and compounds of nitrogen are not of importance to this discussion. Because of its inert character, nitrogen can be used to protect chemicals, reagents, and pharmaceuticals from air oxidation by displacing the air in reaction vessels and containers.

Uses. The physiological aspects of nitrogen are discussed in Chapter 12. Nitrogen is used as an inert atmosphere to retard oxidation in oxidation-sensitive products—cod liver oil, olive oil, multiple vitamin preparations, etc. It is also used to replace air in containers for parenterals and solutions for topical application. Such usage must be declared on the product label. Nitrogen is also used to retard oxidation in the qualitative test for carbon monoxide.

Water

Water is an interesting and, although very common, an extremely unusual liquid. Its properties are very different from those of hydrides of elements immediately beside or below oxygen in the periodic table, i.e., hydrogen sulfide (H_2S), ammonia (NH_3), or hydrogen fluoride (HF). Water's uniqueness is due to its ability to form strong hydrogen bonds with other water molecules or other electronegative or -positive ions or molecules. Water's physical and chemical properties are one of the determining factors for life and living processes.

Properties of Water. These determining factors become more obvious when water's biological and chemical properties are examined in terms of their biological significance. For example, the density of ice is less than the density of liquid water between 0° and 4° C. The reason is that a given

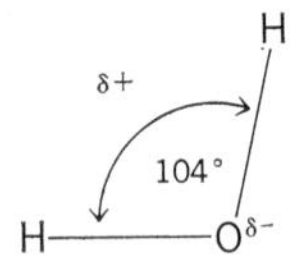

Fig. 4–2.

weight of water contracts in volume when heated from 0° to 4° C due to a partial collapse of hydrogen bonds of the ice crystal. The significance of this property is that ice remains on the surface rather than sinking, thereby acting as an insulating layer and holding the temperature of a lake or stream fairly constant. Also, the ice will melt readily at the first warming trend. If ice sank, it would not melt until the water itself began to warm up, and would require considerable energy due to water's high specific heat. Specific heat is that quantity of heat, expressed in calories, required to raise 1 g of substance 1° C. The specific heat of water at 14.5° C is 1 calorie. In other words, 1 calorie of energy is required to raise the temperature of 1 g of water from 14.5° C to 15.5° C. This property makes water a good insulator and is one of the reasons warm-blooded animals are able to maintain their rather constant internal body temperature.

Water has a high dielectric constant, which is extremely important to its capacity as a solvent and stabilizer of structure. The dielectric constant is related to the separation of charge within the water molecule as a result of its nonlinear "bent" shape (angle of 104°). The negative center may be considered as being localized at the oxygen end of the molecule and the electron-deficient (positive) center at the hydrogen end (Fig. 4–2).

The combination of water being the predominant constituent of the body (60–70%), having a high dielectric constant, being a small molecule, and readily forming hydrogen bonds makes water the solvent of choice in most liquid dosage forms. Because it is the predominant constituent of the body, problems of toxicity, metabolism, and excretion of the solvent do not normally have to be considered when water is used as a solvent. Having a high dielectric constant allows water to solvate ions readily and to stabilize macromolecules (proteins, etc.). On the other hand this solvation may be undesirable, as water can catalyze certain reactions and cause stability problems. Two common pharmaceutical examples are the hydrolysis of aspirin to salicylic and acetic acids and the oxidation of ferrous (Fe^{+2}) cation to ferric (Fe^{+3}) cation. The small molecular size allows water to penetrate into the free space within a crystal and to break the ionic bonds. Large solvent molecules, which are not able to penetrate into the crystal's free space, can attack only on the surface of the crystal.

From a chemical point of view, water may be considered unreactive. It barely dissociates [rx (lxxix)]. It is a very poor reducing agent.

$$2H_2O \; \rightleftharpoons \; H_3^+O + OH^- \quad K_w = 1.008 \times 10^{-14} \; 25° C \qquad \text{(lxxix)}$$

Water can oxidize any element able to displace one or both of its protons as hydrogen gas. The rusting of iron is an example [rx (lxxx)]:

$$2Fe + 3H_2O \rightarrow Fe_2O_3 + 3H_2 \uparrow \qquad \text{(lxxx)}$$

Water can act as an acid [rx (lxxxi)] or a base [rx (lxxxii)]:

$$\underset{\text{Base}}{Na^+ + {}^-O-\overset{\displaystyle O}{\overset{\|}{C}}-CH_3} + \underset{\text{Acid}}{HOH} \;\rightleftharpoons\; Na^+ + OH^- + HO-\overset{\displaystyle O}{\overset{\|}{C}}-CH_3 \quad \text{(lxxxi)}$$

$$\underset{\text{Acid}}{CH_3-\overset{\displaystyle O}{\overset{\|}{C}}-OH} + \underset{\text{Base}}{HOH} \;\rightleftharpoons\; CH_3-\overset{\displaystyle O}{\overset{\|}{C}}-O^- + H_3O^+ \quad \text{(lxxxii)}$$

Pure water is a tasteless, odorless, clear liquid which is colorless in small quantities, but greenish blue in deep layers. When the general public mentions the taste of drinking water, it is additives either from nature or man that give water flavor. For centuries mineral spas or baths were thought to have certain health-giving properties. There is no basis for these beliefs; actually, all natural water is in a sense mineral water due to suspended or colloidal solids such as dirt, silt, and rust, bacteria, and dissolved electrolytes producing the following ions: Na^+, K^+, Ca^{+2}, Mg^{+2}, Fe^{+3}, Sr^{+2}, Ba^{+2}, HCO_3^-, CO_3^{-2}, Cl^-, SO_4^{-2}, NO_3^-, etc. Mineral waters have been designated according to their most important medicinal constituent. The following is a representative list:

Alkaline waters usually contain appreciable quantities of sodium and magnesium sulfates, together with some sodium bicarbonate. Apollinaris, Vichy, and the waters from the Capon Springs (W. Va.) are examples.

Carbonated waters are those which have been charged with carbon dioxide under pressure while in the earth. They usually effervesce upon coming to the surface. Such waters contain calcium and magnesium bicarbonates. Springs in Colorado and Yellowstone Park (Wyoming) yield waters of this type. Artificial carbonated waters may be made by charging waters under pressure with carbon dioxide.

Chalybeate waters contain iron in solution or in suspension and are characterized by a ferruginous taste. Upon exposure to the atmosphere the iron is usually precipitated as hydroxide or oxide. Spring and well waters containing iron are very common.

Lithia waters, as a rule, do not contain appreciable quantities of lithium ions. If present, it occurs in the form of the carbonate or chloride.

Saline waters are sometimes called "purgative waters" and contain relatively large amounts of magnesium and sodium sulfates with sodium chloride. Springs located at Saratoga Springs, N.Y., and the Blue Lick Springs in Kentucky are examples of these waters.

Sulfur waters contain hydrogen sulfide. These waters (e.g., the waters from White Sulphur Springs, W. Va. and Richfield Springs, N.Y.) deposit sulfur upon exposure to the atmosphere.

Siliceous waters occur in Yellowstone Park and Iceland and contain very small quantities of soluble alkali silicates.

Hardness of Water. The dissolved impurities (including minerals) in all ordinary potable waters amount to very little, usually less than 0.1 of 1%. One of the chief factors that determines the value of a water for domestic and commercial purposes is its *hardness*. This property is occasioned by the presence in solution of varying amounts of calcium, iron, and/or magnesium salts, which convert ordinary soap (water soluble sodium and/or potassium salts of high molecular weight fatty acids) into water insoluble calcium, iron, and/or magnesium salts of the fatty acids [rx (lxxxiii)]:

$$2CH_3(CH_2)_xCO_2Na + CaSO_4 \rightarrow [CH_3(CH_2)_xCO_2]_2Ca \downarrow + Na_2SO_4 \quad \text{(lxxxiii)}$$

Water soluble soap	Hardening agent	Water insoluble soap

x = 14 (Palmitate)

x = 16 (Stearate)

Only when all the hardening substances have been precipitated as a water insoluble curd by the above mechanism will the soap begin to lather. It is apparent that hard water is undesirable in many respects and, therefore, methods of removing hardness have received much attention. Water may possess temporary or permanent hardness or both.

Temporary hardness (bicarbonate hardness) is caused by the presence in the water of soluble calcium or magnesium bicarbonates and can be removed (softened) by boiling or addition of a source of hydroxide. These bicarbonates are formed by the action of water charged with carbon dioxide percolating through limestone deposits to cause the following reaction to occur [rx (lxxxiv)]:

$$CaCO_3 + H_2O + CO_2 \rightarrow Ca(HCO_3)_2 \quad \text{(lxxxiv)}$$

Water insoluble	Water soluble

Permanent hardness is caused by the presence in solution of the sulfates, chlorides, or hydroxides of calcium and/or magnesium. These objectionable salts cannot be removed by boiling or addition of a source of hydroxide.

Temporary hard water may be softened by the following procedures:

1. *Boiling.* The carbon dioxide that has held the insoluble calcium and magnesium carbonates in solution as bicarbonates is driven off by boiling and the insoluble normal carbonate is precipitated [rx (lxxxv)]:

$$Ca(HCO_3)_2 + heat \rightarrow CaCO_3 \downarrow + H_2O + CO_2 \uparrow \qquad (lxxxv)$$

This method, of course, is not especially satisfactory when large volumes of "softened" water are needed. It is also the reason for the development of scale on boilers, pipes, tea pots, etc.

2. *Clark's Lime Process.* Clark, in 1941, suggested that slaked lime, in quantities just sufficient to react with the bicarbonate ion, be added to the water, thus precipitating the normal carbonate [rx (lxxxvi)]:

$$Ca(HCO_3)_2 + Ca(OH)_2 \rightarrow 2CaCO_3 \downarrow + 2H_2O \qquad (lxxxvi)$$

Care must be taken not to add too much slaked lime since this will impart a new, permanent hardness to the water which is more difficult to remove than the original.

3. *Addition of Soluble Alkali Carbonates or Hydroxides.* Sodium carbonate when added to temporary hard water will precipitate calcium carbonate [rx (lxxxvii)]:

$$Ca(HCO_3)_2 + Na_2CO_3 \rightarrow CaCO_3 \downarrow + 2NaHCO_3 \qquad (lxxxvii)$$

This manner of softening water was, at one time, familiar to the housewife through the use of washing soda $(Na_2CO_3 \cdot 10H_2O)$ as a water softener in the laundry and is returning to use as a replacement of phosphates in detergents. However, due to its high alkalinity in solution (pH 11.6), the detergent label must carry a warning statement if the detergent contains sodium carbonate.

By furnishing hydroxide ions, sodium hydroxide, or borax (sodium borate), on the other hand, converts the bicarbonates to slaked lime, $Ca(OH)_2$, [rx (lxxxviii)] which reacts with the calcium or magnesium bicarbonate still remaining [rx (lxxxvi)]. The mechanism by which sodium

$$2NaOH + Ca(HCO_3)_2 \rightarrow Ca(OH)_2 + 2NaHCO_3 \qquad (lxxxviii)$$

borate forms hydroxide ions is shown in rx (lxxxix) and rx (xc) and summarized in rx (xci):

$$Na_2B_4O_7 + 3H_2O \rightleftarrows 2NaBO_2 + 2H_3BO_3 \qquad (lxxxix)$$

$$2NaBO_2 + 4H_2O \rightleftarrows 2NaOH + 2H_3BO_3 \qquad (xc)$$

$$Na_2B_4O_7 + 7H_2O \rightleftarrows 2NaOH + 4H_3BO_3 \qquad (xci)$$

4. *Addition of Ammonia.* Household ammonia is also used for softening water [rx (xcii)].

$$Ca(HCO_3)_2 + 2NH_3 \rightarrow CaCO_3 \downarrow + (NH_4)_2CO_3 \qquad \text{(xcii)}$$

5. *Polyphosphate Chelation.* The detergent industry has been using phosphates and metaphosphate polymers for years. These soften water by two methods. The hydroxide anions [generated by the basic phosphate (PO_4^{-3}) anion] remove calcium bicarbonate by forming calcium hydroxide [rx (lxxxviii)] which then reacts with more calcium bicarbonate to form a calcium carbonate precipitate [rx (lxxxvi)]. The sodium metaphosphate $(NaPO_3)_n$ or Graham's salt chelates the divalent cation, thus making it unavailable for any further reaction.

6. *Chelation by the Zeolite (Permutit) Process.* Artificial zeolites were introduced by Gans (1910) as a means of softening both temporary and permanently hard waters. The process is based on an ion exchange reaction for softening water. Zeolite is a sodium aluminum silicate (said to be $Na_2H_6Al_2Si_2O_{11}$), and may be simply represented by the symbol, Na_2Zeol. By passing the hard water through a zeolite column, an exchange of "water-hardening" cations is made for "nonhardening" sodium cations [rx (xciii)]:

$$Na_2Zeol + Ca(HCO_3)_2 \rightarrow CaZeol + 2NaHCO_3 \qquad \text{(xciii)}$$

When the calcium zeolite reaches a high enough concentration, the column can no longer exchange calcium ion for sodium ion efficiently. To restore its activity a strong solution of sodium chloride is allowed to flow through the inactivated zeolite, reconverting it to the sodium form [rx (xciv)]:

$$CaZeol + 2NaCl \rightarrow Na_2Zeol + CaCl_2 \qquad \text{(xciv)}$$

7. *Deionized or Demineralized Water.* One of the most common methods for softening both types of hard water is an outgrowth of the desire to obtain water approximating distilled water in purity without going through the wasteful and expensive distillation procedure. Water may be deionized for one to ten percent of the cost of preparing it by distillation. Practically all the procedures previously mentioned leave some chemical in the water, although the water may be "softened" in the sense that fatty acids will not precipitate as insoluble salts. The presence of salts in water is undesirable in many manufacturing processes. The development of commercial *resinous ion exchangers* in 1935 made possible the removal of both cations and anions from water. The removal of salts from water by this process consists of two steps. The first step is a removal of the cations by passing the water through a hydrogen exchange resin (HResin or cation exchange

resin) which converts any salt to the corresponding acid by giving up a hydrogen ion in exchange for the metal ion [rx (xcv)]:

$$2HResin + Ca(HCO_3)_2 \rightarrow Ca(Resin)_2 + 2H_2CO_3 \ (H_2O + 2CO_2 \uparrow) \quad (xcv)$$

The second step accomplishes the removal of all anions (now in the form of acids) by passage of the water through another resin (usually an amine-formaldehyde resin or anion exchange resin) whereby the —NH₂ groups neutralize the acids [rx (xcvi)]:

$$Resin \ NH_2 + HCl \rightarrow Resin—NH_3^+ + Cl^- \qquad (xcvi)$$

Usually the process is carried out with four columns (two of each type resin, alternately and in series) because the highly ionized cations (Na^+ and K^+) are removed first, as are the anions of the highly ionized acids (Cl^- and SO_4^{-2}). Incompletely ionized carbonic acid is not completely removed in a single passage through the anion resin bed, because the weak acids are not removed until after the strong acids have been eliminated.

The water which finally issues from the deionizer compares very favorably with distilled water in purity and can be produced at a lower cost. Indeed, the U.S.P. XV modified the name and monograph of the traditional *Distilled Water* to permit the use of deionized water (see below) under the title of *Purified Water*. Ion exchange resins will only remove ionic impurities such as sodium, potassium, calcium, magnesium, chlorides, sulfates, bicarbonates, carbonates, hydroxides, etc. Demineralized water is not suitable when bacteria, pyrogens, organic impurities, etc. are objectionable. When sterility is necessary in pharmacies, hospitals, and biological plants, deionized water is not recommended. Nevertheless, distilled

TABLE 4–11. COMPARISON OF THE EFFECTIVENESS OF DISTILLATION AND DEMINERALIZATION PROCEDURES

	Distillation	*Demineralization*
Strong Electrolytes	Good	Good (one stage can equal several distillations)
Weak Electrolytes	Poor	Good (even silicates, carbonates, borates, and phenolates can be removed)
Nonelectrolytes	Poor, depends on volatility	Ineffectual
Sterility	Sterile (pyrogen-free after repeated distillation)	Possible, but requires care

From Modern Inorganic Pharmaceutical Chemistry, by C. A. Discher, John Wiley and Sons, New York, p. 146, 1964. By permission.

water is no guarantee of bacteria-free water. It has been demonstrated that *Pseudomonas aeruginosa*, a major causative agent of hospital-acquired infections, can grow relatively quickly in distilled water obtained in hospitals and achieve high cell contaminations which remain stable for long periods of time.[8] Water must be distilled several times before it can be considered sterile. Table 4–11 summarizes the advantages and disadvantages of distillation versus demineralization.

Permanent hard water may be softened by the following procedures:

1. *Addition of Soluble Carbonates.* By adding soluble carbonates (e.g., Na_2CO_3 or washing soda) to the hard water the insoluble carbonates of calcium and magnesium are precipitated [rx (xcvii)]:

$$MgSO_4(\text{or } CaSO_4) + Na_2CO_3 \rightarrow MgCO_3(\text{or } CaCO_3) \downarrow + Na_2SO_4 \quad (xcvii)$$

2. *Polyphosphate Chelation.* The chemistry is the same as that of temporary hard water.

3. *Zeolite Process.* As previously indicated, this process applies equally well to temporary and permanent hard waters [rx (xcviii)]:

$$Na_2Zeol + CaSO_4 \rightarrow CaZeol + Na_2SO_4 \quad (xcviii)$$

4. *Deionized or Demineralized Water.* The resinous ion exchangers also soften both types of water, inasmuch as all anions and cations are removed irrespective of the salts. The removal of $CaSO_4$ from permanently hard water is carried out by the typical two-step mechanism [rxs (xcix) and (c)]:

$$2HResin + CaSO_4 \rightarrow Ca(Resin)_2 + H_2SO_4 \quad (xcix)$$

$$2Resin\ NH_2 + H_2SO_4 \rightarrow (Resin\ NH_3^+)_2 \cdot SO_4^{-2} \quad (c)$$

Because of the importance in pharmaceutical manufacturing, compounding, and testing water, the U.S.P. has defined five grades of water ranging from ordinary pure water to water suitable for intravenous injection. Their purity requirements are summarized in Table 4–12.

Official Waters

Water, U.S.P. XVIII (H_2O; Mol. Wt. 18.02)

Water U.S.P. is a clear, colorless, odorless liquid which has specifications for pH, zinc, other heavy metals, foreign volatile matter, total solid content, and bacteriological purity. It is official as a solvent and is used to make several official solutions, tinctures, and extracts.

Purified Water U.S.P. XVIII

Purified Water U.S.P. is water obtained by distillation or by ion exchange treatment. It occurs as a clear, colorless, odorless liquid and *is not intended*

TABLE 4-12. PURITY REQUIREMENTS FOR OFFICIAL WATERS

	pH	Oxidizable Substances	Total Solids	Bacteriological Purity
Water U.S.P. XVIII	5–8	None	1000 ppm	U.S.P.H.S. requirements
Purified Water U.S.P. XVIII	5–7	Pink color remains in 100 ml containing 0.1 ml of 0.1 N KMnO$_4$	10 ppm	U.S.P.H.S. requirements
Water for Injection, U.S.P. XVIII	5–7	Pink color remains in 100 ml containing 0.1 ml of 0.1 N KMnO$_4$	10 ppm	Pass pyrogen test
Bacteriostatic Water for Injection, U.S.P. XVIII	4.5–7	No test due to added bacteriostatic additive	40 ppm corrected for bacteriostatic additive	Pass pyrogen test
Sterile Water for Injection, U.S.P. XVIII	5–7	Pink color remains in 100 ml containing 0.2 ml of 0.1 N KMnO$_4$	40 ppm: 30-ml vial or smaller; 30 ppm: 30- to 100-ml vial; 20 ppm: container larger than 100 ml	Pass pyrogen test

for parenteral administration. It is called for in the preparation of most U.S.P. test reagents. There are specifications for pH, chloride, sulfate, ammonia, calcium, carbon dioxide, heavy metals, oxidizable substances, total solids, and bacteriological purity. It must be labeled to indicate the method of preparation. It is the water of choice for extemporaneous compounding.

Water for Injection, U.S.P. XVIII

Water for Injection U.S.P. is water purified by distillation and occurs as a clear, colorless, odorless liquid. It contains no added substance. It meets all the specifications called for in the *Purified Water* monograph with the exception of bacteriological purity. Instead, *Water for Injection* must pass a pyrogen test. *Water for Injection is intended for use as a solvent for the preparation of parenteral solutions.* The finished preparation should then be sterilized. *For parenteral solutions that are prepared under aseptic conditions and are not sterilized by appropriate filtration or in the final container, first render the Water for Injection sterile and thereafter protect it from microbial contamination.* The U.S.P. requires that *Water for Injection* be stored at a temperature below 4° C or above 37° C, the range in which microbial growth occurs. *Water for Injection* would be used by the large scale pharmaceutical manufacturers. A pharmacy preparing extemporaneous parenterals would more likely use *Bacteriostatic Water for Injection* or *Sterile Water for Injection.*

Bacteriostatic Water for Injection, U.S.P. XVIII

Bacteriostatic Water for Injection U.S.P. is sterile water for injection containing one or more suitable antimicrobial agents and occurs as a clear, colorless liquid, odorless or having the odor of the antimicrobial substance. The U.S.P. warns: *Use Bacteriostatic Water for Injection with due regard for the common compatibility of the antimicrobial agent or agents it contains with the particular medicinal substance that is to be dissolved or diluted.* Benzyl alcohol is a common bacteriostatic agent. Most of the specifications such as pH, chloride, total solids, and oxidizable substances have been modified to make allowance for the bacteriostatic agent. It does have specifications for antimicrobial agents, pyrogens, and sterility. *Bacteriostatic Water for Injection* is stored in single-dose or in multiple-dose containers, preferably of Type I or Type II glass, of not larger than 30-ml size. The antimicrobial agent or agents and the concentrations must be on the label.

Bacteriostatic Water for Injection was developed in response to a need for easily stored sterile water that could be used to compound small volumes of extemporaneous parenterals for intramuscular injection. The bacteriostatic agents prevent the use of this water for intravenous administration.

Sterile Water for Injection, U.S.P. XVIII

Sterile Water for Injection U.S.P. is water for injection sterilized and suitably packaged. It is a clear, colorless, odorless liquid which contains no antimicrobial agent. This water is the most difficult to prepare. It would appear from examination of Table 4–12 that the specification for oxidizable substances and total solids is less strict as compared with *Water, Purified Water,* and *Water for Injection.* This is a good example of a case where manufacturing limitations have required the modification of a monograph. By official definition, *Water for Injection* has been distilled; this would mean that the specifications for oxidizable substances and total solids should be minimal. The same argument would hold for *Purified Water* prepared by distillation. For purified water prepared by ion exchange treatment, the specifications for oxidizable substances and total solids serve as a check for traces of resin or bacterial contamination that may have been on the resin bed. In contrast, the tests in *Sterile Water for Injection* serve as a check for material extracted from the closures during sterilization and storage. The reason for the range of total solids permitted as an inverse function of volume is that the closure surface-to-volume ratio increases as the volume of the container decreases.[9]

Sterile Water for Injection may be stored in single-dose containers, preferably of Type I or Type II glass, of not larger than 1000-ml size. The label must indicate that no antimicrobial or other substance has been added, and that it is not suitable for intravascular injection without its first having been made approximately isotonic by the addition of a suitable solute. *Sterile Water for Injection* is used for the extemporaneous compounding of parenterals for either intravenous or intramuscular injection.

Pharmaceutically Acceptable Glass

The chemistry of glass is extremely complicated. The word "glass" is a generic term referring to vitreous material (material which softens gradually over a temperature range rather than melting sharply). Most commercial glasses are vitreous silicates with some type of additive which confers special properties. Boron decreases the coefficient of expansion in Pyrex® glass. Potassium gives a brown light-resistant glass. The rare earths selectively absorb light of certain wavelengths.

Glass may be considered as sodium silicate (Na_4SiO_4). Aqueous solutions will slowly become alkaline upon standing for prolonged times in soft glass containers [rx (ci)].

$$4Na^+ + SiO_4^{-4} + 3H_2O \rightarrow H_3SiO_4^- + 4Na^+ + 3OH^- \qquad \text{(ci)}$$

The compendia usually specify the type of glass container or a suitable buffer if the drug is base-sensitive. The waters for injection must be stored in glass of Types I or II as rx (ci) is greatly speeded up during heat sterilization. Both the U.S.P. and N.F. define the same four glass types and their alkalinity limits. Table 4–13 summarizes this information.

TABLE 4-13. GLASS TYPES AND TEST LIMITS

Type	General Description*	Type or Test	Limits Size,† ml	Limits 0.02 N Acid, ml
I	Highly resistant, borosilicate glass	Powdered Glass	All	1.0
II	Treated soda-lime glass	Water Attack	100 or less	0.7
			Over 100	0.2
III	Soda-lime glass	Powdered Glass	All	8.5
NP	General purpose soda-lime glass	Powdered Glass	All	15.0

* The description applied to containers of this type of glass which are usually available.
† Size indicates the overflow capacity of the container; U.S.P. XVIII, p. 925; N.F. XIII, p. 796.

Certain drugs must be protected from light. There are numerous types of containers, colorless, opaque, and colored, available for packaging. The opaque container would be the superior light-protective container, but, except for ointments, clear containers are preferred for dispensing, if for no other reason than that the patient likes to see what he is buying, the amount left in the bottle, and ease of pouring.

There are four types of clear containers available: colorless, green, blue, and amber. Since most light-catalyzed degradation is due to ultraviolet radiation, this is the area of the electromagnetic spectrum that must be excluded. For the most part blue glass does not confer much protection. Green glass may or may not, depending upon the type and intensity of green. Amber glass will usually screen out ultraviolet radiation very effectively and is usually the color recommended for protection from light. Both the U.S.P. and N.F. have identical standards for percent transmission of light in the 290 to 450 nm region, dependent upon the thickness of the glass. It is assumed that radiation below 290 nm has been screened out by the atmosphere.

References

1. Vanderwerf, C. A. Acids, Bases, and the Chemistry of the Covalent Bond. New York: Reinhold Publishing, 1961.
2. Pearson, R. G. The principle of hard and soft acids and bases. J. Amer. Chem. Soc., **85**:3533, 1963.
3. Pearson, R. G. Hard and soft acids and bases, HSAB, Part I. J. Chem. Ed., **45**:581, 1968.
4. Connors, K. A. A Textbook of Pharmaceutical Analysis. New York: John Wiley and Sons, p. 13, 1967.
5. Gearien, J. E., and Grabowski, B. F.: Methods of Drug Analysis, Philadelphia: **Lea & Febiger**, p. 7, 1969.

6. Edwards, J. O., Morrison, G. C., Ross, V. F., and Schultz, J. W.　The structure of the aqueous borate ion.　J. Amer. Chem. Soc., **77**:266, 1955.
7. Wolff, I. A., and Wasserman, A. E.　Nitrates, nitrites, and nitrosamines.　Science, **177**:15, 1972.
8. Favero, M. S., Carson, L. A., Bond, W. W., and Petersen, N. J.　*Pseudomonas aeruginosa:* growth in distilled water from hospitals.　Science, **173**:836, 1971.
9. Morecomb, F. A.　U.S.P. Convention, Inc.　Personal communication to John H. Block, 1967.

5
Major Intra- and Extracellular Electrolytes

"If possible, all additions of drugs to parenteral solutions should be made by a pharmacist who has been specially trained to perform this function."[1]

The body's fluids are solutions of inorganic and organic solutes. The concentration balances of the various components are maintained in order for the cells and tissues to have a constant environment. In order for the body to maintain this internal homeostasis, there are regulatory mechanisms which control pH, ionic balances, osmotic balances, etc. There are also a large number of products under the general heading of replacement therapy which can be used by the physician when the body itself is unable to correct an electrolyte imbalance due to a change in the composition of its fluids. These products include electrolytes, acids and bases, blood products, carbohydrates, amino acids, and proteins. Electrolytes used for replacement and acid-base correction therapy will be discussed in this chapter. Cations with a very specific biochemical function or found only in trace amounts will be described in the following chapter. Table 5–1 summarizes the inorganic ions associated with body metabolism.

The electrolyte concentration will vary with a particular fluid compartment. The three compartments are: (1) intracellular fluid (45–50% of body weight); (2) interstitial fluid (12–15% of body weight); and (3) plasma or vascular fluid (4–5% of body weight). The term "extracellular fluid" includes both interstitial and vascular fluids. These three compartments are separated from each other by membranes that are permeable to water and many organic and inorganic solutes. They are nearly impermeable to macromolecules such as proteins, and are selectively permeable to certain ions such as Na^+, K^+, and Mg^{++}. The end result is that each fluid compartment has a distinct solute pattern (see Table 5–2). Also, the solution in each compartment is ionically balanced. Thus, sodium and chloride are found in the plasma and interstitial fluids while potassium, magnesium, and phosphate (as phosphate esters, HPO_4^{-2}, and $H_2PO_4^-$) are found in the intracellular fluid.

TABLE 5-1. INORGANIC IONS ASSOCIATED WITH
BODY METABOLISM.[a]

Element[b] and Total Amount in Human Body	Best Food Sources	RDA[c] 1968	Absorption and Metabolism	Principal Metabolic Functions	Clinical Manifestations of Deficiency
Sodium (Na^+) 1.8 g/kg	Table salt, salty foods, animal foods, milk, baking soda, baking powder, some vegetables	About 3–5 g[d]	Readily absorbed, extracellular, excreted in urine and sweat; aldosterone increases reabsorption in renal tubules	Buffer constituent, acid-base balance, water balance, osmotic pressure, CO_2 transport, cell membrane permeability, muscle irritability	Dehydration; acidosis; tissue atrophy; excess leads to edema, hypertension
Potassium (K^+) 2.6 g/kg	Vegetables, fruits, whole grains, meat, milk, legumes	About 1.5–4.5 g[d]	Readily absorbed, intracellular; excreted by kidney	Buffer constituent, acid-base balance, water balance, CO_2 transport, membrane transport, neuromuscular irritability	Acidosis; renal damage
Calcium (Ca^{++}) 22 g/kg	Milk, milk products, fish bones (cooked)	0.8 g	Poorly absorbed (20%–40%) according to body need; absorption aided by vitamin D, lactose, acidity; hindered by excess fat, phytate, oxalate; excreted in feces; parathyroid hormone mobilizes bone Ca^{++}	Formation of apatite in bones, teeth; blood clotting; cell membrane permeability; neuromuscular irritability	Rickets (child), poor growth; osteoporosis (adult); hyperexcitability
Phosphorus (PO_4^{-3}) 12 g/kg	Milk, milk products, egg yolk, meat, whole grains, legumes, nuts	0.8 g	Readily absorbed; excreted by kidney	Constituent of bones, teeth; constituent of buffers; constituent of ATP, NAD, FAD, etc.; constituent of metabolic intermediates, nucleoproteins, phospholipids, phosphoproteins	Osteomalacia (rare); renal rickets; cardiac arrhythmia

TABLE 5–1.   INORGANIC IONS ASSOCIATED WITH
BODY METABOLISM (continued)

Element[b] and Total Amount in Human Body	Best Food Sources	RDA[c] 1968	Absorption and Metabolism	Principal Metabolic Functions	Clinical Manifestations of Deficiency
Magnesium (Mg^{++}) 0.5 g/kg	Chlorophyl, nuts, legumes, whole grains	350 mg	Absorbed; competes with Ca^{++} for transport	Cofactor for PO_4-transferring enzymes; constituent of bones, teeth; decreases neuromuscular irritability	Magnesium-conditioned deficiency, muscular tremor, choreiform movements, confusion; vasodilatation, hyperirritability
Iron (Fe^{++} or Fe^{+++}) 75 mg/kg	Liver, meats, egg yolk, green leafy vegetables, whole grains, enriched bread and cereals	10 mg male; 18 mg female	Absorbed according to body need; aided by HCl, ascorbic acid	Constituent of hemoglobin, myoglobin, catalase, ferredoxin, cytochromes; electron transport, enzyme cofactor	Anemia, hypochromic; pregnancy demands; excess $\rightarrow$ hemochromatosis
Iodine (I^-)	Seafoods, iodized salt	140 μg male; 100 μg female	Concentrates in thyroid; transported as PBI	Constituent of thyroxin, triiodothyronine; regulator of cellular oxidations	Endemic (simple) goiter (hypothyroidism); cretinism
Zinc (Zn^{++}) 28 mg/kg	Liver, pancreas, shellfish; widely distributed in animal and plant tissue	10–15 mg[d]	1–2 mg absorbed; phytate decreases absorption	Constituent of insulin, carbonic anhydrase, carboxypeptidase, lactic dehydrogenase, alcohol dehydrogenase, alkaline phosphatase	Anemia; stunted growth; hypogonadism in male
Copper (Cu^{++}) 2 mg/kg	Liver, kidney, egg yolk, whole grains	2–3 mg[d]	Limited absorption; transport by ceruloplasmin; stored in liver; excretion via bile	Formation of hemoglobin (increases iron utilization); constituent of oxidase enzymes (tyrosinase, cytochrome oxidase, ascorbic acid oxidase)	Hypochromic anemia; excessive hepatic storage in Wilson's diesase

TABLE 5–1. INORGANIC IONS ASSOCIATED WITH
BODY METABOLISM (continued)

Element[b] and Total Amount in Human Body	Best Food Sources	RDA[c] 1968	Absorption and Metabolism	Principal Metabolic Functions	Clinical Manifestations of Deficiency
Cobalt (Co^{++}) 3 mg	Liver, pancreas, mushrooms	1–2 mg[d]	Limited absorption; stored in liver; excretion via bile	Constituent of vitamin B_{12}	Anemia in animals; deficiency as vitamin B_{12} → pernicious anemia; excess → polycythemia
Manganese (Mn^{++}) 20 mg	Liver, kidney, wheat germ, legumes, nuts	3–9 mg[d]	Stored in liver, mitochondria, and bone; excreted via bile	Cofactor for number of enzymes— arginase, carboxylase, kinases, etc.	Unknown in man; in animals → decreased glucose tolerance, perosis, congenital ataxia
Molybdenum (Mo) 5 mg	Liver, kidney, whole grains, legumes, leafy vegetables	Trace[d]	Readily absorbed; excreted in urine and bile	Constituent of xanthine oxidase, aldehyde oxidase	Unknown
Chromium (Cr^{+++})	Liver, animal and plant tissue	Trace[d]		Involved in carbohydrate utilization	Unknown; deficiency in diabetes claimed; decreased glucose tolerance in rats; possible relation to cardiovascular disease
Selenium (Se)	Liver, kidney, heart	Trace[d]	Excreted in urine	Constituent of factor 3; acts with vitamin E to prevent liver necrosis and muscular dystrophy in animals; inhibits lipid peroxidation	Unknown; excess → alkali disease in cattle, sheep
Chloride (Cl^-) 50 mEq/kg	Animal foods, table salt	Intake 5–10 g as NaCl[d]	Rapid absorption; excreted in urine; high renal threshold; not stored	Electrolyte, osmotic balance; gastric HCl; acid-base balance	Hypochloremic alkalosis (pernicious vomiting)

TABLE 5–1. INORGANIC IONS ASSOCIATED WITH
BODY METABOLISM (continued)

Element[b] and Total Amount in Human Body	Best Food Sources	RDA[c] 1968	Absorption and Metabolism	Principal Metabolic Functions	Clinical Manifestations of Deficiency
Fluoride (F)	Seafoods, some drinking water	1 mg[d] (1 ppm in drinking water)	Easily absorbed; excreted in urine; deposited in bones and teeth	Constituent of fluoroapatite— tooth enamel	Dental caries, osteoporosis; excess (5–8 ppm in water) → mottled enamel
Sulfur (SO₄⁻²)	Plant and animal proteins as Cysteine and Methionine	2–3 g[d]	Derived from metabolism of Cys and Met; excreted in urine	Constituent of proteins, mucopolysaccharides, heparin, thiamine, biotin, lipoic acid; detoxication	Cystinuria; cystine renal calculi

[a] From Biochemistry, 8th ed., by J. M. Orten and O. W. Neuhaus, St. Louis: C. V. Mosby, pp. 436–439, 1970. By permission.

[b] The inorganic elements included are those for which evidence exists that they are *essential* for man. Other elements not included but present in the human body in trace amounts, for which there is fragmented evidence for some biochemical function, include cadmium, lithium, nickel, vanadium. Other elements present in human tissues in trace amounts as incidental constituents of no known significance include Ag, Au, Al, As, Br, Pb, Rb, Si, Ti, B. The amounts of the element present in the entire human body are averages from the literature (Dairy Council Digest, *39*, 26, 1968). They are expressed as grams or milligrams per kilogram of body weight (*fat-free basis*) or as milligrams in the entire body.

[c] Recommended dietary allowance (RDA) per day, established by the Food and Nutrition Board, National Research Council, 1968. The values given are for a normal adult male, 22 years of age.

[d] An estimated value is given if no RDA value has been established. The estimated value is the average daily dietary intake of a normal adult.

As noted from Table 5–2, the concentrations of individual ions are expressed by mEq/l (milliequivalents/liter) rather than by weight/volume (w/v). Dosages of individual ions are also expressed in mEq/l. The concentration of the salt supplying the required ion will usually be expressed in a w/v unit. It is imperative that a pharmacist be able to understand and to interconvert the mEq/l and w/v expressions with facility.

The equivalent weight is a means of expressing combining power. It is obtained by dividing the atomic or molecular weight by the valence. It is quite easy to calculate the mEq/l concentration from Eq. 1.

$$mEq/l = \text{mg of substance}/l \div Eq. \text{ wt.} = \qquad \text{(Eq. 1)}$$

$$\text{mg of substance}/l \div \frac{\text{Mol. wt.}}{\text{Eq./mole}}$$

TABLE 5–2.   FLUID ELECTROLYTE CONCENTRATIONS

Cations	Plasma	Interstitial Fluid	Intracellular Fluid
Na^+	142 mEq/l	145 mEq/l	10 mEq/l
K^+	4	4	160
Ca^{+2}	5	3	—
Mg^{+2}	3	2	35
Totals	154	154	205
Anions			
HCO_3^-	27 mEq/l	30 mEq/l	8 mEq/l
Cl^-	103	115	2
HPO_4^{-2}	2	2	140
SO_4^{-2}	1	1	—
Organic Acids	5	5	—
Protein	16	1	55
Totals:	154	154	205

From Fluid and Electrolytes: Some Practical Guides to Clinical Use.   Chicago: Abbott Laboratories, pp. 10–11.   By permission.

TABLE 5–3.   SIMPLIFIED METHOD OF CONVERSION BETWEEN MILLIGRAMS AND MILLIEQUIVALENTS*

Ion or Salt	To obtain mEq/liter from mg per liter† Divide the number of mg by	To obtain mg per liter from mEq/liter Multiply the number of mEq by
Na^+	23.0	23.0
K^+	39.1	39.1
Ca^{+2}	20.0	20.0
Mg^{+2}	12.2	12.2
HCO_3^-	61.0	61.0
Cl^-	35.5	35.5
HPO_4^{-2}	48.0‡	48.0‡
SO_4^{-2}	48.0‡	48.0‡
NaCl	58.5	58.5

* From Electrolyte solution calculations, by P. J. Wurdack and J. L. Colaizzi, J. Amer. Pharm. Ass., **NS10** (7): 413, 1970.   By permission.

† Conversion factors for multivalent ions or compounds already are based on their valences; thus, there is no need to divide or multiply any factor by the valence of the ion or compound again.

‡ Values for phosphate and sulfate have been calculated on the basis of the complete (HPO_4^{-2}) and (SO_4^{-2}) radicals, and not on the basis of phosphorus and sulfur.

Equation 2 permits the calculation of the weight of salt necessary to yield the required number of mEq.

$$\text{mg/liter} = (\text{mEq/l})\ (\text{Eq. wt.}) = (\text{mEq/l}) \left(\frac{\text{Mol. wt.}}{\text{valence}}\right) \quad \text{(Eq. 2)}$$

Table 5–3 gives the Eq. wt. for the common ions and salt.
Below are some sample problems.

Problem 1.

Calculate the number of mEq of NaCl in one liter of a 0.9% w/v solution.
0.9% NaCl = 9 g NaCl/l
Substituting into Eq. 1:

$$\text{mEq/l} = \frac{9000\ \text{mg}}{58.5} = 153.8\ \text{mEq NaCl/l}$$

Since the number of eq. of Na^+ must equal the number of equivalent of Cl^-, the concentrations of each ion are 153.8 mEq Na^+/l and 153.8 mEq Cl^-/l.

Problem 2.

Calculate the amount of salt necessary to make a solution that contains 153 mEq/l each of Na^+ and Cl^-.
Using Eq. 2:

$$\text{mg/l} = (153)\ (58.5) = 9000\ \text{mg/l} = 9\ \text{g/l}$$

Problem 3.

What is the weight of calcium chloride ($CaCl_2 \cdot 2H_2O$) needed to prepare a liter of solution containing 9 mEq Ca^{+2}/l?
Using Eq. 2:

$$\text{mg/l} = \frac{(9)\ (147)}{2} = 661.5\ \text{mg}$$

Problem 4.

What will be the chloride content of a solution containing 661.5 mg $CaCl_2 \cdot 2H_2O$/l?
Using Eq. 1:

$$\text{mEq/l} = \frac{(661.5)\ (2)}{147} = 9\ \text{mEq Cl}^-\text{/l}$$

Note that in Problems 3 and 4, the valence form is 2 because of divalent calcium. In other words, each mole of calcium chloride contains two chlorides.

Problem 5.

Compare the amount of total iodine and iodide in 0.3 ml of *Strong Iodine Solution* U.S.P. XVIII with the amount of total iodide in 0.3 ml of *Potassium Iodide Solution* N.F. XIII.

Strong Iodine Solution U.S.P. XVIII consists of:

$$\text{KI} \quad 100 \text{ g/l and}$$
$$\text{I}_2 \quad 50 \text{ g/l}$$

$$\text{mEq I}^-/\text{l (from KI)} = \frac{(100,000) \ (1)}{166} = 603 \text{ mEq I}^-/\text{l}$$

$$\text{mEq I/l (from I}_2) = \frac{(50,000) \ (2)}{254} = 393 \text{ mEq I/l}$$

Total mEq of iodine and iodide per liter = 603 + 393 = 996 mEq. Therefore a 0.3-ml dose will contain 0.2988 mEq of total iodine and iodide.

Potassium Iodide Solution N.F. XIII consists of:

$$\text{KI} \quad 1000 \text{ g/l}$$

$$\text{mEq I}^-/\text{l} = \frac{(1,000,000) \ (1)}{166} = 6030 \text{ mEq I}^-/\text{l}$$

Therefore, a 0.3-ml dose will contain 1.809 mEq I$^-$ and furnishes six times as much iodide (iodine) when used in thyroid therapy.

By calculating the dose of an ion in mEq, it is easier to compare and contrast two similar preparations.

Major Physiological Ions

Chloride. Chloride is the major extracellular anion and is principally responsible for maintaining proper hydration, osmotic pressure, and normal cation-anion balance in the vascular and interstitial fluid compartments. Because chloride is the major extracellular anion, chloride salts are a preferred dosage form providing, of course, other pharmaceutical and toxicological criteria are met.

Food is the main source of chloride, with the anion being almost completely absorbed from the intestinal tract. Chloride is removed from the blood by glomerular filtration and possibly is reabsorbed by the kidney tubules (see the discussions on kidney and acid-base balance). Hypochloremia can be caused by: (1) salt-losing nephritis (inflammation of the kidney) associated with chronic pyelonephritis (inflammation of the kidney and its pelvis), leading to a probable lack of tubular reabsorption of chloride; (2) metabolic acidosis such as found in diabetes mellitus and renal failure, causing either excessive production or diminished excretion of acids leading to the replacement of chloride by acetoacetate and phosphate; and (3) prolonged vomiting with loss of chloride as gastric hydrochloric acid.

Hyperchloremic conditions are seen in dehydration, decreased renal blood flow found with congestive heart failure, severe renal damage, and excessive chloride intake.

The chloride ion, as such, has practically no pharmacological activity. When administered as the ammonium salt it may be used as a urinary acidifier.

Phosphate. Phosphate (as HPO_4^{-2}) is the principal anion of the intracellular fluid compartment. Its biochemistry is very complex and a detailed discussion is beyond the scope of this book. However, it can be summarized as follows:

1. Hexoses are metabolized as phosphate esters.
2. The phosphoric acid anhydride linkage is the body's means of storing potential chemical energy as adenosine triphosphate (ATP).
3. The $HPO_4^{-2}/H_2PO_4^-$ is an important buffer system, both biochemically (see physiological acid-base balance discussion) and pharmaceutically (see Chapter 4).
4. Phosphorus is essential for proper calcium metabolism (see discussion of calcium metabolism).
5. Phosphorus is essential for normal bone and tooth development since it is a component of hydroxyapatite, the main calcium salt found in bones and teeth (see Chapter 11).

Phosphates are not considered to have any specific pharmacological action. Thus, many drugs in which the cation is the active ion are administered as phosphate salts providing other pharmaceutical criteria are met. In those cases where phosphate therapy is necessary, oral administration can be difficult. Only the dihydrogen phosphate ($H_2PO_4^-$) anion will be absorbed from the intestines. Phosphate is found in this form only in the acid stomach and in the upper part of the duodenum. At pH 6.8 it will exist as a 1:1 ratio of $HPO_4^{-2}/H_2PO_4^-$. Because the monohydrogen phosphate anion is so poorly absorbed, its salts find use as saline cathartics (see Chapter 8).

The nomenclature of the phosphate acids and their salts is confusing. Most phosphate salts of pharmaceutical concern and phosphate esters of biochemical interest are derived from phosphoric acid, commonly represented as H_3PO_4, but more accurately as $PO(OH)_3$ (see structural formula, below). This acid is also known as orthophosphoric acid. The prefix

$$
\begin{array}{c}
O \\
\uparrow \\
HO-P-OH \\
| \\
OH
\end{array}
$$

"ortho" indicates the most highly "hydroxylated" known form of the acid (or its salt or ester).

Metaphosphoric acid refers to the product resulting when a molecule of orthophosphoric acid *intra*molecularly loses the equivalent of one molecule of water [rx (i)]:

$$H_3PO_4 \rightarrow HPO_3 + H_2O \tag{i}$$

Structurally metaphosphoric acid is represented as:

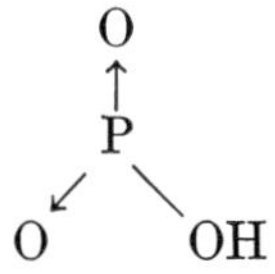

$$
\begin{array}{c}
O \\
\uparrow \\
P \\
\swarrow \quad \searrow \\
O \qquad OH
\end{array}
$$

Sodium metaphosphate (Graham's Salt), a polymer, is a commonly used water softening agent (see Chapter 4).

Pyrophosphoric acid refers to phosphoric acid anhydride [rx (ii)]:

$$2H_3PO_4 \rightarrow H_4P_2O_7 + H_2O \tag{ii}$$

Structurally, pyrophosphoric acid can be represented as:

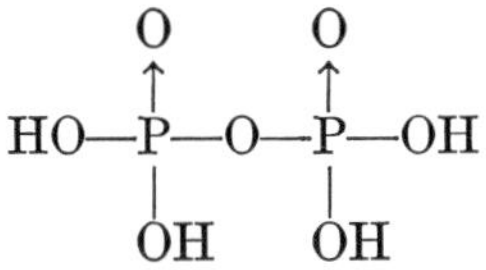

$$
\begin{array}{ccc}
O & & O \\
\uparrow & & \uparrow \\
HO\!-\!P\!-\!O\!-\!P\!-\!OH \\
| & & | \\
OH & & OH
\end{array}
$$

and is the equivalent of an *inter*molecular dehydration of two molecules of orthophosphoric acid. The so-called "energy-rich" phosphate bond referred to in most biochemistry textbooks is an anhydride linkage between two or three phosphoric acids, as found in adenosine triphosphate (ATP). Approximately 8000 calories/mole of energy are released during the hydrolysis of a pyrophosphate bond.

To complicate nomenclature matters even further, the naming of the salts derived from orthophosphoric acid has not been universally standardized. It is important that the pharmacist realize this when ordering and dispensing the specific salts. The nomenclature is summarized in Table 5–4.

Serum phosphate levels usually correlate with serum calcium values. Whenever calcium concentrations are not within normal range, serum phosphate will either be too high (hyperphosphatemia) or too low (hypophosphatemia). Hyperphosphatemia may be found in hypervitaminosis D (which increases intestinal phosphate absorption along with calcium), renal failure due to the inability to excrete phosphate into the urine, and hypo-

TABLE 5-4.   NOMENCLATURE OF THE PHOSPHATES

Formula	Preferred Name	Other Commonly Used Names	Name in Compendia
NaH₂PO₄	Sodium dihydrogen phosphate	Monobasic sodium phosphate; primary sodium phosphate	Sodium biphosphate
Na₂HPO₄	Sodium monohydrogen phosphate	Dibasic sodium phosphate; secondary sodium phosphate	Sodium phosphate
Na₃PO₄	Sodium phosphate	Tribasic sodium phosphate; tertiary sodium phosphate	

From Modern Inorganic Pharmaceutical Chemistry by C. A. Discher, John Wiley and Sons, New York, p. 122, 1964.   By permission.

parathyroidism.   In the latter case, the lack of parathyroid hormone permits renal tubular reabsorption of phosphate which results in decreased urinary phosphate and a rise in serum phosphate.   A complication of hyperphosphatemia is the formation of phosphatic urinary calculi (a form of kidney stone) with resultant possible kidney damage.   Basic aluminum carbonate, $Al(OH)CO_3$ (Basaljel®), is used to remove dietary phosphate by excreting it in the feces as slightly soluble aluminum phosphate.

Hypophosphatemia may be seen in: vitamin D deficiency (rickets) probably caused by decreased intestinal calcium absorption; hyperparathyroidism, in which increased levels of parathyroid hormone further inhibit renal tubular phosphate reabsorption, resulting in increased urinary phosphate excretion and decreased serum phosphate; lack of phosphate reabsorption by the kidney tubule from other causes (infection, cancers, etc.); and possible long-term aluminum hydroxide gel antacid therapy.[2] In the latter, the aluminum hydroxide gel forms insoluble aluminum phosphate salts from dietary phosphate therapy, thus preventing the absorption of dietary phosphate from the intestinal tract.   This would only be a problem with long-term, continuous usage of aluminum hydroxide gel.

Dietary hypophosphatemia is not normally considered a problem, since a person eating a reasonably balanced diet should receive more than an adequate amount of phosphate.   Hypophosphatemia has been reported in patients receiving their entire nutritional and caloric requirements by intravenous hyperalimentation (intravenous administration of a solution meeting the patients' essential caloric and nutritional needs).   Hyperalimentation solutions usually contain protein hydrolysate or a balanced amino acid mixture, glucose, a multivitamin mixture, and the appropriate salts, containing calcium, magnesium, sodium, potassium, iron, chloride, bicarbonate, phosphate, and sulfate ions.   By withholding any ion, it is possible to induce an artificial deficiency state.   When phosphate is with-

held the resulting hypophosphatemia causes marked alterations in erythrocyte metabolism, resulting in decreases in erythrocyte glucose-6-phosphate
(G-6-P), fructose-6-phosphate (F-6-P), 3-phosphoglyceric acid (3-PG),
2-phosphoglyceric acid (2-PG), phosphoenolpyruvate (PEP), 2,3-diphos-
phoglycerate (2,3-DPG), and adenosine triphosphate (ATP). There is an
increase in total triose phosphate (glyceraldehyde-3-phosphate, dihydroxy
acetone phosphate, etc.). The decreased erythrocyte levels of 2,3-DPG and
ATP cause an increase in the oxygen affinity of the red blood cell. It is
hypothesized that serum inorganic phosphate has an important regulatory
role in erythrocyte glucose metabolism; this regulation may occur at the
glyceraldehyde-3-phosphate dehydrogenase step.[3]

Phosphates have been used as tonics, but this use has no validity.
Phosphate sodas of various flavors were a popular soft drink at one time.
They were made by mixing the appropriate flavored syrup, a diluted phosphoric acid solution, and sodium bicarbonate. The reaction between the
acid and sodium bicarbonate produced the carbonation [rx (iii)]:

$$H_3PO_4 + 2NaHCO_3 \rightarrow Na_2HPO_4 + 2H_2CO_3$$
$$\hookrightarrow 2H_2O + 2CO_2 \uparrow \quad \text{(iii)}$$

Bicarbonate. Bicarbonate (HCO_3^-) is the second most prevalent anion
in the extracellular fluid compartments. Along with carbonic acid, it
functions as the body's most important buffer system (see physiological
acid-base balance). A lack of bicarbonate causes metabolic acidosis and
an excess causes metabolic alkalosis.

Sodium. Sodium is the principal cation in the extracellular fluid
compartments. This ion is responsible for maintaining normal hydration
and osmotic pressure. Normally, more than adequate amounts of sodium
are contained in the daily diet with nearly complete absorption from the
intestinal tract. Excess sodium is excreted by the kidneys, which make
them the ultimate regulator of the sodium content of the body. Approximately 80–85% of the sodium in the glomerular filtrate is reabsorbed.
This reabsorption is under hormonal control that is still not completely
understood. It has been theorized that renin, a proteolytic enzyme released
by the kidney, cleaves a linear protein and forms angiotensin I. Angiotensin I is then cleaved to form the octapeptide angiotensin II, which
stimulates the adrenal cortex to increase its secretion of aldosterone, and is
effective in increasing the reabsorption of sodium.[4] A prostaglandin has
also been implicated in the hormonal control of tubular reabsorption of
sodium.[5]

Conditions causing hyponatremia (low serum sodium level) are: (1) extreme urine loss, such as seen in diabetes insipidus (a disease of pituitary
origin as contrasted with diabetes mellitus, which is caused by deficient
insulin secretion by the β-cells of the islets of Langerhans in the pancreas);

(2) metabolic acidosis, in which sodium is excreted; (3) Addison's disease, with decreased excretion of the antidiuretic hormone, aldosterone; (4) diarrhea and vomiting; and (5) kidney damage. Hypernatremia (increased serum sodium level) is found in: (1) hyperadrenalism (Cushing's syndrome) with increased aldosterone production; (2) severe dehydration; (3) certain types of brain injury; and (4) excess treatment with sodium salts.

There is a good correlation between sodium content (as sodium chloride) of the tissues and hypertension. If, for some reason, the body is unable to eliminate sodium and the concentration starts to increase, water is retained in the tissues to maintain osmotic balance. Edema results and, outwardly, the patient can take on a puffy appearance with swelling, particularly of the lower extremities. The buildup of fluids puts an added burden on the heart which may be aggravated if the heart is also diseased. Treatment includes low salt diets, diuretics, cardiotonic drugs, and combinations of each. In temporary conditions such as pregnancy, elimination of salt and highly salted foods will greatly reduce the edema and concurrent weight problems. Sodium-free salt substitues (Neocurtasal®, Co-Salt®) can be used to enhance the flavor of food. Neocurtasal® contains a mixture of potassium chloride, glutamic acid, potassium glutamate, calcium silicate, and tribasic calcium phosphate ($Ca_3[PO_4]$) as pouring agents, and potassium iodide. Co-Salt® is a blend of choline, potassium chloride, ammonium chloride, and tricalcium phosphate.

Potassium. Potassium is the major intracellular cation, present in a concentration approximately 23 times higher than the concentration of potassium in the extracellular fluid compartments. This concentration differential is maintained by an active transport mechanism. During transmission of a nerve impulse, potassium leaves the cell and sodium enters the cell. It is currently thought that an active transport mechanism reestablishes the concentration differential after transmission of the nerve impulse. This active transport mechanism has been called the *sodium-potassium pump*.

Potassium in the diet is rapidly absorbed. Any excess potassium is rapidly excreted by the kidneys. Potassium salts have been used for their diuretic action because of this efficient excretion of potassium by the kidneys, since a certain volume of fluid (urine) will be excreted in order to keep the potassium salt in solution.

Both elevated and low serum potassium levels can be serious to the patient. Hypopotassemia (hypokalemia) causes changes in myocardial function, flaccid and feeble muscles, and low blood pressure. It can occur from vomiting, diarrhea, burns, hemorrhages, diabetic coma, intravenous infusion of solutions lacking in potassium (a dilution effect), overuse of thiazide diuretics, and alkalosis. The latter occurs due to the movement of potassium into the cell as protons move out of the cell into the proton-deficient extracellular fluid. Notice that, in this situation, it is possible for a low serum potassium level and elevated cellular potassium level to occur concurrently.

Hyperpotassemia (hyperkalemia) is less common and usually occurs during certain types of kidney damage. If the kidney is functioning properly, the body can eliminate excess potassium readily. In certain acidotic conditions, interference with the sodium and potassium proton exchange can result in potassium retention (see physiological acid-base balance discussion).

The heart appears to be particularly sensitive to potassium concentrations. In hypopotassemia there are alterations in the electrocardiogram (ECG) and distinct histological changes in the myocardium. An increase in potassium levels also results in changes in the electrocardiogram and causes the heart muscle to become flaccid with possible cessation of heart beat (potassium arrest). It is thought that potassium may be displacing calcium in the cardiac muscle, since a decrease in calcium will produce a similar pattern in the heart muscle and may explain why calcium gluconate is effective in hyperpotassemic conditions.

Because potassium is the major intracellular cation, serum potassium levels may not be a sensitive enough measure of the body's potassium levels. There is a current effort at developing whole body counts of potassium by measuring levels of potassium-40.[6]

Calcium. The biochemistry of calcium is very complex. Ninety-nine percent of body calcium is found in the bones. The remaining calcium is largely found in the extracellular fluid compartments. The biochemical functions of calcium, as well as the hormonal control of serum calcium levels, are very complicated, with many of the details yet to be elucidated. While a detailed discussion of calcium biochemistry is beyond the scope of this book, a brief overview of its absorption, hormonal control, and function will be presented.

Calcium is absorbed from the upper part of the small intestine where the intestinal contents are still acidic, and it exists as ionized water soluble salts. As the intestinal contents become neutral to alkaline, calcium is precipitated as the dibasic phosphate ($CaHPO_4$), carbonate, oxalate, and sulfate salts, and as insoluble calcium soaps. The fatty acid portion of the soaps comes from lipase-catalyzed hydrolysis of dietary triglycerides.

The actual absorption of calcium across the intestinal membranes is controlled by the parathyroid hormone and a metabolite of vitamin D. It is currently believed that cholecalciferol (vitamin D_3) is hydroxylated at the C-25 position in the liver and then at the C-1 position in the kidneys. This activated metabolite, 1,25-dihydroxycholecalciferol, may function as a gene activator, causing the synthesis of a calcium-binding protein which transfers the calcium cation across the intestinal wall.[7]

A report has appeared that epileptic children on anticonvulsants with a minimal daily vitamin D intake (140–290 I.U. daily) showed reduced calcium levels. The anticonvulsants are postulated to increase vitamin D metabolism by microsomal enzyme induction. Increased vitamin D intake appears to correct the problem.[8]

The intestinal absorption and serum level of calcium are also heavily influenced by phosphate. Low dietary phosphorus intake (a relatively rare occurrence) will restrict the amount of calcium that is absorbed. Furthermore, under the influence of parathyroid hormone (see previous phosphate discussion), renal tubular reabsorption of phosphorus is inhibited. This results in increased excretion of phosphorus in the urine, causing a fall in the blood level of phosphorus and a rise in the blood calcium level. Increased serum phosphorus levels will lower serum calcium levels. The administration of phosphorus salts, both oral and I.V., has been used with some success in the treatment of hypercalcemia. The mechanism is not completely clear. The calcium-lowering effect is apparently not due to increased urinary excretion, since renal excretion of calcium decreases during therapy. There is some evidence for extraskeletal calcification.[9-12] There is also evidence that lactose plays a role in calcium absorption with lactose-deficient patients having a higher incidence of osteoporosis. This is an attractive hypothesis since milk, which is one of the main sources of dietary calcium, also contains lactose. Most lactose-intolerant patients avoid milk. In addition, feeding of lactose to a lactose-intolerant patient induces a greater negative calcium balance.[13]

Calcium absorption and distribution are also under a complex hormonal control: parathyroid hormone (PTH) and the recently discovered calcitonin (thyrocalcitonin). Removal of the parathyroid glands causes a severe tetany (muscle twitchings, cramps, convulsions) to develop due to a sharp drop in serum calcium levels and a rise in serum phosphorus levels. Urinary calcium and phosphorus excretion is also diminished. Parathyroid hormone (PTH) controls the blood calcium and phosphate levels by acting on both the kidney and bone. Administration of PTH raises the blood calcium, lowers blood phosphorus, and increases the elimination of both in the urine. In man there is an initial decrease in calcium urinary excretion followed by a sustained increased excretion as serum calcium levels increase.[14] PTH also causes the migration of calcium from bone if this element is not available in sufficient amounts in the food, and increases the phosphatase activity of serum. Its action on bone is attributed to increasing the activity of the osteoclast cells which will cause the breakdown of bone tissue and release of calcium. Blood calcium levels control the secretory activity of the parathyroid gland: decreased blood calcium increases parathyroid secretion and increased blood calcium decreases parathyroid secretion.

The other hormone involved in calcium control is calcitonin, or thyrocalcitonin. It was first reported in 1962. The hormone is secreted by the perifollicular C cells of the mammalian thyroid gland. Calcitonin acts directly on bone and indirectly in kidney. The action on bone is due to an inhibition of calcium resorption. The mechanism is complex, but it appears that calcitonin reduces the number of osteoclast cells.[15]   In this respect it opposes the parathyroid hormone. This property has led to the experimental use of calcitonin in certain bone wasting diseases (see discussion below). It has not been completely determined if calcitonin actually promotes bone formation.

In the kidney calcitonin increases the urinary excretion of phosphate by an indirect effect. Because calcitonin produces hypocalcemia, parathyroid hormone is released causing increased urinary phosphate excretion.

As with parathyroid hormone, serum calcium levels control the secretion of calcitonin with increased calcium levels causing calcitonin secretion. Calcitonin has been used in the treatment of hypercalcemia due to hyperparathyroidism and vitamin D intoxication.[16]

Functionally, 99% of all body calcium is supportive, being found in bone as hydroxyapatite. The remaining ionic calcium is involved in neurohormonal functions, blood clotting, muscle contraction, and possibly in other biochemical processes. Calcium is necessary for the release of acetylcholine from preganglionic nerve endings. Its action in muscle contraction has been strongly associated with cyclic AMP.[17]   Evidence has been published that calcium cations give rise to muscle contractions; the muscle becomes flaccid when calcium is removed or displaced. The deleterious effects of hyperpotassemia on the heart may be due to excessive potassium displacing calcium from the cardiac muscle.

Calcium is essential for blood clotting. Citrate is added to whole blood to complex the blood calcium and thereby prevent clot formation in the collected blood.

From all of the above discussion, it should be obvious that either a hypo- or hypercalcemic condition is serious and the cause should be found or, at the very least, the condition treated symptomatically. Hypercalcemia is found in hyperparathyroidism, hypervitaminosis D, and some bone neoplastic diseases. Symptoms include fatigue, muscle weakness, constipation, anorexia (loss of appetite), and cardiac irregularities. If the condition persists, calcium salts may be deposited in the kidneys and blood vessels. There are many methods of reducing the intestinal absorption of calcium, ranging from the precipitation of calcium as insoluble sulfate or phosphate salts to complexation with ethylenediamine tetraacetic acid (EDTA). Recent reports indicate that cellulose phosphate is effective in reducing intestinal calcium hyperabsorption.[19]

Hypocalcemia can be caused by hypoparathyroidism, vitamin D deficiency, osteoblastic metastasis (spreading bone cancer), steatorrhea (fatty stool), Cushing's syndrome (hyperactive adrenal cortex), acute pancreatitis, and acute hyperphosphatemia. If the serum calcium levels fall enough, hypocalcemic tetany can result.

Associated with the above condition are disorders in bone metabolism. Bone is a dynamic tissue involving constant exchange of calcium and phosphate ions with the body fluids. As has already been described, much of this exchange is under hormonal control. Bone, in addition to providing structural support, is also a storage tissue for calcium. At night, when a person goes for 12 hours or more without food intake, resorption of the bone occurs in order to maintain blood calcium levels.

A serious condition of bone degeneration commonly associated with aging is osteoporosis, which is a reduced volume of bone tissue per unit volume of anatomical bone. As the condition progresses, the bones become weaker and more fragile. Broken hips due to the bone's inability to support body weight are commonly seen in the elderly with this disease. There have been several hypotheses put forth as to probable causes: (1) decreased calcium absorption due to diet or some problem associated with intestinal calcium absorption; (2) vitamin D deficiency and/or inability to hydroxylate vitamin D, thus reducing the levels of the active metabolite 1,25-dihydroxy-cholecalciferol; (3) increased sensitivity to parathyroid hormone, particularly in postmenopausal women; and (4) bone dissolution as a possible mechanism to buffer the fixed acid load from dietary protein.[20–22] The suggested treatments vary as widely as the postulated causes. Treatments include increased calcium and vitamin D intake, increased phosphate such as sodium monohydrogen phosphate (Na_2HPO_4), administration of sodium fluoride, and administration of calcitonin.[23,24]

Hypothesis four, bone dissolution acting as a buffer, is particularly interesting in terms of the continuing debate on what constitutes a proper diet. This hypothesis leads to the conclusion that a vegetarian will have less bone loss than will an omnivore (meat eater), since the omnivore's diet will potentially be more acidic following digestion and absorption into the body of the amino acids from the protein. Of course, the vegetarian also consumes protein, but the amount is nowhere near the amount consumed by an omnivore. A recent study reported that the bone density of vegetarians was significantly greater when compared with age- and sex-matched controls.[22] If this study can be confirmed, it would appear that omnivores should increase their dietary calcium consumption probably by suitable dairy products. On the other hand, this study did not compare calcium intake between the two groups. In order to obtain adequate amounts of protein, many vegetarians will consume greater quantities of dairy products as compared to omnivores, thereby having a greater calcium intake than omnivores.

Another bone disease due to faulty calcium metabolism is Paget's disease, characterized by an initial phase of decalcification and softening of the bone followed by calcium deposition with resultant thickening and deformity. It is estimated that 2% of all adults over age 40 are affected.

Treatment with phosphate salts and/or calcitonin is in the trial stage of development.[25,26]

Magnesium. Magnesium is the second most plentiful cation in the intracellular fluid compartment and the fourth most abundant cation in the body. A healthy, lean, 70-kg adult male has about 2000 mEq of magnesium, as compared to 2600 mEq of calcium, 3400 mEq of sodium, and 3000 mEq of potassium.[27] Fifty percent of total body magnesium (10–20 g) is combined with calcium and phosphorus in bone. It is an essential component of many of the enzymes involving phosphate metabolism which also require adenosine triphosphate (ATP). Magnesium is also apparently indispensable for protein synthesis and for the smooth functioning of the neuromuscular system.[28]

Negative magnesium balance is more widespread than most people generally realize. Causes include malnutrition, dietary restrictions, chronic alcoholism, faulty absorption or utilization, gastrointestinal diseases, medications, and parathyroid hormone (PTH) imbalances.[29–31] The clinical significance of magnesium depletion varies widely both in symptomatology and in biochemistry, depending upon the overall health of the individual. Because the body has tissue reserves of magnesium, it may take several weeks before the blood levels start to show a decrease. Indeed, serum magnesium levels may be misleading. A urinary magnesium determination utilizing an intravenous magnesium test dose may be more accurate. Normal subjects will promptly excrete 90% of the test dose while depleted subjects may retain 40% of the magnesium in the test dose.[32]

Symptoms of magnesium deficiency include personality changes after depletion of three or four months' duration, failure to gain weight properly, and cardiac disturbances.[33] Hypomagnesemia and alkalosis have been correlated with the withdrawal symptoms of the chronic alcoholic.[34–36]

Magnesium cation has a definite pharmacological action. Magnesium salts when injected intramuscularly or intravenously have a powerful general anesthetic action which resembles that produced by chloroform. This depressant action affects the cellular portion of the neuron and the neuromuscular junction. An excess of magnesium decreases the amount of transmitter substance, acetylcholine, liberated at the end plate. The blocking effect of magnesium ions at the neuromuscular junction is in certain respects similar to that of curare, which depresses the sensitivity of the muscle to intraarterially injected acetylcholine and to acetylcholine applied directly to the single end plate. Calcium ions relieve the block produced by magnesium ions and restore output of acetylcholine from nerve endings. For this reason soluble magnesium salts (usually magnesium sulfate) have been used as central nervous system depressants in obstetrics, convulsant states, and for the symptoms of tetanus.

Magnesium ion is not readily absorbed from the gastrointestinal tract because its absorption is retarded by alkaline media. Most of the absorption takes place in the acid medium of the duodenum. Due to the slow absorption of magnesium ions, a saline laxative action occurs upon the ingestion of any water soluble magnesium compound.

In any case, intravenous (or intramuscular) injection of magnesium sulfate for any purpose should not be carried out on patients with impaired kidney function.[37] The toxic possibilities in the presence of deficient kidney function should also be kept in mind in the case of oral use of magnesium salts.

Electrolytes Used for Replacement Therapy

Sodium Replacement

Sodium Chloride, U.S.P. XVIII (NaCl; Mol. Wt. 58.44)

Sodium Chloride U.S.P. occurs as colorless cubic crystals or as a white, crystalline powder having a saline taste. It is freely soluble in water, and slightly more so in boiling water; it is soluble in glycerin and slightly soluble in alcohol.

It is the salt of the extracellular fluids, and its uses range from replacement therapy and manufacture of isotonic solutions to a flavor enhancer. In order to be isotonic, a salt solution should be 0.9% w/v. Ordinary table salt does not meet U.S.P. standards. Table salt usually contains water insoluble salts such as silicates, which prevent caking and add to the product's pouring qualities, and potassium iodide for the prevention of goiter (see Chapter 6).

Isotonic solutions are used as wet dressings, for irrigating body cavities or tissues, and as injections when fluid and electrolytes have been depleted in isotonic proportions. Buildup of excessive extracellular fluid due to administration of isotonic sodium chloride may lead to both pulmonary and peripheral edema. Hypotonic solutions are administered for maintenance therapy when patients are unable to take fluid and nutrients orally for one to three days. Dextrose (glucose) is usually the caloric source.

Hypertonic injections are used when there is loss of sodium in an excess of water. These injections should be given slowly in small volumes (200–400 ml).[38] Orally administered hypertonic sodium chloride solutions which will induce vomiting have been recommended for household poisoning accidents. At least one death has been reported in a 2-year-old child who was given this type of first aid treatment. The child's serum sodium was sharply elevated.[39]

Usual Dose: Oral, 1 g three times a day.

Intravenous infusion, 1 liter of a 0.9% solution.

Topically to wounds and body cavities, as a 0.9% solution for irrigation.

TABLE 5–5

Percent Dextrose	Percent Sodium Chloride	mEq/l	Injection, ml Sodium Chloride
5	0.11	18.8	250, 500, and 1000
5	0.2	34.2	250, 500, and 1000
5	0.225	38.5	250, 500, and 1000
3.3	0.3	51.3	500 and 1000
5	0.3	51.3	250, 500, and 1000
5	0.33	56.4	250, 500, and 1000
2.5	0.45	76.9	150, 250, 500, and 1000
5	0.45	76.9	250, 500, and 1000
10	0.45	76.9	1000
2.5	0.9	153.8	250, 500, and 1000
5	0.9	153.8	150, 250, 500, and 1000
10	0.9	153.9	500 and 1000
20	0.9	153.9	500

Usual Dose Range: Oral, 3 to 6 g daily.
Occurrence:
Sodium Chloride Injection, U.S.P. XVIII
  Contains 0.9% NaCl.
  Category: Fluid and electrolyte replenisher; irrigation solution.
  Usual Dose: Intravenous infusion, 1 liter.
Bacteriostatic Sodium Chloride Injection, U.S.P. XVIII
  Contains 0.9% NaCl.
  Category: Sterile vehicle.
Sodium Chloride Solution, U.S.P. XVIII
  Contains 0.9% NaCl.
  Category: Isotonic vehicle.
Sodium Chloride Tablets, U.S.P. XVIII
  Category: Electrolyte replenisher.
  Usually available as 600-mg, 1-, and 2.25-g tablets.
Dextrose and Sodium Chloride Injection, U.S.P. XVIII
  Injections available, see Table 5–5.
  Category: Fluid, nutrient, and electrolyte replenisher.
  Usual Dose: Intravenous, 1 liter.
Sodium Chloride and Dextrose Tablets, N.F. XIII
  Usually available as tablets of 200 mg (3.42 mEq) sodium chloride
  and 450 mg dextrose.
  Category: Electrolyte and nutrient replenisher.
Mannitol and Sodium Chloride Injection, U.S.P. XVIII
  Injections available, see Table 5–6.
  Category: Diuretic.
  Dose: Intravenous infusion, 50 to 200 g daily.

TABLE 5–6

Percent Mannitol	Percent Sodium Chloride	mEq/l	Injection, ml Sodium Chloride
5	0.3	51.3	500 and 1000
10	0.3	51.3	500 and 1000
15	0.45	76.9	150 and 500
20	0.45	76.9	250 and 500

Fructose and Sodium Chloride Injection, N.F. XIII

Contains 10% fructose and 0.9% NaCl.

Category: Fluid, nutrient, and electrolyte replenisher.

Usual Dose: Intravenous and subcutaneous, as required.

Ringer's Injection, U.S.P. XVIII

Contains 0.86% NaCl (147 mEq/l Na, 4 mEq/l K, 4.5 mEq/l Ca, 155.5 mEq/l Cl).

Category: Fluid and electrolyte replenisher.

Usual Dose: Intravenous infusion, 1 liter.

Lactated Ringer's Injection, U.S.P. XVIII

Contains 0.6% NaCl (130 mEq/l Na, 4 mEq/l K, 2.7 mEq/l Ca, 109.7 mEq/l Cl, 27 mEq/l lactate).

Category: Systemic alkalizer; fluid and electrolyte replenisher.

Usual Dose: Intravenous infusion, 1 liter. It is contraindicated in alkalosis conditions.

Potassium Replacement

Potassium Chloride, U.S.P. XVIII (KCl; Mol. Wt. 74.56)

Potassium Chloride U.S.P. occurs as colorless, elongated, prismatic, or cubical crystals, or as a white, granular powder. It is odorless, has a saline taste, and is stable in air. *Potassium Chloride* U.S.P. is freely soluble in water and even more so in boiling water, giving solutions that are neutral to litmus. It is insoluble in alcohol.

Potassium Chloride is the drug of choice for oral replacement of potassium, preferably as a solution. It is irritating to the gastrointestinal tract and solutions must be well diluted. The U.S.P. requires that the tablets must be enteric-coated, but several authorities do not recommend the use of tablets because of the possible occurrence of small bowel ulceration and because the tablet's absorption is undependable. The enteric-coated potassium-containing diuretics have been withdrawn from the market because they have produced intestinal ulceration. Intravenous injections may be used if the patient is unable to take potassium orally or if hypopotassemia is severe. When injections are used, serum potassium concentration, electrocardiograms, and urinary output should be monitored. Although marketed as a concentrate, potassium chloride is given alone as an isotonic solution, in an isotonically balanced sodium chloride solution, or as 500 ml of 5% glucose (dextrose) solution containing 40 mEq of potassium.[40] Precipitates containing mostly silica have been reported in admixtures of potassium chloride and 5% dextrose in water, attributed to leaching of the glass. The silica precipitates were not found in separate solutions of potassium chloride or dextrose.[41]

In addition to hypopotassemia, potassium chloride is indicated in the treatment of familial periodic paralysis (a recurring, rapidly progressive, flaccid paralysis), Meniere's syndrome (disease of the inner ear which includes dizziness and noise in the ear), and as an antidote in digitalis intoxication. The latter results from either large doses of digitalis administered together with diuretics or the cumulative effect of maintenance doses taken over long periods of time. The symptoms range from gastrointestinal disturbances to arrhythmias. While potassium administration is the treatment of choice for digitalis intoxication, it is not a specific antidote and can potentiate some of the cardiac complications of digitalis intoxication. Potassium chloride is also used as an adjunct to drugs used in the treatment of myasthenia gravis (a progressive, severe muscle weakness).

When given as a dilute oral solution, potassium chloride is mixed with fruit or vegetable juices to mask the saline taste. Tomato juice (regular or low sodium) has been recommended as an effective masking agent.[42]

Potassium therapy is contraindicated in patients with impaired renal function with oliguria (diminished urine output), acute dehydration, hyperpotassemic conditions exacerbated (increased severity of a disease) by potassium, such as myotonia congenita (tonic muscle rigidity and spasm), adynamia episodica hereditaria (periodic weakness or paralysis of skeletal muscle), and in patients receiving potassium-sparing drugs.

Usual Dose: 1 g four times a day.

Usual Dose Range: 500 mg to 8 g daily.

Occurrence:

Potassium Chloride Injection, U.S.P. XVIII

Available as concentrates: 1.5 g in 10 ml; 3 g in 12.5 and 20 ml; 4.5 and 6 g in 30 ml.

Potassium Chloride Tablets, U.S.P. XVIII

Available as enteric-coated tablets containing 300 mg or 1 g. (See discussion above concerning potassium chloride tablets.)

Ringer's Injection, U.S.P. XVIII

Contains 0.03% KCl (147 mEq/l Na, 4 mEq/l K, 4.5 mEq/l Ca, 155.5 mEq/l Cl).

Category: Fluid and electrolyte replenisher.

Usual Dose: Intravenous infusion, 1 liter.

Lactated Ringer's Injection, U.S.P. XVIII

Contains 0.03% KCl (130 mEq/l Na, 4 mEq/l K, 2.7 mEq/l Ca, 109.7 mEq/l Cl, 27 mEq/l lactate).

Category: Systemic alkalizer; fluid and electrolyte replenisher.

Usual Dose: Intravenous infusion, 1 liter.

Lactated Potassic Saline, Injection, N.F. XIII

Contains 0.026% KCl (121 mEq/l Na, 35 mEq/l K, 103 mEq/l Cl, 53 mEq/l lactate).

Available in volumes of 150, 250, 500, and 1000 ml.

Category: Fluid and electrolyte replenisher.

Usual Dose Range: Intravenous and subcutaneous, 40 to 80 ml/kg of body weight per day, given slowly over a period of 4 to 12 hours.

Potassium Gluconate, N.F. XIII (Mol. Wt. 234.25)

$$\begin{array}{c} CO_2^-K^+ \\ | \\ HCOH \\ | \\ HOCH \\ | \\ HCOH \\ | \\ HCOH \\ | \\ CH_2OH \end{array}$$

Potassium Gluconate N.F. occurs as a white to yellowish white, crystalline powder or as granules. It is odorless and has a slightly bitter taste. It is stable in air, and its solutions are slightly alkaline to litmus. *Potassium Gluconate* is freely soluble in water. It is practically insoluble in dehydrated alcohol, ether, benzene, and chloroform.

Potassium gluconate is claimed to be less irritating and easier to use so as to mask potassium's saline taste. Since chloride ion is often needed to allow complete potassium replacement, potassium chloride is the drug of choice in treatment of hypopotassemia. (See the discussion of potassium chloride for indication and contraindication of potassium therapy.)

Category: Electrolyte replenisher.

Usual Dose: The equivalent of 10 mEq of potassium four times daily.

Occurrence:

Potassium Gluconate Elixir, N.F. XIII (Kaon® Elixir)

Available as an elixir containing 4.68 g of potassium gluconate in each 15 ml, equivalent to 20 mEq of potassium.

Potassium Gluconate Tablets, N.F. XIII (Kaon® Tablets)

Available as sugar-coated tablets containing 1.17 g of potassium gluconate equivalent to 5 mEq of potassium.

Calcium Replacement

Calcium Chloride, U.S.P. XVIII ($CaCl_2 \cdot 2H_2O$; Mol. Wt. 147.02)

Calcium Chloride U.S.P. occurs as white, hard, odorless fragments or granules which are deliquescent. It is freely soluble in water, alcohol, and boiling alcohol and very soluble in boiling water. It is irritating to the veins and should be injected slowly. Rapid injection may cause cutaneous burning sensation, peripheral vasodilation, and fall in blood pressure. It is contraindicated in hypocalcemia associated with renal insufficiency. Calcium chloride is used as the calcium source in many commercially available electrolyte replacement and maintenance solutions.

Occurrence:

Ringer's Injection, U.S.P. XVIII

Contains 0.033% $CaCl_2 \cdot 2H_2O$ (147 mEq/l Na, 4 mEq/l K, 4.5 mEq/l Ca, 155.5 mEq/l Cl).

Category: Fluid and electrolyte replenisher.

Usual Dose: Intravenous infusion, 1 liter.

Lactated Ringer's Injection, U.S.P. XVIII

Contains 0.02% $CaCl_2 \cdot 2H_2O$ (130 mEq/l Na, 4 mEq/l K, 2.5 mEq/l Ca, 109.7 mEq/l Cl, 27 mEq/l lactate).

Category: Systemic alkalizer; fluid and electrolyte replenisher.

Usual Dose: Intravenous infusion, 1 liter.

Calcium Gluconate, U.S.P. XVIII (Mol. Wt. 430.88)

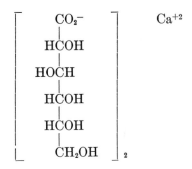

Calcium Gluconate U.S.P. occurs as white crystalline, odorless, tasteless granules or powder which is stable in air. Its solutions are neutral to litmus. It is sparingly (and slowly) soluble in water, freely soluble in boiling water, and insoluble in alcohol and in many other organic solvents.

Calcium gluconate is considered by many to be the treatment of choice for hypocalcemia because it is nonirritating when given orally and intravenously. There is some risk of abscess formation when given intramuscularly.

Usual Dose: Oral, 1 g three or more times a day.

Intravenous, 1 g one or more times a day.

Usual Dose Range: Oral or intravenous, 1 to 15 g daily.

Occurrence:

Calcium Gluconate Injection, U.S.P. XVIII (97 mg Calcium Gluconate/ml)

Usually available in 10-ml ampules.

Calcium Gluconate Tablets, U.S.P. XVIII

Usually available as 500-mg and 1-g tablets.

Calcium Lactate, N.F. XIII (Mol. Wt. [anhydrous] 218.22)

$$\left(\underset{\mathrm{CH_3CHCO_2^-}}{\overset{\mathrm{OH}}{|}} \right)_2 Ca^{+2} \cdot xH_2O$$

Calcium Lactate N.F. occurs as white, almost odorless, granules or powder. The pentahydrate is somewhat efflorescent and at 120 ° C becomes anhydrous. Calcium lactate pentahydrate is soluble in water and practically insoluble in alcohol. It is claimed to be a nonirritating calcium salt for oral calcium replacement therapy.

Usual Dose Range: 1 to 5 g three times a day.

Occurrence:

Calcium Lactate Tablets, N.F. XIII

Usually available as 300- and 600-mg (calcium lactate pentahydrate) tablets.

Dibasic Calcium Phosphate, N.F. XIII ($CaHPO_4$; Mol. Wt. 136.06; calcium monohydrogen phosphate; secondary calcium phosphate; often incorrectly called *dicalcium phosphate*)

Dibasic Calcium Phosphate N.F. occurs as a white, odorless, tasteless powder and is stable in air. It is practically insoluble in water, soluble in diluted hydrochloric and nitric acids, and insoluble in alcohol.

Dibasic calcium phosphate is given orally as a source of calcium and phosphorus in pregnancy and in lactation and calcium deficiency states. Although water insoluble, in the acid stomach it is converted to the soluble monobasic calcium phosphate ($Ca[H_2PO_4]_2$) and calcium chloride [rx (iv)].

$$2CaHPO_4 + 2HCl \rightarrow Ca(H_2PO_4)_2 + CaCl_2 \qquad \text{(iv)}$$

It will remain in solution in the upper part of the duodenum where the gastrointestinal contents are still acidic, and the calcium cation can be absorbed across the intestinal wall.

Usual Dose: 1 g three times a day.

Usual Dose Range: 1 to 5 g.

Tribasic Calcium Phosphate, N.F. XIII (a variable mixture of phosphates equivalent to not less than 90% $Ca_3(PO_4)_2$; tricalcium phosphate; tertiary calcium phosphate)

Tribasic Calcium Phosphate N.F. occurs as a white, odorless, and tasteless powder, which is stable in air. It dissolves readily in dilute hydrochloric and nitric acids, and is insoluble in alcohol and almost insoluble in water. Like dibasic calcium phosphate, it forms soluble calcium salts in the acid stomach and is used as an oral source of calcium and phosphorus. Because it "consumes" protons when dissolving, it is also used as an antacid (see discussion of calcium-containing antacids in Chapter 8).

Usual Dose Range: 1 to 5 g three times a day.

Parenteral Magnesium Administration

Magnesium Sulfate, U.S.P. XVIII ($MgSO_4 \cdot 7H_2O$; Mol. Wt. 246.47)
(The chemistry is discussed in Chapter 8 in the section on saline cathartics.)

Magnesium sulfate, when injected, has been used as a central nervous system depressant in the treatment of eclampsia (convulsions and coma), in severe cases of hypomagnesemia in the presence of tremors and seizures, and in magnesium-deficient alcoholics.[34–36] Overtreatment can cause respiratory paralysis and cardiac depression.[37] Intravenous injection of a calcium salt can be used to counteract magnesium intoxication. Magnesium sulfate should not be administered to patients with renal insufficiency.

Category: Anticonvulsant; cathartic.

Usual Dose: Anticonvulsant—Intramuscular, 1 g in a 25 to 50% solution.
　　Intravenous, 4 g in a 10% solution.
　　Cathartic—Oral, 15 g.

Usual Dose Range: Parenteral, 1 to 10 g daily; oral, 10 to 30 g daily.

Occurrence:
　　Magnesium Sulfate Injection, U.S.P. XVIII (available as 1 g in 2 ml, 2 g in 20 ml, 15 g in 30 ml)

Physiological Acid-Base Balance

Before discussing the electrolytes used in acid-base balance, an overview of the means used by the body to maintain physiological pH will be presented. Acids (either carbonic from carbon dioxide and lactic from anaerobic metabolism) are constantly being produced during metabolism. Since most metabolic reactions occur only within a very narrow pH range (7.38–7.42), the body utilizes several efficient buffer systems.

Two of the major buffer systems in the body are bicarbonate/carbonic acid (HCO_3^-/H_2CO_3) found in the plasma and kidneys and monohydrogen phosphate/dihydrogen phosphate ($HPO_4^{-2}/H_2PO_4^-$) found in the cells and kidneys. Also in the red blood cells is the hemoglobin (Hb) buffer system which is the most effective single system for buffering the carbonic acid produced during metabolic processes. For each millimole of oxygen that dissociates from hemoglobin 0.7 millimole H^+ is removed [rx (vi)].

Carbon dioxide (the acid anhydride of carbonic acid) is continuously produced in the cells. It diffuses from the cells into the plasma where a small portion is dissolved, and another small portion reacts with the water to form carbonic acid. The increased carbonic acid is buffered by plasma proteins. Most carbon dioxide enters the erythrocytes where it either rapidly forms carbonic acid by the action of carbonic anhydrase [rx (v)], or combines with hemoglobin. The tendency to lower the pH of

$$CO_2 + H_2O \xrightarrow{\text{carbonic anhydrase}} H_2CO_3 \qquad \text{(v)}$$

the erythrocytes due to increased concentration of carbonic acid is compensated by hemoglobin [rx (vi)]. The bicarbonate anion then diffuses

$$H_2CO_3 + K^+ + HbO_2^- \rightarrow K^+ + HCO_3^- + HHb + O_2 \qquad \text{(vi)}$$

out of the erythrocyte and chloride anion diffuses in. This has been named the *chloride shift*. The bicarbonate in the plasma, along with the plasma carbonic acid, now acts as an efficient buffer system. The normal HCO_3^-/H_2CO_3 ratio is $27/1.35$ mEq/l ($20/1$) corresponding to pH 7.4. (See Chapter 4 for a discussion of the use of the Henderson-Hasselbalch equation for calculating pH's.) In the lungs, there is a reversal of the above processes due to the large amounts of oxygen present. Oxygen combines with the protonated deoxyhemoglobin, releasing protons. These combine with bicarbonate, forming carbonic acid, which then dissociates to carbon dioxide and water [rx (vii)]. The carbon dioxide is exhaled by the lungs. Thus, by regulating the breathing, it is possible for the body to exert a partial control on the HCO_3^-/H_2CO_3 ratio.

$$O_2 + HHb + K^+ + HCO_3^- \rightarrow K^+ + HbO_2^- + H_2CO_3$$
$$ \hookrightarrow H_2O + CO_2 \uparrow \quad \text{(vii)}$$

The phosphate buffer system is also effective in maintaining physiological pH. At pH 7.4 the $HPO_4^{-2}/H_2PO_4^-$ ratio is approximately $4:1$. In the kidneys, the pH of the urine can drop to pH 4.5–4.8, corresponding to $HPO_4^{-2}/H_2PO_4^-$ ratios of $1:99$ to $1:100$.

The steps for acid excretion in the kidneys occur as follows (Fig. 5–1):
1. Sodium salts of mineral and organic acids are removed from the plasma by glomerular filtration.

FIG. 5–1.   *Top:* Mobilization of hydrogen ions in proximal tubule.   *Middle:* Secretion of hydrogen ions in distal tubule.   *Bottom:* Production of ammonia in distal tubule. (From Current Diagnosis and Treatment, by M. A. Krupp and M. J. Chatton, Lange Medical Publications, Los Altos, CA, p. 18, 1971. By permission.

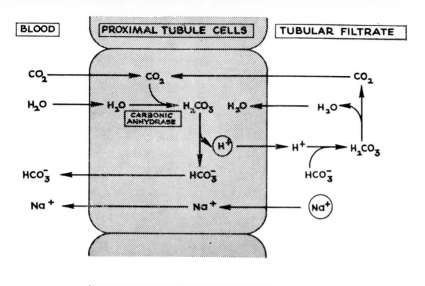

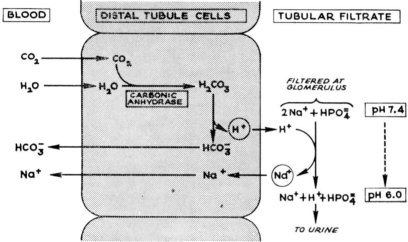

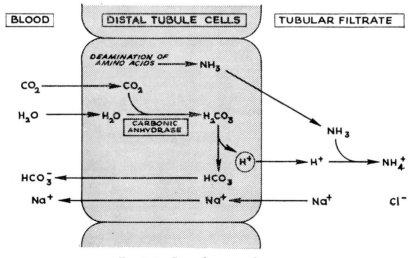

Fig. 5–1. Legend on opposite page.

197

2. Sodium is preferentially removed from the renal filtrate or tubular fluid and, in the tubule cells, reacts with carbonic acid formed by the carbonic anhydrase-catalyzed reaction of carbon dioxide and water [rxs (v) and (viii)]. This is sometimes called the Na^+–H^+ exchange.

$$Na^+ + H_2CO_3 \rightarrow Na^+ + HCO_3^- + H^+ \tag{viii}$$

3. The sodium bicarbonate returns to the plasma (eventually being removed in the lungs as carbon dioxide) and the protons enter the tubular fluid, forming acids of the anions that originally were sodium salts ($H_2PO_4^-$, lactate, etc.).

Potassium excretion is very complex and is not completely understood. It is filtered from the plasma, but it is then removed from the tubular filtrate by active reabsorption in the proximal tubules. Potassium is then secreted back into the tubular fluid by the distal tubular cells. This distal secretion of potassium involves an exchange process by which sodium is reabsorbed from the tubular fluid and potassium is secreted into the tubular fluid. At the same time there is the already described Na^+–H^+ exchange, which results in a competition by potassium and hydrogen ions for the sodium ion in the distal tubular fluid. Thus, a change in sodium or hydrogen ion concentrations can have a pronounced effect on the amount of potassium being secreted by the distal tubular cells.

Potassium excretion will be decreased when (1) the amount of sodium reaching the distal tubule is low or (2) the proton secretion by the kidney tubule is increased. In the former case, a low sodium concentration in the distal tubule fluid impedes the Na^+–K^+ exchange, thereby reducing the amount of potassium being excreted. In the latter case, increased proton secretion will lead to the protons "winning" the competition for sodium, thereby facilitating the Na^+–H^+ exchange at the expense of the Na^+–K^+ exchange.

Similarly, alterations in the potassium concentration will have an effect on proton excretion into the tubular fluid. When total body potassium is high, there is a passage of protons from the cells (including the kidney tubule cells) into the extracellular fluid, causing an intracellular alkalosis and possibly an extracellular acidosis. The resulting intracellular alkalosis causes protons to be retained by the kidney tubule cells, resulting in an inhibition of the Na^+–H^+ exchange and, since there will be less competition for sodium by the protons, the Na^+–K^+ exchange will be facilitated, causing increased potassium excretion.

When total body potassium is low, the intracellular fluid is acidic due to the passage of protons into the potassium-depleted cells (including the kidney tubule cells), resulting in an intracellular acidosis and possible extracellular alkalosis. Since there is now increased hydrogen ion concentration in kidney tubule cells, potassium excretion will be inhibited as the excess hydrogen ions compete for the sodium ions. In other words, the Na^+-H^+ exchange will be facilitated at the expense of the Na^+-K^+ exchange. This can result in an acidic urine even though the extracellular fluids are proton-deficient (metabolic alkalosis) due to the fact that protons are leaving the extracellular fluids and entering the cells as replacement for loss of cellular potassium.

The formation of ammonia from protein and amino acid metabolism is another means of removing protons (Fig. 5–1). The ammonia is secreted by the tubule cells into the tubular filtrate where it combines with the protons from carbonic acid to form ammonium ions. The "consumption" result is that protons and toxic ammonia are excreted, physiologic pH is maintained, and sodium is reabsorbed from the tubular filtrate.

Due to a variety of causes, the body's acid levels may increase and/or alkali levels decrease below normal, causing acidosis, or the acid levels may decrease and/or alkali levels increase above normal, causing alkalosis. The terms acidosis and alkalosis refer to the pH dropping slightly below 7.38 or increasing slightly above 7.42, respectively. If the body can restore the pH back to 7.35–7.45 by alterations in respiration and kidney function, it is referred to as a *compensated* metabolic acidosis or alkalosis. On the other hand, if, after the body takes corrective action, the buffer ratios are still not back to their normal ranges, it is said to be an *uncompensated* acidosis or alkalosis. Rather than uncompensated, *partially compensated* might be better terminology since the patient will recover, given

TABLE 5-7. COMPENSATORY MECHANISMS

Condition (causes)	Buffer System	Respiratory Function	Renal Function
Metabolic acidosis—primary HCO_3^- deficit (diabetic acidosis, renal failure, diarrhea)	HCO_3^-/H_2CO_3	Hyperventilation causing increased excretion of H_2CO_3 as CO_2	Increased acid excretion by Na^+-H^+ exchange, increased NH_3 formation, and HCO_3^- reabsorption
Metabolic alkalosis—primary HCO_3^- excess (administration of excess alkali, vomiting, potassium ions)	HCO_3^-/H_2CO_3	CO_2 retention causing increased H_2CO_3 concentration	Decreased Na^+-H^+ exchange, decreased NH_3 formation, and reabsorption of HCO_3^-
Respiratory acidosis—primary H_2CO_3 excess (cardiac disease, lung damage, drowning)	Hemoglobin and protein	Increased CO_2 excretion through the lungs	Same as metabolic acidosis
Respiratory alkalosis—primary H_2CO_3 deficit (fever, hysteria, anoxia, salicylate poisoning)	Same as metabolic alkalosis	Same as metabolic alkalosis	Same as metabolic alkalosis

time and the ability to respond to proper supportive treatment. Table 5–7 summarizes the types of acid-base imbalances and compensatory mechanisms.

Electrolytes Used in Acid-Base Therapy

Metabolic acidosis is treated with the sodium salts of bicarbonate, lactate, acetate, and occasionally citrate. Administration of bicarbonate increases the HCO_3^-/H_2CO_3 ratio when there is a bicarbonate deficit. Lactate, acetate, and citrate ions are normal components of metabolism and will be degraded to carbon dioxide and water by the tricarboxylic acid cycle (TCA cycle, citric acid cycle, Krebs cycle). The carbon dioxide, by the action of carbonic anhydrase, will form bicarbonate and thereby reduce the bicarbonate deficit.

Metabolic alkalosis has been treated with ammonium salts. Its action is in the kidneys where it retards the $Na^+–H^+$ exchange. For reasons discussed under the discussion of ammonium chloride, its use is rarely recommended.

Sodium Acetate, N.F. XIII ($CH_3CO_2Na \cdot 3H_2O$; Mol. Wt. 136.08)

Sodium Acetate N.F. occurs as colorless, transparent crystals, as a white granular crystalline powder or as white flakes. It is odorless or has a faint, acetous odor. It is efflorescent in warm, dry air. *Sodium Acetate* N.F. is very soluble in water and is soluble in alcohol.

It is claimed that sodium acetate, which is metabolized to carbon dioxide and then to bicarbonate, can be used as an effective buffer in metabolic acidosis.[43] It approaches sodium bicarbonate in its ability to restore blood pH and plasma bicarbonate in patients suffering from metabolic acidosis of acute cholera (a disease involving severe diarrhea resulting in the loss of electrolytes).[44] Uremic acidosis (acidic urine) has been corrected by infusion of sodium acetate giving results comparable to equivalent amounts of sodium bicarbonate infusion.[45] Sodium acetate solutions are also easier to sterilize than sodium bicarbonate solutions (see sodium bicarbonate discussion).

Usual Dose: 1.5 g.

Potassium Acetate, N.F. XIII (CH_3CO_2K; Mol. Wt. 98.15)

Potassium Acetate N.F. occurs as colorless, monoclinic crystals, or as a white crystalline powder. It has a saline and slightly alkaline taste. *Potassium Acetate* N.F. deliquesces on exposure to moist air. It is very soluble in water and is freely soluble in alcohol. It is categorized by the N.F. as an alkalizer. Along with potassium citrate and bicarbonate, it is found in Potassium Triplex®. All the precautions mentioned for potassium chloride apply to potassium acetate.

Usual Dose: 1 g.

Sodium Bicarbonate, U.S.P. XVIII ($NaHCO_3$; Mol. Wt. 84.01; sodium acid carbonate; sodium hydrogen carbonate; baking soda; bicarbonate of soda)

Sodium Bicarbonate U.S.P. occurs as a white, crystalline powder which is stable in dry air, but slowly decomposes in moist air. Its solutions, when freshly prepared with cold water without shaking, are alkaline to litmus. The alkalinity increases as the solutions stand, are agitated, or are heated. It is soluble in water and insoluble in alcohol.

When heated, the salt loses water and carbon dioxide and is converted into the normal carbonate [rx (ix)]:

$$2\,NaHCO_3 \rightleftharpoons Na_2CO_3 + H_2CO_3$$
$$\longrightarrow H_2O + CO_2\uparrow \qquad\qquad (ix)$$

The above decomposition takes place when the dry salt or a solution is heated. It accounts for one of the major difficulties in attempting to sterilize either the dry salt or its solutions, since the sodium carbonate solution which remains is much more alkaline than the bicarbonate solution and consequently is dangerous to use parenterally. Both the U.S.P. and the British Pharmacopoeia (B.P.) recognize an injection of sodium bicarbonate which is a sterile solution of *Sodium Bicarbonate* in *Water for Injection*. Although the U.S.P. is silent on the matter, the B.P. states that the solution can be sterilized by bacteriologic filtration or by autoclaving. The latter process is performed by passing carbon dioxide through the solution for one minute and then placing the solution in gas-tight containers for the autoclaving process. This has the effect of exerting an equilibrium control on reaction (ix) and causing the reaction to be reversed. After two hours of cooling at room temperature, the solution is assayed by titration to make sure that no decomposition has taken place, and, if so, the pH can be adjusted by passing in CO_2. Another procedure that is sometimes used is to weigh reagent grade sodium bicarbonate aseptically into warm *Sterile Water for Injection*. Although there is a definite possibility of having a nonsterile solution as a result, this method has been used with few untoward results.

Sterilization of a sodium bicarbonate solution also may be effected by heating in an open vessel and then resaturating the cooled solution with sterile carbon dioxide.

Because sodium bicarbonate injections are often used in emergency situations (see later discussion), attempts have been made to prepare 7.5% sodium bicarbonate injections in disposable syringes in small lots, store them under refrigeration for 60–90 days or at room temperature for 7–30 days, and date each lot.[46,47] It would also appear that the pH change accompanying the conversion of sodium bicarbonate to sodium carbonate should be checked for each lot over the period of storage, since the time that a sodium bicarbonate solution will remain acceptable is quite variable.

Another characteristic reaction of bicarbonate (also carbonate) salts is that carbon dioxide is liberated when they are treated with acids [rx (x)].

$$NaHCO_3 + HA \rightarrow NaA + CO_2 \uparrow + H_2O \text{ (HA = any acid)} \qquad (x)$$

The liberated CO_2 bubbling through the liquid is termed *effervescence*. Effervescent tablets and salts make use of the reaction of sodium bicarbonate with acids (usually organic acids, e.g., tartaric acid, citric acid, etc.), because in the dry state the bicarbonate and acid do not react, whereas when introduced into water a vigorous evolution of CO_2 takes place. However, it is well to remember that the reaction can take place in moist air and may account for incompatibilities in dry prescription mixtures of sodium bicarbonate with acetylsalicylic acid (aspirin) or other acidic substances.

Aqueous solutions of sodium bicarbonate are slightly alkaline (pH of about 8.2) as a result of hydrolysis of the bicarbonate ion [rx (xi)]. Sodium

$$HCO_3^- + H_2O \rightleftarrows H_2CO_3 + OH^- \qquad (xi)$$

bicarbonate is so slightly alkaline that it fails to turn phenolphthalein red. This fact constitutes a distinguishing test between sodium bicarbonate and sodium carbonate, because the carbonate ion in the latter salt is so extensively hydrolyzed that the solution is quite alkaline [rx (xii); pH is about 11.6].

$$CO_3^{-2} + H_2O \underset{\longleftarrow}{\overset{\longrightarrow}{\rightleftharpoons}} HCO_3^- + OH^- \qquad (xii)$$

Because sodium bicarbonate is the principal bicarbonate of drug use, a discussion of its actions will serve as a guide to considerations of other bicarbonates. Sodium bicarbonate may be considered from the standpoint of two relationships: (1) its relationship to the body economy as a buffer component and (2) its therapeutic and miscellaneous uses.

1. The normal acid-base balance of the plasma is maintained by three mechanisms working together: the buffers of the body fluids and red blood cells; the pulmonary excretion of excess carbon dioxide; and the renal excretion of either acid or base, whichever is in excess.

Although there are other buffer systems in the plasma [e.g., (1) the mono-hydrogen/dihydrogen biphosphate system (see earlier discussion) and (2) the proteins], the bicarbonate/carbonic acid system is by far the most important plasma buffer. This buffer system involves an equilibrium between sodium bicarbonate and carbonic acid. At a given pH, the ratio of the concentrations of the two substances is constant. While workings of the system are complex in detail, they are simple in principle. If an excess of acid is liberated in the body, it is neutralized by some of the sodium bicarbonate [rx (xiii)]. The excess carbonic acid decomposes into water

$$H^+ + NaHCO_3 \rightarrow Na^+ + H_2CO_3 \qquad (xiii)$$

and carbon dioxide and the latter is excreted by the lungs until the normal bicarbonate/carbonic acid ratio is achieved [rx (xiv)]. If an excess of

$$H_2CO_3 \rightarrow H_2O + CO_2 \uparrow \qquad (xiv)$$

alkali occurs in the body, it combines with carbonic acid to form bicarbonate, and more carbonic acid is formed from carbon dioxide and water to restore the balance. Since carbon dioxide is an end product of the metabolism of all types of foodstuff, there is always an abundant supply upon which to draw.

2. Sodium bicarbonate is used in medicine principally for its acid-neutralizing properties. It is used: (a) to combat gastric hyperacidity (see antacid discussion); (b) to combat systemic acidosis; and (c) for miscellaneous uses.

Oral administration of the drug causes a lessening of the acidity of the urine or may even produce an alkalinization. This effect has been used during the administration of certain drugs to increase their effectiveness or lessen the possibility of their crystallizing in the kidneys or urinary tract. Notable among these drugs was sulfanilamide and its related drugs which, in an earlier day, were often prescribed with sodium bicarbonate for the purpose of preventing, if possible, the deposition of crystals of the acetylated sulfa drug in the kidney with consequent mechanical injury to that organ. Experience seems to indicate that large amounts of fluids ingested during the sulfa treatment are more apt to prevent this so-called "crystalluria" than is sodium bicarbonate and, in fact, modern sulfa drugs avoid this problem because the acetylated forms are quite soluble. In addition to the use of sodium bicarbonate with drugs, changing the pH of the reaction of the urine alternately from acid to alkaline has been used in the treatment of certain types of urinary tract infections.

Occasionally, it is found that the simultaneous administration of sodium bicarbonate with other drugs inhibits the activity of the administered drug. Such a therapeutic incompatibility is found in the case of sodium bicarbonate and sodium salicylate which have been prescribed in equivalent amounts, the sodium bicarbonate being administered to alleviate the gastric discomfort attendant upon oral sodium salicylate administration. The administration of sodium bicarbonate and sodium salicylate simultaneously in equal amounts greatly retards the rise in the serum salicylate level, in contrast to sodium salicylate alone which rather quickly brings up the salicylate level. Furthermore, if a satisfactory salicylate level in the blood is reached by sodium salicylate alone, it is found that oral administration of sodium bicarbonate will markedly reduce the level. This has been related to an increased renal excretion in alkaline urine. Sodium bicarbonate, by increasing gastric pH, has been shown to interfere with tetracycline absorption.[48]

Sodium bicarbonate is administered parenterally and orally as the current drug of choice to combat systemic acidosis. A dramatic use in this respect is in the treatment of methyl alcohol poisoning. Inasmuch as the metabolite of methyl alcohol is formic acid, the intravenous use of large quantities of sodium bicarbonate is credited with the saving of many lives. Sodium bicarbonate's effect is used mainly to increase the alkali reserve of the blood and to replace sodium ion in cases of clinical dehydration. Sodium bicarbonate injection has been recommended for the acidotic condition found in drowning victims even before clearing the victim's airways and administering oxygen.[49] Because it could easily be given in excessive amounts, causing an alkalosis, its intelligent use requires that the alkali reserve of the blood be determined before its administration. Since it requires a laboratory analysis to determine the alkali reserve of the blood, and because it is so difficult to sterilize sodium bicarbonate, several other sodium salts have been suggested as replacements, e.g., sodium lactate, sodium acetate, etc. These salts are claimed to be superior because the organic portion of the molecule is oxidized in the tissues to yield essentially sodium bicarbonate, with the advantage that these salts can be sterilized and do not tend to cause alkalosis.

Occurrence:

Sodium Bicarbonate Injection, U.S.P. XVIII

> *Usual Dose:* Intravenous infusion, 500 ml of a 1.4% solution.
> Available as 1% in 20 ml; 1.4% in 500 ml; 5% in 500 ml; 7.5% in 50 ml; 8.4% in 30 ml.

Sodium Bicarbonate Tablets, U.S.P. XVIII

Potassium Bicarbonate, U.S.P. XVIII ($KHCO_3$; Mol. Wt. 100.12: potassium acid bicarbonate)

Potassium Bicarbonate U.S.P. occurs as colorless, transparent, monoclinic prisms or as a white, granular powder which is odorless, and is stable in air. Its solutions are neutral or alkaline to phenolphthalein T.S. *Potassium Bicarbonate* U.S.P. is freely soluble in water and practically insoluble in alcohol. It is officially classified as an electrolyte replenisher and is a component, along with potassium acetate and citrate, of Potassium Triplex® and oral effervescent potassium replacement solutions (K-Lyte®). The chemical properties are the same as those of sodium bicarbonate and the restrictions the same as with any potassium salt. Potassium bicarbonate has been used as an antacid for people who must restrict their sodium intake, but there is risk of hyperpotassemia with prolonged use.

Usual Dose: 1 g four times a day.

Usual Dose Range: 500 mg to 8 g daily.

Sodium Biphosphate, N.F. XIII ($NaH_2PO_4 \cdot H_2O$; Mol. Wt. 137.99) (The chemistry is described under saline cathartics in Chapter 8).

Sodium Biphosphate N.F. is classified as a urinary acidifier since the excess dihydrogen phosphate anion is excreted by the kidneys.

Usual Dose: 600 mg four times a day.

Usual Dose Range: 500 mg to 1 g one to six times a day.

Sodium Citrate, U.S.P. XVIII (Mol. Wt. [anhydrous] 258.07)

$$CH_2CO_2^-Na^+$$
$$|$$
$$HOCCO_2^-Na^+$$
$$|$$
$$CH_2CO_2^-Na^+$$

Sodium Citrate U.S.P. occurs as colorless crystals, or as a white, crystalline powder. It may be either anhydrous or contain two moles of water of hydration. The label must indicate the physical form. As the hydrous salt, sodium citrate is freely soluble in water and very soluble in boiling water. It is insoluble in alcohol.

Sodium citrate is official for its use as an anticoagulant for whole blood. It chelates serum calcium, thereby removing one of the components of blood clotting. Citrates are used for the chelation of other cations, e.g., Benedict's Solution and *Ferrous Sulfate Syrup* N.F. XIII (see discussion on complexation in Chapter 1). Citric acid and its salts are also used as buffering agents (see Chapter 2).

Because citrate, a component of the tricarboxylic acid or Krebs cycle, is rapidly metabolized to carbon dioxide and then to bicarbonate, sodium citrate is used in chronic acidosis to restore bicarbonate reserve. It also has a diuretic effect due to increased body salt concentration. The kidney excretes this extra salt. Patients who must restrict their sodium intake should be careful about using sodium salts for any diuretic effect.

Usual Dose: Oral: Adults, 1 to 2 g every two to four hours as required.

Occurrence:

Anticoagulant Citrate Dextrose Solution, U.S.P. XVIII

Anticoagulant Citrate Phosphate Dextrose Solution, U.S.P. XVIII

Potassium Citrate, N.F. XIII (Mol. Wt. 324.42; tripotassium citrate)

$$CH_2CO_2^-K^+$$
$$|$$
$$HOCCO_2^-K^+ \cdot H_2O$$
$$|$$
$$CH_2CO_2^-K^+$$

Potassium Citrate N.F. occurs as transparent crystals or as a white, granular powder. It is odorless, has a cooling saline taste, and is deliquescent when exposed to moist air. *Potassium Citrate* N.F. is freely soluble in water and almost insoluble in alcohol.

It is classified as an alkalizer and has the same contraindications and precautions as potassium chloride. Potassium citrate is a component of Potassium Triplex® along with potassium acetate and bicarbonate.

Usual Dose: 1 g.

Sodium Lactate ($CH_3CHOHCO_2Na$)

Sodium lactate is commercially available as a mixture with water containing 70–80% sodium lactate. Although the lactate obtained during anaerobic glycolysis has the L-configuration, the commercial preparations are mostly racemic mixtures. Sodium lactate is official as *Sodium Lactate Injection* U.S.P. XVIII. This is a sterile solution of lactic acid in water for injection which has been neutralized with sodium hydroxide. It must be between pH 6.0 and 7.3 to assure that overtitration has not occurred. *Sodium Lactate Injection* U.S.P. is available in a 1/6 M solution in 150-, 250-, and 1000-ml containers as a fluid and electrolyte replenisher used in the treatment of metabolic acidosis.

Usual Dose: Intravenous, 1 liter of a 1/6 molar solution.

Ammonium Chloride, U.S.P. XVIII (NH_4Cl; Mol. Wt. 53.49; ammonium muriate, sal ammoniac, salmiac)

Ammonium Chloride U.S.P. occurs as colorless crystals or as a white, fine or coarse crystalline powder which has a cool, saline taste, and is somewhat hygroscopic. *Ammonium Chloride* U.S.P. is freely soluble in water and in glycerin, and even more soluble in boiling water. It is sparingly soluble in alcohol.

The ammonium cation falls into certain pharmacological categories: (1) acid-base equilibrium of the body; (2) diuretic effect; and (3) expectorant effect.

1. Ammonium ion plays a rather important role in the maintenance of the acid-base equilibrium of the body, particularly in combating acidosis.

Most of the ammonia found in the kidneys comes from the deamination of glutamine and other amino acids. By its ability to excrete ammonia (essentially ammonium ion) the kidney saves base (e.g., sodium) for the body by substituting the ammonium cation for the sodium cation in the compound being excreted (Fig. 5–1).

2. The diuretic effect of ammonium chloride has been extensively studied. The effect is produced by conversion of the ammonium cation (a so-called "labile" cation) to urea [rx (xv)] with consequent formation of a proton (H^+) and a chloride ion (Cl^-), i.e., the equivalent of hydrochloric acid. The hydrogen ion reacts with the body buffers, mainly bicarbonate (HCO_3^-) [rx (xvi)], to form CO_2 and the net effect is, therefore, a displacement of the bicarbonate ion by a chloride ion. The fact that

bicarbonate (usually considered as $NaHCO_3$) is lost indicates that the alkali reserve of the body has been reduced below normal, i.e., an acidosis

$$2NH_4Cl + CO_2 \xrightarrow[\text{steps}]{\text{several}} CO(NH_2)_2 + 2H^+ + 2Cl^- + H_2O \qquad \text{(xv)}$$

$$H^+ + HCO_3^- \rightleftarrows H_2CO_3$$
$$\hookrightarrow H_2O + CO_2 \uparrow \qquad \text{(xvi)}$$

has been induced. It is well to note that this does not imply any significant change in blood pH. The actual diuresis is brought about by the increased chloride load presented to the tubule, causing a state of affairs where all of it cannot be reabsorbed and some thus escapes with an equivalent amount of cation (mostly Na^+) and an isoosmotic equivalent of water.

Termination of the diuretic action is believed to depend upon the acidosis produced, which calls into play a renal defense of the acid-base pattern in which the kidney converts urea back to ammonia and also produces H^+ in exchange for Na^+. Thus, the net effect becomes one of excreting increasing amounts of NH_4^+ and Cl^- (i.e., NH_4Cl) until the amount of NH_4Cl excreted becomes equivalent to the amount ingested for the diuretic action. Diuretic action then ceases after a few days of therapy, making further administration of ammonium chloride inefficient.

3. The expectorant action of the ammonium salts is probably due to local irritation, which in turn is due to a salt action, but this is merely a postulation. However, the ammonium salts have been used extensively in the treatment of coughs associated with a thick viscous sputum. The action of the ammonium salt is to thin out and perhaps to increase the quantity of the mucus. Ammonium chloride and ammonium carbonate particularly have been used in cough preparations.

Ammonium Chloride is admitted to U.S.P. XVIII as a systemic acidifier. The 1971 A.M.A. Drug Evaluations cautions against its use for metabolic alkalosis, as it may perpetuate potassium and sodium depletion.[50] In order to maintain ionic balance in an alkalosis condition, the kidney tubule cells are already excreting sodium and potassium cations in an attempt to conserve protons. In addition, the increased chloride load in the tubule (see above) resulting from administration of ammonium chloride causes increased sodium and potassium excretion, as there must be a cation excreted for each chloride anion excreted.

Ammonium chloride is contraindicated in patients with impaired hepatic or renal function because of the risk of ammonia toxicity. Intravenous injection must be given slowly to permit metabolism of ammonium ions by the liver and to avoid ammonia toxicity.

Ammonium chloride has been used to potentiate mercurial diuretics or to correct hypochloremic alkalosis that occurs after prolonged use of these mercurial compounds.

Usual Dose (Sytemic Acidifiers): Oral, 1 to 2 g four times a day. Intravenous infusion, 500 ml of a 2% solution over a period of three hours.
Usual Dose Range: Oral, 4 to 12 g daily. Intravenous, 100 ml to 1 liter of a 2% solution daily.
Occurrence:

Ammonium Chloride Injection, U.S.P. XVIII
   Usually available as 160 mg in 30 ml; 600 mg in 100 ml; 10.7 g in 500 ml; 21.4 g in 1000 ml.

Ammonium Chloride Tablets, U.S.P. XVIII
   Usually available as 500-mg and 1-g enteric-coated tablets.

Electrolyte Combination Therapy

Infusions. In short-term therapy, such as following surgery, infusion of a standard glucose and saline solution may be adequate; however, when deficits are severe or protracted, solutions containing additional electrolytes are usually required. While combinations compounded according to the needs of each individual patient would be ideal, this usually is not feasible from the standpoint of cost and sterility. Furthermore, there is a broad selection of commercial electrolyte infusion solutions with differing amounts of electrolytes available. One of these will usually fit the electrolyte needs of a patient.

These combination products can be divided into two groups: fluid maintenance and electrolyte replacement. Maintenance therapy with intravenous fluids is intended to supply normal requirements for water and electrolytes to patients who cannot take them orally. All maintenance solutions should contain at least 5% dextrose. This minimizes the buildup of those metabolites associated with starvation: urea, phosphate, and ketone bodies. In addition to dextrose, the general electrolyte composition of maintenance solutions is 25–30 mEq/l Na, 15–20 mEq/l K, 22 mEq/l Cl, 20–23 mEq/l HCO_3 (or equivalent amounts of lactate or acetate), 3 mEq/l Mg, and 3 mEq/l P.

Replacement therapy is needed when there is heavy loss of water and electrolytes, as in prolonged fever, severe vomiting, and diarrhea. There are usually two types of solutions used in replacement therapy: a solution for rapid initial replacement and a solution for subsequent replacement. The electrolyte concentrations in solutions for rapid initial replacement more or less resemble the electrolyte concentrations found in the extracellular fluids. Some may have larger amounts of potassium. Typical concentration ranges are 130–150 mEq/l Na, 4–12 mEq/l K, 98–109 mEq/l Cl, 28–55 mEq/l HCO_3 (or equivalent amounts of lactate, acetate, or gluconate), 3–5 mEq/l Ca, and 3 mEq/l Mg. Electrolyte concentrations in subsequent replacement solutions are 40–121 mEq/l Na, 16–35 mEq/l K, 30–103 mEq/l Cl, 16–53 mEq/l HCO_3 (or equivalent amounts of lactate or acetate), 0–5 mEq/l Ca, 3–6 mEq/l Mg, and 0–13 mEq/l P. The wide variance allows the clinician to select a solution which best fits the electrolyte and acid-base needs of the patient.[51]

Official Combination Electrolyte Infusions

Ringer's Injection, U.S.P. XVIII

Each liter contains 8.6 g sodium chloride, 0.3 g potassium chloride, and 0.33 g calcium chloride (as the dihydrate). This is equivalent to 147 mEq/l Na, 4 mEq/l K, 4.5 mEq/l Ca, and 155.5 mEq/l Cl. It is usually available in 500- and 1000-ml injections.

Usual Dose: Intravenous infusion, 1 liter.

Lactated Ringer's Injection, U.S.P. XVIII

Each 100 ml contains 600 mg sodium chloride, 310 mg sodium lactate, 30 mg potassium chloride, and 20 mg calcium chloride (as the dihydrate). These are equivalent to 130 mEq/l Na, 4 mEq/l K, 2.7 mEq/l Ca, 109.7 mEq/l Cl, and 27 mEq/l lactate. It is usually available as 150-, 250-, 500- and 1000-ml injections.

Usual Dose: Intravenous infusion, 1 liter.

Oral Electrolyte Solutions. Orally administered electrolyte solutions are available. Two (Lytren® and Pedialyte®) are used to supply water and electrolytes in amounts needed for maintenance as soon as intake of usual foods and liquids is discontinued, and before serious fluid losses or deficits occur. They are also given to replace mild to moderate fluid losses due to diarrhea and other conditions associated with excessive fluid loss or deficient fluid intake. These solutions are not intended to promote the total water requirements of the individual. If additional liquid is needed, water or other nonelectrolyte fluids should be given.

There have been sodium chloride and sodium chloride plus dextrose tablets available for years to replace the salt lost through excessive perspiration. An oral sodium chloride sustained release preparation has been reported.[52] These have been used by people working in hot (either due to weather or industrial environment) areas who could suffer from the hyponatremic water intoxication syndrome (cramps, fever, and confusion) if they replaced their fluid loss by water only. For many individuals, adding additional salt to their food will provide the additional sodium chloride that is needed.

In the last few years, oral electrolyte solutions have been introduced, originally for athletes, to replace fluid and electrolytes lost through excessive perspiration as well as to quench thirst. A liter of these preparations (Gatorade®, Sportade®, and Bike Half-Time Punch Mix®) will provide the equivalent of 1–2 salt tablets. There appears to be no real need for the potassium in these preparations.[53] Table 5–8 lists the electrolyte composition of several of these oral preparations.

TABLE 5-8. ORAL ELECTROLYTE PREPARATIONS

Preparation	Electrolytes (mEq/l)									Carbohydrate
	Na^+	K^+	Ca^{+2}	Mg^{+2}	Citrate	SO_4^{-2}	Cl^-	P	Lactate	
Lytren®	25	25	4	4	15	4	30	5	4	280 cal/l (dextrose)
Pedialyte®	30	20	4	4			30		28	192 cal/l (50 g sucrose)
Sodium Chloride and Dextrose Tablets, N.F. XIII	3.4/tab						3.4/tab			450 mg dextrose/tab
Sodium Chloride Tablets, U.S.P. XVIII	10.3, 17, and 38/tab						10.3, 17, and 38/tab			
Gatorade®	21	3								50 g dextrose/l
Sportade®	27.4	14								7 g glucose/l and 40 g sucrose/l
Bike Half-Time®	24	2								1 g glucose and 76 g sucrose/l

References

General Review

Searcy, R. L. Diagnostic Biochemistry. New York: McGraw-Hill, 1969.

Tietz, N. W. Fundamentals of Clinical Chemistry. Philadelphia: W. B. Saunders, pp. 612–697, 1970.

1. A.M.A. Drug Evaluations, 1st ed. Chicago: American Medical Association, pp. 119, 1971.
2. Lotz, M., Zisman, E., and Bartter, F. C. Evidence for phosphorus-depletion syndrome in man. New Eng. J. Med., **278**: 409, 1968.
3. Travis, S. F., Sugerman, H. J., Rubery, R. L., Dudrick, S. J., Delivoria-Papadopoulos, M., Miller, L. D., and Oski, F. A. Alterations of red-cell glycolytic intermediates and oxygen transport as a consequence of hypophosphatemia in patients receiving intravenous hyperalimentation. New Eng. J. Med., **285**:763, 1971.
4. Davis, B. B., and Knox, F. G. Current concepts of the regulation of urinary sodium excretion—a review. Amer. J. Med. Sci., **259**:373, 1970.
5. PGA vs. essential hypertension. Med. World News, **11**:20F, Oct. 23, 1970.
6. Edmonds, C. J., and Jasani, B. Total-body potassium in hypertensive patients during prolonged diuretic therapy. Lancet, **2**:8, 1972.
7. Chem. Eng. News, **46**:41, Nov. 11, 1968.
8. Hunter, J., Maxwell, J. D., Stewart, D. A., Parsons, V., and Williams, R. Altered calcium metabolism in epileptic children on anticonvulsants. Brit. Med. J., **4**:202, 1971.
9. Salter, F. J. Inorganic phosphates in the treatment of hypercalcemia. Drug Intel. Clin. Pharm., **4**:4, 1970.
10. Eisenberg, E. Effect of intravenous phosphate on serum strontium and calcium. New Eng. J. Med., **282**:889, 1970.
11. Goldsmith, R. S. Multiple effects of phosphate therapy. New Eng. J. Med., **282**:927, 1970.
12. Hebert, L. A., Lemann, J., Petersen, J. R., and Lennon, E. J. Studies of the mechanism by which phosphate infusion lowers serum calcium concentration. J. Clin. Invest., **45**:1886, 1966.
13. Condon, J. R., Nassim, J. R., Millard, F. J. C., Hilbe, A., and Stainthorpe, E. M. Calcium and phosphorus metabolism in relation to lactose tolerance. Lancet, **1**: 1027, 1970.
14. Pechet, M. M., Bobadilla, E., Carroll, E. L., and Hesse, R. H. Regulation of bone resorption and formation. Amer. J. Med., **43**:696, 1967.
15. Foster, G. V., Doyle, F. H., Bordier, P., and Matrajt, H. Effect of thyrocalcitonin on bone. Lancet, **2**:1428, 1966.
16. West, T. E. T., Joffe, M., Sinclair L., and O'Riordan, J. L. H. Treatment of hypercalcaemia with calcitonin. Lancet, **1**:675, 1971.
17. Rasmussen, H. Cell communication, calcium ion, and cyclic adenosine monophosphate. Science, **170**:404, 1970.
18. Hoyle, G. How is Muscle Turned On and Off? Sci. Amer., **223**:85, April 1970.
19. Biochemical undoing of common calculi. Med. World News, **13**:17, June 21, 1972.
20. Nordin, B. E. C. Clinical significance and pathogenesis of osteoporosis. Brit. Med. J., **1**:571, 1971.
21. Hassain, M., Smith, D. A., and Nordin, B. E. C. Parathyroid activity and postmenopausal osteoporosis. Lancet, **1**:809, 1970.
22. Ellis, F. R., Holesh, S., and Ellis, J. W. Incidence of osteoporosis in vegetarians and omnivores. Amer. J. Clin. Nutr., **25**:555, 1972.
23. Osteoporosis and fluoride therapy. Brit. Med. J., **3**:660, 1970.
24. Prevention and treatment of postmenopausal and "senile" osteoporosis. Med. Lett., **11**:101, 1969.
25. Halting bone resorption in Paget's. Med. World News, **12**:16, July 10, 1970.

26. Woodhouse, N. J. Y., Reiner, M., Bordier, P., Kalu, D. N., Fisher, M., Foster, G. V., Joplin, G. F., and MacIntyre, I. Human calcitonin in the treatment of Paget's bone disease. Lancet, **1**:1139, 1971.
27. Harris, I., and Wilkinson, A. W. Magnesium depletion in children. Lancet, **2**:735, 1971.
28. Wacker, W. E. C. The biochemistry of magnesium. Ann. N.Y. Acad. Sci., **162**:717, 1969.
29. Caddell, J. L. Magnesium deficiency in protein–calorie malnutrition: A followup study. Ann. N.Y. Acad. Sci., **162**:874, 1969.
30. Seelig, M. S. Electrographic patterns of magnesium depletion appearing in alcoholic heart disease. Ann. N.Y. Acad. Sci., **162**:906, 1969.
31. Jones, J. E., Shane, S. R., Jacobs, W. H., and Flink, E. B. Magnesium balance studies in chronic alcoholism. Ann. N.Y. Acad. Sci., **162**:934, 1969.
32. Med. World News, **13**:12M, Feb. 4, 1972.
33. Belgrade, M. L. Magnesium metabolism. J. Appl. Nutr., **21**:48, 1969.
34. Levitt, A. E. Alcoholism: Costly health problem for U.S. industry. Chem. Eng. News, **49**:25, Sept. 13, 1971.
35. Mendelson, J. H., Ogata, M., and Mello, N. K. Effects of alcohol ingestion and withdrawal on magnesium studies of alcoholics: Clinical and experimental findings. Ann. N.Y. Acad. Sci., **162**:918, 1969.
36. Wolfe, S. M., and Victor, M. The relationship of hypomagnesemia and alkalosis to alcohol withdrawal symptoms. Ann. N.Y. Acad. Sci., **162**:973, 1969.
37. Flink, E. B. Therapy of magnesium deficiency. Ann. N.Y. Acad. Sci., **162**:901, 1969.
38. A.M.A. Drug Evaluations, 1st ed. Chicago: American Medical Association, p. 124, 1971.
39. DeGenaro, F., and Nyban, W. L. Salt—a dangerous "antidote." J. Pediat., **78**: 1048, 1971.
40. Potassium formulations and therapy. Med. Lett., **11**:77, 1969.
41. Kramer, W., Tanja, J. J., and Harrison, W. L. Precipitates found in admixtures of potassium chloride and dextrose 5% in water. Amer. J. Hosp. Pharm., **27**:548, 1970.
42. Cobey, J. C. How to make HCl solutions tasty. New Eng. J. Med., **248**:395, 1971.
43. Cash, R. A., Toha, K. M. M., Nalin, D. R., Huq, Z., and Phillips, R. A. Acetate in the correction of acidosis secondary to diarrhea. Lancet, **2**:302, 1969.
44. Watten, R. H., Gutman, R. A., and Fresh, J. W. Comparison of acetate, lactate, and bicarbonate in treating the acidosis of cholera. Lancet, **2**:512, 1969.
45. Eliahou, H. E., Feng, P. H., Weinberg, U., Iaina, A., and Reisin, E. Acetate and bicarbonate in the correction of uraemic acidosis. Brit. Med. J., **4**:399, 1970.
46. Hicks, C. I., Gallardo, J. P. B., and Guillory, J. K. Stability of sodium bicarbonate injection stored in polypropylene syringes. Amer. J. Hosp. Pharm., **29**:210, 1972.
47. Deluca, P. P., and Kowalsky, R. J. Problems arising from the transfer of sodium bicarbonate injection from ampuls to plastic disposable syringes. Amer. J. Hosp. Pharm., **29**:217, 1972.
48. Barr, W. H., Adir, J., and Garrettson, L. Decrease of tetracycline absorption in man by sodium bicarbonate. Clin. Pharmacol. Ther., **12**:779, 1971.
49. Sodium bicarb. proves a lifesaver in near-drownings. Med. World News, **10**:20, Nov. 21, 1969.
50. A.M.A. Drug Evaluations, 1st ed., Chicago: American Medical Association, pp. 127–128, 1971.
51. Parenteral water and electrolyte solutions. Med. Lett., **12**:77, 1970.
52. Clarkson, E. M., Curtis, J. R., Jewkes, R. J., Jones, B. E., Luck, V. A., deWardener, H. E., and Phillips, N. Slow sodium: An oral slowly released sodium chloride preparation. Brit. Med. J., **3**:604, 1971.
53. Gatorade and other oral electrolyte solutions. Med. Lett., **11**:71, 1969.

6
Essential and Trace Ions

The ions to be discussed in this chapter are restricted to those that, based on current knowledge, have specialized biochemical functions. They are not found in the general electrolyte replacement preparations. At least two of them, iron and iodide, show a defined deficiency syndrome. Specific deficiency symptoms for some of the others are seen in animals other than man.

Iron

Iron is present in some form wherever respiration occurs in higher animals. It is essential to the elementary metabolic processes in the cell. In the respiratory chain, iron functions as an electron carrier. Iron is responsible for the transport of molecular oxygen in higher organisms. Both of these functions depend on the ability of iron to exist in coordination compounds in different states of oxidation and bonding.

Table 6–1 shows the distribution of iron in blood and in other tissues, and its function in these tissues.

TABLE 6–1. BODY COMPONENTS CONTAINING IRON*

Occur-rence	Iron Bound as	Mode of Linkage	Function	Iron Content Total†	Iron Content % of Body Iron
Blood system	Hemoglobin	Heme	O_2 transport	3 g	65.4
	Plasma	Transferrin	Fe transport	4 mg	0.1
Tissues	Functional iron (myoglobin, cell hemes)	Heme	Cell respiration	650 mg	13.9
	Storage iron	Ferritin Hemosiderin	Iron pool Detoxication	} 1 g	} 21.5
				4.65 g	100.0

* From Problems of iron metabolism with special reference to biochemical, physiological, and clinical aspects, by W. Keiderling and H. P. Wetzel, Angew. Chem. [Eng.] **5**: 633, 1966. By permission.
† Approximate values.

Most of the iron found in the body is associated with two types of protein: (1) hemoproteins and (2) iron storage and/or transport proteins. (1) *Hemoproteins* are those iron-containing proteins responsible for respiration (i.e., respiratory enzymes) and for carrying oxygen. Cytochrome c is an example of a respiratory enzyme in which iron is complexed in a porphyrin (i.e., heme) ring system which is in turn covalently bonded to the protein portion of the molecule. It is of interest that the porphyrin system of cytochrome c differs from the porphyrin of hemoglobin in the side chains of the molecule. In addition, the heme portion of the cytochrome c molecule is shielded by the protein chain to prevent oxygen from interacting with iron. The iron functions as an electron carrier and can be present as ferrous (Fe^{+2}) or ferric (Fe^{+3}) as it picks up or donates an electron in the process of electron transfer. Other oxidative enzymes containing iron are catalase and the peroxidases.

The other group of hemoproteins, represented by myoglobin and hemoglobin, stores and/or transports oxygen. Hemoglobin consists of four protein chains, each of which contains a heme unit of a porphyrin ring and ferrous iron. The iron complexes molecular oxygen by utilizing a vacant orbital which can be used by a pair of nonbonding electrons from oxygen. The release and uptake of molecular oxygen are influenced by the oxygen tension, pH, carbon dioxide concentration, and the presence of 2,3-diphosphoglycerate. Patients suffering from iron-deficiency anemia have a decreased capability for transporting oxygen.

(2) *Iron storage and/or transport proteins* are the body's method of handling these requirements. Ferritin and hemosiderin are iron storage proteins found in the liver, spleen, and bone marrow. Ferritin is a water soluble, crystallizable iron protein built up from apoferritin and micelles of a colloidal ferric hydroxide-phosphate complex. Although the iron in ferritin is stored in the Fe^{+3} form it is incorporated and released in the Fe^{+2} form. Hemosiderin, on the other hand, is water insoluble and is considered by some to be a dehydrated ferritin.[1] (See discussion of copper in this chapter for the postulated role of ceruloplasm in iron oxidation-reduction.) The major iron transport protein of blood plasma is a glycoprotein known as *transferrin* (siderophilin). It binds two atoms of ferric iron per molecule so tightly that, for all practical purposes, there is no free plasma iron. Transferrin releases iron to the red cell precursor by attaching to a receptor on the surface of the developing red blood cell.[2]

The body is very efficient at recycling the iron obtained from broken-down red blood cells. As a result, daily iron requirements are low: 10–12 mg for a male and 12–18 mg for a female for both of whom only 5 to 10% is absorbed. Iron is lost in sloughed-off cells, hemorrhaging, and in menstrual flow, bile, and other secretions. For a male this amounts to an iron loss of 0.6 mg per day. A nonpregnant female loses 1.2 to 1.8 mg of iron per day after puberty due to menstrual flow. During pregnancy, the iron loss increases to 3 to 4 mg per day due to placental iron transport. Iron supplementation is common during pregnancy.

For most people, diet is the source of replacement iron but it turns out that simply because a food is "rich" in iron, it may not be a good source of iron. In general, iron in liver and muscle is better absorbed than is the iron in eggs and leafy vegetables and, in particular, the iron in wheat, corn, and black beans is relatively unavailable to the body.[3,4] This may be explained by the observation that hemoglobin iron is well absorbed because it is still bound in the porphyrin ring. Other iron complexes, such as ferritin, are poorly absorbed because the protein must first be digested by the gastrointestinal proteases before absorption can occur.[5] Once released from its bound state in food, the inorganic iron can either be complexed by sugars, ascorbic acid, citric acid, and amino acids from the meal or, in their absence, be complexed by a gastric mucopolysaccharide. The net result is that iron is prevented from being precipitated as the pH of the gastric contents increases in the small intestine. It is still not known how iron is released from these complexes to be absorbed across the intestinal wall.[6] Since the dietary need for iron is limited, the body possesses a control mechanism for the absorption of iron across the intestinal wall. Most iron absorption occurs in the upper portion of the duodenum where the contents are still

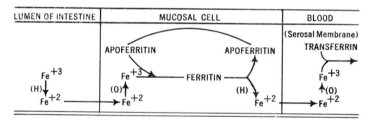

FIG. 6–1. *The Mucosal Block Hypothesis.* Iron absorption regulated and controlled by availability of apoferritin. From Saltman, P. The role of chelation in iron metabolism. J. Chem. Ed., **42**:683, 1965—with some modification.

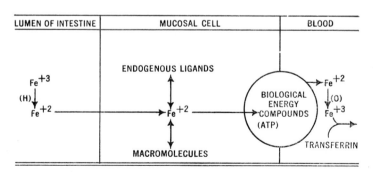

FIG. 6–2. *The Active Transport Hypothesis.* Regulation and control of iron absorption by regulating the amount of cellular energy available for active transport. From Saltman, P. The role of chelation in iron metabolism. J. Chem. Ed., **42**:683, 1965—with some modification.

acidic. Iron transport into the intestinal mucosa is facilitated by ascorbic acid, hydrochloric acid, fructose, and other organic molecules which tend to hold iron in a soluble ferrous state. The presence of phosphate and oxalate causes the precipitation of insoluble iron salts. There have been three postulates put forth as explanations of the control of intestinal iron absorption: (1) the mucosal block hypothesis (Fig. 6–1); (2) the active transport hypothesis (Fig. 6–2); and (3) the iron-chelate hypothesis (Fig. 6–3).

The *mucosal block* postulation suggests that dietary or administered iron is reduced to the ferrous form, which diffuses into the mucosal cell where it is reoxidized and then combined with apoferritin (which is being continually formed and destroyed) to form stable ferritin, the iron-carrying protein. As ferritin it crosses the cell and is released to be reduced again to ferrous iron for diffusion across the serosal cell membrane (membrane covering the intestines) and eventual reoxidation to ferric iron and combination with the iron-transport protein, transferrin. In this form it is transferred to the liver for storage or to the bone marrow for use in heme synthesis for erythrocyte production. The key to the mucosal block hypothesis is the assertion that only small amounts of ferritin can be formed in any one cell. Once the full complement of ferritin is obtained for a cell it can no longer pick up iron, no matter what the concentration is in the lumen of the intestine. Further absorption occurs only in cells that do not have their full amount of ferritin or if the ferritin unloads its iron through the serosal membrane to regenerate apoferritin. There are numerous arguments against this hypothesis, among which are the facts that no maximum limit of absorption has been demonstrated, that increased amounts (although smaller percentages) of iron are absorbed from larger doses, that unphysiologic amounts of iron are required to show the blocking effect, and that nonferritin-bound iron is found in intestinal mucosal cells.

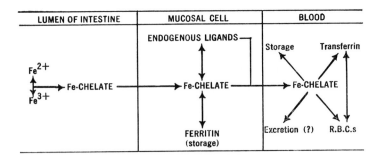

Fig. 6–3. *The Iron-Chelate Hypothesis.* Primary control of iron absorption resides in the presence of endogenous or exogenous chelating agents which are able to bind either Fe^{2+} or Fe^{3+} to form soluble low molecular weight complexes. From Rogers' Inorganic Pharmaceutical Chemistry, 8th ed., by T. O. Soine and C. O. Wilson, Lea & Febiger, Philadelphia, p. 612, 1967. By permission.

The second hypothesis may be termed the *active transport* mechanism.[7,8] As in the mucosal block mechanism, ferrous iron enters the mucosal cell by diffusion, where it combines with endogenous low molecular weight ligands or is stored as ferritin.   To cross the serosal membrane into the blood a specific transport system intimately linked to adenosine triphosphate (ATP) has been suggested.   The control of iron entry into the blood occurs by this active transport system.   Once past the serosal membrane the events are the same as postulated by the mucosal block hypothesis.   Although attractive, this hypothesis fails in that there has been no demonstration that iron movement across the serosal membrane is dependent upon metabolic energy. Perhaps one of the most telling arguments against this hypothesis is the fact that iron movement is not affected by an anaerobic condition, whereas other known active transport processes (e.g., Na^+) are vitally affected.

The third mechanism proposes that primary control is exerted by exogenous or endogenous ligands or chelating agents, which can bind either oxidation state of iron to form low molecular weight complexes capable of passively diffusing through the mucosal cell membrane from the intestine.[9] No prior reduction to ferrous iron is postulated.   Within the cell the iron can be transferred to other endogenous ligands or stored as ferritin.   Diffusion across the serosal membrane is viewed as occurring with either the original chelating material or with some endogenous ligand.   Once across the membrane the iron is transported in the chelated form to depot cells where it is transferred to transferrin.   The sequence following this is the same as in the other hypotheses.   The major attributes of this theory are that no redox reactions or metabolic energy requirements are directly involved.   The hypothesis needs experimental proof.   Indeed, there is doubt that low molecular weight chelates are present in the gastrointestinal mucosal cell.[10]   The suggestion that both ferrous and ferric ions are equally complexed for diffusion into the mucosal cell could be difficult to rationalize in face of the large body of empirical fact indicating the superior absorption of ferrous iron over ferric.   On the other hand, much of the "superiority" of ferrous iron over ferric iron may be explained by the poorer solubility of ferric salts as compared to ferrous salts.

Recent *in vivo* studies using rat intestine and ^{59}Fe-labeled ferrous sulfate have shown that newly absorbed iron appears in distinct separable fractions within the duodenal mucosal cells.   Within one hour, 90% of the iron is incorporated into ferritin in the normal animal.   Mucosal iron distribution is different in the iron-deficient animal, with only minimal incorporation of iron into ferritin until the major proportion of iron, unbound as a free salt, has been transferred into the body.   Then the level of iron found in ferritin in the mucosal cell approaches that of the normal animal.   The data from the above experiment tend to disprove the mucosal block theory, in that the binding of iron by ferritin does not significantly inhibit mucosal transfer of physiological doses of iron. Rather, the distribution of newly absorbed iron in the mucosal cell during the early phase of absorption is merely a function of the initial rate of uptake and amount of apoferritin or partially saturated ferritin molecules present in the cell at the time of exposure to iron. The initial rate of uptake appears to be at least partially dependent upon the body's iron requirements at the time of exposure to iron. Also, because of the rapid cellular turnover of the gastrointestinal mucosa, the state of its iron content can be quickly varied, depending upon the conditions existing at the time of production of new mucosal cells. This implies that the gastrointestinal mucosal cell is a labile iron pool whose iron content reflects the immediate needs of the animal for iron.[10] The above data and conclusions most fit the postulated iron-chelate mechanism except that there are no low molecular weight ligands involved. If nothing else, it should be obvious that any investigation of gastrointestinal iron absorption must be carried out under very carefully defined and controlled conditions in terms of the iron needs of the test animal.

Another problem to be answered is how the body communicates to the intestine that additional iron is needed. None of the above postulates and data fully explains control of gastrointestinal uptake in terms of the body's needs. Contrary to earlier work, a pancreatic secretion probably has no direct control on iron absorption.[11] The response of the intestine to iron depletion is more rapid when erythropoiesis (red blood cell production) is stimulated than when storage sites in the liver are reduced. This has led to the concept that as transferrin becomes depleted of iron during red blood cell formation, transferrin returns to the intestinal wall to pick up more iron, thereby "signaling" the intestine to absorb more iron.[12]

Occasionally some aspect of the control mechanism fails and a condition known as pathological hemochromatosis can develop, a rare disease found predominantly in the male. In this disease the body absorbs iron in quantities greater than the amounts being excreted, leading to siderosis (the deposition of iron in the tissues). The cause is not known since the mechanism regulating iron absorption is unknown. Indeed, as pointed out in the above discussion, there are a number of factors affecting iron absorption which possibly make hemochromatosis a family of disorders. Excess iron deposited in the liver causes cirrhosis, in the pancreas causes diabetes mellitus, and in the skin produces a bronzed appearance. Phlebotomy (bleeding) is the treatment of choice.[13]

A person deficient in iron will become anemic. Anemia is a general term for a condition in which circulating red blood cells are deficient in number, or deficient in total hemoglobin content, per unit of blood volume. The net result is a lower oxygen-carrying capacity by the blood. Anemias can be classified by etiology or by description of the red blood cell.

Anemias can be caused by excessive blood loss or destruction, or they may be due to decreased blood formation. Excessive blood loss can be

TABLE 6–2. IRON BALANCE

	Iron Requirements mg	Dietary Iron mg	Required Absorption %
Men	0.9 (0.6–1.2)	15	6 (4–8)
Menstruating Women	1.3 (0.7–2.5)	10	13 (7–25)
Pregnant Women	2.5 (2.0–5.0)	10	25 (20–50)

From Iron-deficiency anemia, by C. A. Finch, Amer. J. Clin. Nutr., **22**:512, 1969. By permission.

caused by hemorrhaging, menstrual flow, and bleeding ulcer. Blood destruction is caused by hemolytic agents (drug therapy, infections, and toxins) or defective hemoglobins (sickle cell anemia, thalassemias). Anemia due to decreased blood formation can be caused by deficiencies of key materials (cobalamin, folic acid, iron, pyridoxine), infections, renal insufficiency, malignancy, and marrow failure.[14] Treatment ranges from transfusions and treatment of the pathological condition causing the anemia to administration of the deficient material necessary for red blood cell formation. Supplemental iron therapy is indicated if there is inadequate iron intake, excessive loss of iron due to blood loss, or inadequate absorption of iron. Iron-deficiency anemia is widespread throughout the world and is more common in women and infants than in men. In particular, the pregnant or menstruating female has sharply increased iron requirements, meaning that she must absorb a greater percentage of her dietary iron (Table 6–2).

Compounding the problem in females is the trend in recent years for the American woman to reduce her caloric intake.[15] Furthermore, food prepared in the so-called developed countries under sanitary conditions in aluminum, stainless steel, glass, ceramic, and plastic utensils may contain less iron than in those countries where food is prepared in iron vessels. It is suggested that one third of dietary iron may be due to extrinsic contamination when iron vessels are used in food preparation.[16,17] Whether this extrinsic iron is in a form suitable for absorption has not been ascertained.

Since iron-deficiency anemia is not uncommon throughout the entire U.S. population, there have been proposals that the iron content of enriched flours and breads be raised to certain minimum levels. In terms of iron-deficiency anemia, the population can be divided into three groups: infants, men, and women.

The infant's iron stores become exhausted due to: (1) a very low iron intake in the first few months of life (see discussion below) which is needed to maintain the iron stores present at birth; and (2) high iron requirements which are required for the rapid growth of the first year of life. The problem is nutritional and can be prevented by fortifying those foods commonly eaten by an infant in its early life (milk formulas, dried milk, infant foods).

The cause of iron deficiency in the adult male has traditionally been considered to be non-nutritional since a male should have no trouble obtaining adequate amounts of iron by eating a balanced diet. Rather, some pathological process may be responsible, most frequently gastrointestinal bleeding. This means that, for the male, iron-deficiency anemia usually requires a clinical investigation of its cause. Nevertheless, U.S. Army studies of the iron nutritional status of male recruits indicate that a significant percentage of U.S. males may have below normal iron stores, due to not eating a proper diet as civilians.[18]

The adult female's iron requirements fit into a third category. Her diet may be adequate in terms of proteins, vitamins, and caloric requirements, but may still need additional iron due to her normally higher iron requirements (menses and pregnancy). Even if she does not develop iron-deficiency anemia, her iron stores may still be inadequate causing later health problems.[18]

For the above reasons, the Food and Drug Administration, with the concurrence of the American Medical Association's Council on Foods and Nutrition, has formally proposed that suitable iron salts be added to enriched wheat flour, farina (starch), bread, buns, and rolls. There has been some opposition expressed that such increased enrichment could be potentially harmful to the largely male segment of the population suffering from hemochromatosis (discussed above).[20,21] This is a rare disease, and many physicians have never seen a case. Most of the time hemochromatosis can only be confirmed at autopsy when the affected organs are examined. Currently, it is concluded that it is infeasible to carry out definitive clinical studies to determine whether moderate enrichment of cereal-based foods increases the prevalence or the severity of pathological hemochromatosis. The problem is that there is no satisfactory experimental animal model, and the disease in humans is rare. Also, the amounts proposed for the iron enrichment program are only in quantities capable of reducing the occurrence of iron-deficiency anemia and improving inadequate iron stores. There would not be enough iron present to treat already developed iron-deficiency anemia. In general, the new regulations would increase the daily iron intake by 20 to 25%.[19] Nevertheless, restraint will have to be exercised in the fortification of foods with iron and research in the field of iron metabolism will have to be expanded.*

Iron-deficiency anemia in infants and preschool children is common among the poor. The full-term infant has about 250 mg of iron, of which about 150 mg are bound by hemoglobin, a value found even when the mother is iron-deficient. This store of iron is sufficient to prevent anemia during the first few months of life. This is an important protective mecha-

---

* February 1974, the F.D.A. stayed the proposed regulation pending further study.

nism when one considers that the largely milk diet (human or cow) consumed during the first few months of the newborn's life supplies relatively little iron.[22]

In the second six months of life, the infant needs 0.9 mg of iron each day. Much of this is supplied by infant foods and proprietary milk formulas. However, infants of poor income families are traditionally fed whole milk which is less expensive, but contains little iron. Whole milk is difficult to fortify since iron hastens the development of rancidity and imparts an objectionable color change when the milk is used for cooking or in coffee.[23] There is also evidence that there is a factor in fresh cow's milk (homogenized and pasteurized) which interferes with iron absorption by inducing an allergic response in the intestinal tract of some infants and, thereby, impedes iron absorption across the intestinal wall. Use of evaporated milk, a soybean substitute, or human milk usually returns to normal those clinical symptoms localized in the intestinal tract that are normally associated with iron deficiency.[24,25] In other words, many gastrointestinal symptoms normally associated with iron-deficiency anemia in infants may be due to an allergy to fresh milk rather than to the iron deficiency. A change in the dairy part of the diet along with iron administration may be indicated for these children.

Iron deficiencies can also occur in patients being maintained on hemodialysis.[26,27] Evidence has also been published that febrile (feverish) illness will depress iron absorption, independently of the patient's nutritional state.[28]

An iron compound used for replacement or supplemental therapy must meet two requirements. It must (1) be biologically available and (2) be nonirritating. There is considerable disagreement in the literature as to which iron salts are sufficiently available biologically. Usually water soluble ferrous sulfate is the standard to which other iron salts are compared. If the test subjects have sufficient body iron stores, so little iron will be absorbed that little difference will be noted between iron preparations. On the other hand, it is apparently difficult to gather a sufficient number of iron-deficient humans who are definitely anemic at any one time in order to obtain statistically meaningful results, because test animals are usually utilized. The problem then becomes one of finding a test animal whose intestinal absorption of iron closely resembles that of man.

There is also the problem of testing the iron salt in a realistic food situation. In the case of breads and other bakery products, the iron salt is either carefully blended into the flour used to prepare the bakery product or is incorporated into the dough at the bakery. During the baking process, a significant percentage of the soluble ferrous salts is probably oxidized to the poorly soluble ferric salts. This may explain why comparisons between ferrous sulfate and other iron salts can show little differences in bioavailability.

In a carefully designed evaluation of iron salts commonly used for food fortification, day-old chicks and weanling rats were made anemic on iron-deficient diets and fed test diets containing an iron salt. Their hemoglobin regeneration was then measured. Salts which were 75 to 100% as available as ferrous sulfate for hemoglobin regeneration included the dihydrogen ferrous salt of EDTA, ferric ammonium citrate, ferric choline citrate, ferric citrate, ferric glycerophosphate, ferric sulfate, ferrous ammonium sulfate, ferrous chloride, ferrous fumarate, ferrous gluconate, and ferrous tartrate. Salts which were 25 to 75% as available as ferrous sulfate included ferric chloride, ferric pyrophosphate, ferric orthophosphate, and reduced iron. Iron salts found to be mediocre sources of iron were ferric oxide, ferrous carbonate, and sodium iron pyrophosphate.[29]

When iron supplements are prescribed, the oral route is the method of choice. The ferrous salts are absorbed better, with the incidence of untoward effects being approximately the same with ferric as with ferrous compounds. Iron is an astringent (protein precipitant) and can irritate the gastrointestinal tract. For this reason, iron is usually given after a meal when there is food in the stomach although better absorption occurs when taken between meals. Sustained-release iron dosage forms have been utilized as a means of minimizing the irritant properties of iron. There is some question as to whether these sustained-release iron preparations are advantageous, since most iron absorption occurs in the duodenum and the unabsorbed iron is apt to pass beyond this area.

Part of the problem of comparing iron-containing products is measuring a clinically significant hematopoietic response to iron. For example, molybdenized ferrous sulfate sustained-release capsules can be shown to give a greater percentage increase of serum iron levels over fasting levels as compared to a sustained-release exsiccated ferrous sulfate capsule.[30] A comparison of equivalent doses of molybdenized ferrous sulfate sustained-release capsules with *Ferrous Sulfate* U.S.P. capsules in a year-long study showed nearly the same results when measuring hematocrit and hemoglobin concentrations. However, there definitely was more incidence of side effects, leading to the stoppage of therapy with the conventional *Ferrous Sulfate* U.S.P. capsules.[31] A three-way double-blind study involving a sustained-release ferrous sulfate tablet, an equivalent amount of a conventional ferrous sulfate tablet, and a placebo showed no statistically significant difference in hemoglobin levels or hematocrit levels between the two ferrous sulfate preparations. However, the conventional ferrous sulfate preparation had a much higher incidence of side effects as compared to the placebo.[32] In contrast to the above, a comparison of serum iron level per unit time of administering a sustained-release dosage form containing ferrous sulfate or fumarate with that of an equivalent amount of ferrous sulfate or fumarate in a conventional dosage form showed only one sustained-release product to have better intestinal absorption properties.

When ferrous sulfate is given in divided doses it does not give a sustained serum iron response. This can be attributed to a decrease in the rate of iron absorption occurring when a large dose is followed by a second and third dose. A good correlation can be obtained from *in vitro* dissolution data and the physiologic availability of iron. In other words, some of the sustained-release preparations have less side effects because less iron is available. Even though these products give good hematocrit and hemoglobin value improvement, they do not necessarily show a good increase in serum iron levels. Finally, it would appear that sustained-release iron products cannot give a sustained serum iron response due to the observation that the rate of iron absorption decreases once the iron reaches a certain level in the mucosal cells.[33] It appears that there is still much evaluation work to be done in this area.

Parenteral iron preparations are indicated only in those conditions where either iron absorption is defective (steatorrhea, partial gastrectomy) or the iron salt may be irritating (ulcerative colitis, peptic ulcer). Except for these types of indications, there seems to be no real advantage for using parenteral iron because it can cause vomiting and allergic reactions. It does not seem to correct iron-deficiency anemia any more rapidly than oral therapy.[34] The total dose for parenteral administration must be calculated carefully. One of the standard formulas is:

patient's weight (kg) $\times$ [normal hemoglobin value (g%) $-$
patient's hemoglobin value (g%)] $\times$ 2.5 = total milligrams of iron needed

This is one of several formulas. For each parenteral iron product, it is advisable to use the formula suggested by the manufacturer.

Overdosage of oral iron is serious and can cause death, particularly in young children. Ingestion of 10 to 15 300-mg ferrous sulfate tablets may be lethal to a child, with a mortality rate near 50%.[35,36] The human lethal dose is considered to be 150 to 200 mg iron/kg body weight.[37] Iron poisoning progresses in three to four stages. Stage one begins 30 to 40 minutes after ingestion and includes gastrointestinal distress due to the astringent action of ionized iron, developing into cardiovascular collapse, shock, and possible death in six hours. In stage two recovery seems apparent and may continue for 10 to 14 hours. Stage three may then develop with a recurrent cardiovascular collapse, Cheyne-Stokes respiration (cyclical, rythmical variation in intensity), convulsions, metabolic acidosis, shock, coma, liver damage, and possible death occurring in one to three days. If the patient survives, stage four may occur one to two months later with gastrointestinal complications (scarring, pyloric obstruction) due to the necrotizing effect (cell death) of the iron.[38]

Treatment usually includes gastric lavage and administration of salts (sodium bicarbonate and sodium dihydrogen phosphate) to form insoluble iron salts. Oral administration of deferoxamine will prevent iron absorption. Provided kidney damage has not occurred, deferoxamine can be given parenterally to chelate the iron and allow it to pass out in the urine. However, chelated feroxamine has been reported to have similar toxicities to that of iron.[39] Peritoneal dialysis has also been tried with poor results.[40] A recent report suggests lavage of the stomach with a phosphate salt to bind all unabsorbed iron that was not expelled by emesis. Chelation by deferoxamine mesylate (Desferal® mesylate) is indicated only when there is unbound iron demonstrated in the serum.[41] With the above in mind, it is important to warn the patient to be especially careful about keeping iron supplements away from children.

Orally administered iron has been shown to interfere with the absorption of tetracycline, oxytetracycline, methacycline, and oxycycline, presumably by forming a 1:1 chelate.[42] The reverse inhibition is also true: a tetracycline will inhibit iron absorption.[43]

Oral ferrous salts may aggravate gastrointestinal diseases such as peptic ulcer, regional enteritis, and ulcerative colitis. There is no currently approved rationale for the coadministration of copper salts.

There are three officially approved iron salts available for the oral administration of iron. One of the three, ferrous sulfate, is the most widely used salt and is the standard for comparison with the other two salts, ferrous fumarate and ferrous gluconate. The latter are claimed to be less irritating to the gastrointestinal tract than ferrous sulfate. Keeping in mind that it is the iron which causes gastric irritation, this claim is valid only when comparing doses which contain equivalent amounts of iron. Thus 200 mg of ferrous fumarate contain 60 mg of iron; 300 mg of ferrous sulfate (as the heptahydrate) contain 60 mg of iron; but 300 mg of ferrous gluconate contain only 35 mg of iron. The dose of ferrous gluconate would have to be doubled before comparisons of gastric irritation with the other two iron preparations would have any meaning.

Official Iron Products

Ferrous Fumarate, U.S.P. XVIII (Mol. Wt. 169.9)

$$\begin{array}{c} HC\!-\!C\!=\!O \\ \| \qquad \diagdown \\ O\!=\!C\!-\!CH \qquad O \\ | \qquad \diagup \\ O\!-\!\!-\!\!-\!\!-\!\!Fe \end{array}$$

Ferrous Fumarate U.S.P. occurs as a reddish orange to red-brown, odorless powder which may contain soft lumps that produce a yellow streak when crushed. It is slightly soluble in water and very slightly soluble in alcohol. *Ferrous Fumarate* U.S.P. dissolves in dilute hydro chloric acid with the precipitation of fumaric acid [rx (i)]:

$$FeC_2H_2(CO_2)_2 + 2HCl \rightarrow Fe^{-2} + 2Cl^- + C_2H_2(CO_2H)_2 \qquad (i)$$

There is little evidence to substantiate the claim that it is less irritating than ferrous sulfate if one administers equivalent doses of iron. One of the useful attributes of this salt is its resistance to oxidation on exposure to air. In this respect it may be superior to both ferrous sulfate and ferrous gluconate. Even on exposure to a hot humid atmosphere over an extended period of time, there is little conversion to the ferric form.

Usual Dose: 200 mg (the equivalent of 60 mg of elemental iron) two or three times a day.

Usual Dose Range: 200 to 600 mg daily.

Occurrence:

Ferrous Fumarate Tablets, U.S.P. XVIII (Ircon®, Toleron®)
Usually available as 200-mg tablets.

Ferrous Gluconate, N.F. XIII (Mol. Wt. 482.18)

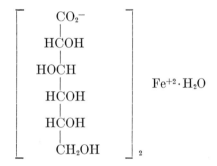

Ferrous Gluconate N.F. occurs as a yellowish gray or pale greenish yellow, fine powder or as granules having a slight odor resembling that of burnt sugar. Its 1 in 20 solution is acid to litmus. *Ferrous Gluconate* N.F. is soluble in water with slight heating and practically insoluble in alcohol.

Ferrous gluconate was introduced in 1937. Compared to some of the preparations used in those days (ferrous carbonate or Blaud's Pills) ferrous gluconate was a great improvement as it has good bioavailability. It is doubtful if it is any less irritating than ferrous fumarate or sulfate when equivalent doses of iron are administered (see discussion above).

Usual Dose: 300 mg three times a day (equivalent to 35 mg of iron/300 mg dose).

Usual Dose Range: 200 to 600 mg.

Occurrence:

Ferrous Gluconate Tablets, N.F. XIII (Fergon®)

Ferrous Sulfate, U.S.P. XVIII ($FeSO_4 \cdot 7H_2O$; Mol. Wt. 278.02; iron sulfate, green vitriol, iron vitriol)

Ferrous Sulfate U.S.P. occurs as pale, bluish green crystals or granules which are odorless. It has a saline, styptic taste, and is efflorescent in dry air. *Ferrous Sulfate* U.S.P. oxidizes readily in moist air to form brownish yellow basic ferric sulfate [approximate formula $Fe_4(OH)_2(SO_4)_5$]. For this reason the U.S.P. carries the italicized warning: Note—*Do not use Ferrous Sulfate that is coated with brownish yellow basic ferric sulfate.*

A 1 in 10 solution of *Ferrous Sulfate* U.S.P. is acid to litmus, having a pH of about 3.7. It is freely soluble in water, very soluble in boiling water, and insoluble in alcohol.

Ferrous sulfate is the most widely used oral iron preparation and is considered as the drug of choice for treating uncomplicated iron-deficiency anemia. It can be irritating to the gastrointestinal mucosa due to the astringent action of soluble iron, but it is probably no more irritating than any other iron salt when equivalent doses are used. Although there are a large number of ferrous sulfate tablets on the market, they can vary as to the potential bioavailability of the iron. This is usually due to the tablet adjuvants.[44]

Usual Dose: 300 mg, the equivalent of 60 mg of elemental iron, two or three times a day.

Usual Dose Range: 300 mg to 1 g daily.

Ferrous Sulfate Tablets, U.S.P. XVIII (Feosol®, Fer-In-Sol®)

Usually available as 300-mg tablets of $FeSO_4 \cdot 7H_2O$ or the equivalent amount of a dried ferrous sulfate, approximately 200 mg.

Ferrous Sulfate Syrup, N.F. XIII

It contains 40 g $FeSO_4 \cdot 7H_2O$ in 1000 ml. Sucrose (825 g) acts as a stabilizing agent, holding the iron in the ferrous state, and the citric acid (2.1 g) tends to act as an acid buffer. There is not enough citric acid present to complex the ferrous ion.

Dried Ferrous Sulfate, U.S.P. XVIII ($FeSO_4 \cdot xH_2O$)

It is primarily the monohydrate, although some tetrahydrate may be present. It is used in making tablets because a smaller amount may be used.

Iron Dextran Injection, U.S.P. XVIII (Imferon®)

Iron Dextran Injection U.S.P. is a sterile, colloidal solution of ferric hydroxide [$Fe(OH)_3$] complexed with partially hydrolyzed dextran (glucose polymer) of low molecular weight, in *Water for Injection.* The pH will be between 5.2 and 6.5. Prior to mixing, it is a dark brown, slightly viscous liquid.

It is for intramuscular injection only. The dose must be calculated very carefully using the tables and formulas supplied by the manufacturer. It is used only in confirmed cases of severe iron-deficiency anemia where oral therapy is contraindicated or ineffective, or if the patient cannot be relied upon to take oral medication. It should not be used in a prophylactic manner. Anaphylactic (severe allergic response) reactions, including three deaths, have been reported. Reports of carcinogenicity in animal studies do not appear to extend to humans, and the possibility is considered unlikely. A recent study indicates that iron dextran is effective in iron-deficiency anemia only when the bone marrow iron stores are depleted. In iron-deficiency anemia where there is adequate although unavailable bone marrow iron, iron dextran will not give an optimal response.[45]

Usual Dose: Intramuscular, the equivalent of 100 mg of iron once a day.

Usual Dose Range: Intramuscular, the equivalent of 50 to 250 mg of iron daily to every other day.

Iron Sorbitex Injection, U.S.P. XVIII (Jectofer®)

Iron Sorbitex Injection U.S.P. is a sterile solution of a complex of iron, sorbitol, and citric acid that is stabilized with the aid of dextrin and an excess of sorbitol. The pH is between 7.2 and 7.9. By itself, iron sorbitex is a dark brown, clear liquid.

Iron Sorbitex Injection U.S.P. is to be administered by the intramuscular route only. It is indicated for patients who either cannot or will not take adequate amounts of orally administered iron or for patients in whom orally administered iron is ineffective. Examples of the latter would be idiopathic steatorrhea, also called nontropical sprue (chronic diarrhea). The concurrent administration of oral iron is contraindicated. To date there have been no antigenic properties reported. The patient's urine can become dark on standing due to the formation of iron sulfide. The manufacturer's supplied dosage tables and formulas should be followed carefully.

Usual Dose: Intramuscular, the equivalent of 100 mg of iron once a day.

Usual Dose Range: 100 to 200 mg daily.

Nonofficial Iron Preparations

Dextriferron (Astrafer®)

This product, introduced in 1958, is a colloidal aqueous solution of ferric hydroxide which has been complexed with partially hydrolyzed dextrin. It is used for intravenous administration only. Immediate reactions have included flushing of face, nausea, vomiting, headache, abdominal pain, profound hypotension, collapse, and severe anaphylactoid responses. Delayed effects such as chills, fever, and sensations of stiffness in the arms, legs, or face have been reported. The dose should be calculated carefully.

Usual Dose: Intravenous. *Adults*—the equivalent of 30 mg elemental iron initially, with increments of 20 to 30 mg daily to a maximum daily dose of 100 mg until hemoglobin levels return to normal. *Small adults and children*—daily dose should not exceed 50 mg. *Infants*—daily dose should not exceed 20 mg.

Ferrocholinate (Chel-Iron®, Ferrolip®)

$$H_2O \quad\quad ^-O_2CCH_2 \quad\quad\quad CH_2CH_2OH$$
$$H_2O\text{---}Fe^{+2}\text{---}O\text{---}C\text{---}CH_2CO_2^- \ \ ^+N(CH_3)_3$$
$$H_2O \quad\quad\quad ^-O_2C$$

Ferrocholinate is a ferric iron chelate preparation available as pediatric drops, liquid, and tablets. It is claimed to be less toxic than ferrous sulfate or gluconate. The iron is claimed to be readily bioavailable.[29]

Other combinations of iron, choline, and citric acid that have been prepared for pharmaceutical use include a 1:2:2 chelate and 2:3:3 chelate respectively.

Ferric Ammonium Citrate, N.F. XI

The iron in ferric ammonium citrate is biologically available, at least in laboratory animals.[21]

Green Ferric Ammonium Citrate, N.F. XI

At one time this product was the preparation of choice for intramuscular injection of iron. It differs from ferric ammonium citrate by containing more ammonium citrate and seems to be biologically available.

Ferric Cacodylate, N.F. X $(Fe[(CH_3)_2AsO_2]_3)$

This salt has been used as a source of iron and, because of the arsenic, in the treatment of leukemias. Its efficacy is doubtful.

Ferric Chloride $(FeCl_3 \cdot 6H_2O)$

This has been part of *Ferric Chloride Solution* N.F. XI, *Ferric Chloride Tincture* N.F. XI (Iron Tincture), *Iron and Ammonium Acetate Solution* N.F. X (Basham's Mixture), *Ferric Citrochloride Tincture* N.F. XI, and *Iron, Quinine, and Strychnine Elixir* (which has *Ferric Citrochloride Tincture* as a component). In laboratory animals, ferric chloride is only partially bioavailable.[29]

Ferric Hypophosphite, N.F. X $(Fe[PH_2O_2]_3 \cdot xH_2O)$

Soluble Ferric Phosphate, N.F. XI $(FePO_4 +$ sodium citrate)

Ferric Pyrophosphate $(Fe_4[P_2O_7]_3 \cdot 9H_2O)$

Ferric Glycerophosphate, N.F. X $(Fe_2[C_3H_5(OH)_2PO_4]_3)$

The phosphate salts of iron have been shown to have poor bioavailability in laboratory animals.[29]

Saccharated Ferric Oxide, N.F. VII (A mixture of sucrose and ferric hydroxide and ferric oxide; soluble ferric oxide, eisenzucker)

Ferric oxide (Fe_2O_3) has poor iron bioavailability in laboratory animals.[29]

Ferrous Carbonate $(FeCO_3)$

This salt has been official as *Ferrous Carbonate Mass* N.F. X, *Ferrous Carbonate Pills* N.F. XI (Chalybeate Pills, Blaud's Pills, and Ferruginous Pills), and *Saccharated Ferrous Carbonate* N.F. X. Ferrous carbonate has poor iron bioavailability in laboratory animals.[29]

Copper

Like iron, copper is indispensable for normal metabolism in man and most animals. Unlike iron, it is currently believed that most of the population obtains adequate amounts of copper from food, water, and cooking utensils. Thus, copper supplements are probably not necessary. The adult human is estimated to contain about 2 mg/kg of copper, distributed mostly in enzymes and other proteins. The average daily intake is estimated at 2–5 mg per day.

Figure 6–4 illustrates the distribution of copper in the body. Copper is solubilized in the acidic stomach and is absorbed from the stomach and upper small intestine. It is not known if the body regulates the absorption of copper, but current evidence indicates that regulation is accomplished by excretion of excess copper. Of the approximately 5 mg of copper in the daily diet, 30% is absorbed. Of the amount absorbed, 80% is excreted through the bile, 16% is emptied directly back into the intestine through the gut walls, and only 4% is excreted in the urine.[46] From the intestine, copper moves into the blood serum where it exists first as a copper-albumin complex. It then goes to the liver where the copper becomes part of the copper protein ceruloplasmin. At equilibrium 93% of serum copper is in ceruloplasmin and 7% in albumin- and amino acid-bound fractions. Ceruloplasmin copper is not released until the protein is catabolized. The albumin copper is probably the means used to transport copper to the liver, red blood cell, bone marrow, kidneys, and tissues. It appears that amino acid-bound copper is the form of copper for movement across membranes.

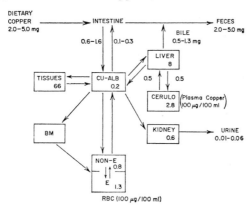

FIG. 6–4. Schematic representation of some metabolic pathways of copper. The numbers in the boxes refer to milligrams of copper in the pool. The numbers next to the arrows refer to milligrams of copper traversing the pathway each day. CU-ALB, direct-reacting fraction; CERULO, ceruloplasmin; NON-E, nonerythrocuprein; E, erythrocuprein; BM, bone marrow; RBC, red blood cell. From Cartwright, G. E. and M. M. Wintrobe; Amer. J. Clin. Nutr., 14: 224, 1964; and from Rogers' Inorganic Pharmaceutical Chemistry, 8th ed., by T. O. Soine and C. O. Wilson, Lea & Febiger, Philadelphia, p. 322, 1967.

In the liver, the copper is either stored, incorporated into ceruloplasmin, or excreted in the bile. Sixty percent of the copper in the red blood cell is found in the copper protein erythrocuprein and 40% as a labile non-erthyrocuprein fraction. The erythrocuprein copper may have come from the copper in the bone marrow during formation of the normoblasts (a red blood cell precursor). The level of copper in the kidneys is higher than one would predict from the small amount excreted in the urine.[47]

Several roles in body metabolism have been attributed to copper. One of these has to do with hemoglobin formation. Despite an adequate supply of iron, copper is required to prevent anemic conditions. There have been three roles postulated for copper: (1) copper could facilitate iron absorption; (2) copper could be stimulatory to the enzymes in the heme and/or globin biosynthetic pathways; and (3) copper could be involved in mobilization of stored iron, preparative to the incorporation of iron into the hemoglobin molecule. Data concerning the role of copper in iron absorption are conflicting. There is no evidence that copper plays any direct role in heme or globin biosynthesis. There is evidence that copper does have a function in mobilization of stored iron.[47] A recent postulate is that ceruloplasmin is really a ferrioxidase enzyme. It oxidizes ferrous iron to ferric for binding by transferrin.[48] Transferrin then releases its iron to the red cell precursor. If the postulated role of ceruloplasmin is correct, a copper deficiency would lead to an effective lack of iron available to the newly forming red blood cell.

Copper also is important in oxidative phosphorylation (ATP production by cellular respiration). Copper is a constituent of cytochrome oxidase, the terminal oxidase in the electron transport mechanism from which high-energy phosphate bonds are derived (see Fig. 12–1).

Copper is associated with the formation of aortic elastin. It may be that copper is necessary for amine oxidase activity and may play a role in the formation of crosslinkages of elastin. Copper deficiency in pigs can result in a weakening of the aorta and other blood vessels.

Copper is also a component of tyrosinase, an enzyme responsible for conversion of tyrosine to the black pigment, melanin. A copper deficiency in animals may cause loss of hair color which can be attributed to reduced tyrosinase activity. Albinism is associated with either an absence of or an inactive form of tyrosinase.

Most studies of copper deficiency have been done with animals, since the condition is rare in humans for the reasons stated in the opening paragraph. Indeed, certain livestock diseases can be attributed to copper deficiency. Nevertheless, human deficiencies have been identified in severely malnourished infants with chronic diarrhea, particularly those infants using modified cow's milk preparations that contain very little copper. Early symptoms can include neutropenia (neutrophile leukocyte deficiency), with or without iron-deficiency anemia. In more protracted cases, a regenerative anemia and scurvy-like bone lesions have appeared. It may be that premature infants with small livers (low store of copper) fed the standard formulas would be more likely to develop a copper deficiency.[49,50]

More common, although still rare (4 or 5 per million), is Wilson's disease, a condition of excess copper storage. Wilson's disease is of genetic origin, being transmitted by an autosomal recessive gene. Patients have increased copper levels in liver, brain, kidney, and cornea. The symptoms are hepatic cirrhosis, brain damage and demyelination, and kidney defects. They can be reversed by placing the patient on negative copper balance. This is usually done by diet and by use of the chelating agent penicillamine. The diet should contain less than 1.1–1.5 mg of copper per day. Low copper menus are available.[51] Penicillamine remains the drug of choice, although carbacrylamine resin and potassium sulfide have been used to remove dietary copper by causing its excretion in the feces.[52] It is postulated that penicillamine mobilizes copper by reductive chelation.[53]

Metabolically, Wilson's disease is characterized by an unusually low serum ceruloplasmin, which leads to a decrease in total blood copper. Urinary copper excretion is increased. Unfortunately, the correlation between severity of Wilson's disease and ceruloplasmin levels does not always hold. Some patients have normal ceruloplasmin levels. Heterozygotes can have low ceruloplasmin levels and not have any other symptoms.[47] A recent explanation is that patients with Wilson's disease have defective metallothionein, a metal-binding protein, in which the protein binds copper more tightly than in normal individuals. It is also possible that the ceruloplasmin is also defective and does not bind copper as well. The end result is that copper is stored in certain tissues and is not transported.[54]

A potential hepatotoxicity of copper in people undergoing long-term hemodialysis has been reported. The problem is that tap water, particularly if it is soft, will tend to dissolve copper from the water pipes. As a result, it can contain up to 1 ppm copper which, if present in the dialysis fluid, can concentrate in the patient's liver. In hard water areas, the copper pipes will usually be coated with a layer of salts that have precipitated from the water.[55] It seems advisable that water used as a dialysis fluid be completely deionized prior to use in order that its mineral content be known with certainty. In addition, copper tubing and fittings are being removed from dialysis machines.[56]

There are no preparations official for the administration of copper for a copper deficiency. Copper preparations are used topically as fungicides and astringents. The only official copper salt, *Cupric Sulfate* N.F. XIII, is official as an antidote for phosphorus poisoning and is discussed in Chapter 12. In addition to its topical uses, copper sulfate is used as a fungicide and algicide in agriculture in combination with lime (Bordeaux Mixture) and in swimming pools in conjunction with chlorine (see Chapter 9).

Copper sulfate is also the essential component of (1) Fehling's solution and (2) Benedict's solution, both of which are used for the determination of reducing sugars, principally glucose. A positive test is the production of a cuprous oxide precipitate [rx (ii)]:

$$Cu^{+2} + \text{sugar} \xrightarrow{\quad OH^- \quad} \underset{\text{red}}{Cu_2O} \downarrow + \text{sugar acids} \qquad (ii)$$

Herman von Fehling developed Fehling's solution in 1850. Solution 1, the copper sulfate solution, and Solution 2, the alkaline potassium sodium tartrate solution, are prepared separately and mixed in equal volumes at the time of analysis. When a portion of Solution 1 is mixed with an equal volume of Solution 2, the copper sulfate and sodium hydroxide probably react to form cupric hydroxide, which immediately is complexed by the potassium sodium tartrate into a soluble blue-colored compound. It is generally believed today that the solution probably contains a sequestered form of the cupric ion, a soluble cupric-tartrate ion that prevents the cupric ion *per se* from reacting with the hydroxide ions present, but which does not prevent the cupric ions from being reduced (see sequestering agents).

$$
\begin{array}{c}
\text{H} \\
| \\
\text{O}-\text{C}-\text{COONa} \\
\diagup \qquad\qquad | \\
\text{Cu} \qquad\qquad\qquad\quad \cdot 2H_2O \\
\diagdown \qquad\qquad | \\
\text{O}-\text{C}-\text{COOK} \\
| \\
\text{H} \\
\text{(soluble)}
\end{array}
$$

The copper ion in the solution is readily converted to cuprous oxide (Cu_2O) when heated with a few drops of a solution containing reducing sugars, e.g., diabetic urine. In testing urine it is found that traces of sugar in the urine (0.1%) will only cause a bluish-green coloration of the solution without a precipitate and that with increasing amounts of sugar the color changes gradually to red, accompanied by a precipitate (10% sugar and over).

Benedict's solution is used both as a qualitative reagent and as a quantitative solution. The qualitative solution contains crystalline copper sulfate, sodium citrate, monohydrated sodium carbonate, and distilled water. The cupric ion is combined with the citrate ion to form cupric citrate, which is insoluble except in alkali citrates. It is used for much the same purpose as Fehling's solution—namely, to detect sugar in urine, and the color changes are very similar. It has an advantage over Fehling's solution in that the single solution is stable, whereas Fehling's solutions A and B must be combined at the time of use.

Benedict's solution is less alkaline than Fehling's solution and utilizes citrate, which is a better sequestering agent. In Benedict's test the reagent must be boiled before a reduction by sugar will take place.

The qualitative reagent should not be confused with the quantitative solution, which has its own specific composition. The quantitative solution is less commonly used than the qualitative, but it is useful for determining the sugar content of urine more accurately than is possible with the qualitative solution.

Zinc

The zinc ion is widely distributed in the body. Biochemically, it is associated with certain metalloenzymes. These include alcohol dehydrogenase, lactic dehydrogenase, malate dehydrogenase, D-lactate-cytochrome c reductase, glyceraldehyde-3-phosphate dehydrogenase, glutamic dehydrogenase, aldolase, carbonic anhydrase, alkaline phosphatase, carboxypeptidase, and neutral protease. It has not been ascertained in all cases if zinc is essential for the mechanism of action for each enzyme, such as is the case for carbonic anhydrase and carboxypeptidase. Zinc is also bound to RNA, stabilizing the secondary and tertiary structures.

Zinc is an essential dietary mineral, but currently it is felt that there is little justification for including it in mineral supplements. No minimum daily requirements have been established although 10 to 15 mg daily have been estimated. Foods rich in zinc include meat, milk, fish, nuts, and legumes. It is conceivable that a person on a vegetable diet may not receive adequate amounts of zinc because phytic acid (Fig. 6–5), found in vegetable proteins such as soybean, combines with zinc and decreases its absorption. This may be clinically significant as there has appeared to be a correlation between zinc deficiency and phytate content of bread in an Iranian village and city.[57]

Zinc deficiency is associated with impaired growth, parakeratosis (a thickened, scaly, inflamed skin), and retarded sexual maturation. Studies involving zinc supplementation for supposedly zinc-deficient population groups have yielded mixed results. Part of the problem may be that people who have zinc-deficient diets usually have diets deficient in other essential

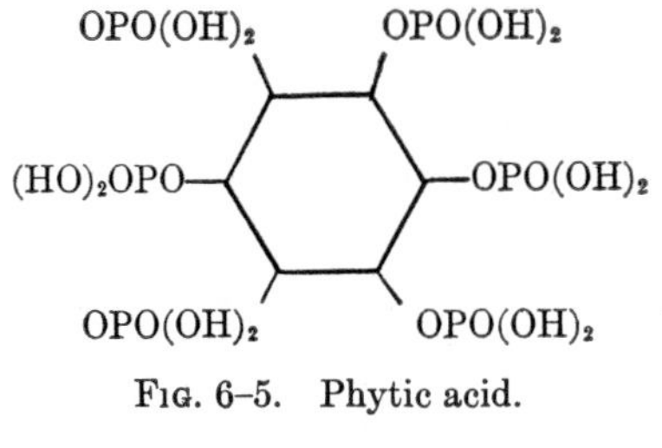

Fig. 6–5. Phytic acid.

nutrients.[58] Low plasma zinc levels are found in alcoholic cirrhosis (progressive liver disease), other types of liver disease, active tuberculosis, indolent (painless) ulcers, uremia, myocardial infarction, nontuberculous pulmonary infection, Down's syndrome (mongolism), cystic fibrosis with growth retardation, growth-retarded Iranian villagers, pregnancy, and in women taking oral contraceptives. No conditions have been observed with higher than normal plasma-zinc concentrations. Whether the lower plasma-zinc levels contribute to all the disease states listed above, or whether the lower levels are a result of the condition, must still be determined.[59]

There is some question as to what tissue or fluid should be measured in order to determine the degree of zinc nutriture. Zinc levels of human hair have been determined, but no correlation with zinc levels of red blood cells and plasma has been obtained.[60,61]

Zinc toxicity has resulted from the ingestion of acid food kept in a galvanized metal container and from industrial workers inhaling zinc oxide. Typical symptoms include chills, fever, malaise, coughing, salivation, and headache.[62] A 16-year-old boy who ingested 12 g of metallic zinc in two doses (4 to 8 g) was observed to have a very lethargic condition. Chelation therapy with dimercaprol promoted dramatic clinical improvement and a fall in blood zinc levels.[63]

The use of oral zinc sulfate has been suggested for wound healing. Oral doses of zinc sulfate (220 mg three times daily) have been reported to greatly increase healing following surgery.[64] In an uncontrolled study, the same dosage regimen was given to six elderly patients with bedsores, with a reported increase in the rate of healing.[65] A controlled study reported a dramatic increase in the healing of leg ulcers using the same dosage regimen of zinc sulfate, with the treated group being healed in an average of 32 days as compared to 77 days for the control group.[66] A similar double-blind study reported that oral zinc sulfate (220 mg three times daily) increased the healing rate of sickle cell leg ulcers three times faster than in the placebo group.[67] The only side effect reported has been mild diarrhea. A study using rats on controlled diets showed that zinc would increase the rate of wound closure only if the animal was zinc-deficient. Zinc supplementation had no measurable effect on wound healing if the animal had a normal zinc level.[68] Zinc sulfate did not have a direct accelerator effect on either cellular proliferation or collagen biosynthesis in a human skin fibroblast culture.[69] It would appear that a study of zinc requirements in man during wound healing as compared to normal individuals should be performed to determine exactly what phase of wound healing is benefitted by zinc sulfate. It is known that corticosteroid therapy in burn treatment causes serum zinc depletion.[70]

Currently, zinc sulfate is official as a topical astringent. It is more fully discussed in Chapter 9.

Chromium

Chromium is probably an essential trace element. It can be shown that chromium is necessary for optimal growth of experimental animals. In larger quantities it is toxic. There seems to be relatively less chromium in tissues of people living in the United States as compared to those in foreign countries. Chromium levels are also higher in infants than adults. Whether these differences are clinically significant is still to be determined. Chromium, however, seems to play some role in glucose tolerance. Chromium supplementation has been shown to improve or to normalize the impaired glucose tolerance of some diabetics, old people, and malnourished children, but not of others. While low chromium states do exist, they are not universal and definitely do not constitute the sole cause for impairment of glucose tolerance.[71–73]

The pharmacodynamic actions of chromium salts, chromates, and dichromates are very similar. They are destructive to tissue, regardless of whether applied topically or administered orally. When taken internally they produce a characteristic nephritis and glycosuria. Persons exposed to "chromate dust" develop deep ulcers of the skin and nasal mucosa that heal very slowly.

Manganese

The total manganese content of an adult human has been estimated to be 10 to 20 mg, with the highest concentration occurring in bone and liver as well as in the pituitary, pineal, and lactating mammary glands. Manganese functions in many metalloproteins as a nonspecific cation. It also is associated with ribonucleic acid and may play a role in protein synthesis, oxidative phosphorylation, fatty acid metabolism, and cholesterol synthesis.

No deficiency state has been found in humans. The minimum daily requirements have been estimated at 3–9 mg. In other mammals, manganese deficiency is characterized by defective growth, bone abnormalities, reproductive dysfunction, central nervous system malfunctions, and disturbances in fat and lipid metabolism.[74]

Excessive manganese intake can lead to chronic manganism (manganese poisoning). It is found mostly in the mining villages of Chile. Symptoms include mental disturbances, progressive bradykinesia (abnormal slowness of movement), asthenia (weakness), paresis (incomplete paralysis), dysarthria (imperfect articulation in speech), dystonia (disordered muscle tone), and disturbances of gait. Chest hair will usually contain quite a high level of manganese, and scalp hair, which normally has no manganese, will also contain manganese. In many ways, manganism is similar to Parkinson's disease. Interestingly, the use of levodopa (L-dihydroxyphenylalanine) has been successful in relieving many of the symptoms.[75,76]

Manganese salts which were official at one time include a citrate, $Mn_3(C_6H_5O_7)_2$, a glycerophosphate, $MnC_3H_5(OH)_2PO_4$, and a hypophosphite, $Mn(PH_2O_2)_2 \cdot H_2O$. They were used as "tonics," since it was known that manganese was an essential trace element. There is no known current therapeutic rationale for the administration of manganese.

Molybdenum

The requirements for molybdenum in human diet have not been established. It is present in all plant and animal tissues. The largest amounts (up to 3 ppm) are found in liver, kidney, bone, and skin. Molybdenum has been found associated with flavin-dependent enzymes.[74]

The only current use of molybdenum today is as the oxide, which together with ferrous sulfate in a specially coprecipitated complex is marketed in the ratio of 3 mg oxide/195 mg ferrous sulfate in the form of tablets, capsules, and drops as a hematinic preparation (Mol-Iron®).

Selenium

Until recently selenium has been considered toxic when taken internally. Because it is below sulfur in the periodic table, there have been attempts to replace sulfur with selenium in pharmacologically active and metabolically important compounds but, for the most part, these have been too toxic for systemic use. Selenium antimetabolites have been suggested as antitumor agents, but they are not effective when nontoxic doses are used. The oral administration of large doses of selenium salts produces intestinal irritation and interference with the functioning of small blood vessels and blood-forming organs.

There is accumulating evidence that selenium is essential in trace amounts in animals and possibly in man. Selenium prevents a liver necrosis in rats. It may be of therapeutic value in the treatment of kwashiorkor, a protein deficiency affecting large numbers of children and infants in South America, Africa, and Asia. Selenium has been implicated in cellular respiration and as an antioxidant in conjunction with vitamin E.[74] There is a report that the increase in dental cavities in children is proportional to the amount of selenium in their diet.[77]

Currently, selenium is official as *Selenium Sulfide* N.F. XIII, a suspension for the treatment of seborrheic dermatitis of the scalp (dandruff). It is described in Chapter 9.

Sulfur

Sulfur is widely distributed throughout the body as sulfhydryl groups of cysteine, disulfide linkages in protein from cystine, and sulfate salts and esters found in mucopolysaccharides and sulfolipids. Dietary sulfur comes from these same groupings found in plant and animal foodstuffs. The minimum daily requirements are 2–3 g. Currently there seems to be no need for dietary supplements of sulfur.

Sulfur has been used therapeutically since antiquity and, during that time, certain well defined uses emerged. These are: (1) cathartic action; (2) parasiticide in scabies; (3) stimulant in alopecia; (4) fumigation; and (5) miscellaneous skin diseases. In addition, (6) sulfides have been used for many years as depilatories. The cathartic action of sulfur is described in Chapter 8 and the topical uses in Chapter 9.

Iodine (Iodide)

Iodide is an essential ion necessary for the synthesis of the two hormones produced by the thyroid gland, triiodothyronine (T_3) and thyroxine (T_4).

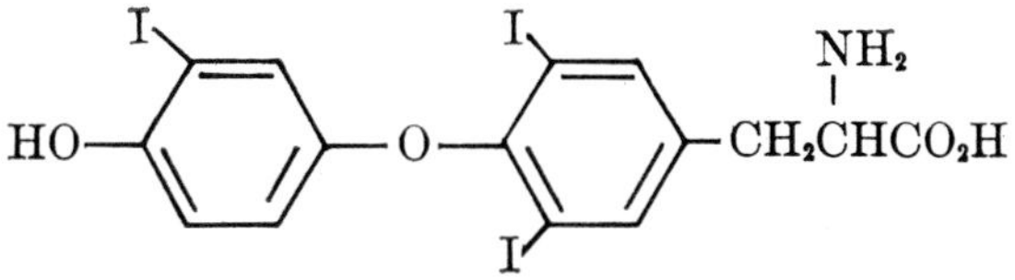

Triiodothyronine (T_3)

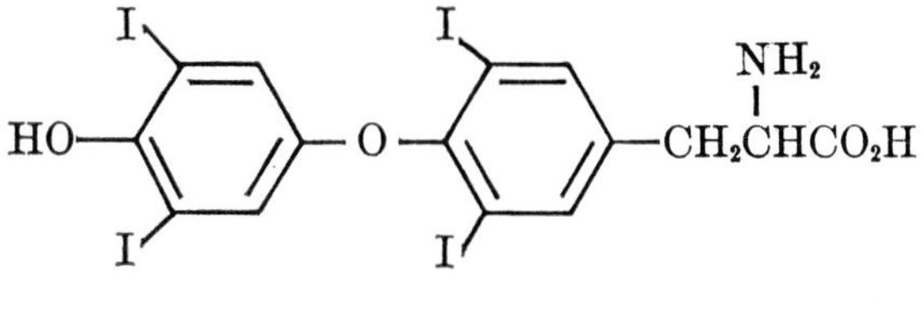

Thyroxine (T_4)

Internally iodine or iodide can be administered, since iodine is reduced to iodide in the intestinal tract. For solubility reasons it is more common to administer an iodide salt.

Iodine can be discussed from two standpoints: (1) its biochemical role in thyroid hormone formation; and (2) its pharmacological action as a fibrolytic agent, expectorant, and bactericidal agent.

The usual daily iodine requirement for an average man is approximately 140 micrograms and for an average female about 100 micrograms. Lack of sufficient iodine in the diet results in an enlargement of the thyroid gland, known as simple or colloid goiter. It is characterized by a swelling at the neck. The enlargement of the thyroid gland is a compensatory mechanism whereby the body attempts to make up for the hormone deficiency by increasing the size of the gland. Endemic goiter is almost always the result of a dietary deficiency of iodine. There are certain geographical areas, including locations in the United States, in which the soil is deficient in iodine. Today adequate amounts of iodine are easily insured by the use of iodized table salt containing 0.01% potassium iodide.

Iodine is an essential constituent of the thyroid hormones, thyroxine and triiodothyronine, as already described. In the normal utilization of iodine for the synthesis of these hormones it may be considered as being in the iodide form, which is then oxidized to iodine for incorporation into tyrosine to form monoiodotyrosine and diiodotyrosine. These are then coupled to form the triiodo- and tetraiodothyronines. The oxidation step from iodide to iodine is not well understood and is thought by many to involve a peroxidase system. Tracer studies with radioactive iodine have shown that iodine (as iodide) is incorporated into the thyroid gland solely to form the thyroid hormones. In comparing goitrous glands with normal glands it is always found that there is a lesser iodine and hormone content in the former. The size of the thyroid gland has been shown to be, in general, inversely proportional to the iodine content of the gland. Also, it has been well established that the iodine content of the thyroid gland is roughly proportional to the iodine intake.

When iodine is administered, its uptake is governed by three principal factors: (1) character of the local thyroid tissue, because abnormal adenomatous (tumorous) thyroid tissue has a slower uptake of iodide and a lower content of iodine than normal tissue; (2) blood level of inorganic iodide, because a high level keeps the iodine at a high level in the colloid, thus using up only a small part of the administered iodide: and (3) thyrotropin (a hormone secreted by the anterior pituitary) level in the blood, because the thyrotropin content has a direct bearing on the complete utilization of iodine in the formation of the iodinated hormones, and thyrotropin also controls the release of thyroid hormone from the thyroid gland. Paradoxically, the administration of excessive amounts of iodide will inhibit the incorporation of the element into the iodinated hormones provided other factors are not involved.

Actually, the amount of thyroid hormone released is a reflection of the true state of the thyroid gland, mere storage of the thyroid hormone in the colloid not being truly indicative. In this connection it is interesting to note that the effect of such antithyroid drugs as propylthiouracil is not to inhibit iodine uptake but rather to block the enzyme system that iodinates the amino acid precursors. The actual nature of the active thyroid hormone itself has not been determined. It is known that the hormone is not thyroglobulin but is, nevertheless, a protein combination of some kind with triiodothyronine and thyroxine. Indeed, thyroid hormone activity has been induced in such ordinary proteins as casein and plasma protein by *in vitro* treatment with iodine.

Iodide has been used therapeutically as: an ameliorating (improving) agent in hyperthyroidism; a fibrolytic agent in syphilis, leprosy, sporotrichosis, blastomycosis, and actinomycosis; an expectorant; and, finally, as an "alterative" (archaic term for a drug that reestablishes the health of the individual).

In hyperthyroidism it is well established that in virtually all cases a moderate drop in metabolic level can be demonstrated following iodide therapy. About 6 mg per day are the optimal dosage, even though the total daily prescribed dose usually amounts to nearly 500 mg iodine. The effect is not upon the systemic action of thyroid hormone, but appears to be brought about by a slower release of hormone. The slower release of thyroid hormone takes place even though the follicular colloid of the thyroid gland becomes more highly iodinated. A secondary result of iodide medication in hyperthyroidism is the involution of the gland. The vascularity is reduced, the gland becomes firmer, and the quantity of bound iodine increases. The changes are those that would be expected if the excessive stimulus to the gland had somehow been removed. This characteristic is not possessed by propylthiouracil and makes iodide therapy an invaluable adjunct to therapy by propylthiouracil, as well as in preparation for thyroidectomy. It has been demonstrated that administration of propylthiouracil, although controlling the formation and release of thyroid hormone, will not bring about involution of the gland as does iodide and may possibly bring on hyperplasia (abnormal cell increase). On the other hand, although administered iodide relieves the symptoms of hyperthyroidism, the effect is not permanent and in time the symptoms return, often in an exaggerated form. For this reason, iodine is used in thyrotoxic crises (thyroid crises) and to decrease the vascularity of the thyroid gland in the preparation of the thyrotoxic patient for surgery.

As a fibrolytic agent, iodide in large doses often causes striking recessions of the gummatous formations in late secondary and tertiary syphilitic infections, although it is becoming increasingly difficult to find cases that have progressed to these stages in this day of antibiotics. The iodide, apparently, has no curative properties nor does it influence resistance in any way. In actinomycosis and other fungal infections beneficial results are often brought about with relatively small doses, although in blastomycoses large doses are the rule. Iodides remain the drug of choice for the treatment of sporotrichosis.

As expectorants the iodides have had a long medical history, having been used in asthma, chronic bronchitis, and the late stages of acute bronchitis. Actually, although they are still popular today, it has not been demonstrated conclusively that they are effective in changing either the character or the amount of sputum. Allergists, in particular, feel that there is a definite liquefaction of tenacious mucus associated with the aftermath of an asthmatic attack. The effect may possibly be ascribed to reflex action from the saline taste of salivary secretions, the iodide ion appearing promptly in these secretions following administration.

Administration of iodides in amounts which exceed a certain level in the body often brings about certain irritative phenomena to the skin and mucous membranes. This is termed *iodism*, and is exhibited by coryza (head cold), rashes, headache, conjunctivitis, laryngitis, and the like. For this reason iodides should not be given in cases of acne. However, iodides are indicated when skin infections such as seborrhea require "opening up." Gastrointestinal effects are characterized by nausea, vomiting, and diarrhea. Discontinuation of the medication together with forcing of fluids is indicated in such cases. Administration of sodium chloride may also aid in the more rapid elimination of iodide.

Iodide therapy has traditionally been contraindicated in tuberculous patients because of fibrolytic effect on granulomatous tissue. It is thought that iodides may cause the spread of the disease. In contrast, others have advocated the use of iodide in the treatment of tuberculosis in order to improve the accessibility of the antitubercular chemotherapeutic agents to the mycobacteria.

Official Iodine Products

Iodine, U.S.P. XVIII (I; At. Wt. 126.90)

Iodine U.S.P. occurs as heavy, grayish black plates or granules, having a metallic luster and a characteristic odor. It is very slightly soluble in water, freely soluble in carbon disulfide, chloroform, carbon tetrachloride, and ether, soluble in alcohol and solutions of iodides, and sparingly soluble in glycerin.

Occurrence:
> **Strong Iodine Solution,** U.S.P. XVIII
> **Iodine Tincture,** U.S.P. XVIII (see Chapter 9)
> **Iodine Solution,** N.F. XIII (see Chapter 9)
> **Iodine Ampules,** N.F. XIII (see Chapter 9)

Strong Iodine Solution, U.S.P. XVIII (Lugol's Solution)

Strong Iodine Solution U.S.P. contains 5 g of iodine and 10 g of potassium iodide per 100 ml total volume. It is a transparent liquid having a deep brown color and odor of iodine.

Category: Source of iodine.

Usual Dose: 0.1 to 0.3 ml three times a day.

Usual Dose Range: 0.1 to 3 ml daily.

Potassium Iodide, U.S.P. XVIII (KI: Mol. Wt. 166.01)

Potassium Iodide U.S.P. occurs as hexahedral crystals, either transparent and colorless or somewhat opaque and white, or as a white granular powder. It is slightly hygroscopic and its solutions are neutral or alkaline to litmus. *Potassium Iodide* U.S.P. is very soluble in water and even more so in boiling water, freely soluble in glycerin, and soluble in alcohol.

Potassium iodide has a wide variety of uses, all based on its being an iodide salt. It is the iodide source in table salt, being present as 0.01%. It is used in larger doses for hyperthyroidism. Potassium iodide is official as an expectorant. It is also used as an antifungal agent and, by taking advantage of its expectorant action, as an antitussive agent.

Usual Dose: Expectorant, 300 mg four times a day.

Usual Dose Range: 300 mg to 2 g daily.

Occurrence:

Potassium Iodide Solution, N.F. XIII
Strong Iodine Solution, U.S.P. XVIII

Potassium Iodide Solution, N.F. XIII

Potassium Iodide Solution N.F. contains 100 g potassium iodide in each 100 ml of solution. It is a clear, colorless, and odorless liquid, with a characteristic, strongly salty taste. It is neutral or alkaline to litmus and has a specific gravity of about 1.70. Sodium thiosulfate is added (500 mg/l) to reduce any liberated iodine that may form when potassium iodide solution stands for periods of time [rx (iii)]:

$$2Na_2S_2O_3 + I_2 \rightarrow Na_2S_4O_6 + 2NaI \qquad \text{(iii)}$$

Potassium Iodide Solution N.F. has the same uses as the parent salt.

Usual Dose: Iodine supplement; expectorant, 0.3 ml; sporotrichosis treatment, oral, 0.6 ml three times daily, amount increased each day by 0.06 ml at each dose until maximal tolerated dose is reached. Cure requires six to eight weeks' treatment.

Sodium Iodide, U.S.P. XVIII (NaI; Mol. Wt. 149.89)

Sodium Iodide U.S.P. occurs as colorless, odorless crystals, or as a white crystalline powder. It is deliquescent in moist air, and develops a brown tint upon decomposition [rx (iv)]. *Sodium Iodide* U.S.P. is very soluble in water and freely soluble in alcohol and glycerin. It is used as a source of iodine, mostly for hyperthyroid conditions, and to solubilize iodine in *Iodine Tincture* U.S.P. and *Iodine Solution* N.F.

$$4I^- + O_2 + 2H_2O \rightarrow 2I_2 + 4OH^- \qquad \text{(iv)}$$

Usual Dose: Oral, 300 mg two to four times a day. Intravenous infusion, 1 g.

Usual Dose Range: Oral, 300 mg to 2 g daily. Intravenous infusion, 1 to 3 g daily.

Occurrence:

Iodine Tincture, U.S.P. XVIII (see Chapter 9)
Iodine Solution, N.F. XIII (see Chapter 9)
Iodine Ampules, N.F. XIII (see Chapter 9)

References

1. Keiderling, W., and Wetzel, H. P. Problems of iron metabolism with special reference to biochemical, physiological, and clinical aspects. Angew. Chem. [Eng.], 5:633, 1966.
2. Fletcher, J., and Huehns, E. R. Function of transferrin. Nature, 218:1211, 1968.
3. Bothwell, T. H., and Charlton, R. W. Absorption of iron. Ann. Rev. Med., 21:145, 1970.
4. Layrisse, M., Martinez-Torres, C., and Roche, M. Effect of interaction of various foods on iron absorption. Amer. J. Clin. Nutr., 21:1175, 1968.
5. Hussain, R., Walker, R. B., Layrisse, M., Clark, P., and Finch, C. A. Nutritive value of food iron. Amer. J. Clin. Nutr., 16:464, 1965.
6. Jacobs, A., and Miles, P. M. Intraluminal transport of iron from stomach to small-intestinal mucosa. Brit. Med. J., 4:778, 1969.
7. Wheby, M. S., Jones, L. G., and Crosby, W. H. Studies on iron absorption. Intestinal regulatory mechanisms. J. Clin. Invest., 43:1433, 1964.
8. Dowdle, E. B., Schachter, D., and Schenker, H. Active transport of Fe^{59} by everted segments of rat duodenum. Amer. J. Physiol., 198:609, 1960.
9. Saltman, P. The role of chelation in iron metabolism. J. Chem. Ed., 42:682, 1965.
10. Sheehan, R. G., and Frenkel, E. P. The control of iron absorption by the gastrointestinal mucosal cell. J. Clin. Invest., 51:224, 1972.
11. Crosby, W. H. Control of iron absorption by intestinal luminal factors. Amer. J. Clin. Nutr., 21:1189, 1968.
12. Crosby, W. H. Intestinal response to the body's requirement for iron. J.A.M.A., 208:347, 1969.
13. MacDonald, R. A. Hemochromatosis: A perlustration. Amer. J. Clin. Nutr., 23:592, 1970.
14. The Merck Manual of Diagnosis and Therapy, 11th ed. Rahway, N.J.: Merck Sharp and Dohme Research Laboratories, p. 29, 1966.
15. Finch, C. A. Iron-deficient anemia. Amer. J. Clin. Nutr., 22:512, 1969.
16. Monsen, E. R., Kuhn, I. N., and Finch, C. A. Iron status of menstruating women. Amer. J. Clin. Nutr., 20:842, 1967.
17. Butterworth, C. E., Jr. Iron "undercontamination"? J.A.M.A., 220:581, 1972, and references cited therein.
18. Finch, C. A., and Monsen, E. R. Iron nutrition and the fortification of food with iron. J.A.M.A., 219:1462, 1972.
19. Council on Foods and Nutrition. Iron in enriched wheat flour, farina, bread, buns, and rolls. J.A.M.A., 220:855, 1972.
20. Krikker, M. A. Fortification of food with iron. New Eng. J. Med., 283:206, 1970.
21. Should we triple the iron in bread? Med. World News, 13:6, 1972.
22. Iron in infancy. Brit. Med., J., 2:728, 1972.
23. Iron-fortified formulas for infants. Med. Lett., 13:65, 1971.
24. Woodruff, C. W., and Clark, J. L. The role of fresh cow's milk in iron deficiency, I. Albumin turnover in infants with iron deficiency anemia. Amer. J. Dis. Child., 124:18, 1972.
25. Woodruff, C. W., Wright, S. W., and Wright, R. P. The role of fresh cow's milk in iron deficiency, II. Comparison of fresh cow's milk with a prepared formula. Amer. J. Dis. Child., 124:26, 1972.
26. Edwards, M. S., Pegrum, G. D., and Curtis, J. R. Iron therapy in patients on maintenance haemodialysis. Lancet, 2:491, 1970.
27. Brozovich, B., Cattell, W. R., Cottrall, M. F., Gwyther, M. M., McMillan, J. M., Malpas, J. S., Salsbury, A., and Trott, N. G. Iron metabolism in patients undergoing regular dialysis therapy. Brit. Med. J., 1:695, 1971.
28. Beresford, C. H., Neale, R. J., and Brooks, O. G. Iron absorption and pyrexia. Lancet, 1:568, 1971.
29. Fritz, J. C., Pla, G. W., Roberts, T., Boehne, J. W., and Hove, E. L. Biological availability in animals of iron from common dietary sources. J. Agr. Food Chem., 18:647, 1970.

30. Mouratoff, G. J., and Batterman, R. C. Serum iron absorption tests of a sustained-release form of oral iron. J. New Drugs, 1:157, 1961.
31. Posner, L. B., and Wilson, F. An evaluation of sustained-release molbydenized ferrous sulfate in iron-deficiency anemia of pregnancy. J. New Drugs, 3:155, 1963.
32. Webster, J. J. Treatment of iron deficiency anemia in patients with iron intolerance: Clinical evaluation of a controlled-release form of ferrous sulfate. Curr. Ther. Res., 4:130, 1962.
33. Middleton, E. J., Nagy, E., and Morrison, A. B. Studies on the absorption of orally administered iron from sustained-release preparations. New Eng. J. Med., 274:136, 1966.
34. McCurdy, P. R. Oral and parenteral iron therapy, a comparison. J.A.M.A., 191:859, 1965.
35. Iron Poisoning, J.A.M.A., 198:1303, 1966.
36. Clark, W. M., Jurow, S. S., Walford, R. L., and Warthen, R. O. Ferrous sulfate poisoning. Amer. J. Dis. Child., 88:220, 1954.
37. Reissman, K. R., Coleman, T. J., Budai, B. S., and Moriarty, L. R. Acute intestinal iron intoxication, I. Iron absorption, serum iron and autopsy findings. Blood, 10:35, 1955.
38. Eickholdt, T. H. Pharmacology of Iron. Canad. J. Pharm., 4:1, 1969, and references cited therein.
39. Whitten, C. F., Gibson, G. W., Good, M. H., Goodwin, J. F., and Brough, A. J. Studies in acute iron poisoning. 1. Desferrioxamine in the treatment of acute iron poisoning: Clinical observations, experimental studies, and theoretical considerations. Pediatrics, 36:322, 1965.
40. Lavender, S., and Bell, J. A. Iron intoxication in an adult. Brit. Med. J., 2:406, 1970.
41. Fischer, D. S., Parkman, R., and Finch, S. C. Acute iron poisoning in children: The problem of appropriate therapy. J.A.M.A., 218:1179, 1971.
42. Neuvonen, P. J., Gothoni, G., Hackman, R., and Björksten, K. Interference of iron with the absorption of tetracyclines in man. Brit. Med. J., 4:532, 1970.
43. Greenberger, J. J., Ruppert, R. D., and Cuppage, F. E. Inhibition of intestinal iron induced by tetracycline. Gastroenterology, 53:590, 1967.
44. Blezek, C. E., Lach, J. L., and Guillory, J. K. Some dissolution aspects of ferrous sulfate tablets. Amer. J. Hosp. Pharm., 27:533, 1970.
45. Davies, A. G., Beamish, M. R., and Jacobs, A. Utilization of iron dextran. Brit. Med. J., 1:146, 1971.
46. Cartwright, G. E., and Wintrobe, M. M. Copper metabolism in normal subjects. Amer. J. Clin. Nutr., 14:224, 1964.
47. Dowdy, R. P. Copper metabolism. Amer. J. Clin. Nutr., 22:887, 1969, and references cited therein.
48. Freiden, E. Ceruloplasmin, a link between copper and iron metabolism. Nutr. Rev., 28:87, 1970.
49. Graham, G. G. Human copper deficiency. New Eng. J. Med., 285:857, 1971.
50. Al-Rashid, R. A., and Spangler, J. Neonatal copper deficiency. New Eng. J. Med., 285:841, 1971.
51. Lawler, M. R., and Jelang, M. A. Recipes for low-copper diets. J. Amer. Diet. Ass., 57:420, 1970.
52. Strickland, G. T., Blackwell, R. Q., and Whatten, R. H. Metabolic studies in Wilson's disease: Evaluation of efficacy of chelation therapy in respect to copper balance. Amer. J. Med., 51:31, 1971.
53. Peisach, J., and Blumberg, W. E. A mechanism for the action of penicillamine in the treatment of Wilson's disease. Molec. Pharmacol., 5:200, 1969.
54. Protein defective in Wilson's disease. Chem. Eng. News., 49:29, April 26, 1971.
55. Bloomfield, J., Dixon, S. R., and McCredie, D. A. Potential hepatotoxicity of copper in recurrent hemodialysis. Arch. Intern. Med., 128:555, 1971.
56. Klein, W. J., Metz, E. N., and Price, A. R. Acute copper intoxication, a hazard of hemodialysis. Arch. Intern. Med. (Chicago), 129:578, 1972.

57. Reinhold, J. G. High phytate content of rural Iranian bread: A possible cause of human zinc deficiency. Amer. J. Clin. Nutr., **24**:1204, 1971.
58. Prasad, A. S. A century of research on the metabolic role of zinc. Amer. J. Clin. Nutr., **22**:1215, 1969, and references cited therein.
59. Halsted, J. H., and Smith, J. C., Jr. Plasma-zinc in health and disease. Lancet, **1**:322, 1970.
60. Zinc in hair as a measure of zinc nutriture in human beings. Nutr. Rev., **28**:209, 1970.
61. McBean, L. D., Mahloudji, M., Reinhold, J. G., and Halsted, J. A. Correlations of zinc concentrations in human plasma and hair. Amer. J. Clin. Nutr., **24**:506, 1971.
62. Zinc. Med. Lett., **11**:15, 1969.
63. Murphy, J. V. Intoxication following ingestion of elemental zinc. J.A.M.A., **212**:2119, 1970.
64. Pories, W. J., Henzel, J. H., Rob, C. G., and Strain, W. H. Acceleration of wound healing in man with zinc sulphate given by mouth. Lancet, **1**:121, 1967.
65. Cohen, C. Zinc sulphate and bedsores. Brit. Med. J., **2**:561, 1968.
66. Husain, S. L. Oral zinc sulphate in leg ulcers. Lancet, **1**:1069, 1969.
67. Serjeant, G. R., Galloway, R. E., and Gueri, M. C. Oral zinc sulphate in sickle-cell ulcers. Lancet, **2**:891, 1970.
68. Sandstead, J. H., Lanier, U. C., Jr., Shephard, G. H., and Gillespie, D. D. Zinc and wound healing: Effects of zinc deficiency and zinc supplementation. Amer. J. Clin. Nutr., **23**:514, 1970.
69. Waters, M. D., Moore, R. D., Amato, J. J., and Houck, J. C. Zinc sulfate failure as an accelerator of collagen biosynthesis and fibroblast proliferation. Proc. Soc. Exp. Biol. Med., **138**:373, 1971.
70. Flynn, A., Pories, W. J., Strain, W. H., Hill, O. A., Jr., and Fratianne, R. B. Rapid serum-zinc depletion associated with corticosteroid therapy. Lancet, **2**:1169, 1971.
71. Schroeder, H. A. The role of chromium in mammalian nutrition. Amer. J. Clin. Nutr., **21**:230, 1968, and references cited therein.
72. Mertz, W. Chromium occurrence and function in biological systems. Physiol. Rev., **49**:165, 1969, and references cited therein.
73. Gürson, C. T., and Saner, G Effect of chromium on glucose utilization in marasmic protein calories malnutrition. Amer. J. Clin. Nutr., **24**:1313, 1971.
74. Prasad, H. S., Oberleas, D., and Rajasekaran, G. Essential micronutrient elements: Biochemistry and changes in liver disorders. Amer. J. Clin. Nutr., **23**:581, 1970, and references cited therein.
75. Mena, I., Court, J., Fuenzalida, S., Papavisiliou, P. S., and Cotzias, G. C. Modification of chronic manganese poisoning. New Eng. J. Med., **282**:5, 1970.
76. Rosenstock, H. Z., Simons, D. G., and Meyer, J. S. Chronic manganism: Neurologic and laboratory studies during treatment with levodopa. J.A.M.A., **217**:1354, 1970.
77. Consumption of small amounts of selenium in foods increases the prevalence of caries. Chem. Eng. News, **48**:37, June 29, 1970.

7

Nonessential Ions

The ions discussed in this chapter are currently considered nonessential, even though some of them have a beneficial pharmacological action in appropriate dosage. Particularly with fluoride, it may be shown in the future that this anion is essential with minimum daily requirements. Several of the ions discussed in this chapter have a toxicological action, perhaps in levels found in the environment.

Fluoride

Fluorides are widely used today for their anticariogenic action (inhibition of dental cavity development). The postulated mechanisms and dosage forms are discussed in detail in Chapter 10. This discussion will be restricted to an overview of fluoride biochemistry and the use of fluorides in other diseases involving calcium metabolism.

When fluoride is taken orally, approximately 95% is absorbed, and the balance is excreted in the feces. About 50% of the ingested fluoride is excreted in the urine. This figure seems to be fairly independent of the amount of fluoride ingested. While some fluoride is excreted in the perspiration, the balance is retained in the bone and is released very slowly to be excreted in the urine.[1]

There is considerable controversy concerning the toxic doses of fluoride, usually in the form of sodium fluoride. There does not seem to be any harm in the quantities used to fluoridate public water supplies (see Chapter 10). Indeed, there seems to be an excellent margin of safety between the effective dose of 2.2 mg sodium fluoride per day for the prevention of caries and the acute lethal dose of 4 g.

Sodium fluoride, in toxic doses, is a general protoplasmic poison which inhibits enzyme activity. A double-blind study of human subjects ingesting a controlled daily dose of 5 mg fluoride ion (equivalent to an average daily intake of water fluoridated at 5 ppm) for three months showed a 20% reduction in alkaline phosphatase activity. In areas just beginning fluoridation, based on a daily intake of 1 ppm fluoride, there was an initial 16% drop in serum alkaline phosphatase activity followed by a return to normal within 8 to 22 weeks. There was no noticeable effect on the subjects' health. It must be kept in mind that serum alkaline phosphatase is formed in the liver and has no specific function as such in the serum. Thus, it is postulated that fluoride causes a transient decreased production of enzyme for which the body soon compensates.[2]

In higher doses fluoride is definitely toxic. A condition known as wine fluorosis, caused by wine to which fluoride has been added illegally in concentrations of 15–75 ppm to retard fermentation, can cause fluoride-induced arthritic changes. Calcification of the periarticular (near a joint) ligaments is responsible for restricted movement in hands, shoulders, and hip joints. Twenty-six individuals living near a fluoride-emitting factory and eating fluoride-contaminated food grown in the area were reported to have preosteosclerotic (before abnormal bone hardening) arthritis. There is also a report that crippling fluorosis can be treated by decreasing dietary calcium.[3] Bone diseases resulting from the use of fluoridated water while on hemodialysis are more fully discussed in Chapter 10. It must be kept in mind that hemodialysis presents special problems regarding the maintenance of proper serum ion concentrations. It is now believed that deionized water should be used during hemodialysis.

An interesting report has appeared correlating goiter in Himalayan villages (Nepal) independently with fluoride content and hardness of water in each village. The iodine intake was low in each village. There was a positive relationship found between incidence of goiter and fluoride content of the drinking water. Animals also show a positive relationship between goiter and fluoride intake.[4] However, goiter is not common in areas in the United States in which the drinking water contains natural fluoride in amounts several times greater than that of the drinking water in the Himalayan villages. This can probably be explained by the fact that, in the United States, there is adequate daily iodine intake due to the use of iodized salt (see Chapter 6).

All the above severe toxic symptoms involve calcium metabolism. It is difficult to pinpoint the mechanism of these effects. For example, a short-term (22–42 days) feeding of 20.6 mg sodium fluoride per day to human subjects showed no statistical change in intestinal absorption of calcium and calcium balances, nor was there any change in phosphorus or nitrogen balances.[5] Similar results were obtained in a feeding experiment involving three patients with osteoporosis given 88 to 100 mg sodium fluoride per day for 2.5–3 months.[6]

Nevertheless, reports that fluoride reduces the prevalence of osteoporosis (loss of bone calcium) continue to appear.[7] An early study comparing populations living in a high-fluoride area (4 to 5.8 ppm fluoride in the water supply) and low-fluoride area (0.15 to 0.3 ppm) reported that reduced bone density and collapsed vertebrae, especially in women, were substantially higher in the low-fluoride area. Visible calcification of the aorta, particularly in men, was significantly higher in the low-fluoride area.[9] The implication here is that fluoride facilitates or retains calcium deposited in the hard tissues (teeth and bones) and not in the soft tissues, the aorta in this example. Also, the subjects in this study had ingested fluoride for long periods of their lifetimes.

As already pointed out, short-term studies involving the feeding of large amounts of sodium fluoride have shown little effect on calcium or phosphorus balances. Use of ^{47}Ca as a tracer to measure skeletal metabolism showed no significant change in the size of the exchangeable calcium pools or in the accretion (growth) rate. The implication from this study is that fluoride cannot be considered to be an effective treatment for osteoporosis, at least not for short time intervals.[7]

Sodium fluoride has been given to patients with multiple myeloma. One of the most incapacitating aspects of this disease is bone damage, which usually is resistant to present modes of therapy. However, doses up to 50 mg of sodium fluoride administered four times a day for at least six months produced no subjective or objective improvement.[8]

There is some evidence that patients with bone wasting diseases may have impaired conversion of cholecalciferol to 1,25-dihydroxycholecalciferol, currently believed to be necessary for intestinal calcium absorption (see calcium discussion in Chapter 5).

Bromides

Bromides were first introduced into medicine by Locock in 1853 for their antiepileptic effect. Their introduction into therapy was based on the theory that epilepsy might be caused by an overabundance of amorousness (sexual enjoyment) and, since bromides were considered to be a sexual depressant, the depressant action should be therapeutically desirable. Although the hypothesis was incorrect, the fact was that bromides did produce a beneficial depression of the convulsions associated with epilepsy. Later, in 1864, Behrend utilized them in certain cases of sleeplessness. At one time the bromides were very widely used, much as aspirin is today. Early studies on bromides were done with potassium bromide and some confusion arose as to which ion of the salt contributed the sedative properties. This confusion was dispelled when it became apparent that other bromide salts had the same activity. Likewise, a depressant effect on the heart was noted with potassium bromide, but this was later shown to be due to the potassium content.

Administration of small doses (0.5 to 2 g) of a bromide (e.g., KBr) serves to cause a depression of the central nervous system. Larger doses (4 to 8 g) depress all reflexes and cause a narcotic type of effect. The use of repeated small doses for a sedative effect depends on the abovementioned depression of the central nervous system. The actual mechanism of sleep

9

production is not known, but sleep is induced by diminishing the susceptibility of the patient to external stimuli.

Bromides have been a standard form of medication for epileptiform
seizures but have yielded to the organic antiepileptics (diphenylhydantoin,
phenobarbital, etc.). Bromides' usefulness in epilepsy depends on their
ability to depress the motor areas of the brain, an effect brought about by
large doses. Bromides do not have a cleancut, highly selective action on
the motor areas alone, but also depress the sensory functions to some
extent. This makes it impossible to avoid some degree of sedation at all
times if an effective level is to be maintained in an epileptic. Nevertheless,
bromides, usually as the sodium or potassium salt, may still be useful in
grand mal seizures in children in whom other drugs prove unsuitable.[10]

Bromides are rapidly absorbed and are excreted principally in the urine.
Repeated doses tend to cause accumulation with a consequent replacement
of chloride ion by accumulated bromide ion. On the other hand, administration of sodium chloride tends to hasten elimination of bromide. The
distribution of bromide ion in the body is virtually the same as that of
chloride ion, a fact that makes highly selective action by bromides unlikely.

The use of bromides is attended by the possibility of bromism (poisoning
by bromides). The occurrence of bromism is not infrequent, in spite of the
fact that cautions have been emphasized in both the medical and pharmaceutical journals. One of the reasons for this is the uncontrolled sale of
bromide-containing proprietaries. Another reason is that physicians
themselves are lax in keeping a careful watch on patients who are taking
bromides over long periods of time. The early signs of bromide intoxication
include insomnia and restlessness, as well as dizziness, weakness, and
headache. It is easy to see where physicians might have been led to increase
the dose in order to combat the very condition they were causing. A skin
rash known as bromide acne often occurs with bromism but is not as
common a finding as one would be led to believe from the older literature.
Bromide psychosis (mental disease) may also be induced by long-continued
use of bromides. Death from bromides is rare, and the symptoms, as a
rule, recede upon discontinuation of the drug. However, treatment of
bromism may be carried out by the administration of substantial doses of
sodium chloride (6 g daily in divided doses) or, where sodium intake is to
be limited, ammonium chloride may be used instead.

Today there is no bromide salt or combination product that is official.
The product on the market associated with bromide therapy, Bromo-
Seltzer®, contains 162.5 mg potassium bromide per capful (maximum of
6 capfuls per day). The official dose of potassium bromide was 1 g daily,
which would be equivalent to taking the maximum dose of 6 capfuls per
day. Most of Bromo-Seltzer's® ability to relieve "nervous tension" is
probably due to the phenacetin, caffeine, and acetaminophen present.

Nevertheless, bromism is observed today due to long-term chronic administration of bromide products (Three Bromides Elixir, Nervine®, etc.).[11]

Lithium

The lithium ion is a depressant to the central nervous system and to circulation. The ion also has a diuretic action. Its toxic nature led to the discontinuance of lithium chloride as a component of salt substitutes used by patients on salt-free diets. Indeed, the toxic effects of lithium are aggravated by a reduction in sodium intake.

Lithium is readily absorbed from the intestine and accumulates in the body. The extent of lithium accumulation is dependent upon the sodium intake. A decreased intake of sodium accelerates lithium accumulation and accentuates its toxic effects. Conversely, lithium intoxication is treated by withholding the lithium salt and providing an adequate sodium intake.

Lithium salts have been advocated at different times as central nervous system depressants. At one time lithium bromide was official because of the depressant action of the bromide anion (lithium was not considered a depressant at that time). The supposed advantage of administering lithium bromide was that there was more bromide available per unit weight of lithium bromide since lithium (At. Wt. 7) is so much lighter than sodium (At. Wt. 23), ammonium (Mol. Wt. 18) or potassium (At. Wt. 39).

Indeed, the current use of lithium carbonate in manic-depressive disorders was due more to serendipity than to a careful screening of potential drugs. The discoverer, J. F. J. Cade, was actually looking for a cation to form a very soluble urate salt during some studies to determine whether uric acid enhanced urea toxicity in guinea pigs. Lithium urate is very water soluble. He found that both lithium urate and lithium carbonate protected the animals against the convulsant death associated with toxic doses of urea. The next step was to determine what effect the lithium salts would have on animals without urea being coadministered. The animals became lethargic. Cade then concluded that lithium salts should be investigated in mania because of their sedative effect and in epilepsy because of their anticonvulsant action.[12]

The manic-depressive reaction is characterized by extremes in emotion and behavior. In the manic phase, the individual is excited with the excitement being mild, acute, or delirious. The patient becomes hyperactive, ranging from simple overactivity to sustained and frenzied "busyness." His behavior may become bizarre, but, unless he is also paranoid, he is not apt to harm anyone. At the other extreme is the depression phase which can be mild, acute, or stuporous. The patient's behavior ranges from being downhearted and/or hypochondriac (preoccupied with disease) to all physical activity becoming a great exertion. As the depression recedes, the danger of suicide increases.

The prognosis is dependent upon age of onset and whether the first attack was manic or depressive. Until the advent of lithium carbonate, treatment has included the use of phenothiazine tranquilizers and electroshock therapy. Hospitalization is usually required during either phase.

Lithium carbonate (Li_2CO_3; Eskalith®, Lithonate®, Lithane®) is administered orally in doses of 300 to 600 mg three times a day to manic patients. The lower dose range is used for elderly patients in whom renal clearance is likely to be reduced. A phenothiazine tranquilizer usually is administered also for the first few days since it takes three to ten days for lithium to become effective. Lithium carbonate should be discontinued if a satisfactory response is not obtained in 14 days.

Since lithium is toxic, serum lithium levels should be monitored. A satisfactory range is 0.5 to 1.5 mEq Li/l (see Chapter 5 for a discussion of mEq definitions and calculations) with the upper level for acute manic phases, and about 1 mEq/l for maintenance. Due to its toxic effects, and because it is excreted by the kidneys, if blood levels of lithium are elevated lithium carbonate is contraindicated in patients with impaired renal function. Lithium can induce a diabetes insipidus (increased urination without concurrent glucosuria) condition by apparently interfering with the action of vasopressin.[13] Also, since lithium toxicity increases with a decrease in sodium intake, patients on salt-restricted diets or those who are receiving diuretics should be monitored carefully.[14,15]

The effectiveness of lithium carbonate in manic reactions is well documented. It corrects the patient's mood without the mental confusion that is usually observed following electroconvulsive therapy. It is not yet clear how effective lithium is for the depressive phase nor how effective it is as a prophylaxis to prevent recurring attacks.[15]

Discontinuance studies have indicated that lithium prophylaxis is a useful means of preventing recurring attacks of either phase.[17,18] In these types of investigations, the subjects are switched from drug to placebo in double-blind fashion. The incidence of relapse was found to be greater while the subjects were on the placebo. Patients have been maintained on lithium carbonate prophylaxis for an average of 40 months at a dose that produces a plasma lithium level of between 0.6 and 1.4 mEq/l. The mean number of admissions to a hospital for episodes of depression and/or mania during the investigational period was 0.55, compared with 3.36 during a period of similar duration before lithium prophylaxis. Time spent in the hospital dropped from a group average of 26.9 weeks to 3.5 weeks.[19]

Not all patients will accept lithium carbonate prophylaxis. Three reasons have been given as to why some patients reject it: (1) creative patients (artists, painters, etc.) feel that the drug interferes with their earning a living by interfering with their creativity; (2) the hypomanic phase is pleasurable; and (3) the patient refuses to accept the fact that he has a chronic mental illness.[20]

The mechanism of lithium's action is still to be determined. There are conflicting reports about whether lithium is retained in greater amounts during manic episodes than during normal states.[21,22] There is some evidence that lithium affects sodium, potassium, magnesium, and calcium balance.[16,23] Current postulates concerning the mode of lithium's actions involve alteration in the metabolism of the neurotransmitters, norepinephrine and serotonin.[16,31]

Several toxic effects are being noticed now that lithium carbonate is being more widely used in a larger number of people. Lithium carbonate can affect thyroid function, causing myxedema (deficient thyroid function), decreased protein-bound iodine levels, and increased iodine uptake.[15,24] Lithium carbonate reversibly alters the electrocardiogram pattern.[25] There has been a proposal published that lithium may actually reduce atherosclerotic heart disease.[26] Lithium passes the placental barrier, resulting in the fetus having elevated serum lithium levels.[27] For this reason, pregnant women on lithium therapy should be monitored very carefully.[28]

Since lithium is found naturally in certain public water supplies, the question arises as to whether populations drinking water containing lithium might not have a lowered incidence of manic-depressive reactions. A comparative survey of cities near and distant from lithium deposits showed that the highest lithium concentration was 0.1 mEq/l, a dose too small for any psychopharmacologic effect.[29] On the other hand, a report of a Texas study showed that as the lithium content of public water increased, mental hospital admissions decreased.[30] It must be kept in mind that the highest concentration of lithium reported was 160 μg/l (0.023 mEq/l) which is well below that found in the previous study. On the other hand, the previous study[29] did not report any clinical criteria for its conclusion. If these Texas investigations should be confirmed, the controversy concerning addition of a lithium salt to public water supplies could well make the fluoridation issue insignificant in comparison.

Gold

The early use of gold salts in medicine was of an empirical nature, based largely on legend and folklore. Koch's discovery in 1890 that gold cyanide was effective *in vitro* against the tubercle bacillus may be said to mark the beginning of modern gold therapy. This discovery led investigators to use various gold salts in the treatment of many diseases believed to be tubercular in origin. The current use of gold (chrysotherapy) in rheumatoid arthritis is based on the early belief that this disease was an atypical form of tuberculosis.

Therapeutic gold compounds are administered by intramuscular injections. The gold rapidly enters the plasma where it remains bound to albumin for several days. Gold is usually administered on a weekly basis, since plasma-gold levels remain fairly constant for a prolonged period. Orally administered gold is poorly and erratically absorbed.

Gold is toxic. It is slowly excreted by the kidney and will accumulate in the body. Regular determinations of plasma gold levels must be run for the patient on gold therapy. Gold toxicity involves the skin and mucous membranes, joints, blood, kidney, liver, and nervous tissue. Much of the time, cessation of gold administration and supportive treatment are adequate to remove the toxic effects. If gold toxicity is severe, dimercaprol can be used to remove the accumulated gold from the body.

Gold is used primarily in the treatment of rheumatoid arthritis. This progressive, very painful, and crippling disease is poorly understood. With the possible exception of the pig, there are no good animal models of the disease by which it can be studied and drugs evaluated. Current hypotheses are based on an autoimmune mechanism, but the triggering agent is unknown. The disease is a chronic inflammatory disorder that mainly attacks joints and their surrounding structures (muscles, tendons, and other connective tissue). The joints of the hands, wrists, feet, ankles, knees, elbows, and hips are mostly afflicted. Although joints are the primary sites of the disease, it can attack the heart, lung, kidney, and other organs.

The triggering of the inflammatory response involves the release of a group of hydrolytic enzymes known as the lysosomal enzymes, which are contained in membrane sacs called lysosomes and are found in the polymorphonuclear leukocytes, liver cells, and the cells of the synovial membranes of joints. These hydrolases include at least 20 different enzymes, including acid phosphatases, β-glucuronidases, lipases, proteases, sulfatases, and esterases. Together this group of enzymes hydrolyzes phosphate esters, lipids, carbohydrates, proteins, nucleic acids, sulfate esters, etc. Lysosomal enzymes normally function as part of the inflammatory process by breaking down the debris resulting from injury and infection.

It has been postulated that in rheumatoid arthritis some factor triggers the continual release of these enzymes, causing the breakdown of normal synovial membranes, cartilage, muscle, and bone. In advanced cases, the cartilage may be completely destroyed and fibrous tissue may grow out of the exposed bone ends. Eventually the fibrous tissue may become calcified, resulting in the fusion of the joint.[32-34] Because of the severe pain and discouraging prognosis, the arthritic patient is very susceptible to quackery. It is important that the pharmacist continually encourage the arthritic patient to follow the drug regimen prescribed for him and to see his physician on a regular basis.

Drug treatment includes the salicylates, with aspirin in high doses being the drug of choice, corticosteroids, nonsteroidal antiinflammatory agents, gold compounds, and certain antimalarials. The corticosteroids bring about dramatic relief. Unfortunately they cannot be used on a chronic basis. It has been postulated that most of these agents act by stabilizing the lysosomal membranes, thereby reducing the enzymatic breakdown of the joint tissues. Recently it has been suggested that aspirin and possibly other nonsteroidal antiinflammatory agents act by blocking the synthesis of certain prostaglandins that may take part in the inflammatory process.[35]

There is some question as to whether the drugs used in rheumatoid arthritis actually stop the progressive joint destruction. Some of these agents, gold included, may give symptomatic relief only.

Gold is administered beginning with small doses (see specific compounds for doses) and increasing the dose until a total of 1 g has been administered. There is disagreement as to whether a second course of treatment should be instituted if the first course does not produce any beneficial results. Furthermore, it has not been decided if a patient in remission should receive maintenance gold treatment, or if treatment should be stopped until the disease begins again.

Gold has also been used in nondisseminated lupus erythematosus but is contraindicated in disseminated lupus. Gold should not be given to individuals with renal disease, a history of infectious hepatitis, skin or blood disorders, diabetes, pregnancy, hypertension, or congestive heart failure. Prior to each injection it is recommended that a complete blood count and urinalysis be performed and the patient questioned about signs of toxicity (rash, pruritus).

Official Gold Compounds

Aurothioglucose, U.S.P. XVIII ($C_6H_{11}AuO_5S$; Mol. Wt. 392.18; [1-thio-D-glucopyronosato] gold)

Aurothioglucose U.S.P. is an odorless or nearly odorless yellow powder which is stable in air. An aqueous solution is unstable on long standing. The U.S.P. permits the addition of not more than 5.0% of sodium acetate as a stabilizing agent. The pH of its 1 in 100 solution is about 6.3. *Aurothioglucose* U.S.P. is freely soluble in water, and practically insoluble in acetone, alcohol, chloroform and ether. It is administered as a suspension in oil.

Usual Dose: Intramuscular, 10 mg, increased to 25 mg, and then to 50 mg per week to a total of 750 mg; then in decreasing amounts. If a patient has improved and no toxic effects have developed, a dose of up to 50 mg may be given at three- to four-week intervals. Children of 6 to 12 years can receive about one fourth the adult dose.[36]

Usual Dose Range: 10 to 50 mg weekly.

Occurrence:

Aurothioglucose Injection, U.S.P. XVIII (Solganol®)

Contains 100% of the labeled amount as a sterile suspension in a suitable vegetable oil with suitable thickening agents. Solganol® uses sesame oil with 2% aluminum monostearate. It is available in concentrations of 50 and 100 mg/ml.

Gold Sodium Thiomalate, U.S.P. XVIII ($C_4H_3AuNa_2O_4S \cdot H_2O$; Mol. Wt. 408.09; [Disodium mercaptosuccinato]gold)

$$\underset{\text{Au—S—CHCO}_2\text{Na}}{\overset{\text{CH}_2\text{CO}_2\text{Na}}{|}} \quad \cdot H_2O$$

Gold Sodium Thiomalate U.S.P. is a white to yellowish white, odorless, fine powder which is affected by light. It is very soluble in water, and insoluble in alcohol, ether, and most organic solvents.

Usual Dose: Intramuscular, 10 mg, increased to 25 mg, and then to 30 mg per week to a total of 750 mg; then in decreasing amounts.

Usual Dose Range: 10 to 50 mg weekly.

Occurrence:

Gold Sodium Thiomalate Injection, U.S.P. XVIII (Myochrysine®)

Contains 100% of the labeled amount as a sterile solution in *Water for Injection.* It is available as 10, 25, 50, and 100 mg in 1 ml and 500 mg in 10 ml.

Nonofficial Gold Products

Gold Sodium Thiosulfate, N.F. XII ($Na_3Au(S_2O_3)_2 \cdot 2H_2O$; Mol. Wt. 526.22)

Aurothioglycanide ($C_6H_5NHCOCH_2SAu$; Lauron®)

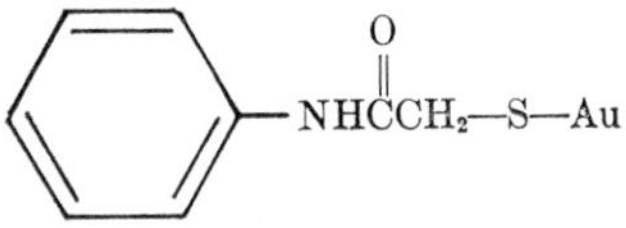

Arsenic

Arsenicals have had a long history in medicine and in crime. Arsenic compounds injure or destroy all cells and are known as protoplasmic poisons. Arsenic reacts with sulfhydryl groups of protein and simple molecules (thioglycollic acid, glutathione, lipoic acid). A more complete discussion is found in Chapter 9.

There has been controversy as to whether an arsenical is active as trivalent or pentavalent arsenic. The body is fully capable of reducing a pentavalent arsenical or oxidizing a trivalent arsenical. Historically, there has been a therapeutic separation between organic trivalent and pentavalent arsenicals, but each has certain advantages. For example, the trivalent congeners of pentavalent compounds appear to be more potent trypanosides, but the pentavalent arsenicals seem to penetrate the cells of the parasite selectively as compared to the cells of the host.

Arsenic as a potassium arsenite solution (Fowler's solution—a solution of arsenic trioxide and potassium bicarbonate) has been used for leukemia because it lowers leukocyte count. It has also been used for psoriasis. There is some reason now to believe that inorganic arsenicals may be carcinogenic. This is based on observations dating back to 1888 that the ingestion of arsenic predisposed a person to skin and visceral cancers. On the other hand, attempts at inducing a carcinoma in animals by administering various arsenic salts using various methods were found to induce malignant changes only if repeated applications of a contact irritant, such as croton oil, were also used. However, most of these animal experiments were not carried out over the three- to ten-year period of chronic arsenic ingestion required for a premalignant condition of the skin to develop (see discussion below).[37]

Arsenic was a popular homicidal poison because embalming fluids once contained arsenic, making its detection impossible once the victim was embalmed. The acute toxic dose of the trivalent inorganic salt is about 100 mg. Symptoms begin with gastric pain developing into severe vomiting and diarrhea, and there may be severe skeletal muscle cramps. The fluid loss can cause shock. Death may not occur until 24 hours after ingestion of the poison. The time course is affected by the dose and the presence of food in the stomach. Chronic poisoning is difficult to detect as the initial symptoms (weakness, anorexia, occasional nausea and vomiting, and diarrhea or constipation) are the same as many other disorders. Treatment involves gastric lavage, saline cathartics, and use of the chelating agent dimercaprol.

More common than the macabre has been chronic arsenic poisoning from industrial and drinking waters. In the former, workers in factories producing fruit sprays, sheep dips, and weed-killers were exposed to sodium or potassium arsenite and possibly to the very dangerous gas, arsine (AsH_3). Certain water supplies have been found to contain arsenic from either natural sources or industrial pollution. In addition to the symptoms already described, patients ingesting arsenic chronically for two to five years may develop a hyperpigmentation which later can form a "rain-drop" appearance. If ingestion continues for a period of three to ten years, a permanent keratosis (horny or callus-like growth) can develop. These keratoses later tend to undergo malignant change.[37]

There has been the traditional belief that one can build up a tolerance to arsenic by ingesting small amounts daily. Whether or not this is true is debatable. It has been stated that those population groups habitually eating arsenic (arsenic eaters) are ingesting a poorly soluble form of arsenic. Others state the long-continued use of arsenic leads to a resistance of the intestinal mucosa to the inflammatory action of arsenic, resulting in less absorption of the arsenic.

Arsenicals are currently used in the treatment of trypanosomiasis and amebiasis (see Chapter 9).

Antimony

Antimony compounds differ from arsenicals by being less readily absorbed and by producing topical irritation. Otherwise, their actions closely resemble those of the arsenicals. They are more caustic than the arsenicals, causing papular (skin elevation) eruptions which develop into vesicular (saclike) and pustular (filled with pus) sores. The irritant action is also exerted upon the gastrointestinal mucosa, resulting in an emetic action. The salivary and bronchial glands are reflexly stimulated. Antimony compounds have been used in cough preparations as expectorants (antimony potassium tartrate in brown mixture). Pentavalent organic antimonials are used for protozoal infections (see Chapter 9).

Aluminum

Soluble aluminum compounds are astringent and antiseptic. Several of the soluble aluminum salts are used by the cosmetic industry as deodorants because of their mild astringent action (see Chapter 9). The insoluble aluminum compounds are mostly used as nonsystemic antacids since the aluminum cation is not absorbed across the intestinal wall (see Chapter 8).

Silver

Silver ion, in common with other heavy metals, is a protein precipitant. The action of silver ion on tissue ranges from antiseptic, astringent, and irritant to corrosive, as the concentration of free silver ion increases. Silver products are used topically and are discussed in detail in Chapter 9.

Silver ion in sufficient concentration is corrosive to the mucosa of the digestive tract. The internal use of silver salts results in little or no systemic action because of the readiness with which silver is precipitated. The toxic dose of silver nitrate has been stated to be about 10 g, although survival has been noted with larger doses.

Whenever silver preparations are used for long periods of time they can cause discoloration of the skin, called *argyria*. The color ranges from gray to one suggesting marked cyanosis. Part of the pigment may be silver sulfide (Ag_2S), but it is also partly metallic sulfur resulting from the reduction of silver in the tissues. Since this reduction is facilitated by light (as in photographic emulsion), those portions of skin exposed to light are more likely to become discolored.

Argyria is irreversible, although it is said that injection of 6% sodium thiosulfate and 1% potassium ferricyanide subcutaneously will remove the color. This treatment requires numerous small injections in the affected area. Chelating agents are not effective since argyria involves free rather than ionized silver.

Barium

The barium cation is extremely toxic systemically due to its muscle stimulating action. The symptoms of barium poisoning are the result of this muscle stimulating action. For example, stimulating gastrointestinal musculature causes vomiting, severe colic (abdominal pain), diarrhea, and hemorrhage; stimulation of the cardiovascular musculature causes spasm of the arterioles, leading to hypertension and cardiac arrhythmias. There may be other muscle tremors, respiratory failure, and convulsion. Death may be caused by cardiac arrest.

Treatment of barium poisoning consists of precipitation of insoluble barium sulfate by oral administration of sodium or magnesium sulfate, followed by gastric lavage. Sodium sulfate may also be administered intravenously.

Barium chloride has been used in complete heart block (heart beat stops) since it is a powerful stimulant of cardiac muscle but because of its low therapeutic index, other drugs and procedures are preferred. The only official barium salt is *Barium Sulfate* U.S.P. XVIII. This water insoluble salt is used as a radiopaque in x-ray studies of the gastrointestinal tract (see Chapter 11). The U.S.P. requires that the title be written out in full when used in prescriptions to avoid possible confusion with poisonous barium sulfide or sulfite.

Even though barium sulfate is not considered to be absorbed across the intestinal mucosa because it is water insoluble, a certain amount appears to be absorbed. Electrocardiographic monitoring of patients undergoing barium enema roentgenographic examinations showed that 46% had electrocardiographic changes during the procedure. The changes occurred more often in the elderly and in patients with heart disease. None of the patients died or developed definite myocardial damage directly attributable to barium enema. Further, no elevations in serum creatine phosphokinase and α-hydroxybutyric dehydrogenase could be attributed to the barium enemas.[38]

Cadmium

The toxicity of the cadmium ion has been known for a long time. Recently it has been publicized due to diseases resulting from possible cadmium contamination. A local disease described by Japanese scientists, known as itai-itai (ouch-ouch), is now believed to have been caused by drinking river water contaminated by cadmium which came from a zinc, lead, and cadmium mine upstream from the village. Later, as pollution abatement at the mine was put into effect, the incidence of the disease decreased. Although soluble zinc and lead salts were also present in the water and could have contributed to the symptoms, itai-itai still is considered to be one of the best examples of cadmium poisoning. The disease has a high fatality incidence, with women of 50 to 60 years of age who had borne many children especially at risk. The symptoms include severe bone pain, waddling gait, amino-aciduria, glycosuria, severe osteomalacia (bone softening), and multiple pathological fractures.[39,40]

More recently, cadmium in drinking water has been suggested as a contributing factor towards heart disease. This is based on the observation that soft water may dissolve the cadmium, which is found with the zinc used to galvanize pipes. Thus, in hard water areas, there would be a protective layer of insoluble calcium salts lining the pipes. There does seem to be a higher incidence of heart disease in soft water areas.[41,42] If this hypothesis is correct, there should also be a lower incidence of heart disease in newer housing areas where copper tubing has been used.[43]

Cadmium accumulates in the kidneys. Of the estimated 200 to 400 mg of cadmium normally ingested each day, about 3 μg is retained in the body, with 1 μg bound in the kidneys. By the time an individual is 30 years old, about 10 mg of cadmium could have accumulated in the kidneys. This observation has led some investigators to suggest that cadmium is implicated in hypertension.[44]

An important source of cadmium may be cigarette smoke. In comparing the total cadmium content of the kidneys, liver, and lungs of subjects at their death, nonsmokers had a mean cadmium level of 6.63 mg as compared to 15.80 mg for smokers. Age and occupation were not significant variables, but number of cigarettes was significant.[45] Much investigative work remains to be done to determine if the cadmium in cigarette smoke contributes to health problems attributed to cigarette smoking.

The insoluble salt, cadmium sulfide, is used in a shampoo (Capsebon®) for the treatment of dandruff (see Chapter 9).

Lead

Lead and its compounds have had a long medical history. Although recognized for centuries as being toxic, it has been widely used industrially, in food and beverage processing, and in medicine. Plumbism (lead poisoning) at one time was an occupational hazard in the manufacture of storage batteries and pottery. Many commercial painters became ill or died in the early twentieth century from the use of paints containing pigments made from lead. These same paints are now affecting children who live in older housing and ingest the paint chips (see later discussion of pica). The colonial American was exposed daily to lead in pewter plates and vessels and inadequately glazed pottery, to lead salts used to sweeten wine, and to lead tubing in stills. This chronic diet of lead was the cause of an intestinal cramping that was given various names: dry gripes, dry bellyache, colica pictonium, and Poitou colic.[46] Lead salts were used topically as an astringent, and at one time, there were lead salts and solutions that were official for this purpose.

Today, lead poisoning is a well known hazard in our society. Indeed, it is becoming uncomfortably well recognized as sources of lead, ranging from older paint and automobile fumes to earthenware utensils, cocktail glasses, and moonshine whiskey, are continually being exposed.[47]

Depending upon the form of lead, just about any route of administration is possible. While oral lead is generally absorbed slowly and excreted reasonably well, postmortem analyses show that lead does accumulate in the body. Inorganic lead cannot pass through intact skin, but it will be absorbed through abraded skin. Thus, lead solutions used as astringents could be absorbed systemically. Organic lead, such as tetraethyl lead, can penetrate the skin rapidly. This would be more of an industrial problem.

While subcutaneous administration of lead may seem unusual, lead poisoning from an embedded bullet, particularly lead shot, does occur. Thus the trite phrase heard in the western movie about giving the victim a dose of lead poisoning when shooting him has some truth in it.

It is believed that most lead poisoning occurs by the oral route. It has been estimated that an individual ingests about 0.3 mg of lead daily from food. Of this only about 8 to 12% is absorbed. This amount can be much greater in the child living in an older home who eats peeling paint chips containing lead, a phenomenon known as *pica* (referring to the magpie and its indiscriminate eating habits). In theory this type of lead poisoning should be disappearing since titanium dioxide-containing paints began replacing leaded interior paints in the 1940s. But in older cities, particularly in the slums, the apartments have not been repainted, and the paint is peeling.[48] Another source has been toys painted with leaded exterior pigments. It is important that a parent read the label of a paint can before using it on any interior surface or on any object that a child is apt to place in his mouth. While several cities have laws forcing landlords to repaint windowsills and place wallboard over the older painted walls, enforcement and inspection have been lax. Other sources of oral lead have included improperly glazed earthenware vessels which have contained acidic fluid or beverages and lead-containing moonshine whiskey in which the still was made from discarded automobile radiators.[49,50]

At one time, lead poisoning from inhalation was an industrial problem. There is recent evidence that the lead in gasoline fumes may be forming lead aerosols in the atmosphere with the greatest concentration near the ground in the auto-congested cities, where there is considerable start-and-stop driving. The small child may be more susceptible since he is closer to the source (the exhaust pipe).[51-54]

Once absorbed, the lead can be found initially in the erythrocytes and soft tissues. In the latter, the kidneys contain the most lead, with the liver second. The lead becomes redistributed over time and is found in bone, teeth, and hair. Once deposited in the bone, lead is considered nontoxic until it is mobilized again. Thus, even though the environmental source of lead is removed, the possibility of lead poisoning persists due to the lead that is stored in the body.

It is difficult to determine a person's exposure to lead since the sources can be so diffuse. The U.S. Public Health Service has issued guidelines for the detection and treatment of lead poisoning in children. The guidelines assume that the normal range of blood lead is 15 to 40 μg/100 ml of whole blood. Concentrations of 40 μg/100 ml or over found on two separate occasions are to be considered as evidence suggesting excess lead absorption, past or present. Blood lead values between 50 to 70 μg/100 ml should be regarded as evidence of possible lead poisoning, with further evaluation required. Children with blood lead values of 80 μg/100 ml or higher should be considered to have lead poisoning, regardless of the presence or absence of symptoms or other laboratory findings.[55] Because lead is stored in the bones and teeth, measuring the lead levels in deciduous (baby) teeth has been proposed as a means of determining a child's exposure to lead.[56]

There is still much work to be done in identifying and removing sources of lead in the environment. The levels described above are arbitrary and based on surveys of blood lead levels. Clinical evidence is lacking as to just what blood lead levels are toxic. There may be undetected metabolic disturbances occurring at the lower blood lead levels that may not become obvious until the child becomes older.

Another reason for identifying and removing sources of lead poisoning is that the child treated for lead poisoning is very susceptible to the toxic effects of lead upon reexposure. Children who had mild encephalopathy (discussed below) from lead poisoning often had permanent neurological damage upon reexposure to lead.[57]

While lead may be considered a protein precipitant by combining with the cysteine sulfhydryl groups of protein, chronic lead poisoning manifests itself by inhibition of heme synthesis. Figure 7–1 outlines heme biosynthesis. Steps 1, 5, 6, and 7 occur in the mitochondria, and steps 2, 3, and 4 in the cytoplasm. The enzymes at steps 2 (aminolevulinic acid dehydrase) and 6 (Fe^{+2}-protoporphyrin chelatase) are inhibited by lead. Lead may also act at steps 1 (δ-aminolevulinic acid synthetase) and 5 (coproporphyrinogen oxidase). It must be kept in mind that heme is being synthesized not only in the newly forming red blood cells, but also in nearly every

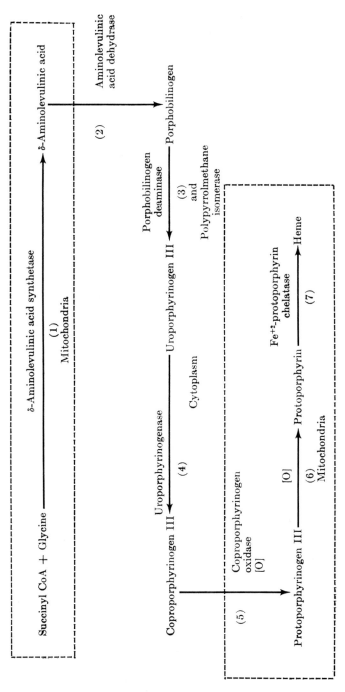

FIG. 7-1. Heme biosynthesis.

Reactions within dotted lines occur in the mitochondria; outside the dotted lines in the cytoplasm.

cell in the body, since heme is also an essential constituent of the respiratory pigments, the cytochromes. There is increased excretion of δ-aminolevulinic acid (ALA) and coproporphyrin III (an oxidized product of coproporphyrinogen). This interference in heme synthesis in the immature red cell can lead to a reversible lead anemia.

The most serious of the symptoms of lead poisoning is lead encephalopathy.[58] It is more common in children than adults. There is brain damage with a fatality rate of about 25%, and about 40% of the survivors will have mental retardation, EEG (electroencephalogram) abnormalities, cerebral palsy, seizures, optic atrophy, etc.

There can be renal damage characterized by a scarring and shrinking of kidney tissue. Chronic lead nephropathy has been seen in heavy moonshine drinkers. It may be that the damaged renal tubular cells do not reabsorb phosphorus, leading to a hypophosphatemia (low blood phosphate level). As the phosphate is mobilized from the bone, the lead present in the bone may also be released, thereby aggravating the condition.[49] Treatment of chronic lead poisoning is based on the use of chelating agents to remove the accumulated lead from the erythrocytes and soft tissues. Dimercaprol and calcium disodium edetate are used initially, followed by penicillamine for follow-up treatment.

Acute lead poisoning from oral ingestion can be treated by administering sodium or magnesium sulfate to precipitate the lead, followed by gastric lavage.

Mercury

Mercury and its compounds have had a long history in medicine. While not as medically important today, certain mercury salts are still official for topical use (see Chapter 9). Actually, metallic mercury is relatively nontoxic as such since it is the mercurous (Hg^+) and mercuric (Hg^{+2}) cations that are toxic. Mercury has been administered in doses as large as 16 oz for intestinal obstruction. As long as gut motility was restored, the patients recovered without mercury poisoning.

Mercury vapor, however, is toxic. It is believed that mercury as finely divided vapor is more readily oxidized and is absorbed through bronchi. Chronic mercury poisoning from inhaling mercury vapor is believed to have occurred in early scientists working with mercury, in industrial situations, and in people living near industrial plants emitting mercury vapor in the air.[59,60] Nevertheless, except for the occasional careless laboratory worker, poisoning by mercury vapor, particularly from industrial sources, is probably rare today.

Mercury poisoning by soluble inorganic mercury salts can be avoided by adhering to a strict dosage schedule. Even mercuric chloride ($HgCl_2$;

mercury bichloride, corrosive sublimate), which can cause bloody diarrhea and death by kidney failure, has been used orally for treatment of syphilis. In most cases, there were no incidences of mercury poisoning when taken in moderate dosages.[59] The mercury salts used topically are, for the most part, water insoluble and can be considered nontoxic for short-term use on intact skin.

In contrast with the above observation, alkylated mercurials, commonly called organic mercurials, are very toxic and are the cause of most modern reports of mercury poisoning. An example is noted in farmers eating grain seed originally intended for planting that had been treated with organic mercurial fungicides. Persons eating animals fed grain treated with mercurial fungicides have also developed mercury poisoning.[59,60]

In a recently reported case, three members of a family of nine continually eating mercury-contaminated pork over a three-month period were hospitalized for mercury poisoning. Nearly two years later, the 8-year-old girl was still blind and unable to speak more than occasionally mumbling one or two syllables; the 13-year-old boy could feed himself and walk without assistance, but was still nearly blind; and the 20-year-old girl was able to walk with the aid of walker frame, speak slowly, and attend school, but still had difficulty seeing. Finally the mother, who was in the early stage of pregnancy during the time that the contaminated pork was eaten, gave birth to a male infant who developed convulsions and who, by the age of one year, still could not sit up or see. No mercury was ever detected in the amniotic fluid. It is not completely understood why the other members of the family did not become ill, but it is believed that it was related to the short exposure to the contaminated pork as compared to other reported incidents of mercury poisoning.[61]

Mercury poisoning due to ingestion of contaminated grains and other foodstuffs can be prevented by proper food inspection, labeling, and, if necessary, outright banning of mercurial fungicides. It must be pointed out, however, that the chances of developing the severe symptoms described above from eating food from commercial sources are slight. Even if a contaminated animal should be slaughtered commercially, its parts would be distributed over the meat counter and would probably last for only one or two meals in any given household. Nevertheless, while this should be kept in mind, it should not be used as an excuse for relaxed vigilance against mercury contamination of foodstuffs.

More insidious is mercury entering the food chain. Plants and animals tend to concentrate mercury, with marine algae being especially efficient. Much of the industrial mercury "escaping" from factories making chlorine, paper, mercury lamps, batteries, electrical appliances, and other products does not lie inertly in the mud, as was once thought, but is converted to dimethyl mercury by the anaerobic bacteria in lakes and bay bottoms. The dimethyl mercury rises through the food chain from microorganisms to algae to smaller fishes, and then to larger fishes. The classic example used to illustrate this means of mercury poisoning is that of the fishermen and their families living around Minamoto Bay in Japan. The fish and shell-fish taken from the bay had high concentrations of mercury. The source of mercury was traced to the effluent from a factory.[59,60] Recently samples of commercial swordfish and tuna were reported to have unacceptably high levels of mercury. Usually the feeding grounds of the fish correlated with an industrial source of mercury. On the other hand, it has been suggested that mercury has been present in fish for many years and, interestingly, in fish where there is no known source of mercury contamination. The leaching of mercury from natural mineral deposits should probably be investigated. There is disagreement as to what upper levels should be permitted in foodstuffs.[62-64]

The toxic effects of mercury, like those of lead and arsenic, are due to its combining with protein sulfhydryl groups. Indeed, the other common name for sulfhydryl groups, mercaptans, is based on the ability of sulfur atoms to bind or capture mercury. Once absorbed, the mercuric cation concentrates mostly in the kidney with lower concentrations in the liver, blood, bone marrow, and other tissues. It is excreted by the kidney and colon. There seems to be a steady state between mercury absorption and excretion which is adjustable based on the intake of mercury.

Mercury poisoning can be divided into two types, acute and chronic. Acute poisoning usually occurs by ingestion of a soluble mercuric salt. Vomiting usually results, which can often empty the stomach of the salt. If large amounts of the mercuric salt were ingested, such severe damage to the intestinal mucosa could occur that shock and death may result. Systemically, mercury affects several organs, with the kidney being profoundly affected. Diuresis from suppression of tubular reabsorption is followed by additional renal damage. The combination of vomiting, diarrhea, and diuresis leads to a fluid and electrolyte imbalance.

Treatment includes gastric lavage, use of a reducing agent such as sodium formaldehyde sulfoxylate ($Na[HOCH_2SO_2]$) to reduce the mercuric cation forming less soluble mercurous salts, and use of chelating agents such as dimercaprol or penicillamine.

Chronic mercury poisoning can occur from industrial exposure, eating of foods contaminated with mercury, and long-term exposure to topical mercurials. The symptoms mimic other disorders. In the case of the family eating pork contaminated with mercury, it took over a month before mercury was suspected, and this only after a more detailed investigation of the family's surroundings was undertaken.[61] Chronic mercury poisoning affects the central nervous system causing behavioral and personality changes, decreased visual acuity, tremors, insomnia, and ataxia. It is more difficult to treat than acute mercury poisoning and consists of removing the source of mercury, administering chelating agents, and providing symptomatic treatment. N-acetyl-D,L-penicillamine has recently been recommended as a superior chelating agent as compared with dimercaprol in chronic mercury poisoning.[65]

Mercury and its salts have enjoyed several uses in therapeutics. Almost all therapeutic applications may be looked upon as modifications of the principal action of the mercuric ion: combining with the sulfhydryl groups of proteins. Currently mercurials are used as diuretics, antiseptics, parasiticides, and fungicides. At one time mercurials were used in the treatment of syphilis, but they have yielded first to arsenicals and bismuth compounds and now to antibiotics. Calomel (Hg_2Cl_2; mercurous chloride) was, at one time, the cathartic of choice. Its action is discussed in Chapter 8. The remaining official mercurials are admitted to the compendia as diuretics (see below) or as antiseptics, parasiticides, and fungicides (Chapter 9).

The diuretic action of mercury salts, inorganic and organic, is due to a direct renal effect by mercuric ion. It is postulated that mercurial diuretics, by reacting with protein sulfhydryl groups, inactivate specific enzymes of the renal tubules, thus preventing sodium ion reabsorption in the proximal tubule and thereby bringing about a sodium and water diuresis. The organic mercurials are used almost to the exclusion of the inorganic salts, as they are more selectively distributed to the tubule. Comparable doses of inorganic salts (mercuric chloride) can be dissipated by reaction with sulfhydryl groups throughout the body and, therefore, would also cause more undesirable toxic reactions.

The chief use of mercurial diuretics is to rid the body of excess fluid caused by cardiac edema. It should be emphasized that the mercury compounds are used simply as potent diuretics and have no curative action whatsoever except that incidental to the diuretic action. Acid-forming salts such as ammonium chloride (see Chapter 5) potentiate the diuretic action of mercurials, and the alkalizing salt sodium bicarbonate (see Chapter 5) inhibits the diuretic action.

Disadvantages of organic mercurial diuretics include such poor absorption from the gastrointestinal tract that most must be administered parenterally, with possible chronic mercury poisoning resulting from prolonged use. Their advantages include: little potassium loss and little alteration of the electrolyte balance of the body fluids; no significant change in carbohydrate metabolism, eliminating the risk of onset of diabetes; and no interference with uric acid excretion, eliminating the risk of hyperuricemia.

Official Mercury Products

Meralluride, U.S.P. XVIII ($C_{16}H_{22}HgN_6O_7$; Mol. Wt. 610.98; Mercuhydrin®)

Meralluride U.S.P. occurs as a white to slightly yellow powder which is slowly affected by light. Its saturated solution is acid to litmus. Meralluride is slightly soluble in water, and soluble in hot water and in glacial acetic acid.

> *Usual Dose:* Parenteral, 1 ml of the injection equivalent to 39 mg of mercury and 43.6 mg of anhydrous theophylline (48 mg of hydrous theophylline) one or two times a week.
>
> *Usual Dose Range:* 1 to 2 ml.
>
> *Occurrence:*
>
> **Meralluride Injection,** U.S.P. XVIII

Sodium Mercaptomerin, U.S.P. XVIII ($C_{16}H_{25}HgNNa_2O_6$; Mol. Wt. 606.01; Thiomerin Sodium®)

Sodium Mercaptomerin U.S.P. occurs as a white, hygroscopic powder or amorphous solid having a characteristic honeycomb structure. Its 1 in 50 solution is neutral or slightly alkaline to litmus. It is freely soluble in water, soluble in alcohol, and slightly soluble in chloroform and in ether. Its solution must be protected from light.

> *Usual Dose:* Parenteral, 125 mg once a day.
>
> *Usual Dose Range:* 25 to 250 mg daily to weekly.
>
> *Occurrence:*
>
> **Sodium Mercaptomerin Injection,** U.S.P. XVIII
>
> **Sterile Sodium Mercaptomerin,** U.S.P. XVIII

Chlormerodrin, N.F XIII ($C_5H_{11}ClHgN_2O_2$; Mol. Wt. 367.20; Neohydrin®)

$$H_2N-C(=O)-N(H)-CH_2-CH(OCH_3)-CH_2-Hg-Cl$$

Chlormerodrin N.F. occurs as a white, odorless, bitter powder. It is sparingly soluble in water, slightly soluble in methanol and absolute alcohol, and practically insoluble in acetone and ether. Chlormerodrin is the one official mercurial diuretic administered orally.

Usual Dose Range: 55 to 110 mg daily.

Occurrence:

Chloromerodrin Tablets, N.F. XIII

Nickel

Nickel compounds have no current medical use. At one time, nickel bromide was suggestd for epilepsy; nickel carbonate and sulfate for tonics and hematinics; and a 1 or 2% solution of nickel sulfate for certain parasitic skin diseases.

Internally, nickel lowers blood pressure and causes nephritis. It has been reported that serum nickel concentrations increased after acute myocardial infarction, suggesting that nickel may be a trace nutrient.[66]

Beryllium

Beryllium and its salts are very toxic. Beryllium salts were widely used in lamp manufacturing during the period of 1940 to 1946. By 1949, the lamp manufacturing industry had eliminated beryllium as a component in fluorescent light tubes because of illnesses developing in their employees. Over twenty years later, workers are still being treated for chronic beryllium poisoning. The diagnosis can be difficult if beryllium exposure is not suspected.[67] No specific antidotes for either acute or chronic beryllium poisoning are known.

Strontium

Strontium salts have been used at various times, ranging from the use of *Strontium Bromide* N.F. X as a sedative to strontium lactate used in the treatment of osteoporosis. The latter use has some rationale as the strontium cation does affect calcium metabolism and distribution. Strontium chloride is used in a dentifrice (Sensodyne®) as a tooth temperature desensitizing agent (see Chapter 10).

Strontium can replace calcium in bone formation and has been used to hasten bone remineralization in diseases such as osteoporosis, mentioned above. Early work indicated that strontium caused the formation of a rachitic bone. Later work showed that if vitamin D, estrogens, and androgens were also administered with strontium, functional bone would be produced. The above two observations can be rationalized by the recent report that strontium may inhibit the synthesis of 1,25-dihydroxycholecalciferol from cholecalciferol (vitamin D_3), which would prevent proper calcium absorption from the intestinal tract.[68] The end result would be a weak bone formed from imperfect apatite (mixed salt of calcium phosphate and hydroxide) crystals. The addition of vitamin D, estrogens, and androgens presumably overcame this inhibition, suggesting that these three agents are probably the effective drugs and strontium has little or no beneficial action.

Strontium 90 (^{90}Sr) is a long-lived radioisotope ($t_{\frac{1}{2}}$ = 28 years) whose presence is greatly feared in radioactive fallout from atomic testing and warfare. It mimics calcium biochemistry and can be found in those plant and animal sources (milk) that are normally considered good sources of calcium. Once ingested, it localizes in the bone, just like calcium, resulting in several foci or sources of ionizing radiation from the ^{90}Sr remaining in the body for up to half the patient's lifetime. Much research has been performed for developing methods to mobilize ^{90}Sr from the bone.

References

1. Spencer, H., Lewin, I., Wistrowski, E., and Samachson, J. Fluoride metabolism in man. Amer. J. Med., **49**:807, 1970.
2. Ferguson, D. B. Effects of low doses of fluoride on serum proteins and a serum enzyme in man. Nature New Biol., **231**:159, 1971.
3. Waldbott, G. L. Biological action of fluoride. Amer. J. Clin. Nutr., **22**:1407, 1969.
4. Day, T. K., and Powell-Jackson, P. R. Fluoride, water hardness, and endemic goitre. Lancet, **1**:1135, 1972.
5. Spencer, H., Lewin, I., Fowler, J., and Samachson, J. Effect of sodium fluoride on calcium absorption and balances in man. Amer. J. Clin. Nutr., **22**:381, 1970.
6. Spencer, H., Lewin, I., Osis, D., and Samachson, J. Studies of fluoride and calcium metabolism in patients with osteoporosis. Amer. J. Med., **49**:814, 1971.
7. Cohn, S. H., Dombrowski, C. S., Hauser, W., and Atkins, H. L. Effects of fluoride on calcium metabolism in osteoporosis. Amer. J. Clin. Nutr., **24**:20, 1971.
8. Harley, J. B., Schilling, A., and Glidewell, O. Ineffectiveness of fluoride therapy in multiple myeloma. New Eng. J. Med., **286**:1283, 1972.
9. Bernsten, D. S., Sadowsky, N., Hegsted, D. M., Guri, C. D., and Stare, F. J. Prevalence of osteoporosis in high- and low-fluoride areas in North Dakota. J.A.M.A., **198**:499, 1966.
10. A.M.A. Council on Drugs. A.M.A. Drug Evaluation, 1st ed. Chicago: American Medical Association, p. 253, 1971.
11. Carney, M. W. P. Five cases of bromism. Lancet, **2**:523, 1971.
12. Cade, J. F. J. Lithium salts in the treatment of psychotic excitement. Med. J. Aust., **2**:349, 1949. Cited in Schoenberg, B. S. Serendipity, lithium and the affective psychoses. J.A.M.A., **207**:951, 1969.
13. Lithium-induced diabetes insipidus. Brit. Med. J., **2**:726, 1972.
14. Lithium for manic-depressive states. Med. Lett., **12**:10, 1970.
15. A.M.A. Council on Drugs. A.M.A. Drug Evaluation, 1st ed. Chicago: American Medical Association, p. 240, 1971.

16. Gershon, S. Lithium in mania. Clin. Pharmacol. Ther., **11**:168, 1970, and references cited therein.
17. Baestrup, P. C., Poulsen, J. C., Schou, M., Thomsen, K., and Amidsen, A. Prophylactic lithium double blind, discontinuation in manic-depressive and recurrent-depressive disorders. Lancet, **2**:326, 1970.
18. Stokes, P. E., Stoll, P. M., Shamoian, C. A., and Patton, M. J. Efficacy of lithium as acute treatment of manic-depressive illness. Lancet, **1**:1319, 1971.
19. Hullis, R. P., McDonald, R., and Allsopp, M. N. E. Prophylactic lithium in recurrent affective disorders. Lancet, **1**:1044, 1972.
20. Polatin, P., and Fieve, R. R. Patient rejection of lithium carbonate prophylaxis. J.A.M.A., **218**:864, 1971.
21. Greenspan, K., Goodwin, F. K., Bunney, W. E., and Durell, J. Lithium ion retention and distribution. Arch. Gen. Psychiat. (Chicago), **19**:664, 1968.
22. Platman, S. R., Rohrlich, J., and Fieve, R. R. Absorption and excretion of lithium in manic-depressive disease. Dis. Nerv. Syst., **29**:733, 1968.
23. King, L. J., Carl, J. L., Archer, E. G., and Castellanet, M. Effects of lithium on brain energy reserves and cations *in vivo*. J. Pharmacol. Exp. Ther., **168**:163, 1969.
24. Luby, E. D., Schwartz, D., and Rosenbaum, H. Lithium-carbonate-induced myxedema. J.A.M.A., **218**:1298, 1971.
25. Demers, R. G., and Heninger, G. R. Electrocardiographic T-wave changes during lithium carbonate treatment. J.A.M.A., **218**:381, 1971.
26. Voors, A. W. Does lithium depletion cause atherosclerotic heart disease?, Lancet **2**:1337, 1969.
27. Wilbanks, G. D., Bressler, B., Peete, C. H., Jr., Cherny, W. B., and London, W. L. Toxic effects of lithium carbonate in a mother and newborn infant. J.A.M.A., **213**:865, 1970.
28. Goldfield, M., and Weinstein, M. R. Lithium in pregnancy: a review with recommendations. Amer. J. Psychiat., **127**:888, 1971. Cited in Clin.-Alert, No. 22, Feb. 16, 1971.
29. Steinberg, J. S., and Rosin, D. A. Lithium content of water in United States cities. J.A.M.A., **211**:1012, 1970.
30. In Texas: The more lithium in tap water, the fewer mental cases. Med. World News, **12**:18, Oct. 15, 1971.
31. Perez-Cruet, J., Tagliamonte, A., Tagliamonte, P., and Gessa, G. L. Stimulation of serotonin synthesis by lithium. J. Pharmacol. Exp. Ther., **178**:325, 1971.
32. Sanders, H. J. Arthritis and drugs. Chem. Eng. News, **46**:52, July 22, 1968.
33. *Ibid.*, p. 52, July 29, 1968.
34. *Ibid.*, p. 46, Aug. 12, 1968.
35. Collier, H. O. J. Prostaglandins and aspirin. Nature, **232**:17, 1971, and references cited therein.
36. A.M.A. Council on Drugs. A.M.A. Drug Evaluation, 1st ed. Chicago: American Medical Association, p. 209, 1971.
37. Black, M. M. Prolonged ingestion of arsenic. Pharm. J., **199**:593, 1967, and references cited therein.
38. Eastwood G. L. ECG abnormalities associated with the barium enema. J.A.M.A., **219**:719, 1972.
39. Cadmium pollution and itai-itai disease. Lancet, **1**:382, 1971.
40. Cadmium in ouch-ouch. Chem. Eng. News, **48**:16, Aug. 10, 1970.
41. Anderson, T. W., Le Riche, W. H., and MacKay, J. S. Sudden death and ischemic heart disease: Correlation with hardness of local water supply. New Eng. J. Med., **280**:805, 1969.
42. Crawford, M. D., Gardner, M. J., and Morris, J. N. Changes in water hardness and local death-rates. Lancet, **2**:327, 1971.
43. Schroeder, H. A. The water factor. New Eng. J. Med., **280**:836, 1969.
44. When metal can mean hypertension. Med. World News, **11**:30, Feb. 20, 1970.
45. Lewis, G. P., Jusko, W. J., Coughlin, L. L., and Hartz, S. Contribution of cigarette smoking to cadmium accumulation in man. Lancet, **1**:291, 1972.

46. Guinee, V. F.   Lead poisoning.   Amer. J. Med., **52**:283, 1972.
47. Dickinson, L., Reichert, E. L., Ho, R. C. S., Rivers, J. B., and Kominami, N.: Lead poisoning in a family due to cocktail glasses.   Amer. J. Med., **52**:391, 1972.
48. Oberle, M. W.   Lead poisoning: A preventable childhood disease of the slums. Science, **165**:991, 1969.
49. Chisolm, J. J., Jr.   Lead poisoning.   Sci. Amer., **224**(2):15, 1971.
50. Klein, M., Namer, R., Harpur, E., and Corbin, R.   Earthenware containers as a source of fatal lead poisoning.   New Eng. J. Med., **283**:669, 1970.
51. Chow, T. J., and Earl, J. L.   Lead aerosols in the atmosphere: Increasing concentrations.   Science, **169**:577, 1970.
52. Lead in air.   Brit. Med. J., **3**:653, 1971.
53. Lead poisoning in the middle class?   Med. World News, **12**:16, Feb. 19, 1971.
54. Rothschild, E. O.   Lead poisoning—the silent epidemic.   New Eng. J. Med., **283**:704, 1970.
55. New policy on lead poisoning in children.   Med. World News, **12**:40I, Oct. 15, 1971.
56. Needleman, H. C., Tuncay, O. C., and Shapiro, I. M.   Lead levels in deciduous teeth of urban and suburban American children. Nature, **235**:111, 1972.
57. Lin-Fu, J. S.   Undue absorption of lead among children—A new look at an old problem.   New Eng. J. Med., **286**:702, 1972.
58. Whitfield, C. L., Ch'ien, L. T., and Whitefield, J. D.   Lead encephalopathy in adults.   Amer. J. Med., **52**:289, 1972.
59. Goldwater, L. J.   Mercury in the environment.   Sci. Amer., **224**(5):15, 1971.
60. McIntyre, A. R.   The toxicities of mercury and its compounds.   J. Clin. Pharmacol. and New Drugs, **11**:397, 1971, and references cited therein.
61. Pierce, P. E., Thompson, J. F., Likosky, W. H., Nickey, L. N., Barthel, W. F., and Hinman, A. R.   Alkyl mercury poisoning in humans.   Report of an outbreak. J.A.M.A., **220**:1439, 1972.
62. Eyl, T. B.   Tempest in a teapot.   Amer. J. Clin. Nutr., **24**:1199, 1971.
63. Dunlap, L.   Mercury: Anatomy of a pollution problem.   Chem. Eng. News, **49**:22, July 5, 1971.
64. Miller, G. E., Grant, P. M., Kishore, R., Steinkruger, F. J., Rowland, F. S., and Guinn, V. P.   Mercury concentrations in museum specimens of tuna and swordfish. Science, **175**:1121, 1972.
65. Kark, R. A. P., Poskanzer, D. C., Bullock, J. D., and Boylen, G.   Mercury poisoning and its treatment with N-acetyl-D,L-penicillamine.   New Eng. J. Med., **285**:10, 1971.
66. Sunderman, F. W., Jr., Nomoto, S., Pradhan, A. M., Levine, H., Bernstein, S. H., and Hirsch, R.   Increased concentrations of serum nickel after acute myocardial infarction.   New Eng. J. Med., **283**:896, 1970.
67. Stoeckle, J. D., Hardy, H. L., and Weber, A. L.   Chronic beryllium disease: Long-term follow-up of sixty cases and selective review of the literature.   Amer. J. Med., **46**:545, 1969, and references cited therein.
68. Omdahl, J. L., and DeLuca, H. F.   Strontium induced rickets: Metabolic basis. Science, **174**:949, 1971.

8
Gastrointestinal Agents*

Inorganic agents used to treat gastrointestinal disorders include: (1) products for altering gastric pH, (2) protectives for intestinal inflammation, (3) adsorbents for intestinal toxins, and (4) cathartic or laxatives for constipation. Most of these products do not require a prescription, which places the responsibility directly on the pharmacist as to who should purchase these items. Many of the above indications are symptoms of a more serious condition and, therefore, the professional pharmacist should inform himself of the uses and limitations of these and all other products he sells and be ready to advise his patients about their utilization.

Acidifying Agents

Achlorhydria is the absence of hydrochloric acid in the gastric secretions. Patients with this condition fall into one of two groups: (1) those who remain free of gastric hydrochloric acid after stimulation with histamine phosphate; and (2) those in whom there is normally a lack of gastric hydrochloric acid but who respond to stimulation by histamine. Patients with the first type of achlorhydria include those with a subtotal gastrectomy, atrophic gastritis (chronic gastritis with atrophy of the mucous membranes and glands), carcinoma of the stomach, or gastric polyps. The second type, in which the patients are initially free of gastric hydrochloric acid but will secrete it upon histamine stimulation, include those with chronic nephritis (inflammation of the kidneys), chronic alcoholism, tuberculosis, hyperthyroidism, pellagra, sprue (a periodic fatty, frothy diarrhea), and parasitic infestations. It is common in otherwise normal individuals after age 50.

Gastric hydrochloric acid functions by killing the bacteria in ingested food and drink, softening fibrous foods, and promoting formation of the proteolytic enzyme, pepsin. The latter is formed by pepsinogen being converted to pepsin when the pH of the gastric contents drops below 6.

---

* At the time this material was sent to the printer, a preliminary monograph for antacids was published in the Federal Register (Vol. 38, No. 65, April 5, 1973, Part II) describing an *in vitro* test for measuring acid-consuming capacity and calling for the development of an *in vivo* means of evaluation. The panel also recommended certain labeling requirements for antacid products sold over the counter.

Thus, a lack of hydrochloric acid could reasonably be expected to cause gastrointestinal disturbances.

The symptoms of achlorhydria can vary with the associated disease, but they generally include mild diarrhea or frequent bowel movements, epigastric (upper middle portion of the abdomen) pain, and sensitivity to spicy foods. Because pepsin possesses its greatest proteolytic activity below pH 3.5 (see pepsin in the antacid discussion which follows), there is usually a lack of pepsin activity, if not lack of the enzyme itself, in the absence of gastric hydrochloric acid. This usually is not considered to be a problem, since the many proteolytic enzymes in the intestinal tract are still present and fully functional. It is not uncommon for patients with achlorhydria to have pernicious anemia due to a lack of intrinsic factor, the protein necessary to carry vitamin B_{12} across the intestinal wall.

In an attempt to relieve the gastrointestinal symptoms caused by achlorhydria, *Diluted Hydrochloric Acid* N.F. has been utilized. (A complete description of *Diluted Hydrochloric Acid* will be found in Chapter 4.) The usual 5-ml dose of *Diluted Hydrochloric Acid* N.F. added to 200 ml of water provides about 15 mEq of acid. In order to avoid exposure of dental enamel to hydrochloric acid, the use of a drinking straw laid well back on the tongue has been recommended or the use of equivalent (but more expensive) products such as glutamic acid hydrochloride (Acidulin®) which is administered in capsules. However, this product provides only 1.7 mEq of hydrochloric acid, and the recent Drug Efficacy Study of the National Academy of Sciences-National Research Council for the Food and Drug Administration concluded that the glutamic acid preparations were ineffective. It must also be kept in mind that it is not clear if the decrease in gastric hydrochloric acid is the cause of any specific symptoms, or if it is a symptom associated with many possible pathological conditions. For this reason it is doubtful if the administration of *Diluted Hydrochloric Acid* with its higher acid content serves any useful clinical or physiological purpose.[1]

Antacids

When the general public is asked why it takes antacids, the answers will include: (1) that uncomfortable feeling from overeating, (2) heartburn, and (3) a growing hungry feeling between meals. Antacids are widely advertised with these reasons given as an indication for purchasing and using a certain product. Unfortunately, few of these antacid advertisements carry any statement recommending that the patient consult with a physician before undertaking self-medication.

The chief indication for administering an antacid is to neutralize excess gastric hydrochloric acid which may be causing pain and possible ulceration. Another objective may be to inactivate the proteolytic enzyme, pepsin. Depending upon the extent of hyperacidity an anticholinergic agent may also be indicated, and, depending on the degree and extent of ulceration, bed rest and surgery may also be required. Thus, those taking antacids on a continuing basis should be under medical supervision. Because the vast majority of antacids (except sodium bicarbonate) are relatively free of serious side effects, and since they mask the pain resulting from an ulcer, it is easy for the patient to indulge in self-medication and not seek further medical attention.

The stomach pH can range from pH 1 when empty to 7 when food is present. The low acid pH is due to the presence of endogenous hydrochloric acid, which is always present under physiological conditions. When hyperacidity develops, the results can range from gastritis (a general inflammation of the gastric mucosa) to peptic ulcer (a specific circumscribed erosion). A peptic ulcer can be located in the lower end of the esophagus (esophageal ulcer), the stomach (gastric ulcer), duodenum (duodenal ulcer), or on the jejunal side of a gastrojejunostomy (surgical formation of a passage between the stomach and jejunum). An esophageal ulcer occurs when the esophageal sphincter is defective, as in hiatal hernia. Patients with this condition will frequently suffer from "heartburn," which is due to gastric acid entering the esophagus either during a belch or upon lying in bed. Frequently, these people obtain relief by sleeping in a bed elevated at the head to reduce the flow of gastric fluid from the stomach into the esophagus. Sleeping on the right side will sometimes be painful because of gastric outflow into the esophagus.

More common than the esophageal ulcer is the gastric ulcer, which occurs in the lesser curvature of the stomach, and most common is the duodenal ulcer, found in the first portion of the duodenum. The etiology of peptic ulcer is unknown. It is not always associated with hyperacidity. Patients with gastric ulcer proximal to the pyloric antrum usually have a lower than normal secretion of acid. Gastric ulcer may be due more to decreased tissue resistance to pepsin than to any gastric hypersecretion. Hyperacidity, however, is associated with duodenal ulcer and benign peptic ulcers are not found in achlorhydric patients. There frequently is a correlation between the emotional makeup of the individual and the incidence of peptic ulcer. The tense individual who contains his emotions tends to have a greater incidence of peptic ulcer than does the person who can "release" his tension. Most peptic ulcers are chronic, rather than acute, with remissions followed by exacerbations. Complications will include hemorrhage, perforation, pyloric obstruction due to scar tissue, and the transformation from a benign to a malignant condition. Fortunately, the latter complication is rare, but it is more likely to happen in gastric rather than duodenal ulcers. On the other hand, differentiation between benign and malignant gastric ulcers is difficult.

Hemorrhage is more common with gastric than with duodenal ulcers and, in this event, it is not uncommon for blood transfusions to be indicated.

Whereas malignancy and hemorrhage are more common with gastric ulcers, perforation is more common with duodenal ulcers. If not treated in time, peritonitis leading to death can result. Surgery is usually indicated.

Depending on the severity and location of an ulcer, treatment will range from diet and antacid and/or anticholinergic therapy to complete bed rest and possible surgery. Frequently, removal of an emotional stress situation is indicated.

Because food generally reduces the gnawing or hunger pain of peptic ulcer, the patient will often be advised to eat smaller meals but to eat more often. The older bland, tasteless diets largely have been abandoned since they do not promote healing of the ulcer any faster than a balanced, standard, more tasty meal. The latter is also much better received by the patient. Stimulants of gastric acid secretion such as coffee and alcohol are usually eliminated, together with irritating foods such as uncooked roughage, spices, and fried foods.

Again, depending on the severity, hospitalization may be necessary. Bed rest seems to speed up healing.[2] This observation would seem to correlate with the concept of emotional stress being a causative factor in gastric ulcers.

Antacid Therapy. Antacids are alkaline bases used to neutralize the excess gastric hydrochloric acid associated with gastritis and peptic ulcers. This in turn inactivates pepsin, which functions optimally at low pH although some antacids may inhibit pepsin, independently of the pH effects. Since gastric hydrochloric acid production is continuous, the administration of antacids is also usually on a continuing basis. As with the chronic administration of any drug, antacid therapy is not without side effects. Of historical interest is acid rebound, which has been viewed as overtitration by the parietal (stomach wall) cells. It is quite normal for the stomach to be at pH 1–2, the optimal pH for the proteolytic enzyme, pepsin. Most antacids commonly used today raise the gastric pH to 4–5, which greatly reduces pepsin's proteolytic action. Theoretically, if the gastric pH is raised too much, acid rebound may occur since, in an effort to maintain a lower pH, the stomach secretes additional hydrochloric acid which consumes the antacid. In addition, the gastric contents, including the antacid in the stomach, will be emptying into the intestine. The result is that an excess of acid is secreted into the stomach, leading to a possible hyperacidic condition which could further aggravate the ulcer. The concept of acid rebound as a clinical problem is not universally accepted and apparently is difficult to document.[3] In a study using human patients with duodenal ulcers, both aluminum hydroxide-magnesium hydroxide mixtures (30 or 60 ml) and sodium bicarbonate (4 or 8 g) failed to stimulate gastric secretion, while calcium carbonate (4 or 8 g) induced gastric secretion. A calcium ion-stimulating effect on the gastrointestinal tract was proposed. The effect was potentiated by food.[4,34]

A second potential problem is systemic alkalosis. If the antacid is sufficiently water soluble and is composed of readily absorbable ions, the antacid may be absorbed and exert its alkaline effects on the body's buffer systems. This is not as much of a current problem, since most antacids in use today are water insoluble (see later discussion). An exception is sodium bicarbonate which is usually found in most households.

A third problem is the sodium content of the antacid. Those patients who are on sodium-restricted diets should be advised of this when an antacid is recommended. Continually updated tables of the sodium content of antacids can be found in the *Handbook of Non-Prescription Drugs.*[5]

A fourth side effect is the local effect in the gastrointestinal tract. Antacids containing calcium and aluminum salts, after being converted to soluble salts by gastric acid, tend to be constipating, while those containing magnesium salts tend to have a laxative effect. It is quite common to market a constipative- and laxative-acting antacid in combination.

So far this discussion has treated antacids with the assumption that they are a valid treatment for gastrointestinal hyperacidity because of their ability to neutralize excess acid. It is currently thought that antacids are effective in the reduction of pain associated with peptic ulcers, even in the absence of excessive acid. The reduction of pain has been attributed to an increase in gastric pH and subsequent buffering, an inhibition of the proteolytic action of pepsin by either adsorption by the antacid or increase of gastric pH, and/or a protective coating action on the ulcerative tissue. Furthermore, there is no substantiated evidence that continued use of an antacid following the cessation of pain is of any benefit in the healing process.[6,7]

Because antacids do relieve the pain of a peptic ulcer, are relatively free of side effects when properly used, and do not prolong the ulcer, it would seem to indicate that antacids are justified if for no other reason than patient comfort.

Because the actual mechanism for their relieving of pain is not known, the evaluation of antacids is difficult. While some evaluations are based on the subjective measuring of pain reduction, most are based on the effect of modifying stomach pH. Although the latter approach would appear to involve a simple procedure, it is difficult to determine whether the comparative *in vitro* results have any clinical significance.

There have been many *in vitro* techniques published.[8-17] They range from adding the antacid to a given amount of hydrochloric acid and measuring the amount of acid consumed to attempts at mimicking what goes on in the stomach. In order to do the latter, additional acid is added at given time intervals in order to mimic the continuing secretion of acid into the stomach. Because the stomach is continually emptying, some procedures call for removing aliquots of the antacid-hydrochloric acid mixture from the reaction flask at given time intervals. Most of the published experimental procedures are not properly designed and often yield results showing the antacid to have a longer duration of action than it actually possesses under *in vivo* conditions.

Since gastric fluid contains proteins, pepsin, and a cathepsin, procedures utilizing various artificial gastric juices have been devised.[8,9] Some of these techniques also measure whether the antacid can inactivate the proteases by increasing gastric pH, by adsorption of the enzyme by the antacid, or by a specific denaturation, usually by the metal cation component of the antacid.

Indeed, there has been much debate in the literature as to whether a specific antacid does inhibit pepsin.[18,19] A plot of pH versus human pepsin stability shows pepsin to be stable up to pH 6. Above this pH, pepsin stability decreases until complete inactivation occurs at pH 8. In contrast, the plot of pH versus activity shows 100% activity at pH 2, 70% activity at pH 4.5 (a plateau between pH 3.5–4.5), and almost no peptic activity at pH 5.5. This means that an antacid that acts solely by neutralizing excess acid would have to raise the pH of the stomach contents to above pH 8 to inhibit pepsin irreversibly. Even at pH 5.5, the inactivation of pepsin is only transient, with 70% of activity returning by pH 4.5.[20]

With the pH stability and activity curves in mind, it is possible to realize why published evaluations of antipeptic activity of antacid products can produce such contradictory results. It is also important that human pepsin be utilized in these tests. A 1971 study concluded that none of the antacids tested possessed any intrinsic antipeptic activity and produced their effect only by alteration of pH. However, the investigator used bovine and porcine pepsin which show maximum activity between pH 1.5–2.5, a sharp fall to 30% activity at pH 3, minimal activity at pH 4, and zero activity at pH 5, as contrasted with human pepsin which has 70% activity at pH 3.5–4.5.[20–21]

A 1972 study that also used porcine pepsin reported that not only many commercial antacid preparations, but also barium sulfate and a commercial kaolin-pectin mixture (Kaopectate®), inactivated pepsin by adsorption of the enzyme onto the surface of the particulate matter. The inactivation was independent of pH. However, the porcine pepsin used in this study remained active at least up to pH 6.[22]

The *in vivo* evaluation of antacids is more difficult since it involves either removing aliquots of the gastric contents at intervals and measuring pH or introducing pH electrodes into the stomach. The latter can be accomplished by using a miniaturized pH meter which the subject swallows, usually as a capsule. It can be held in a certain location and later recovered by attaching a string to it prior to swallowing the meter. The pH changes along the person's gastrointestinal tract can also be followed by simply allowing the meter to be carried along by peristaltic movement. X-rays give the exact location of the pH meter at any given time. The pH readings are usually transmitted by a radio signal.[23–25]

While no antacid is "ideal," there have been certain criteria that have been developed. These are:

1. The antacid should not be absorbable or cause systemic alkalosis.
2. The antacid should not be a laxative or cause constipation.
3. The antacid should exert its effect rapidly and over a long period of time.
4. The antacid should buffer in the pH 4–6 range.
5. The reaction of the antacid with gastric hydrochloric acid should not cause a large evolution of gas.
6. The antacid should probably inhibit pepsin.

Antacid Products

Sodium Bicarbonate as an Antacid

Sodium Bicarbonate U.S.P. (baking soda) is a highly water soluble antacid with a very rapid onset of action but relatively short duration. It can cause a sharp increase in gastric pH up to or above pH 7. Because of the evolution of carbon dioxide [rx (i)] in the presence of acid, sodium bicarbonate can cause belching and flatulence.

$$NaHCO_3 + HCl \rightarrow NaCl + CO_2 \uparrow \ + H_2O \qquad \text{(i)}$$

It is readily absorbed and sodium retention can result with continued use. While little harm will probably result with occasional use, sodium bicarbonate is definitely not indicated for patients needing antacid therapy for even limited periods of time. It will inhibit the absorption of tetracycline from the gastrointestinal tract.[26]

Sodium Bicarbonate, U.S.P. XVIII (the physical and chemical properties are discussed in Chapter 5)

Usual Dose: 300 mg to 2 g four times daily.

Usual Dose Range: 300 mg to 16 g daily.

Sodium bicarbonate will be found in many effervescent antacid preparations. However, rather than being used as an antacid, its main function is to react with an acid (citric, tartaric, etc.) with the evolution of carbon dioxide. The result is a "sparkling" flavor in preparations which would otherwise have a flat saline taste.

Aluminum-Containing Antacids. The aluminum-containing antacids are widely used. They are nonsystemic and buffer in the pH 3–5 region. Because of liberation of astringent aluminum cations, they tend to be constipating. Most cause increased fecal phosphate excretion due to formation of insoluble aluminum phosphate in the intestinal tract.

Aluminum Hydroxide

Aluminum hydroxide is recognized by the current U.S.P. in two physical forms plus one dosage form. (1) *Aluminum Hydroxide Gel* U.S.P. XVIII (Amphogel®) is a white viscous suspension, from which small amounts of clear liquid may separate on standing. The U.S.P. permits the inclusion of suitable flavoring and antimicrobial agents. It has a pH between 5.5 and 8.0. (2) *Dried Aluminum Hydroxide Gel* U.S.P. XVIII is not a typical gel but is a white, odorless, tasteless, amorphous powder insoluble in water and alcohol but soluble in dilute mineral acids and solutions of fixed alkali hydroxides. Both forms are assayed in terms of their aluminum oxide (Al_2O_3) content and their acid-consuming capacity. The dried gel is also official as dried *Aluminum Hydroxide Gel Tablets* U.S.P. XVIII (Amphogel® Tablets), a convenient dosage form.

Both the gel and dried gel are popular antacids. They possess many of the properties sought in the ideal antacid although their onset of action is slower. This varies with the dosage form and age of the product. The gel with its more finely divided particle and its ability to be completely wetted has a more rapid onset of action than the various dried gel tablets. The onset of action of the latter will vary, depending on whether the patient must first chew the tablet or simply suck on the tablet. If chewing is required, then the degree of chewing will determine how finely divided will be the particles.

A problem with the gels is that of a loss of antacid properties on aging. This is more of a problem with the dried gel than with the liquid suspension and seems to be related to the manufacturing process. The rate of loss of antacid action is dependent upon the pH used to precipitate the gel. This pH dependency determines the concentration of anions from the buffer in the gel structure. The most acid reactive gels are those in which the concentration of a monovalent anion, such as chloride or bicarbonate, approaches 1 mole per mole aluminum or those in which a bivalent anion, such as sulfate, approaches 0.5 mole per mole of aluminum.[27]

Physical data (appearance, viscosity, x-ray diffraction pattern, differential thermal analysis) indicate that a change in gel structure occurs during the aging process.[28]

The anions left in the gel during the precipitation process apparently stabilize the gel. There are aluminum hydroxide gel products on the market using polyhydroxyl hexitol adjuncts, which are claimed to stabilize the gel.[29]

The aluminum hydroxide gels are nonabsorbable and exert little, if any, systemic effect. If the gel is formed by precipitation in a carbonate/bicarbonate system, there may be some evolution of carbon dioxide when the carbonate and bicarbonate anions react with the gastric hydrochloric acid, but there have been no reports of patient discomfort.

From a chemical standpoint, the aluminum hydroxide gels are ideal buffers in the pH 3–5 region due to their amphoteric character [rxs (ii)–(iv)].

		Conjugate	*Conjugate*
Base	*Acid*	*Acid*	*Base*

$$[Al(H_2O)_3(OH)_3] + H_3O^+ \rightleftarrows [Al(H_2O)_4(OH)_2]^+ + \quad H_2O \quad \text{(ii)}$$

$$[Al(H_2O)_4(OH)_2]^+ + H_3O^+ \rightleftarrows [Al(H_2O)_5(OH)]^{+2} + \quad H_2O \quad \text{(iii)}$$

$$[Al(H_2O)_5(OH)]^{+2} + H_3O^+ \rightleftarrows [Al(H_2O)_6]^{+3} \quad + \quad H_2O \quad \text{(iv)}$$

Two important facts must be kept in mind from this sequence. First, the hydrated aluminum hydroxide in reaction (ii) will not go into solution unless the medium is strongly acid or alkaline. Second, the hydrated aluminum cation, the conjugate acid of reaction (iv) is an acid of pK_a 4.85. Thus, reaction (iv) is at 50% completion at pH 4.85 and will reverse as the pH increases. The end result is that aluminum hydroxide gel should hold the gastric pH at just about the desired level.

Another factor in favor of aluminum hydroxide gel may be its adsorbent properties. *In vitro*, it apparently adsorbs pepsin.[18,19] It can also interfere with the adsorption of other drugs, and caution should be exercised in the coadministration of the gel and other drugs. On the other hand, there are reports that aluminum hydroxide gel's acid-reacting property is inhibited by pepsin as compared with other aluminum-containing antacids.[8,30]

Because the water soluble astringent salt, aluminum chloride, is the product of the reaction with hydrochloric acid, aluminum hydroxide gel can cause constipation and occasionally nausea and vomiting. It eventually forms the insoluble aluminum phosphate salt in the intestinal tract, resulting in increased fecal phosphate excretion. This property has been used in the treatment of phosphatic urinary calculi by retarding phosphate absorption.

For people on a normal diet, there is little danger of phosphate deficiency developing. Nevertheless, if phosphate intake is low, patients receiving large doses of aluminum hydroxide gel for long periods may develop a phosphate deficiency.[31]

Usual Dose: 15 ml four to six times a day; the equivalent of 300 mg of aluminum hydroxide four to six times a day.

Usual Dose Range: 5 to 30 ml up to 12 times daily; the equivalent of 300 mg to 5 g of aluminum hydroxide daily.

Aluminum Phosphate

Aluminum Phosphate is official as *Aluminum Phosphate Gel* N.F. XIII (Phosphagel®). It is a white, viscous suspension from which small amounts of water may separate on standing. It may contain suitable preservatives. The gel has a pH between 6.0 and 7.2. It is assayed in terms of aluminum phosphate ($AlPO_4$) content and must meet specified neutralization rate and acid-consuming capacity criteria.

This nonabsorbable antacid has been used in place of aluminum hydroxide gel where loss of phosphate may be a problem to the patient. Since aluminum phosphate gel is regenerated in the intestine, endogenous phosphate is spared. Otherwise the adsorptive and astringent properties are much the same as those of aluminum hydroxide gel.

The acid-consuming ability of aluminum phosphate is based on the release of phosphate anion [rxs (v)–(vii)].

$$AlPO_4 \rightleftarrows Al^{+3} + PO_4^{-3} \qquad\qquad (v)$$
$$\text{Solid} \qquad\quad \text{Solution}$$

$$PO_4^{-3} + H_3O^+ \rightarrow HPO_4^{-2} + H_2O \qquad\qquad (vi)$$
$$\text{Base} \qquad \text{Acid} \quad \text{Conjugate} \quad \text{Conjugate}$$
$$\text{Acid} \qquad \text{Base}$$

$$HPO_4^{-2} + H_3O^+ \rightleftarrows H_2PO_4^- + H_2O \qquad\qquad (vii)$$
$$\text{Base} \qquad \text{Acid} \quad \text{Conjugate} \quad \text{Conjugate}$$
$$\text{Acid} \qquad \text{Base}$$

Aluminum phosphate is very water insoluble and will only go into solution as phosphate anion is consumed by gastric acid. With the dihydrogen phosphate anion having pK_a 7.21 (6.86) and the monohydrogen phosphate anion having pK_a 2.12, the pH of the stomach will be somewhere in between.

Usual Dose: 15 to 30 ml every two hours.

Usual Dose Range: 15 to 45 ml.

Dihydroxyaluminum Aminoacetate ($NH_2CH_2COO—Al(OH)_2 \cdot xH_2O$; Mol. Wt. (anhydrous) 135.05; (glycinato)dihydroxyaluminum)

Dihydroxyaluminum aminoacetate is recognized by the current National Formulary in two physical forms and one dosage form. (1) *Dihydroxyaluminum Aminoacetate* N.F. XIII is a white, odorless powder with a faintly sweet taste. It is insoluble in water and organic solvents but does dissolve in dilute mineral acids and in solutions of fixed alkalies. The N.F. permits the presence of small amounts of aluminum oxide and glycine (aminoacetic acid). (2) *Dihydroxyaluminum Aminoacetate Magma* N.F. XIII (Robalate® Suspension) is a white, viscous suspension from which small amounts of water may separate on standing but may be readily reformed upon shaking. Also official is *Dihydroxyaluminum Aminoacetate Tablets* N.F. XIII (Robalate® Tablets). All dihydroxyaluminum aminoacetate preparations are assayed in terms of aluminum oxide content; all must meet specified acid-consuming capacity and acid-neutralizing capacity criteria; and all give a pH between 6.5 and 7.5. Dihydroxyaluminum aminoacetate is manufactured by reacting aluminum isopropoxide with glycine.

This nonabsorbable product is claimed to act more promptly than the aluminum hydroxide gels because the amine group of glycine reacts with the gastric acid forming the protonated amine [rx (viii)].

$$NH_2CH_2CO_2Al(OH)_2 + H_3O^+ \rightarrow NH_3^+CH_2CO_2Al(OH)_2 + H_2O \quad (viii)$$

This is followed by reaction with the two hydroxyls and the glycine carboxyl group resulting in a prolonged buffering action [rx (ix)].

$$NH_3{}^+CH_2CO_2Al(OH)_2 + 3H_3O^+ \rightarrow Al^{+3} + NH_3{}^+CH_2CO_2H + 3H_2O \quad (ix)$$

Presumably dihydroxyaluminum aminoacetate would show the same amphoteric properties as aluminum hydroxide gel [rxs (ii)–(iv)] and reaction (ix) would not go to completion.

The efficacy of this nonsystemic antacid is apparently not affected by aging and appears to be superior in that regard to dried aluminum hydroxide gel.[30,32] The precautions associated with the use of aluminum hydroxide gel, possible constipation and phosphate depletion, would be true for dihydroxyaluminum aminoacetate as well. It and magnesium carbonate make up the antacid component of Bufferin®.

Usual Dose: 500 mg to 1 g four times a day.

Usual Dose Range: 500 mg to 2 g.

Dihydroxyaluminum Sodium Carbonate, N.F. XIII [(HO)$_2$AlOCO$_2$Na· xH$_2$O; Mol. Wt. (anhydrous) 144.00; aluminum sodium carbonate hydroxide)

Dihydroxyaluminum Sodium Carbonate N.F. is a fine, white, odorless powder. It is practically insoluble in water and organic solvents but dissolves in dilute mineral acids with the evolution of carbon dioxide. An aqueous suspension has a pH between 9.9 and 10.2. It is assayed in terms of aluminum oxide and carbon dioxide evolution and must meet specified acid-consuming capacity, acid-neutralizing capacity, and prolonged neutralization criteria. Dihydroxyaluminum sodium carbonate is made by the reaction of aluminum isopropoxide and an aqueous solution of sodium bicarbonate.

This nonabsorbable aluminum antacid is reported to give a rapid onset of action [rx(x)] followed by a more prolonged reaction and a buffering action [rx (xi)].

$$NaOCO_2Al(OH)_2 + H_3O^+ \rightarrow Al(OH)_3 + Na^+ + CO_2 \uparrow + H_2O \quad (x)$$

$$Al(OH)_3 + H_3O^+ \rightarrow \text{see reactions (ii)–(iv)} \quad (xi)$$

Aging apparently has little effect on efficacy. Potential drawbacks to this preparation would be the presence of sodium, evolution of carbon dioxide, and the usual problems associated with the aluminum antacids.

Usual Dose Range: 300 to 600 mg as required.

Occurrence:

Dihydroxyaluminum Sodium Carbonate Tablets, N.F. XIII (Rolaids®)

Basic Aluminum Carbonate Gel (Al(OH)CO$_3$; Basaljel®)

The gel is suggested for the management of phosphatic urinary calculi.

It functions very much like aluminum hydroxide in that dietary phosphate is excreted in the feces as aluminum phosphate and thus is not available for the formation of phosphatic calculi.

Calcium-Containing Antacids. The calcium-containing antacids differ from the aluminum antacids in that their action is dependent upon their basic properties and not on any amphoteric effect. Those used in medicine are poorly soluble salts which will only go into solution if there is acid present to consume the small amount of solubilized salt already in solution. *In vitro* and *in vivo* studies show that the calcium antacids raise the stomach pH to nearly 7.[33]

The calcium-containing antacids, calcium carbonate in particular, are considered by some to be the antacids of choice. They are rapid acting and largely nonsystemic. The latter characteristic must be qualified because, even though this group of antacids does not cause systemic alkalosis, the liberated calcium cation can be absorbed, causing increased serum calcium levels.[34] Although not common, renal failure due to hypercalcemia has been reported.[35] There is also evidence suggesting that the calcium cation may cause an increase in gastric acid secretion.[4,34]

An uncommon but potentially very serious side effect is the milk-alkali syndrome (Burnett syndrome). This can occur during prolonged administration of large doses of sodium bicarbonate or calcium carbonate together with large amounts of milk. Features of this uncommon complication are acute alkalosis, renal insufficiency, hypercalcemia, hyperphosphatemia, and azotemia. The syndrome can be mistaken for hyperparathyroidism (see discussion on calcium biochemistry). Usually the symptoms disappear after stopping the antacid treatments, although the prognosis can be grave depending upon the degree of renal insufficiency. Magnesium and aluminum salts have not been implicated in this syndrome.

The recorded incidence of hypercalcemia ranges from 1 to 25% with patients receiving 40 to 60 g of calcium carbonate per day. (While this dose is greater than the usual U.S.P. dose range, it nevertheless is not excessive when the clinician is trying to hold the gastric fluids at a constant elevated pH.) Patients seem to develop hypercalcemia during the first few days of calcium carbonate therapy or not at all.[36,37]

The calcium antacids tend to be constipating and are usually found in combination with magnesium antacids.

Calcium Carbonate ($CaCO_3$; Mol. Wt. 100.09; precipitated chalk)

Calcium carbonate is official as *Precipitated Calcium Carbonate* U.S.P. XVIII. It is a fine, white, odorless, tasteless, microcrystalline powder which is stable in air. It is practically insoluble in water, but its solubility is increased by the presence of any ammonium salt or carbon dioxide. The presence of any alkali hydroxide reduces its solubility. It is insoluble in alcohol and dissolves with effervescence in diluted acetic, diluted hydrochloric, and diluted nitric acids.

Because of its fast action, calcium carbonate is one of the most popular antacids. Its action is limited by the amount of salt that will go into solution. Thus, as gastric hydrochloric acid consumes the solubilized calcium carbonate, more goes into solution. This process continues until the acid or calcium carbonate is consumed [rx (xii)].

$$\underset{\text{Solid}}{CaCO_3} \;\rightleftarrows\; \underset{\text{Solution}}{Ca^{+2} + CO_3^{-2}}$$

$$\left|\; H_3O^+ \right.$$

$$\longrightarrow H_2CO_3 \rightarrow H_2O + CO_2 \uparrow \qquad \text{(xii)}$$

The calcium cations formed in reaction (xii) and present as the water soluble calcium chloride salt can be either absorbed (see above) or precipitated as the insoluble calcium phosphate salt in the intestine or as insoluble calcium soaps from the hydrolyzed glycerides resulting from digested food.

Because of calcium's constipative effect, most calcium carbonate preparations will be found in combination with a magnesium antacid. They are contraindicated in patients with renal disease, a history of urinary calculi, gastrointestinal hemorrhage, hypertension, or dehydration and electrolyte imbalance due to excessive vomiting or aspiration of gastric contents. The liberation of carbon dioxide may cause discomfort in some patients.

Usual Dose: 1 g four to six times a day.

Usual Dose Range: 1 to 10 g daily.

Occurrence:

Calcium Carbonate Tablets, N.F. XIII

Tribasic Calcium Phosphate, N.F. XIII

Tribasic Calcium Phosphate N.F. is occasionally used as an antacid. A description of the compound will be found under calcium replenishers. The principle of its action is based on its going into solution only in acid media, much the same as calcium carbonate. The phosphate anion then consumes two equivalents of gastric acid [rx (xiii)].

$$\underset{\text{Solid}}{Ca_3(PO_4)_2} \;\rightleftarrows\; \underset{\text{Solution}}{3Ca^{+3} + 2PO_4^{-3}}$$

$$\left|\; H_3O^+ \right.$$

$$\longrightarrow HPO_4^{-2} + H_2O$$

$$\left|\; H_3O^+ \right.$$

$$\longrightarrow H_2PO_4^- + H_2O \qquad \text{(xiii)}$$

Another view is that the phosphate ion (PO_4^{-3}) reacts with the water present in the stomach [rx (xiv)] liberating hydroxide, which then reacts with the gastric hydrochloric acid.

$$PO_4^{-3} + 2H_2O \rightarrow H_2PO_4^- + 2OH^- \qquad \text{(xiv)}$$

Once the pH has risen, no more phosphate anion will be consumed and the remaining calcium phosphate will not dissolve.

Usual Dosage: 1 to 4 g with water six or more times daily; to control severe symptoms, 2 to 4 g every hour may be required.

Magnesium-Containing Antacids. There are a large number of official antacids containing magnesium. With the possible exception of magnesium trisilicate, they all function in the same manner. They are poorly soluble salts which only go into solution as acid consumes the small amount of anion already in solution. As the pH of the stomach approaches neutrality, the rate of dissolution of the magnesium salt slows and stops at neutrality. Thus, it is the anion rather than the magnesium cation that confers the antacid properties.

The magnesium cation causes this group of antacids to be laxatives (see saline cathartics). For this reason, they are usually found in combination with aluminum and calcium antacids in an attempt to equalize the constipative and laxative actions.

Although the magnesium antacids are considered nonsystemic and most of the magnesium is excreted in the feces as insoluble magnesium salts, small amounts of magnesium cation may be absorbed. Since the absorbed magnesium is excreted by the kidneys, the magnesium-containing antacids are contraindicated in patients with impaired renal function. Otherwise, magnesium retention can occur, leading to magnesium poisoning (see magnesium pharmacology).

Magnesium Carbonate, N.F. XIII (approximate formula: $(MgCO_3)_4 \cdot Mg(OH)_2 \cdot 5H_2O$; magnesium carbonate hydroxide)

Magnesium Carbonate N.F. occurs as light, white, friable masses or as a bulky white powder. It is odorless and is stable in air. It is practically insoluble in water, to which it imparts a slightly alkaline reaction. It is insoluble in alcohol but is dissolved by dilute acids with effervescence. *Magnesium Carbonate* N.F. is a hydrated mixture of magnesium carbonate ($MgCO_3$) and magnesium hydroxide [$Mg(OH)_2$], and is assayed in terms of magnesium oxide (MgO).

The antacid properties of magnesium carbonate are due to the carbonate and hydroxide anions reacting with the gastric hydrochloric acid. Due to its very limited solubility, magnesium carbonate dissolves only as carbonate and hydroxide are being consumed. It is usually found in combination with calcium or aluminum antacids and, along with dihydroxyaluminum aminoacetate, is the antacid component of Bufferin®.

Usual Dose Range: 500 mg to 2 g four times a day.

Occurrence:

Magnesium Citrate Solution, N.F. XIII (see saline cathartics)

Magnesium Hydroxide, N.F. XIII [$Mg(OH)_2$; Mol. Wt. 58.32]

Magnesium Hydroxide N.F. is a bulky white powder. It is practically insoluble in alcohol and water but dissolves in dilute acids. Like the other magnesium antacids, magnesium hydroxide goes into solution as the anion (hydroxide) is consumed by the gastric hydrochloric acid [rx (xv)].

$$\begin{array}{l} Mg(OH)_2 \rightleftarrows Mg^{+2} + 2OH^- \\ \quad\text{Solid} \qquad \text{Solution} \mid \\ \qquad\qquad\qquad\qquad\quad 2\ H_3O^+ \\ \qquad\qquad\qquad\qquad\quad\underline{\qquad\qquad} \rightarrow 4H_2O \end{array} \qquad (xv)$$

In a high dose, magnesium hydroxide is used as a laxative. When used as an antacid, it is usually found in combination with calcium or aluminum antacids.

Usual Dose Ranges: Antacid, 300 to 600 mg.

Cathartic, 2 to 4 g.

Occurrence:

Magnesia Tablets, U.S.P. XVIII
Magnesia and Alumina Oral Suspension, U.S.P. XVIII
Magnesia and Alumina Tablets, U.S.P. XVIII
Milk of Magnesia, U.S.P. XVIII

Magnesium Oxide, U.S.P. XVIII (MgO; Mol. Wt. 40.30; magnesia)

Magnesium Oxide U.S.P. occurs as a very bulky white powder known as *light magnesium oxide* or relatively dense white powder known as *heavy magnesium oxide*. The two oxides differ from one another in density; 5 g of light magnesium oxide occupy a volume of approximately 40 to 50 ml, while 5 g of heavy magnesium oxide occupy a volume of approximately 10 to 20 ml. Both are practically insoluble in water and alcohol but soluble in dilute acids.

In the presence of acid, the oxide is converted to the hydroxide [rx (xvi)], and, therefore, the chemistry and pharmacology are the same as those of magnesium hydroxide [rx (xv)].

$$MgO + 2H_3O^+ \rightarrow Mg(OH)_2 + 2H_2O \qquad (xvi)$$

Usual Dose: 250 mg four times a day.

Usual Dose Range: 250 mg to 4 g daily.

Magnesium Phosphate, N.F. XIII [$Mg_3(PO_4)_2 \cdot 5H_2O$; Mol. Wt. 352.93; tribasic magnesium phosphate]

Magnesium Phosphate N.F. is a white, odorless, and tasteless powder which is readily soluble in diluted mineral acids but almost insoluble in water. The chemistry is the same as that of *Tribasic Calcium Phosphate* N.F. [rx (xiii)]. In common with all magnesium antacids, it has a potential laxative action.

Usual Dose: 1 g.

Magnesium Trisilicate, U.S.P. XVIII [$2MgO \cdot 3SiO_2 \cdot xH_2O$; Mol. Wt. (anhydrous) 260.86]

Magnesium Trisilicate U.S.P. is defined as a compound of magnesium oxide and silicon dioxide containing varying proportions of water. Due to methods of manufacture, it is more likely to be a mixture of magnesium metasilicate ($MgSiO_3$) and colloidal silicon dioxide, with varying amounts of water. The final product is a fine white, odorless, tasteless powder, free from grittiness, which is insoluble in water and alcohol but readily decomposed by mineral acids. It is assayed in terms of silicon dioxide and magnesium oxide and must have a definite magnesium oxide/silicon dioxide ratio. As the amount of silicon dioxide increases with respect to magnesium oxide, there is loss in antacid capability.[38,39] Magnesium trisilicate must also meet an acid-consuming capacity requirement.

Just what occurs in the stomach has not yet been completely determined. The end result is a gelatinous mass of colloidal silicon dioxide and/or silicic acid. There is no question that hydrochloric acid is also consumed in the process. If one accepts the magnesium oxide formula, the chemistry will be the same as that of the official magnesium oxide [rx (xvi)]. If the composition of magnesium trisilicate is closer to that of magnesium metasilicate, the chemistry may be more like that shown by reactions (xvii) or (xviii).

$$Mg_2Si_3O_8 + 4H_3O^+ \rightarrow 2Mg^{+2} + H_4Si_3O_8 \downarrow \text{ (or } 3SiO_2 \cdot 2H_2O) + 4H_2O \quad \text{(xvii)}$$

$$(MgSiO_3)_2 \cdot SiO_2 \cdot nH_2O + 4H_3O^+ \rightarrow 2Mg^{+2} + \underbrace{2H_2SiO_3 + SiO_2}_{\text{colloidal mixture}} + nH_2O \text{ (xviii)}$$

The antacid capabilities of magnesium trisilicate can vary considerably (compare the graphs in refs. 8 and 10, and see the discussion in ref. 11). There is also the view that magnesium trisilicate should be considered as a protective and adsorbent. The colloidal silicates could protect the ulcer from further acid and peptic attack, and possibly adsorb the pepsin. There is no good *in vivo* evidence to substantiate this theory of protective action. *In vitro* studies also indicate that magnesium trisilicate is a less effective inhibitor of pepsin as compared to aluminum hydroxide gel. Like all *in vitro* evaluations of antacids, the results are dependent upon the procedure.[18,19,22,43] Although magnesium trisilicate is considered a nonsystemic antacid, studies indicate that silicon dioxide is excreted by the kidneys in direct proportion to the amount absorbed.[44] When first reported, it was considered to meet the requirements of the ideal antacid.[40–42] Now, magnesium trisilicate is usually found only in combination products as a means of overcoming the constipating effect of a calcium- or aluminum-containing antacid.

Usual Dose: 1 g. four times daily.

Usual Dose Range: 1 to 16 g daily.

Occurrence:

Magnesium Trisilicate Tablets, U.S.P. XVIII

Combination Antacid Preparations. Because no single antacid meets all the criteria for an ideal antacid, several products are on the market containing mixtures of antacids. Most of these combination products are an attempt to balance the constipative effect of calcium and aluminum with the laxative effect of magnesium. Some of these products are also a mixture of an antacid with rapid onset of action and one with a supposedly longer duration of action. The latter property is somewhat ambiguous, as duration of action is more a function of dose than an inherent property of the specific ingredient.

Aluminum Hydroxide Gel-Magnesium Hydroxide Combinations (Aludrox®, WinGel®, Maalox®, Creamalin®)

These products are all results of attempts to obtain a product that is not constipative or laxative. The U.S.P. recognizes two dosage forms, a suspension and a tablet. For each dosage form, either aluminum hydroxide gel or magnesium hydroxide may predominate. *Alumina and Magnesia Oral Suspension* U.S.P. XVIII contains the equivalent of 4% aluminum oxide (Al_2O_3) and 2% magnesium hydroxide [$Mg(OH)_2$]. In contrast, *Magnesia and Alumina Oral Suspension* U.S.P. XVIII contains the equivalent of 2.2% aluminum oxide and 3.8% magnesium hydroxide. A similar type of relationship holds for *Alumina and Magnesia Tablets* U.S.P. XVIII and *Magnesia and Alumina Tablets* U.S.P. XVIII.

Suspension

    Usual Dose: 15 ml four to six times a day.

    Usual Dose Range: 5 to 120 ml daily.

Tablets

    Usual Dose: 1 or 2 tablets four to six times a day.

    Usual Dose Range: 1 to 4 tablets up to twelve times daily.

Aluminum Hydroxide Gel-Magnesium Trisilicate Combinations (Gelusil®, Tricreamalate®, Triosgel®)

Although not official, this is one of the more common combinations. In addition to the hoped-for balance of laxative and constipative effects, the magnesium trisilicate is supposed to exert its protective effect. Good double-blind clinical studies are lacking for this combination.

Magaldrate, N.F. XIII ($Al_2H_{14}Mg_4O_{14} \cdot 2H_2O$—approximate formula; monalium hydrate; hydrated magnesium aluminate)

Magaldrate N.F. is a chemical combination of aluminum hydroxide and magnesium hydroxide. It contains the equivalent of 28 to 39% magnesium oxide and 17 to 25% of aluminum oxide. *Magaldrate* N.F. occurs as a white, odorless, crystalline powder which is insoluble in water and alcohol but soluble in dilute solutions of mineral acids. It must fit a definite x-ray diffraction pattern and meet an acid-consuming capacity requirement. *Magaldrate* N.F. is assayed in terms of magnesium oxide and aluminum oxide, and is purported to be a product superior to a simple mixture of aluminum hydroxide gel and magnesium hydroxide.

Usual Dose: 400 to 800 mg as required, preferably taken between meals and at bedtime.

Occurrence:

Magaldrate Oral Suspension, N.F. XIII (Riopan® Suspension)

Magaldrate Tablets, N.F. XIII (Riopan® Tablets)

Simethicone-Containing Antacids (Di-Gel®, Mylanta®)

Because many people with gastric hyperacidity complain of being "gassy," the defoaming agent simethicone (see Chapter 9) has been added to some antacids.

Calcium Carbonate-Containing Antacid Mixtures (Tums®, Titralac®, Ducon®)

Calcium carbonate can be found in combination with aluminum hydroxide gel to yield products that supposedly have a rapid onset with prolonged action. It can also be found with magnesium-containing antacids in an attempt to balance the constipative effect of calcium with the laxative effects of magnesium. Three-part combinations of calcium carbonate, aluminum hydroxide gel, and a magnesium-containing antacid are also available.[45] These products are claimed to have rapid onset, prolonged action, and little effect on bowel motility.

Alginic Acid-Sodium Bicarbonate-Containing Antacid Mixtures (Gaviscon Foamtab®)

This type of preparation was formulated in an attempt to provide symptomatic relief of reflux esophagitis (inflammation of the esophagus due to regurgitation of the stomach contents). The tablet is chewed and as the contents come in contact with water, the alginic acid (hydrophilic colloidal carbohydrate acid obtained from seaweed) reacts with sodium bicarbonate, forming sodium alginate and carbon dioxide. The latter causes the formation of foam within the solution. In the acid environment of the stomach, alginic acid is precipitated in the form of a light, viscous gel which floats on top of the stomach contents. As long as the patient remains upright, the antacid (aluminum hydroxide gel and magnesium trisilicate) contained in the foam remains near the gastroesophageal junction and supposedly is the first material passed into the esophagus when reflux of the gastric contents occurs.[46]

Protectives and Adsorbents

This group of gastrointestinal agents is commonly used for the treatment of mild diarrhea. Diarrhea is a symptom and not a disease. Very briefly, it results when some factor impairs digestion and/or absorption, thereby increasing the bulk of the intestinal tract. This increased bulk stimulates peristalsis, propelling the intestinal contents to the anus. (See saline cathartics for a more complete discussion of intestinal movement.) Diarrhea may be acute or chronic. Acute diarrhea can be caused by bacterial toxins, chemical poisons, drugs, allergy, and disease. The effects of these agents range from tissue damage or irritation to that of causing electrolytes to flow from body fluids into the intestinal tract, thereby increasing the osmotic load of the intestinal tract. Chronic diarrhea can result from gastrointestinal surgery, carcinomas, chronic inflammatory conditions, and various absorptive defects. Frequently the causative factor of acute diarrhea is not found, and the patient shrugs it off as a 24- or 48-hour stomach flu. With chronic diarrhea, there is usually more time to locate the cause.

Diarrhea is a serious condition, particularly for very young or elderly patients. The loss of fluids and electrolytes can quickly lead to dehydration and electrolyte imbalances. It is currently believed that some of the bacterial toxins stimulate the flow of electrolytes into the intestines, thereby increasing the intestinal osmotic load.[47] The antidiarrheal agents described in this chapter will only treat the symptoms and occasionally the cause, but they will not treat the complications. Most products for the treatment of diarrhea will consist of an adsorbent-protective, an anti-diarrheal, and possibly an antibacterial agent. The ideal antidiarrheal agent should act directly on the smooth muscles of the gut to produce a spasm-like effect which decreases peristalsis and increases segmentation. The antibacterials are only effective if there is an actual infection in the intestinal tract or during epidemics previously shown to be caused by a microorganism. The adsorbent-protectives supposedly adsorb toxins, bacteria, and viruses along with providing a protective coating of the intestinal mucosa. They include bismuth salts, special clays, and activated charcoal.

Bismuth-Containing Products. The use of bismuth salts as antidiarrheals seems to be supported chiefly by tradition.[48] Bismuth subcarbonate has also found some use as an antacid. One report attributes what gastric acid neutralizing there is to the presence of soluble alkali as an impurity.[10] Others also report very poor neutralization ability.[8,38,43] The results are mixed as to whether bismuth salts inhibit pepsin independent of pH.[11,20,21,43]

Although the bismuth salts used as antidiarrheals are considered to be water insoluble, a small amount does go into solution. The soluble bismuth cation supposedly exerts a mild astringent and antiseptic action but it is doubtful whether this is clinically significant. Intestinal hydrogen sulfide acts upon the bismuth salts to form bismuth sulfide; hence, the black stools resulting from the oral administration of bismuth-containing preparations.

Bismuth Subnitrate, N.F. XIII (approximate formula: $[Bi(OH)_2NO_3]_4 \cdot BiO(OH)$; basic bismuth nitrate)

Bismuth Subnitrate N.F. occurs as a white, slightly hygroscopic powder which gives an acid reaction using blue litmus paper. It is practically insoluble in water and in alcohol, but is readily dissolved by hydrochloric or nitric acid. It is assayed in terms of bismuth trioxide (Bi_2O_3).

Bismuth subnitrate has a well recognized incompatibility with tragacanth, in which tragacanth precipitates as a hard mass in the presence of the salt. An interesting paper in connection with this incompatibility points out that the difficulty may be overcome by the protective action of sodium biphosphate or trisodium phosphate.[49] These authors feel that because tragacanth is a negative colloid, the adsorption of the positive bismuth ion (without corresponding adsorption of the negative nitrate ions) tends to precipitate the colloid. The use of phosphates is based on their supplying the negative ions lacking, which may then be adsorbed by the tragacanth, stabilizing the colloid.

A rather novel use for bismuth subnitrate is in the form of an x-ray shielding putty.[50]

Bismuth subnitrate apparently can inhibit pepsin.[43] However, its main use is as a component of *Milk of Bismuth* where it probably functions as a mild astringent-protective (see *Milk of Bismuth*).

Occurrence:

Milk of Bismuth, N.F. XIII

Milk of Bismuth, N.F. XIII (bismuth magma; bismuth cream)

Milk of Bismuth N.F. contains bismuth hydroxide and bismuth subcarbonate in suspension in water. It is made by converting bismuth subnitrate to bismuth nitrate $[Bi(NO_3)_3]$ by the addition of nitric acid. Then, by treatment with ammonium carbonate and ammonia solution, bismuth nitrate is converted to bismuth hydroxide and subcarbonate [rxs (xix)–(xxii)].

$$NH_2CO_2NH_4 \cdot NH_4HCO_3 + NH_4OH \leftrightharpoons 2(NH_4)_2CO_3 \qquad \text{(xix)}$$

$$3(NH_4)_2CO_3 + 2Bi(NO_3)_3 \rightarrow Bi_2(CO_3)_3 \downarrow\ + 6NH_4NO_3 \qquad \text{(xx)}$$

$$2Bi_2(CO_3)_3 + H_2O \rightarrow [(BiO)_2CO_3]_2 \cdot H_2O + 4CO_2 \uparrow \qquad \text{(xxi)}$$

$$Bi(NO_3)_3 + 3NH_4OH \rightarrow Bi(OH)_3 \downarrow\ + 3NH_4NO_3 \qquad \text{(xxii)}$$

It is classified by the National Formulary as an astringent and antacid.

Usual Dose: 5 ml.

Bismuth Subcarbonate, U.S.P. XVIII (approximate formula: $[(BiO)_2 CO_3]_2 \cdot H_2O$; basic bismuth carbonate)

Bismuth Subcarbonate U.S.P. is a white or pale yellowish white, odorless, tasteless powder which is stable in air, but is slowly affected by light. It is practically insoluble in water and in alcohol but dissolves completely in nitric acid and in hydrochloric acid, with copious effervescence. *Bismuth Subcarbonate* U.S.P. is assayed in terms of bismuth trioxide (Bi_2O_3).

The pharmacological properties of bismuth are covered in the general discussion above. Although still used in preparation for gastrointestinal disorders, bismuth subcarbonate is admitted to the U.S.P. XVIII as a topical protectant with no internal dose given.

Occurrence:

Milk of Bismuth, N.F. XIII (generated *in situ*)

Nonofficial Bismuth Compounds
Bismuth Subgallate, N.F. X ($C_7H_7BiO_7$; Mol. Wt. 412.13)

Bismuth Subsalicylate, U.S.P. XVI ($HOC_6H_4CO_2BiO \cdot H_2O$; Mol. Wt. 362.11)

This is one of the ingredients of Pepto-bismol®.

Bismuth Ammonium Citrate

This is one of the main ingredients of Ulcerine®, a product claimed to help coat the ulcer base and promote healing. No adequately controlled trials have been reported.[51]

Activated Clays and Other Adsorbents. This group is composed mostly of clays which have excellent adsorbent properties, and most of them are used for that purpose industrially. They appear to have a valid clinical use, at least in mild diarrhea of short duration.

Kaolin, N.F. XIII

Kaolin N.F. is a native hydrated aluminum silicate, powdered and freed from gritty particles. It occurs as a soft, white, or yellowish white powder, or as lumps. It has an earthy or clay-like taste and, when moistened with water, assumes a darker color and develops a marked clay-like odor. *Kaolin* N.F. is insoluble in water, in cold dilute acids, and in solutions of the alkali hydroxides. It is usually found together with the vegetable carbohydrate, pectin (Kaopectate®, Kao-Con®) and used as an adsorbent. Kaolin-containing products have been reported to interfere materially with the intestinal absorption of lincomycin.[52]

Attapulgite (Quintess®, Diamagma®)

Attapulgite is a nonofficial clay found near Attapulgus, Georgia. Its adsorbent properties are increased by heating. While attapulgite can alter the rate of absorption of orally administered drugs containing a tertiary amine moiety, it may not necessarily affect the extent of absorption of the drug.[53]

Activated Charcoal, U.S.P. XVIII

Activated charcoal has been used as an adsorbent in the treatment of diarrhea. It is now a recommended antidote in certain types of poisoning. (See Chapter 12 for a complete discussion.)

Saline Cathartics

Saline cathartics (purgatives) are agents that quicken and increase evacuation from the bowels. Laxatives are mild cathartics. Most can be purchased without a prescription and are a group that has been widely used, abused, and often overpromoted by the manufacturer. Constipation and illness have historically been associated with each other. When the child does not have a bowel movement, the parents are concerned; when the child then defecates, the parents are pleased. Many of the old patent remedies associated with liver disorders, bile flow, crankiness, etc., were laxatives or cathartics. Thus, the general public is receptive to self-administration of cathartics.

The 1971 A.M.A. Drug Evaluations states that cathartics are properly used: (1) to ease defecation in patients with painful hemorrhoids or other rectal disorders, and to avoid excessive straining and concurrent increases in abdominal pressure in patients with hernias; (2) to avoid potentially hazardous rises in blood pressure during defecation in patients with hypertension, cerebral, coronary, or other arterial diseases; (3) to relieve acute constipation; and (4) to remove solid material from the intestinal tract prior to certain roentgenographic studies.[54] Laxatives should only be used for short-term therapy as prolonged use may lead to loss of spontaneous bowel rhythm upon which normal evacuation depends, causing the patient to become dependent on laxatives, the so-called "laxative habit."

Material is propelled through the intestinal tract by wave-like contractions called peristaltic movements. The waves travel at 2 to 25 cm per second carrying the contents through the small intestine in about 3.5 to 4 hours. However, the last of a meal leaves the ileum some eight or nine hours after ingestion, as the products of digestion are held up at the ileocolic (between the ileum and colon) sphincter to allow adequate time for completion of intestinal digestion and absorption.

In the colon there is a strong peristaltic wave which occurs about three or four times a day. One of these massive movements may expel the contents into the rectum, causing the individual to experience a sensation of fullness in the rectum and a desire to defecate. The observation that these movements appear to be correlated with food entering the stomach is given as the reason why most individuals commonly defecate following a meal, usually breakfast. By ignoring the urge to defecate by voluntary contraction of the external sphincter of the anus, the urge rapidly fades. However, regularity or the frequency of defecation varies among individuals, ranging from once every two or three days to up to three bowel movements per day. It is important that the patient consult a physician if he experiences a change that persists in what has been a regular elimination schedule.

Constipation is the infrequent or difficult evacuation of the feces. It may be due to a person resisting the natural urge to defecate, causing the fecal material which remains in the colon to lose fluid and to become relatively dry and hard. Constipation can also be caused by intestinal atony (lack of muscle tension), intestinal spasm, emotions, drugs, and diet. Many times the constipated patient can be helped by eating foods such as prunes containing natural laxatives or foods with large amounts of roughage. If a laxative is needed, there are a large number and variety of types from which to choose.

Basically there are four types of laxatives: (1) stimulant, (2) bulk-forming, (3) emollient, and (4) saline. The stimulant laxatives act by local irritation on the intestinal tract, which increases peristaltic activity. They include phenolphthalein, aloin, cascara extract, rhubarb extract, senna extract, podophyllin, castor oil, 1,8-dihydroxyanthraquinone (danthron), oxyphenisatin, bisacodyl, and calomel (no longer used). The bulk-forming laxatives are made from cellulose and other nondigestible polysaccharides. They swell when wet, with the increased bulk stimulating peristalsis. Included in this group are psyllium seed, methyl cellulose, sodium carboxymethylcellulose, and karaya gum. The emollient laxatives act either as lubricants facilitating the passage of compacted fecal material or as stool softeners. Mineral oil is the main lubricant laxative used, and d-octyl sodium sulfosuccinate, an anionic surface active agent, is the most commonly used stool softener. The discussion in this chapter will be limited to the fourth group, the saline cathartics.

The saline cathartics act by increasing the osmotic load of the gastrointestinal tract. They are salts of poorly absorbable anions and sometimes cations. The body relieves the hypertonicity of the gut by secreting additional fluids into the intestinal tract. The resulting increased bulk stimulates peristalsis. Poorly absorbed anions that are used as saline cathartics are biphosphate ($H_2PO_4^-$), phosphate (HPO_4^{-2}), sulfate, and tartrate. Soluble magnesium salts are cathartic due to the poorly absorbed magnesium cation (see magnesium antacids). The saline cathartics are water soluble and are taken with large amounts of water. This prevents excessive loss of body fluids and reduces nausea and vomiting if a too hypertonic solution should reach the stomach.

The saline cathartics, when taken for brief periods, are relatively free of side effects. Over a longer term, patients on low sodium diets should not use the sodium-containing saline cathartics (sodium biphosphate, sodium phosphate, sodium sulfate, and potassium sodium tartrate). For those with impaired renal function the magnesium salts should be restricted, since some magnesium cation is absorbed. Magnesium has a central nervous system depressant effect (see magnesium pharmacology).

Official Saline Cathartics

Sodium-Containing Products
Sodium Biphosphate, N.F. XIII ($NaH_2PO_4 \cdot H_2O$; Mol. Wt. 137.99;

sodium dihydrogen phosphate; sodium acid phosphate; primary sodium phosphate; sodium phosphate, monobasic)

Sodium Biphosphate N.F. occur as colorless crystals or as a white, crystalline powder. It is odorless and is slightly deliquescent. Its solutions are acid to litmus and effervesce with sodium carbonate.

The acidic properties are due to the acid dihydrogen phosphate anion having pK_a 6.7 [rx (xxiii)]. The pH of a 0.1 M solution at 25° C is 4.5.

$$H_2PO_4^- + H_2O \rightleftharpoons H_3O^+ + HPO_4^{-2} \qquad \text{(xxiii)}$$

Although classified by the National Formulary as a urinary acidifier, it is also used as a cathartic (Phospho-Soda®, Vacuetts®, and Sal Hepatica®). In the latter, sodium biphosphate is also the proton source that reacts with sodium bicarbonate, causing effervescence.

Again, because of its being an acid salt, sodium biphosphate has been used in some baking powders.

Sodium Phosphate, N.F. XIII ($Na_2HPO_4 \cdot 7H_2O$; Mol. Wt. 268.07; dibasic sodium phosphate; disodium hydrogen phosphate; disodium orthophosphate; disodium phosphate; DSP; phosphate of soda; secondary sodium phosphate).

Sodium Phosphate N.F. occurs as a colorless or white granular salt which effervesces in warm, dry air. Its solutions are alkaline, with a 0.1 M solution having a pH of about 9.5. It is freely soluble in water and very slightly soluble in alcohol.

The medical and pharmaceutical professions should be aware of the inconsistencies of phosphate nomenclature when mixing or compounding extemporaneous preparations containing *Sodium Phosphate* N.F. XIII. The sodium phosphate commonly referred to in most chemistry texts is trisodium phosphate (Na_3PO_4), a very basic salt used in harsher cleaning compounds. It is very doubtful if trisodium phosphate would ever be used internally. Most manufacturers, in addition to *The Merck Index*, eighth edition, do not use the National Formulary name of *Sodium Phosphate* for the disodium phosphate salt. Therefore, it is very important that the pharmacist be aware of the synonyms listed for this product. (See phosphate nomenclature discussion in Chapter 5.)

Because of the poor intestinal permeability of the monohydrogen phosphate anion, this product is widely used as a saline cathartic (Fleet Enema®, Phospho-Soda®).

Usual Dose: 4 g.

Usual Dose Range: 4 to 8 g.

Occurrence:

Sodium Phosphate Solution, N.F. XIII

Dried Sodium Phosphate, N.F. XIII ($Na_2HPO_4 \cdot xH_2O$; Mol. Wt. (anhydrous) 141.96; exsiccated sodium phosphate)

Dried Sodium Phosphate N.F. is a nearly anhydrous white powder which readily absorbs moisture. It is freely soluble in water and insoluble in alcohol. It is used as a saline cathartic in *Effervescent Sodium Phosphate* N.F. XIII, which is a mixture of sodium bicarbonate, tartaric acid, and citric acid. When dissolved in water a carbonated solution of sodium phosphate, sodium tartrate, and sodium citrate is obtained (see sodium bicarbonate—effervescent powders). Sodium tartrate is also a saline cathartic (see *Potassium Sodium Tartrate* N.F.), and sodium citrate provides a lemon-like flavor which, along with the carbonation provided by the carbon dioxide, masks the saline taste. While the National Formulary allows the proportions of tartaric acid and citric acid to be varied as long as their combined acidity is equivalent to the acidity indicated in the official formula, it must be remembered that a significant replacement of the tartrate anion by the citrate anion could cause a loss of cathartic action.

Dried Sodium Phosphate N.F. is used because the heptahydrate salt of *Sodium Phosphate* N.F. would cause a premature reaction between the sodium bicarbonate and tartaric and citric acids, yielding a product with a flat, saline flavor.

Usual Dose Range: 2 to 4 g.

Occurrence:

Effervescent Sodium Phosphate, N.F. XIII

Potassium Sodium Tartrate, N.F. XIII ($C_4H_4KNaO_6 \cdot 4H_2O$; Mol. Wt. 282.23; Rochelle salt, Seignette's salt)

$$
\begin{array}{c}
\text{H} \\
| \\
\text{HOCCO}_2\text{K} \\
| \qquad\qquad \cdot\ 4\text{H}_2\text{O} \\
\text{HOCCO}_2\text{Na} \\
| \\
\text{H}
\end{array}
$$

Potassium Sodium Tartrate N.F. occurs as colorless crystals, or as a white crystalline powder having a cooling, saline taste. Because it effloresces slightly in warm, dry air, the crystals are often coated with a white powder. It is freely soluble in water and practically insoluble in alcohol. Since the tartrate anion has very poor intestinal permeability, it is used as a cathartic. The potassium sodium salt is the only official tartrate.

Usual Dose: 10 g.

Magnesium-Containing Products

Magnesium Hydroxide, N.F. XIII (see the antacid discussion of magnesium hydroxide for the physical-chemical description)

Magnesium Hydroxide N.F. is official as both an antacid and cathartic. When used in the antacid dose (300 to 600 mg) it may have some laxative effect. Indeed, magnesium hydroxide is found in combination with calcium carbonate and aluminum hydroxide gel in order to overcome their constipative effects. In larger doses (2 to 4 g), magnesium hydroxide is an antacid saline cathartic due to the low intestinal permeability of the magnesium cation. The solubility is limited to that of magnesium hydroxide which reacts with the hydrochloric acid of the stomach, producing water and soluble magnesium chloride [rx (xxiv)].

$$Mg(OH)_2 + 2HCl \rightarrow Mg^{+2} + 2Cl^- + 2H_2O \qquad \text{(xxiv)}$$

When used as a cathartic, magnesium hydroxide is commonly administered as an emulsion or suspension. *Milk of Magnesia* U.S.P. XVIII (Magnesia Magma) is a 7 to 8.5% w/w suspension of magnesium hydroxide which may also contain 0.1% citric acid and not more than 0.05% of a volatile oil or a blend of volatile oils, suitable for flavoring purposes. It is a white, opaque, more or less viscous suspension from which varying proportions of water usually separate on standing, and has a pH of about 10.

In order to minimize the effects of the alkalinity on the soft glass (see discussion on pharmaceutically acceptable glass), 0.1% citric acid is added. This has the effect of "pulling" additional magnesium cations into solution, forming water soluble magnesium citrate [rxs (xxv)–(xxvi)].

$$Mg(OH)_2 \rightleftarrows Mg^{+2} + 2OH^- \qquad \text{(xxv)}$$
$$\text{Solid} \qquad \text{Solution}$$

$$Mg^{+2} + C_3H_4(OH)(CO_2H)_3 + 2OH^- \rightarrow$$
$$Mg^{+2} + C_3H_4(OH)(CO_2H)(CO_2^-)_2 + H_2O \qquad \text{(xxvi)}$$

As the aqueous suspension stands in contact with the glass container, additional hydroxide is formed from the hydrolysis of the sodium silicate of the glass [rx (xxvii)].

$$Na_4SiO_4 + 3H_2O \rightarrow H_3SiO_4^- + 4Na^+ + 3OH^- \qquad \text{(xxvii)}$$

As the pH increases, the law of mass action requires that reaction (xxv) be reversed and magnesium hydroxide reformed as the excess hydroxide from the glass is consumed. Because of the presence of the citric acid, there are enough magnesium cations available to react with excess hydroxide from the glass container. The end result is a product with a milder, less chalky taste. Addition of a soluble magnesium salt ($MgCl_2$, $MgSO_4$), equivalent to 0.1% citric acid, will have the same effect.[55] The use of hardened glass should eliminate the citric acid requirement.

A blue-colored glass bottle enhances the white character and is usually used to dispense milk of magnesia. The white magma in a clear glass container has a pasty, nonpleasing appearance. Studies on the effect of light on storage indicate that protection from light is unnecessary. It should be stored at temperatures not exceeding 35° C, and should not be permitted to freeze. In addition to breaking the bottle, freezing changes the density and character of the hydroxide so that more precipitation occurs, and the precipitate is coarser and more granular. The altered magma is not unfit for use but is not as pleasant to the taste.

Usual Dose: Antacid, 5 ml four times a day.

Cathartic, 15 to 30 ml.

Usual Dose Range: 5 to 50 ml daily.

Also official as an antacid are *Magnesia Tablets*, U.S.P. XVIII. By taking six to eight 300-mg tablets it is possible to obtain a laxative action.

Magnesium Citrate Solution, N.F. XIII (Citrate of Magnesia; purgative lemonade)

Magnesium Citrate Solution N.F. is made by reacting magnesium carbonate with citric acid, forming a soluble magnesium salt (possibly $MgHC_6H_5O_7$). Following the addition of flavoring agents, sodium or potassium bicarbonate is added to react with the remaining citric acid yielding a palatable, carbonated, lemon-flavored solution. It must be sterilized or pasteurized.

Usual Dose: 200 ml.

Magnesium Sulfate, U.S.P. XVIII ($MgSO_4 \cdot 7H_2O$; Mol. Wt. 246.47; epsom salts)

Magnesium Sulfate U.S.P. occurs as small, colorless crystals, usually needle-like, with a cooling, saline bitter taste. It effloresces in air, and its solutions are neutral to litmus. *Magnesium Sulfate* U.S.P. is freely soluble in boiling water and sparingly soluble in alcohol.

Magnesium sulfate is used orally as a cathartic and parenterally as an anticonvulsant. The latter use is more fully described in Chapter 5. As with the other orally administered magnesium salts, their use is restricted in patients with impaired renal function.

Usual Dose: Cathartic—oral, 15 g.

Anticonvulsant—intramuscular, 1 g in a 25 to 50% solution; intravenous, 4 g in a 10% solution.

Usual Dose Range: Oral, 10 to 30 g daily.

Parenteral, 1 to 10 g daily.

Sulfur as a Cathartic

Sulfur exerts a mild cathartic effect. (See Chapter 9 for a complete discussion of the physical, chemical, and pharmacological properties of sulfur.) The classical effect of "sulfur and molasses" is based largely on this action. The laxative action has been attributed to the reduction of sulfur to the sulfide anion by reducing agents present in the intestinal fluids. The sulfide anion then reacts with gastric acid, forming hydrogen sulfide, a mild intestinal irritant [rx (xxviii)].

$$S^- + 2H_3O^+ \rightarrow H_2S + H_2O \qquad \text{(xxviii)}$$

This action is very gradual unless a finely divided form of sulfur, such as precipitated sulfur, is used.

Nonofficial Saline Cathartics

Sodium Sulfate, N.F. XII ($Na_2SO_4 \cdot 10H_2O$; Mol. Wt. 322.19; Glauber's Salt)

Potassium Phosphate, N.F. XII (K_2HPO_4; Mol. Wt. 174.18; dibasic potassium phosphate; dipotassium phosphate, DKP, dipotassium hydrogen phosphate)

Potassium Bitartrate, N.F. XII ($KHC_4H_4O_6$; Mol. Wt. 188.18; cream of tartar; acid potassium tartrate; potassium hydrogen tartrate). Additionally, this acid salt is a common ingredient of baking powders, used to raise the dough [rx (xxix)].

$$C_2H_4O_2(CO_2H)(CO_2K) + NaHCO_3 \rightarrow$$
$$C_2H_4O_2(CO_2Na)(CO_2K) + H_2O + CO_2 \uparrow \qquad (xxix)$$

Calomel, N.F. XII (Hg_2Cl_2; Mol. Wt. 472.09; mercurous chloride, mild mercury chloride). At one time calomel was one of the most popular cathartics because of its thorough cleansing action, starting in the upper part of the small intestine. Its action is attributed to a disproportionation in the alkaline small intestine, forming free mercury and mercuric cation. The latter is a strong intestinal irritant rather than a saline cathartic. For maximum cathartic effect from a given dose, it is desirable to give the dose in several portions spaced at 20- to 30-minute intervals. This spreads the irritant effect over a longer portion of the intestine than would otherwise be affected by the same dose given at one time. Because of the possibility of mercury poisoning, it is always advisable to follow the administration of calomel by a saline cathartic within six hours.

Calomel enjoyed a widespread popularity at one time because it was thought to stimulate the flow of bile. This belief was based upon the fact that the stools were colored green, and this color was supposedly due to increased bile secretion. The green coloration, however, has been shown to be due to the antiseptic effect of the mercury, which prevents the normal conversion of the bile pigment biliverdin (green) to bilirubin (red) by intestinal bacteria.

References

General Reviews

Piper, D. W. Antacid and anticholinergic drug therapy of peptic ulcer. Gastroenterology, **52**:1009, 1967.
Davenport, H. W. Why the stomach does not digest itself. Sci. Amer., **226**:87, Jan. 1972.
1. Hydrochloric acid therapy in achlorhydria. Med. Lett., **14**:56, 1972.

2. Jones, F. A., Kay, A. W., and MacPherson R. Treatment of peptic ulcer. Brit. Med. J., **1**:754, 1964.
3. Pereira-Lima, J., and Hollander, F. Gastric acid rebound—a review. Gastro-enterology, **37**:145, 1959.
4. Fordtran, J. S. Acid rebound, New Eng. J. Med., **279**:900, 1968.
5. Penna, R. P. *In:* Griffenhagen, G. B., and Hawkings, L. L., eds. Handbook of Non-Prescription Drugs. 1973 ed. Washington, D.C.: American Pharmaceutical Association pp. 10–12, 1973.
6. Medical treatment of peptic ulcer. Med. Lett., **11**:105, 1969.
7. A.M.A. Drug Evaluations, 1st ed. Chicago: American Medical Association, p. 573, 1971.
8. Armstrong, J., and Martin, M. An *in vitro* evaluation of commonly used antacids with special reference to aluminum hydroxide gel and dried aluminum hydroxide gel. J. Pharm. Pharmacol., **5**:672, 1953.
9. Gore, D. N., Martin, B. K., and Taylor, M. P. The evaluation of buffer antacids with particular reference to preparations of aluminum. J. Pharm. Pharmacol., **5**:686, 1953.
10. Brindle, H.: The chemical evaluation of antacids. J. Pharm. Pharmacol., **5**:692, 1953.
11. Discussion following Refs. 8–10. J. Pharm. Pharmacol., **5**:703, 1953.
12. Hefferren, J. J., Schratenbaer, G., and Wolman W. An *in vitro* study of antacids: Methods and modifying factors. J. Amer. Pharm. Ass., Sci. Ed., **45**:564, 1956.
13. Schaub, K. Investigations for the elaboration of an *in vitro* evaluation of antacids which takes into consideration the biologic facts, I. Pharm. Acta Helv., **37**:669, 1962.
14. Schaub, K. Part II. *Loc. cit.*, p. 733.
15. Schaub, K. Research to determine an *in vitro* evaluation method for antacids based on biological conditions, III. Pharm. Acta Helv., **38**:15, 1963.
16. Myhill, J., and Piper, D. W. Antacid therapy of peptic ulcer, I. A mathematical determination of an adequate dose. Gut, **5**:581, 1964.
17. Piper, D. W., and Fenton, B. H. Antacid therapy of peptic ulcer, II. An evaluation of antacids *in vitro*. Gut, **5**:585, 1964.
18. Schiffrin, M. J., and Komarov, S. A. The inactivation of pepsin by compounds of aluminum and magnesium. Amer. J. Dig. Dis., **8**:215, 1941.
19. Piper, D. W., and Fenton, B. H. The adsorption of pepsin. Amer. J. Dig. Dis., **NS6**:134, 1961.
20. Piper, D. W., and Fenton, D. H. pH stability and activity curves of pepsin with special references to their clinical importance. Gut, **6**:506, 1965.
21. Kuruvilla, J. T. Antipeptic activity of antacids. Gut, **12**:897, 1971.
22. Wenger, J., and Sundy, M. Pepsin adsorption by commercial antacid mixtures. In vitro studies. J. Clin. Pharmacol., **12**:136, 1972.
23. Goldstein, F. J., and Packman, E. W. Evaluation of an improved Heidelberg telemetry capsule for the study of antacids. J. Pharm. Sci., **59**:425, 1970.
24. Kunz, H. J., Norby, T. E., and Rogers, C. H. A pH-measuring radio capsule for the alimentary canal. Amer. J. Dig. Dis., **16**:739, 1971.
25. Meldrum, S. J., Watson, B. W., Riddle, H. C., Bown, R. L., and Sladen, G. E. pH profile of gut as measured by radiotelemetry capsules. Brit. Med. J., **2**:104, 1972.
26. Barr, W. H., Adir, J., and Garrettson, L. Decrease of tetracycline absorption in man by sodium bicarbonate. Clin. Pharmacol. Ther., **12**:779, 1971.
27. Hem, S. L., Russo, E. J., Bahal, S. M., and Levi, R. S. Effect of pH of precipitation on antacid properties of hydrous aluminum oxide. J. Pharm. Sci., **59**:317, 1970.
28. Hem, S. L., Russo, E. J., Harwood, R. J., Tejani, B. H., Bahal, S. M., and Levi, R. S.: Kinetics of hydrous aluminum oxide conversion in mixtures of amorphous alumina gels of various acid reactivities. J. Pharm. Sci., **59**:376, 1970.
29. Hinkel, E. T., Fisher, M. P., and Tainter, M. L. A new highly reactive aluminum hydroxide complex for gastric hyperacidity, I. J. Amer. Pharm. Ass., Sci. Ed. **48**:380, 1959; Part II, *loc. cit.*, p. 384.
30. Murphey, R. S. *In vitro* differences between dihydroxy aluminum aminoacetate and dried aluminum hydroxide gel. J. Amer. Pharm. Ass. Sci. Ed., **41**:361, 1952.

31. Lotz, M., Zisman, E., and Bartter, F. C. Evidence for a phosphorus-depletion syndrome in man. New Eng. J. Med., 278:409, 1968.
32. Rossett, N. E., and Rice, M. L. An *in vitro* evaluation of the efficiency of the more frequently used antacids with particular attention to tablets. Gastroenterology, **26**:490, 1954.
33. Harrison, J. W. E., Abbott, D. D., Feinman, J. I., and Packman, E. W. Comparative *in vivo* methods for evaluating antacids in humans. J. Amer. Pharm. Ass., Sci. Ed., **46**:549, 1957.
34. Barreras, R. F. Acid secretion after calcium carbonate in patients with duodenal ulcer. New Eng. J. Med., 282:1402, 1970.
35. Malone, D. N. S., and Horn, D. B. Acute hypercalcaemia and renal failure after antacid therapy. Brit. Med. J., **1**:709, 1971.
36. Wenger, J., Kirsner, J. B., and Palmer, W. L. The milk-alkali syndrome: Hypercalcemia, alkalosis and azotemia following calcium carbonate and milk therapy of peptic ulcer. Gastroenterology, 33:745, 1957.
37. McMillan, D. W., and Freeman, R. B. The milk-alkali syndrome: A study of the acute disorder with comments on the development of the chronic condition. Medicine, 44:485, 1965.
38. Kraemer, M. Magnesium trisilicate N.N.R. Its position among antacids used to treat peptic ulcer. Amer. J. Dig. Dis., 8:56, 1941.
39. Mutch, N. Magnesium trisilicate. Brit. Med. J., **2**:735, 1937.
40. Mutch, N. The silicates of magnesium. Brit. Med. J., **1**:143, 1936.
41. Mutch, N. Synthetic magnesium trisilicate, its action in the alimentary tract. Brit. Med. J., **1**:205, 1936.
42. Mutch, N. Hydrated magnesium trisilicate in peptic ulceration. Brit. Med. J., **1**:254, 1936.
43. Bateson, P. R. A comparative *in vitro* evaluation of a new bismuth salt—bismuth aluminate. J. Pharm. Pharmacol., **10**:123, 1958.
44. Page, R. C., Heffner, R. R., and Frey, A. Urinary excretion of silica in humans following oral administration of magnesium trisilicate. Amer. J. Dig. Dis., 8:13, 1941.
45. Powell, R. L., Westlake, W. J., Longaker, E. D., and Greene, L. C. A clinical evaluation of a new concentrated antacid. I. J. Clin. Pharmacol., **11**:288, 1971; Part II, *loc. cit.*, p. 296.
46. Beckloff, G. L., Chapman, J. H., and Shiverdecker, P. Objective evaluations of an antacid with unusual properties. J. Clin. Pharmacol., **12**:11, 1972.
47. Hirschhorn, N., and Greenough, W. B., III. Cholera. Sci. Amer., **225**:15, 1971.
48. A.M.A. Drug Evaluations, 1st ed. Chicago: American Medical Association, p. 579, 1971.
49. Schmitz, R. E., and Hill, J. S. Bismuth subnitrate-tragacanth incompatibility. J. Amer. Pharm. Ass., Pract. Ed., **9**:493, 1948.
50. X-Ray shielding putty. Amer. Prof. Pharm., **17**:1089, 1951.
51. Ulcerine and treatment of peptic ulcer. Med. Lett., **13**:71, 1971.
52. Wagner, J. G. Design and data analysis of biopharmaceutical studies in man. Canad. J. Pharm. Sci., **1**:55, 1966.
53. Sorby, D. L. Effects of adsorbents on drug absorption. I. Modification of promazine absorption by activated attapulgite and activated charcoal. J. Pharm. Sci., **54**:677, 1965.
54. A.M.A. Drug Evaluations, 1st ed. Chicago: American Medical Association, p. 597, 1971.
55. Doerge, R. F. School of Pharmacy, Oregon State University. Unpublished data.

9
Topical Agents

In this chapter, a wide variety of uses will be discussed for an even wider variety of compounds. The term "topical" places the use of these compounds on body surfaces, as opposed to "systemic," which indicates that the compounds are absorbed into the circulatory system and distributed to various organs and tissues. The pharmacological effects of topical compounds are evidenced primarily at the surface to which they are applied. This is not to say that there is no penetration into deeper tissues or absorption into the general circulation. Indeed, this does occur in many instances with resulting beneficial effects—for example, the penetration of an antiseptic compound into the tissues below the exposed area of a wound aids in the prevention of deep infections. On the other hand, systemic effects of many of these compounds may elicit toxic or allergic manifestations depending upon the amounts absorbed, e.g., topically applied mercury-containing compounds. Because of this possibility, general aspects of systemic toxicity will be discussed in association with appropriate compounds. It should be stated further that topical application of drugs may be accomplished within body cavities that open to the outside (e.g., the oral, vaginal, and colonic cavities). This type of application is done with the expectation that the compound will exert local or surface activity. However, it should be noted that systemic absorption from these areas may be more extensive in comparison to the skin. Indeed, drugs are often applied to these areas as a route of systemic administration (e.g., buccal tablets, suppositories, etc.).

The compounds used topically will be divided into broad categories based on their usual action or use. The categories are: (1) protective, and (2) antimicrobial and astringent compounds. These classifications are broad enough to encompass most of the compounds; however, some of the agents have uses extending beyond the limits of the specific category. It may also be noted that there is a tremendous amount of overlap between categories where the particular use will depend on the area of application, the concentration of the agent, the presence of other compounds in the preparation, and the solubility (e.g., insoluble zinc oxide is a protective and soluble zinc sulfate is an astringent).

Protectives

As the term implies, protectives are substances which may be applied to the skin to protect certain areas from irritation, usually of mechanical origin. Those compounds or substances most appropriate for this purpose are insoluble and chemically inert. Insolubility is a desirable property in that this limits the absorption of the compounds through the skin, makes it difficult to wash them off, and diminishes metallic properties on tissue. Compounds which are chemically unreactive are necessary in order to prevent interactions between the protective substance and the tissue. In other words, ideal protectives are biologically inactive.

Many materials serving the purpose of protectives are also efficient adsorbents useful for adsorbing moisture from the surface of the skin. Since removing moisture tends to lessen mechanical friction and irritation, adsorbent action is an important property of protectives. Protective and adsorbent action is maximized with decreasing particle size. Small particles offer a larger surface area, allowing them to adhere to each other, adhere better to the surface of the skin, and adsorb moisture more efficiently. A fine state of subdivision of the particles also offers a smooth substance which is soothing to apply and aids in preventing irritation due to rubbing or friction.

Protectives are generally applied as dusting powders, suspensions containing the insoluble protective substance, or ointments. It should be remembered that protective and adsorbent substances are available for use internally for gastrointestinal irritations (see Chapter 8). Although the reason for using these is the same, the compounds and preparations are not. Topical protectives and adsorbents are usually applied to areas of the skin which are subject to constant irritation due to moisture and/or friction, or areas which have already become irritated or inflamed due to friction, allergy, and the like. If the area to which the protective is to be applied is abraded and exuding fluid, adsorbent-type protectives should not be used. These substances will mix with the exudate and dry to a crust which adheres to the open tissue. It is also possible that systemic absorption may be enhanced. The properties and uses of inorganic protective compounds and preparations are discussed in the following section.

Protective Products

Talc, U.S.P. XVIII ($3MgO \cdot 4SiO_2 \cdot H_2O$)

Talc U.S.P. is a native, hydrous magnesium silicate, sometimes containing a small proportion of aluminum silicate. The U.S.P. describes it as a very fine, white or grayish white, crystalline powder. It is unctuous, adheres readily to the skin, and is free from grittiness.

Talc is a layered silicate and is the softest mineral known. It has a smooth, greasy feeling to the touch, and in its lump form (steatite) it is known as soapstone. The form most desirable for cosmetic and pharmaceutical purposes is known as foliated talc and has a plate-like structure.

Chemically, talc may be considered to be a hydrated magnesium silicate having the elements illustrated by the formula represented above. This grouping of elements may be rearranged to provide a representation of talc as the magnesium salt of dimetasilicic acid, having the formula $Mg_3H_2(Si_2O_6)_2$. The actual composition of talc is somewhat variable, containing from 28.1 to 31.2% MgO, 57 to 61.7% SiO_2, and 3 to 7% H_2O. As a magnesium polysilicate, it is unreactive to acids and bases, and inert to most other reagents.

Talc is odorless, tasteless, and insoluble in water, dilute acids, and dilute bases. It has very low adsorptive properties, which is an important consideration for its use as a filtering aid, allowing filtration without danger of removing important constituents, e.g., alkaloids, dyes, etc.

USES. In spite of its low adsorptive properties, the inert, unctuous nature of talc makes it a useful lubricating, protective dusting powder. It can be used to prevent irritation due to friction, and to protect areas from further irritation. It has been known for a number of years, however, that when used on broken skin—wounds and surgical incisions—talc can produce sterile abscesses or granulomas (a nodule of inflamed tissue in which granulation is occurring).[1] This problem precludes the use of talc-containing dusting powder on surgical gloves; absorbable dusting powders are now recommended for this purpose (cf. U.S.P. XVIII, p. 839). No problems are associated with the use of talc on the intact skin. Talc is used in preparations which may be perfumed for cosmetic purposes, or medicated with antimicrobial agents, such as boric acid.

The insoluble and chemically inert nature of talc, including its non-adsorptive properties, renders the material useful as a filtering aid. The best particle size for this purpose is 80/100 mesh, i.e., the powder that will pass through a No. 80 sieve but not a No. 100 sieve. Particles finer than this will not be retained by the usual filter papers, and filtered preparations will appear cloudy. Some of the pharmaceutical preparations that employ talc as a filtering or distributing agent are aromatic waters, Magnesium Citrate Solution, aromatic elixir, and Orange Syrup.

Insoluble Zinc Compounds

Zinc Oxide, U.S.P. XVIII (ZnO; Mol. Wt. 81.37)

Zinc Oxide U.S.P. is a very fine, odorless, amorphous, white or yellowish white powder, free from gritty particles. It gradually absorbs carbon dioxide from the air. When freshly ignited, it should contain not less than 99.0% and not more than 100.5% of ZnO. When heated to 400° or 500° C, the oxide develops a yellow color that disappears on cooling. *Zinc Oxide* is insoluble in water and alcohol, and will gradually absorb carbon dioxide from the air to form a basic zinc carbonate [$Zn_2(OH)_2CO_3$].

Chemically, zinc oxide reacts with dilute acids and aqueous solutions of ammonium compounds to form water soluble products. When treated with dilute hydrochloric acid, the oxide forms the Lewis acid, zinc chloride [rx (i)].

$$ZnO + 2HCl \rightarrow ZnCl_2 + H_2O \qquad\qquad (i)$$

Ammonia water and ammonium carbonate T.S. (test solution) form water soluble basic ammonia complexes with zinc oxide [rxs (ii) and (iii)].

$$ZnO + 4NH_4OH \rightarrow Zn(NH_3)_4(OH)_2 + 3H_2O \qquad\qquad (ii)$$

$$ZnO + 2(NH_4)_2CO_3 \rightarrow Zn(NH_3)_4(OH)_2 + 2CO_2 \uparrow + H_2O \qquad (iii)$$

Acidic solutions of zinc oxide exhibit the properties of zinc ion.

USES. *Zinc Oxide* is a mild astringent and a weak antimicrobial compound. The U.S.P. classifies it as an astringent and topical protective. The antimicrobial-astringent action is due to the release of a small amount of zinc ion from hydrolysis in the acidic moisture on the skin [see rx (i)]. It is used as a protective in ointments (*Zinc Oxide Ointment*, U.S.P. XVIII), pastes (*Zinc Oxide Paste*, U.S.P. XVIII), and dusting powders in the treatment of skin ulcerations and other dermatological problems. As a dusting powder, it is frequently found in combination with other protectives or antimicrobial agents, e.g., talc and boric acid. *Zinc Oxide* is the primary ingredient in *Calamine*, U.S.P. XVIII (see below). *Zinc Gelatin*, U.S.P. XVIII is a protective jelly containing 10% ZnO. The pharmacological and biochemical actions of zinc are discussed in Chapter 6.

Calamine, U.S.P. XVIII, (ZnO·xFe$_2$O$_3$).

Calamine U.S.P. is zinc oxide with a small proportion of ferric oxide. After ignition, it contains not less than 98.0% and not more than 100.5% of ZnO. The presence of the ferric oxide [Fe$_2$O$_3$; iron(III) oxide] gives the substance a pink color which varies according to the method of preparation and the amount of ferric oxide present. The material is a fine powder, odorless, and practically tasteless. It is insoluble in water, but almost completely soluble in mineral acids.

The term *calamine*, besides being applied to the official product, is also used to describe the impure, naturally occurring zinc carbonate. The official *Calamine* is obtained by calcination (powdered by heating) of the natural ore. The calcined product is then passed through a 100-mesh sieve to obtain the finely powdered material necessary for good cohesive and adhesive (adhering to skin) properties.

USES. *Calamine* is classified by the U.S.P. as a topical protective. It is used in dusting powders, ointments, and lotions (*Calamine Lotion*, U.S.P.) where it is applied to the skin for its soothing, adsorbent, protective properties. The only real difference between zinc oxide and calamine is the latter's better cosmetic acceptability. Dermatological problems, particularly those involving the exudation of fluids, respond reasonably well to the application of products containing calamine. *Calamine Lotion*, U.S.P. XVIII, is one of the most widely used preparations containing calamine. It contains equal quantities of *Calamine* and Zinc Oxide suspended with the aid of Bentonite Magma in a solution of Calcium Hydroxide. The lotion is a protective with a good drying effect and a mild astringent action (see below). *Phenolated Calamine Lotion*, U.S.P. XVIII, contains 1% liquified phenol which provides a local anesthetic and antipruritic (anti-itching) action.

Zinc Stearate, U.S.P. XVIII

The U.S.P. describes *Zinc Stearate* as a compound of zinc with a mixture of solid organic acids obtained from fats, and consists chiefly of variable proportions of zinc stearate ($[CH_3(CH_2)_{16}CO_2]_2Zn$) and zinc palmitate ($[CH_3(CH_2)_{14}CO_2]_2Zn$). It contains the equivalent of not less than 12.5% and not more than 14.0% of ZnO. It is a fine, white, bulky powder, free from grittiness, and it has a faint characteristic odor. It is unctuous to the touch and readily adheres to the skin.

Zinc stearate, either as the pure chemical or the official preparation, is insoluble in water, alcohol, and ether. It can be hydrolyzed by heating in dilute mineral acids to form a soluble zinc salt and an insoluble oily layer of stearic (and palmitic) acid. Zinc stearate is neutral to moistened litmus paper.

Uses. Zinc stearate has mild astringent and antimicrobial properties. It is employed in dusting powders and ointments as a protective. It has a particular advantage over many products, e.g., talc, in that it is not wetted by moisture. This property makes the material more desirable in dermatological problems where large amounts of fluid are exuded, because it will not form crusty patches over the areas being treated. The inhalation of zinc stearate dust can cause pulmonary inflammation. For this reason, its routine use as a dusting powder for infants and children should be strongly discouraged.

Zinc stearate, as well as magnesium stearate, is widely used as a lubricant in the manufacture of tablets (see Chapter 12).

Titanium Dioxide, U.S.P. XVIII (TiO_2; Mol. Wt. 79.90).

Titanium Dioxide as a Protective

Titanium Dioxide U.S.P. is a white amorphous, odorless, tasteless, infusible powder. A 1 to 10 aqueous suspension of the compound is neutral to litmus paper. It is insoluble in water, hydrochloric acid, nitric acid, and dilute sulfuric acid. It is soluble in hydrofluoric acid, and in hot concentrated sulfuric acid. Fusion of the compound with potassium bisulfate or with alkali carbonates or hydroxides renders it soluble in water.

Reaction of titanium dioxide with hydrogen peroxide in dilute sulfuric acid produces titanium peroxide (TiO_3) which imparts an orange-red color to the solution. This reaction is used in the official identification of the compound.

Uses. The U.S.P. classifies *Titanium Dioxide* as a topical protective. Unlike the protectives discussed above this compound is used primarily for its opacity due to its high refractive index (2.7). This high refractivity makes the compound useful for screening out ultraviolet radiation—hence, the presence of titanium dioxide in various sun creams and sun screen products. As a solar ray protective, it is used in a concentration of 5 to 25% in ointments or lotions. Titanium dioxide is the most efficient protective for this purpose; however, its action is due primarily to opacity. Organic sun screen agents, e.g., *p*-aminobenzoic acid, act by chemically absorbing ultraviolet radiation.[2]

Titanium Dioxide is also used as a white pigment in cosmetics and paints. The very white character of the pigment may be toned down for cosmetic purposes to a cream color through the addition of small or trace amounts of iron(III) compounds (e.g., phosphate) during the manufacturing process.

Aluminum as a Protective Agent

Aluminum (Al; At. Wt. 26.98)

Aluminum salts will be discussed in detail in the section on Astringents. Aluminum is a silver-white metal having a density of 2.7. It is a very active element with a great affinity for oxygen. The metal reacts with atmospheric oxygen to form a superficial layer of aluminum oxide, Al_2O_3, which protects the metal from further oxidation. It is insoluble in water and alcohol and unreactive toward oxidizing acids, such as nitric acid and sulfuric acid. This is due once again to the formation of a protective layer of aluminum oxide. The metal reacts rapidly with dilute and concentrated hydrochloric acid to release hydrogen and form aluminum chloride, $AlCl_3 \cdot 6H_2O$. It also reacts with alkali hydroxides to release hydrogen and aluminate salts, e.g., $NaAlO_2$ or $NaAl(OH)_4$.

Aluminum was official in the U.S.P. XVII, which described it as a very fine, free-flowing, silvery powder, free from gritty or discolored particles. Its official recognition was to provide standards for the metal in the protective preparation *Aluminum Paste*, U.S.P. XVII.

Uses. Aluminum is present in 10% concentration in *Aluminum Paste*

U.S.P. XVII which is prepared using *Zinc Oxide Ointment* as the base. It is used as a protective to prevent irritation around intestinal fistulae (an abnormal opening of the intestinal tract through the skin). A common application is around the colostomy opening after surgery for intestinal cancer. The aluminum protects the skin from the digestive action of intestinal fluids.

Silicone Polymers. These are inert protective substances occurring in liquid form and known generally as silicone oils. They are primarily dimethylsilicone ethers represented by the general structure:

$$CH_3-\underset{\underset{CH_3}{|}}{\overset{\overset{CH_3}{|}}{Si}}-O\left[-\underset{\underset{CH_3}{|}}{\overset{\overset{CH_3}{|}}{Si}}-O\right]_n-\underset{\underset{CH_3}{|}}{\overset{\overset{CH_3}{|}}{Si}}-CH_3$$

One such polymer is known as Dimethicone or Simethicone (see Chapter 8) and is used in ointments and creams for application to the skin as a water repellent and protective against contact irritants. The usual concentration in these preparations is about 30%. Simethicone is also mentioned in Chapter 8 as a gastric protective and antiflatulent.

Silicone oils adhere very well to the skin and exclude contact with the air. For this reason, they should not be applied over broken or abraded skin or wounds requiring drainage. Contact with the eyes should also be avoided. The primary use is as a prophylactic against chemical irritants.

Antimicrobials and Astringents

Antimicrobial Terminology. There are several terms employed in describing antimicrobial activity. Since some of these terms refer to specific aspects of this activity, it is necessary to become acquainted with their definitions and general usage.

Antiseptic. This term is generally applied to any agent which either kills or inhibits the growth of microorganisms, i.e., bacteria, fungi, protozoans, etc. The term is reserved, however, for those agents used against microorganisms growing on man specifically or living tissue in general.

Germicide. The term germicide refers to a more specific action in that it describes agents which kill microorganisms. The *-cide* ending on the word arises from the Latin word *caedere*, which means "to kill." Hence, this ending can be applied to the names of various classes of microorganisms to provide terms for more specific agents, e.g., bactericide, fungicide, amebicide, etc.

Those agents which do not kill microorganisms, but function primarily by inhibiting their growth, can be described by terms using the suffix -*stat* (from the Greek word *stasis*, meaning "standing still"). Therefore, the terms, bacteriostat, fungistat, etc., are employed with compounds having this aspect of activity against the indicated microorganisms.

Disinfectant. This term refers to the same type of activity as the term germicide, above. Its usage differs in that it is applied to those agents most appropriately used on inanimate objects, e.g., instruments, equipment, rooms, etc.

Sterilization. This refers to the use of a disinfectant or other procedure to render an object completely free of microorganisms. This frequently involves the use of chemicals or mechanical processes (e.g., heat) which are much too stringent for use on animal or human tissue.

The terms antiseptic and germicide may be further modified according to their area and type of use. For example, agents may be classified with respect to whether they are used topically or internally. Internal agents may be further subdivided into those that are absorbed (systemic), and those that are not absorbed (nonsystemic). The former compounds are distributed through the circulation and used to treat infections in various organs and tissues, while the latter agents remain and function in the area where they are applied, e.g., the gastrointestinal tract.

Mechanisms of Action. The mechanisms of action of inorganic antimicrobial agents can be divided into three general categories: oxidation, halogenation, and protein precipitation. These represent the primary chemical interactions or reactions that occur between the agent and microbial protein and result in the death of the microbe or inhibition of its growth. It is important to note that in contrast to certain organic compounds known as antibiotics, the sites of action of inorganic antimicrobial compounds are, for the most part, nonspecific. These agents will interact in a similar fashion with all protein, and in high enough concentrations will affect host protein as well as microbial protein.

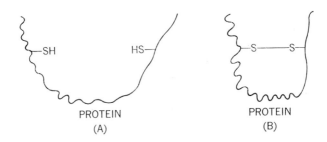

Fɪɢ. 9–1. An illustration of the action of oxidizing agents on protein-containing sulfhydryl groups. (*A*) The protein before oxidation showing the presence of free sulfhydryl groups. (*B*) The protein after oxidation showing the formation of a disulfide bridge between the two—SH groups.

Oxidation. Those compounds capable of functioning as antimicrobial agents through oxidative mechanisms are generally nonmetals and certain types of anions. Most common among these are hydrogen peroxide, metal peroxides, permanganates, halogens (i.e., chlorine and iodine), and certain oxo-halogen anions. The effective oxidative action of these compounds involves the reducing groups present in most proteins, e.g., the sulfhydryl ($-SH$) group in cysteine. An illustration of the reaction between the oxidizing antiseptic and a sulfhydryl-containing protein is shown in Fig. 9–1. Based on the concept that the protein has a specific function in the microorganism, e.g., enzyme, the formation of the disulfide bridge (Fig. 9–1B) will alter the conformation (shape) of the protein and thereby alter its function. The overall change or destruction of function in specific proteins is responsible for the ultimate destruction of the microorganism. Of course, the chemical result of oxidizing the protein is reduction of the antimicrobial agent.

Halogenation. This is a reaction occurring with antiseptics of the hypohalite type and, in particular, hypochlorite, OCl^-. Since these types of compounds can serve as reagents in the chlorination of primary and secondary amides, e.g.,

$$R-\overset{\overset{\textstyle O}{\|}}{C}-NH_2,$$

it is expected that a similar reaction can take place under appropriate conditions with the peptide linkage between the amino acid groups comprising the protein molecule. An example of this is shown in rx (iv).

$$\text{Protein}-\overset{\overset{\textstyle O}{\|}}{C}-\underset{\underset{\textstyle H}{|}}{N}-\text{Protein} \xrightarrow{\;OCl^-\;} \text{Protein}-\overset{\overset{\textstyle O}{\|}}{C}-\underset{\underset{\textstyle Cl}{|}}{N}-\text{Protein} \qquad \text{(iv)}$$

This reaction is ultimately destructive to the function of specific proteins because the substitution of the chlorine atom for the hydrogen produces changes in the forces (hydrogen bonding) responsible for the proper conformation of the protein molecule. As in the oxidative case, the changes in conformation result in destruction of function.

Protein Precipitation. This type of mechanism involves the interaction of proteins with metallic ions having large charge/radius ratios or strong electrostatic fields. This property is available in transition metal cations, including the metals of Groups IB and IIB, e.g., Cu(II), Ag(I), and Zn(II). Aluminum(III) in Group IIIA, due to its charge and small ionic radius, is also an effective protein precipitant. In fact, most metal cations, with the exception of the alkali and alkaline earth metals, will demonstrate protein precipitant activity.

The nature of the interaction is one of complexation in which the various polar groups on the protein act as ligands (see Fig. 9–2). The complexation

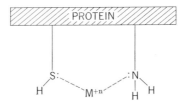

Fɪɢ. 9–2.   A model illustration of the interaction between a metal ion and a protein, resulting in protein precipitation.

of the metal results in a radical change in the properties of the protein or protein precipitant.

The interaction of metal ions with protein is nonspecific, and at sufficient concentration will react with host as well as microbial protein. Certain metals (e.g., mercury, arsenic, and antimony) show some enzyme specificity, and form strong covalent bonds with particular enzyme systems. The presence of the metal "ties up" important functional groups at the active site on the enzyme.

The protein precipitant properties of metal cations can be altered according to the concentration at the site of action. By increasing concentrations, antimicrobial, astringent, irritant, and corrosive properties are successively available.

Astringents.   The application of a very dilute solution of a metal cation to tissue primarily provides a local or surface protein precipitant action. This activity is usually designated as being *astringent.* Being a surface phenomenon, it does not usually result in the destruction of host tissue. Its effect can be observed or felt, however, when applied to skin or mucous membranes. The effect can be generally described as a "shrinkage" or "firming" of the tissue. For example, astringents will cause the constriction of capillaries and small blood vessels, hence, they are used as styptics to stop bleeding from small cuts. The astringent action of aluminum compounds on the gastrointestinal tract, producing constipation, has been mentioned in Chapter 8. Astringents are also used to reduce the volume of exudate from wounds and skin eruptions. They are also found as antiperspirants in deodorants because of their ability to constrict pores and destroy microorganisms that may produce body odors. The concentration of protein precipitants required for this type of activity on host cells is higher than that required for antimicrobial activity.

When applied topically to wounds, astringents may actually stimulate the growth of new tissue. However, higher concentrations will produce an irritant action, and further increases in concentration provide solutions which will have a corrosive effect on contact with tissue. Corrosive effects can be used to advantage in the removal of undesirable tissue, e.g., warts.

Control of Antimicrobial/Astringent Action. As indicated above, inorganic antimicrobial agents are largely nonspecific in their actions on protein. It was further indicated that, at least among the protein precipitants, the action on microbial versus host cells was determined by concentration. In reality, this is the only means available for directing all of these agents toward antimicrobial activity and for minimizing unwanted action on host cells.

In the case of water soluble compounds, control of activity is accomplished by making solutions of the appropriate concentration for the desired use. These concentrations will vary depending upon the area of use—e.g., higher concentrations may be used on the skin than in the eye. Soluble compounds may also be controlled by placing them in a vehicle which would slow their release to the site of action. For certain compounds, this is accomplished by placing them in solutions containing glycerin (e.g., glycerite of hydrogen peroxide) or polyethylene glycol; ointments may also provide controlled release of the antimicrobial agent.

Complexation with a ligand (e.g., Povidone-Iodine) also provides a controlled release of some of these agents, minimizing toxicity and activity at host cells.

Other compounds may be synthesized in an insoluble form and used in suspensions, ointments, or creams for their antimicrobial action because of the slow release of active agent. The insolubility of the compound makes the formulation of preparations somewhat easier from the standpoint of controlling activity. The various dosage forms and concentrations indicated in the following sections are formulated according to the desired activity, e.g., antiseptic, astringent, etc., and the area of application, e.g., skin, eye, ear, mouth, etc.

Antimicrobial/Astringent Products

Oxidative Antimicrobial Agents

Hydrogen Peroxide Solution, U.S.P. XVIII (H_2O_2; Mol. Wt. 34.02)

Hydrogen Peroxide Solution U.S.P. contains, in each 100 ml, not less than 2.5 g and not more than 3.5 g of H_2O_2. Suitable preservatives totaling not more than 0.05% may be added. It is a clear, colorless liquid which may be odorless or may have an odor resembling that of ozone. The solution will usually deteriorate upon standing or upon protracted agitation, and rapidly decomposes when in contact with many oxidizing or reducing substances. It is unstable on prolonged exposure to light, and may decompose suddenly when rapidly heated. The solution is acid to litmus and to the taste, and produces a froth in the mouth. It has a specific gravity of about 1.01.

Pure hydrogen peroxide, at room temperature, is a colorless, syrupy liquid with astringent properties. (Note: It is not used on the skin in this form.) It has a specific gravity of 1.463 at 0° C and, although unstable, it

11

decomposes only very slowly in pure form. When heated to $100°$ C pure hydrogen peroxide decomposes explosively to form water and oxygen. It is miscible in all proportions with water, alcohol, or ether. It is more soluble in ether than in water, and ether will extract it from aqueous solutions.

Chemically, hydrogen peroxide may be considered to be stable in solutions of high purity; however, small amounts of contaminants, e.g., di- and polyvalent ions of chromium, iron, copper, mercury, etc., will catalyze the decomposition of unstabilized solutions. In the absence of impurities, exothermic decomposition to water and oxygen will occur very slowly. Aluminum does not act as a catalyst and, therefore, storage containers may be made of this metal.

Hydrogen peroxide solutions may be stabilized with acids, complexing agents, or adsorbents. Any inorganic or organic acid will stabilize the solution. The compound is a weak acid in aqueous solution, ionizing primarily to form the peroxide ion according to rx (v).

$$H_2O_2 \leftrightarrows 2H^+ + O_2^{-2} \qquad pK_a = 11.62 \qquad \text{(v)}$$

Its acidic strength is less than that of boric acid (see Chapter 4). The above equilibrium provides some indication of the mechanism of stabilization by acids. Any additional hydrogen ion will cause the equilibrium in rx (v) to shift toward hydrogen peroxide.

Alkaline solutions tend to be much less stable; however, the instability is due more to the impurities present in the standard grades of alkalies than to any direct alkaline catalysis.

Complexing or chelating agents are frequently used to stabilize hydrogen peroxide solutions. Some common agents employed for this purpose include acetanilid, quinine sulfate, and 8-hydroxyquinoline (oxinate, see Chapter 1) in concentrations ranging from 0.02 to 0.05%. These compounds will chelate trace amounts of polyvalent metals, thereby making them unavailable to catalyze the decomposition.

Many adsorbents (e.g., alumina and silica) will remove impurities from hydrogen peroxide solutions. Their stabilizing function is then somewhat comparable to the complexing agents. It should be noted, however, that the therapeutic uses of hydrogen peroxide are dependent upon its decomposition by the enzyme catalase (see below), and some of the compounds used to stabilize the solutions will also inactivate the enzyme at the site of action.

Depending upon the chemical environment, hydrogen peroxide will react as either an oxidizing or reducing agent. The oxidation state of oxygen in the peroxide ion, $(O\!-\!O)^{-2}$, is -1. When hydrogen peroxide functions as an oxidizing agent it forms two oxide ions, O^{-2}, requiring two electrons, and resulting in a change of the oxidation state of the oxygen to -2 [rx (vi)].

$$(O_2)^{-2} \xrightarrow[+2e^-]{\text{as oxidizing agent}} 2O^{-2} \qquad \text{(vi)}$$

This type of reaction is most efficient in acidic media to produce water [rx (vii)].

$$H_2O_2 + 2H^+ + 2e^- \rightarrow 2H_2O \qquad \text{(vii)}$$

An example of this type of reaction is illustrated with hydriodic acid in rx (viii).

$$H_2O_2 + 2HI \rightarrow 2H_2O + I_2 \qquad \text{(viii)}$$

The reducing actions of hydrogen peroxide result in the evolution of molecular oxygen. This involves the release of two electrons and a change from the peroxide ion to an oxidation state of zero [rx (ix)].

$$(O_2)^{-2} \xrightarrow[-2e^-]{\text{as reducing agent}} O_2 \qquad \text{(ix)}$$

This type of reaction also occurs best in acidic media and usually involves metallic oxides as shown in rx (x).

$$Ag_2O + H_2O_2 \rightarrow 2Ag + H_2O + O_2 \uparrow \qquad \text{(x)}$$

The official assay of hydrogen peroxide is based on its reducing properties in the decoloration of a standard solution of potassium permanganate [rx (xi)].

$$5H_2O_2 + 2KMnO_4 + 3H_2SO_4 \rightarrow \qquad \text{(xi)}$$
$$K_2SO_4 + 2MnSO_4 + 8H_2O + 5O_2 \uparrow$$

USES. The primary use of *Hydrogen Peroxide Solution*, U.S.P., is as a mild oxidizing antiseptic. This action is produced when the solution comes in contact with open or abraded tissue, exposing the chemical to the enzyme, catalase. This enzyme catalyzes the decomposition of H_2O_2 to water and oxygen [rx (xii)].

$$2H_2O_2 \xrightarrow{\text{catalase}} 2H_2O + O_2 \uparrow \qquad \text{(xii)}$$

In theory, the oxygen acts as an oxidizing agent on bacteria, providing antiseptic action particularly on those organisms obliged to survive through anaerobic metabolism. Hydrogen peroxide will destroy most pathogenic bacteria, e.g., *Escherichia coli*, *Staphylococcus aureus*, and typhoid bacilli.

A major difficulty with hydrogen peroxide, and other peroxides as well, is involved with the rapidity of oxygen release under the influence of catalase. The antiseptic action does not penetrate below the surface to which it is applied, and the surface action is fleeting. The major benefit associated with the use of hydrogen peroxide is the mechanical cleansing action provided by the foaming release of oxygen. This effervescent action aids in the removal of dirt, bacteria, and debris from the surface of a wound or difficult-to-reach areas, e.g., the ear canal.

The official solution containing 3% hydrogen peroxide is often referred to as a 10 volume solution, indicating that 1 ml of the solution will liberate a total of 10 ml of oxygen at standard temperature and pressure. Extending this further, a 6% solution is also a 20 volume solution. Hydrogen peroxide in 6% (20 volume) solutions is used as a hair and fabric bleach. Neither this concentration nor a dilution of it should be used medicinally.

Hydrogen Peroxide Solution is used undiluted for its mild antiseptic and cleansing effects on wounds. When diluted with one part of water, it can be used as a gargle or mouthwash in the treatment of bacterial infections of the throat and mouth (Vincent's angina or Vincent's stomatitis). However, continued use in this manner may lead to irritation of the mouth and tongue, causing hypertrophied filiform papillae or "hairy tongue." Half-strength solutions may also be used as a vaginal douche. When diluted to less than half-strength, hydrogen peroxide tends to lose antiseptic activity.

Urea peroxide (carbamide peroxide, $H_2N—\overset{\overset{\text{O}}{\|}}{C}—NH_2 \cdot H_2O_2$) is a comparatively stable crystalline compound containing 34% H_2O_2. When placed in solution in anhydrous glycerin in concentrations of 4 to 10% (1.5 to 3.75% H_2O_2) and stabilized with 0.1% 8-hydroxyquinoline, a stable antiseptic preparation is produced having more prolonged activity than aqueous solutions of hydrogen peroxide. Products of this type are preferable in the treatment of infections in the oral cavity and ear infections (see ref. 3 and references cited in that article).

A similar preparation known as a glycerite of hydrogen peroxide is a solution containing 1.5% of H_2O_2 in anhydrous glycerin. To minimize the amount of water added to the preparation, 90% hydrogen peroxide is used in making the solution. This preparation has properties similar to those of urea peroxide solutions.

Because of the release of oxygen gas, hydrogen peroxide should never be injected into closed body cavities. In addition to the problem created by having no free exit for the gas, the irritating effects of the compound can cause bleeding and rapid, complete destruction of the H_2O_2 with the accumulation of much gas.

Zinc Peroxide, U.S.P. XVII (ZnO_2; Mol. Wt. 97.37; Medicinal Zinc Peroxide)

Zinc peroxide is, for all practical purposes, the only metal peroxide in present-day use as a topical antibacterial agent. The compound was official in the U.S.P. XVII as a mixture of zinc peroxide, zinc carbonate $(ZnCO_3)$, and zinc hydroxide $[Zn(OH)_2]$. The mixture usually contains about 55% of ZnO_2, and will yield about 9% by weight of oxygen.

It is a white, or faintly yellow, fine, odorless powder which is practically insoluble in water and organic solvents. It will dissolve in dilute acids with decomposition to a soluble zinc salt and hydrogen peroxide [rx (xiii)]. It is unstable at temperatures above 140° C.

$$ZnO_2 + 2HCl \rightarrow ZnCl_2 + H_2O_2 \qquad \text{(xiii)}$$

Although it appears to be insoluble in water, it will slowly hydrolyze to produce zinc oxide and hydrogen peroxide [rx (xiv)].

$$ZnO_2 + H_2O \xrightarrow[\text{Room Temp.}]{\text{Slow}} ZnO + H_2O_2 \qquad \text{(xiv)}$$

This reaction is responsible for its antibacterial action, in that the hydrogen peroxide under the influence of catalase [rx (xii)] will release oxygen at the site of infection, and the remaining zinc oxide will exert an astringent action (see *Zinc Oxide*).

Zinc peroxide as well as other metal peroxides present the same possible chemical incompatibilities as hydrogen peroxide, i.e., warm moist air, heavy metal ions, oxidizing and reducing agents, and alkaline solutions.

USES. Zinc peroxide may be used as an antibacterial and mild astringent in the treatment of infections in various type of wounds. Its effectiveness is greatest on anaerobic organisms, and particularly on those having a strict restriction in their metabolic oxygen requirements. Certain aerobic organisms (e.g., hemolytic streptococci) are also susceptible to the actions of this compound.

Zinc peroxide has been found to be useful in oral infections,[4] e.g., Vincent's stomatitis. A 25% suspension or a direct application of the powder may be used to treat mouth infections caused by anaerobic organisms. Much of the use of this compound has been discontinued in present-day therapeutics.

Other metal peroxides which have been used in the topical therapy of infections include sodium peroxide (Na_2O_2), magnesium peroxide (MgO_2), and calcium peroxide (CaO_2). These compounds enjoy rather limited use in present-day antimicrobial therapy.

Sodium Perborate, $(NaBO_3 \cdot 4H_2O$ or $NaBO_2 \cdot H_2O_2 \cdot 3H_2O$; Mol. Wt. 153.86)

Although no longer official, sodium perborate was official in the N.F. XII as white crystalline granules or as a white powder containing not less than

9% of available oxygen. This corresponds to about 86.5% of $NaBO_3 \cdot 4H_2O$. It is odorless and has a saline taste. The compound is stable in cool, dry air free of carbon dioxide. However, in moist air at temperatures at or above 40° C, it will decompose, with the evolution of oxygen. It is soluble in water to the extent of 1 in 40.

A monohydrate ($NaBO_3 \cdot H_2O$) is known which exhibits greater heat and moisture stability and greater solubility in water. Moist air may convert it to the tetrahydrate form.

The chemical properties of sodium perborate may be viewed as a composite of sodium metaborate ($NaBO_2$), boric acid (see Chapter 4), and hydrogen peroxide. This is emphasized by its slow hydrolysis in aqueous solution to provide sodium metaborate and hydrogen peroxide [rx (xv)].

$$NaBO_3 \cdot 4H_2O \leftrightharpoons NaBO_2 + H_2O_2 + 3H_2O \qquad \text{(xv)}$$

This reaction has led to the possibility of an alternate representation for the composition of sodium perborate, viz., $NaBO_2 \cdot H_2O_2 \cdot 3H_2O$.

The remaining properties of the compound are represented by the alkalinity of its aqueous solutions. The basic nature of the solution is due to the formation of sodium hydroxide from the hydrolysis of the sodium metaborate produced in rx (xv). This type of reaction was discussed in Chapter 4 and is illustrated in rx (xvi).

$$NaBO_2 + 2H_2O \leftrightharpoons NaOH + H_3BO_3 \qquad \text{(xvi)}$$

Because of its hydrolysis products, it is to be expected that the hydrogen peroxide formed in aqueous solution would be unstable at the high pH produced by the hydroxide ions. Therefore, solutions should be freshly prepared at the time of use.

USES. Sodium perborate is used as an antiseptic based solely on the available hydrogen peroxide or oxygen. It is present in preparations used in the treatment of oral infections, e.g., Vincent's angina. Preparations have included a paste formed with glycerin or water and applied orally, or a freshly prepared 2% mouthwash.

It has also been used in powdered or "dry" dentifrice preparations in concentrations of 10 to 20%. There is a danger associated with this type of use in that hydrolysis of particles of the dentifrice lodged between the teeth will produce sodium hydroxide [see rx (xvi)], and local corrosion of enamel or gum tissue. *Aromatic Sodium Perborate*, N.F. XII, a preparation of sodium perborate, saccharin, and peppermint oil, provides a flavored form of sodium perborate for use in these types of preparations.

Sodium perborate is also used as a bleaching agent in nonchlorine laundry bleaches.

Potassium Permanganate, U.S.P. XVIII ($KMnO_4$; Mol. Wt. 158.04)

Potassium Permanganate U.S.P. is an odorless, dark purple crystalline compound. The crystals are almost opaque in transmitted light and of a blue metallic luster in reflected light. The color is sometimes modified by a dark bronze-like appearance. It has a specific gravity of 2.703 and is stable in air.

The compound is soluble in water (1 in 15) and freely soluble in boiling water (1 in 3.5). Concentrated solutions have a deep violet-red color and highly diluted solutions are pink. Solutions have a sweetish astringent taste.

When potassium permanganate is heated to 240° C, it decomposes with the liberation of oxygen, leaving manganese dioxide (MnO_2) and potassium manganate (K_2MnO_4).

Chemically, potassium permanganate is a strong oxidizing agent both in the dry state and in solution. It therefore requires great care in handling. The U.S.P. XVIII makes the following statement: *"Caution—Observe great care in handling Potassium Permanganate, as dangerous explosions may occur if it is brought in contact with organic or other readily oxidizable substances, either in solution or in the dry state."* The compound in the dry state forms explosive mixtures with such materials as charcoal, and will produce a fire when mixed with glycerin. It will also oxidize alcohol.

Acid solutions of potassium permanganate react to reduce the permanganate ion, MnO_4^- (Mn^{+7}) to the manganous ion (Mn^{+2}) with the evolution of oxygen [rx (xvii)].

$$2KMnO_4 + 3H_2SO_4 \rightarrow K_2SO_4 + 2MnSO_4 + 3H_2O + 5[O] \qquad \text{(xvii)}$$

Neutral or alkaline solutions produce a similar reaction with the characteristic brown precipitate of manganese dioxide (MnO_2) [rx (xviii)].

$$2KMnO_4 + H_2O \rightarrow 2MnO_2\downarrow + 2KOH + 3[O] \qquad \text{(xviii)}$$
$$\text{(brown)}$$

The latter reaction is of therapeutic importance since potassium permanganate is usually applied to the skin as a neutral aqueous solution.

An example of the effect of pH on the oxidative reactions of potassium permanganate is illustrated using potassium iodide. In acid solution, iodides are oxidized to molecular iodine [rx (xix)].

$$2KMnO_4 + 10KI + 8H_2SO_4 \rightarrow \qquad \text{(xix)}$$
$$6K_2SO_4 + 2MnSO_4 + 5I_2 + 8H_2O$$

In neutral or alkaline solution, iodides are oxidized to iodates [rx (xx)].

$$2KMnO_4 + H_2O + KI \rightarrow 2MnO_2\downarrow + 2KOH + KIO_3 \qquad \text{(xx)}$$

Potassium permanganate will also oxidize sulfides to free sulfur, ferrous salts to ferric salts, and nitrites to nitrates.

The antibacterial action of potassium permanganate is dependent upon its oxidation of protein or other bioorganic substances. The compound reacts in neutral media in the presence of protein in a manner like that shown in rx (xviii) above. The oxygen released is the effective agent. The manganese dioxide formed as the permanganate is reduced leaves a brown stain on the skin and tissues. The activity of potassium permanganate is not selective; indeed, it oxidizes all organic matter. Therefore, its activity on microorganisms is decidedly diminished in the presence of extraneous organic substances.

Uses. Potassium permanganate solutions are used for both their antibacterial and antifungal actions. However, due to their short duration of action, low penetrating power, and unsightly staining of skin, the use of these preparations has declined. Also, the action is limited to the skin and mucous membranes.

Solutions are used externally in concentrations ranging from 1:500 to 1:15,000 (0.2% to about 0.006%); however, solutions up to 1% may be employed in certain instances. It should be noted that concentrations above 1:5,000 may be irritating to sensitive tissues.

The susceptibility of bacteria to potassium permanganate solutions is variable. Most organisms will be killed in solutions having concentrations between 1:5,000 and 1:10,000; however, some are resistant to concentrations higher than 1:5,000.

Potassium permanganate solutions are primarily used today for skin infections (dermatitis) caused by bacteria and fungi, and for poisonings produced by plant and animal toxins. Wet dressings prepared from a 1:10,000 solution have been used in the treatment of the vesicular (the presence of small blisters or raised areas containing fluid) stage of eczema. Similar dressings have been used in severe poison ivy. Solutions of similar strength have been used in the treatment of the vesicular stage of athlete's foot (tinea pedis) and fungal infections on other portions of the body, e.g., groin (tinea cruris). A.M.A. Drug Evaluations[5] indicate that wet dressings of potassium permanganate are probably ineffective. This may be due to inactivation of the permanganate by the dressing material. Better results may be achieved through direct applications (e.g., bathing the area) of dilute solutions.

Potassium Permanganate Tablets for Solution, U.S.P. XVIII, are available in 60-, 125-, and 300-mg sizes for preparing solutions of appropriate strength. Potassium permanganate crystals are also available. One percent solutions are stable for months and may be prepared to provide a concentrate for dilution at the time of use.

Sodium Hypochlorite Solution, N.F. XIII (Dakin's solution)

Diluted Sodium Hypochlorite Solution, N.F. XIII (modified Dakin's solution)

Hypochlorites in general and sodium hypochlorite (NaOCl; Mol. Wt.

74.44) in particular are very unstable and usually found only in solution. For all practical purposes, solid sodium hypochlorite is not obtainable.

Sodium Hypochlorite Solution N.F. contains not less than 4.0% and not more than 6.0% by weight of NaOCl. It is a clear, pale greenish yellow liquid having an odor of chlorine. The solution is affected by light. It has an alkaline pH coloring red litmus blue; however, the chlorine will later bleach the color. Common household bleach is usually a 4.5 to 5.0% solution of sodium hypochlorite.

Diluted Sodium Hypochlorite Solution N.F. is described as a solution of chlorine compounds of sodium containing, in each 100 ml, not less than 450.0 mg and not more than 500.0 mg (or approximately 0.5%) of NaOCl. The Solution is a colorless or light yellow liquid having a slight odor suggesting chlorine. The Solution is prepared according to a procedure given in the N.F. XIII. Briefly, it involves diluting *Sodium Hypochlorite Solution* with five times the quantity of purified water, and adjusting the pH with a 5% solution of sodium bicarbonate until no color is produced with phenolphthalein. This gives a pH of 8.3 or less; the appropriate chemistry is discussed below. Due to reduced concentration and pH, the resulting solution is the only hypochlorite preparation officially recognized for local application to tissues as an antibacterial. The dilution of household bleach does not normally meet N.F. standards as an antiseptic because of the lack of pH adjustment.

The N.F. XIII classifies *Sodium Hypochlorite Solution* as a disinfectant, and makes the following statement: *"Caution: This Solution is not suitable for application to wounds."* The alkalinity and oxidizing action of this solution is too strong for use on tissues. In addition, the solution dissolves blood clots and delays healing. A diluted form of this is known as Labarraque's solution and consists of *Sodium Hypochlorite Solution* which has been diluted with an equal volume of water (approx. 2.5% NaOCl). This solution is also used primarily as a disinfectant (on *inanimate* objects only).

As indicated above, sodium hypochlorite is quite unstable. Solutions are rapidly decomposed by boiling [rx (xxii)] and by treatment with most acids [rx (xxi)].

$$4NaClO + 4HCl \rightarrow 4NaCl + 4HOCl \rightarrow 2Cl_2\uparrow + O_2\uparrow + 2H_2O \quad \text{(xxi)}$$

Warming the solution will result in the formation of sodium chlorate [rx (xxii)], a compound very low in antibacterial properties.

$$3NaClO \xrightarrow{\text{warm}} 2NaCl + NaClO_3 \quad \text{(xxii)}$$

Solutions of sodium hypochlorite are strong oxidizing agents, as can be shown by the liberation of free iodine from solutions of potassium iodide [rx (xxiii)].

$$2KI + NaClO + H_2O \rightarrow 2KOH + NaCl + I_2 \quad \text{(xxiii)}$$

The antibacterial properties of sodium hypochlorite solutions are due in part to the liberation of chlorine [see rx (iv)], and to the oxidizing action produced by the liberation of oxygen [rx (xxi)]. Sodium hypochlorite solutions are rapidly inactivated in the presence of tissue and bacterial protein due to the chlorination reaction [rx (iv)].

The facility of this action on protein is improved at a pH near 7, which provides higher concentrations of the more active hypochlorous acid, HOCl. The usual pH of sodium hypochlorite solutions is decidedly alkaline, in the range of 10 to 11, due to the hydroxide ions from sodium hydroxide added during their preparation and the hydrolysis of the hypochlorite ion [rx (xxiv)].

$$OCl^- + H_2O \leftrightharpoons HClO + OH^- \qquad \text{(xxiv)}$$

The hypochlorite salt is more stable under alkaline conditions.

In *Diluted Sodium Hypochlorite Solution*, N.F. XIII, the pH is reduced through the addition of sodium bicarbonate solution (see above). This has the effect of reducing the caustic action of the highly alkaline solution on tissues and of increasing the effective concentration of hypochlorous acid. The bicarbonate acts to reduce the hydroxide ion concentration according to the reaction shown in rx (xxv).

$$HCO_3^- + OH^- \leftrightharpoons CO_3^{-2} + H_2O \qquad \text{(xxv)}$$

The equilibrium in the above reaction is maintained toward the right by providing an excess of bicarbonate. The minimum pH available by this method is around 8, which is too low to produce a color in the presence of phenolphthalein (see Indicators in Chapter 4). The removal of hydroxide ions from the solution also increases the concentration of hypochlorous acid relative to hypochlorite according to the equilibrium shown in rx (xxiv), thereby improving the antibacterial properties of the solution. This becomes apparent when the ratio of $[OCl^-]/[HOCl]$ is examined at the more alkaline pH of 11 and at the pH of 8. By applying the Henderson-Hasselbalch equation (see Buffers in Chapter 4) using $pK_a = 7.4$ for HOCl, it is found that at pH $= 11$ the above ratio is approximately 3600 to 1, while at pH $= 8$ the ratio is 3.5 to 1.

Uses. *Sodium Hypochlorite Solution*, N.F. XIII, is useful as a disinfectant and laundry bleach. It is an effective germicidal agent that can be used to disinfect areas, instruments, and utensils which have been exposed to pathogenic organisms. The only solutions of this type available are sold as laundry bleaches (e.g., Hilex®, etc.).

Diluted Sodium Hypochlorite Solution, N.F. XIII, has been used in the past as an antiseptic on pus-forming (suppurating) wounds, and as an irrigation solution for infections inside certain body openings. While the solution is very effective both as an antiseptic and at removing necrotic tissue, it has the disadvantages of dissolving certain type of sutures, and of dissolving blood clots and prolonging clotting time. These properties can lead to secondary hemorrhage, and other antiseptics should be chosen when these are potential problems.

The solution may also be used as a foot bath in the prevention of various fungal infections (athlete's foot, etc.).

The instability of the solution requires storage in tight, light-resistant containers, and the avoidance of excessive heat. The antibacterial effectiveness may be increased by acidifying the solution at the time of use, thereby further increasing the concentration of HOCl. Diluted sodium hypochlorite solutions may be used full strength or diluted 1 to 3.

Diluted Sodium Hypochlorite is available under the official title or as "modified Dakin's solution." Zonite® is a trademarked product containing 1% of sodium hypochlorite which can be used full strength on wounds, diluted 1 to 12 as a mouthwash, or 1 to 30 for feminine hygiene.

OTHER CHLORINE-CONTAINING COMPOUNDS. *Chlorinated Lime* (bleaching powder, chloride of lime) is the product obtained by passing chlorine gas over slaked lime. Among other materials, it contains calcium hypochlorite, calcium hydroxide, and chloride ions. The product is probably best represented by the formula $CaOCl(Cl) \cdot H_2O$. It has the distinct odor of chlorine, and decomposes in air to release hypochlorous acid. It is used as a disinfectant in swimming pools and "sterile" rooms, and as a bleaching agent.

Chloramines are organic amines with one or two chlorine atoms bonded to the nitrogen. When these compounds are dissolved in water they slowly hydrolyze to release hypochlorous acid [rx (xxvi)].

$$R\text{—}NH\text{—}Cl + H_2O \rightarrow R\text{—}NH_2 + HOCl \qquad \text{(xxvi)}$$

Chloramine products are used as disinfectants and in the purification of drinking water supplies. Two products available include chloramine-T [sodium *p*-toluenesulfonchloramide (I)] and *Halazone*, U.S.P. XVIII [*p*-(dichlorosulfamoyl) benzoic acid (II)].

$$SO_2\text{—}NCl^-Na^+ \qquad\qquad SO_2\text{—}NCl_2$$

CH_3	$COOH$
I	II

An important aspect of the relationship between chlorine chemistry and the antimicrobial action of chlorinated compounds is that when chlorine gas is dissolved in neutral or acidic water hypochlorous acid is formed [rx (xxvii)].

$$Cl_2 + H_2O \rightarrow HCl + HOCl \qquad \text{(xxvii)}$$

This is the basis of using chlorine as a disinfectant in swimming pools and water supplies. The antimicrobial action is due to the presence of HOCl.

Iodine Preparations and Compounds

Iodine Solution, N.F. XIII
Iodine Tincture, U.S.P. XVIII

Both *Iodine Solution* N.F. and *Iodine Tincture* U.S.P. contain the same concentrations of ingredients, i.e., in each 100 ml they contain not less than 1.8 g and not more than 2.2 g of iodine (I), and not less than 2.1 g and not more than 2.6 g of sodium iodide (NaI). They differ only in the nature of the solvent, i.e., *Iodine Solution* is aqueous, having been prepared with purified water, and *Iodine Tincture* contains approximately 50% alcohol as the final solvent. Both solutions are transparent, have a reddish brown color, and have the characteristic odor of iodine. *Iodine Tincture* also has the odor of alcohol. The active antimicrobial agent common to both of these preparations is iodine; therefore, some aspects of chemical and physical properties will be considered first.

Iodine, U.S.P. XVIII, (I; At. Wt. 126.90) occurs in the form of heavy, grayish black plates or granules, having a metallic luster and a characteristic penetrating odor. It is very slightly soluble in water (1 in 2950), soluble in alcohol (1 in 12.5) and in solutions of iodides (I^-), and freely soluble in carbon disulfide and chloroform. The solubility in solutions of iodides, e.g., sodium iodide, is due to the formation of the triiodide ion (I_3^-) (see Chapter 3). Both the iodide and triiodide ions are devoid of antibacterial activity, although the latter spontaneously reverts to I_2 and I^-, thus retaining antibacterial potency.

The most notable chemical property of iodine in aqueous solution is that of a mild oxidizing agent. Although less reactive than chlorine in this respect, it is believed that the oxidizing action is mediated through formation of hypoiodous acid [HIO, rx (xxviii)].

$$I_2 + H_2O \leftrightharpoons HI + HIO \rightarrow HI + (O) \qquad \text{(xxviii)}$$

It will oxidize iron to form ferrous iodide [rx (xxix)].

$$Fe + I_2 \rightarrow FeI_2 \qquad \text{(xxix)}$$

For this reason metal spatulas should not be used to handle iodine and balance pans should be protected against pitting by using weighing papers.

Strong oxidizing agents will oxidize iodine to iodate, as illustrated with potassium chlorate in rx (xxx).

$$3I_2 + 5KClO_3 + 3H_2O \rightarrow 5KCl + 6HIO_3 \qquad \text{(xxx)}$$

This demonstrates a possible caution against mixing iodine with other oxidizing antibacterials.

In base, iodine forms both iodide and iodate salts with ammonium hydroxide solution, as shown in rx (xxxi).

$$3I_2 + 6NH_4OH \rightarrow 5NH_4I + NH_4IO_3 + 3H_2O \qquad \text{(xxxi)}$$

This reaction was used in the formation of *Decolorized Iodine Tincture*, N.F. IX, a preparation lacking antibacterial properties, but popular at one time as a "non-staining iodine" preparation.

The biochemistry and physiology of iodine and iodide are discussed in Chapter 6. The antimicrobial action has been known for 100 years. The actual mechanism of its action on microorganisms is not known; however, it is likely to involve oxidation and/or iodination of microbial protein. Since this activity is nonselective, the action is greatly reduced in the presence of other organic material. Unlike chlorine, which is germicidal through the formation of HOCl, free iodine is about six times more effective than hypoiodous acid, HIO.

The antimicrobial properties of iodine preparations have been recently reviewed.[6] Preparations providing free iodine are bactericidal, fungicidal, amebicidal, and virucidal. Iodine is effective in very dilute solutions—e.g., a 1:20,000 concentration will kill most bacteria in one minute and bacterial spores in 15 minutes. The presence of organic material will increase the time required for lethal action, but the solution will remain effective.

The toxicity of preparations containing free iodine is frequently mentioned and somewhat overstated. Iodine is a very active element and is, therefore, easily inactivated by organic materials in the gastrointestinal tract. Very little free iodine is absorbed. Most of the toxicity due to the ingestion of large quantities of iodine is a result of the corrosive action of the element on the gastrointestinal tract, producing abdominal pain, gastroenteritis, and possibly bloody diarrhea. The treatment usually involves gastric lavage with a soluble starch solution, or administration of a 5% sodium thiosulfate solution [rx (xxxii)]. The starch solution forms a complex with the iodine (purple color), thus aiding in its removal from the stomach. Fatalities have resulted from the ingestion of large quantities of a preparation, e.g., 1 to 8 oz of tincture of iodine. Most suicide attempts using tincture of iodine are unsuccessful.

$$2Na_2S_2O_3 + I_2 \rightarrow Na_2S_4O_6 + 2NaI \qquad \text{(xxxii)}$$

There are several preparations which contain iodine for antimicrobial purposes including iodine ointment, phenolated iodine solution (Boulton's solution), and *Iodine Ampuls*, N.F. XIII. The most frequently used preparations, however, are *Iodine Solution*, N.F. XIII, and *Iodine Tincture*, U.S.P. XVIII.

USES. *Iodine Tincture* and *Iodine Solution* are probably the most effective topical antiseptic agents available. They have been used as antiseptics on the skin prior to surgery. There is some indication that *Iodine Tincture* may be more suitable for this purpose, since the alcohol seems to improve the penetration of the iodine due to a "wetting" or spreading effect, as well as providing some additional antibacterial effect.

Both preparations can be diluted with water to provide effective solutions of reduced concentration for application to wounds and abrasions. *Iodine Solution* is preferred for application to wounds because the alcohol in the *Tincture* is very irritating to open tissue and is the entire reason for the stinging when this preparation is applied to wounds. In fact, the A.M.A. Drug Evaluations recommends against the use of Iodine Tincture on tissues.[7] The official solution and tincture are effective against bacterial and fungal infections of the skin. There is virtually no need for antibacterial solutions containing elemental iodine to exceed the 2% concentration of these two preparations. The concentration of *Iodine Tincture* may increase, particularly during household use, due to the evaporation of the alcohol.

Iodine Tincture may be used to disinfect drinking water. Treatment of suspected water supplies with three drops per quart will kill amebae and bacteria in 15 minutes.

Both *Iodine Solution* and *Tincture* are available in official concentrations and can be diluted to 0.5 to 1.0% concentration for application to wounds, and to 0.1% for irrigation. Iodine preparations of any type are contraindicated in patients who have exhibited prior allergy or hypersensitivity to iodine.

Povidone-Iodine, Aerosol, and Solution, N.F. XIII (Poly [1-(2-oxo-1-pyrrolidinyl) ethylene] Iodine Complex; Betadine®, Isodine®)

Povidone-Iodine is a complex of iodine with *Povidone*, N.F. XIII, which is a polymer also known as polyvinylpyrrolidone or PVP. When dried at 105° C to constant weight, the *Povidone-Iodine* complex contains not less than 9.0% and not more than 12.0% of available I (iodine).

The complex is a yellowish brown amorphous powder, and has a slight characteristic odor. Its aqueous solution is acid to litmus. It is soluble in water and in alcohol, and practically insoluble in organic solvents.

Both the *Aerosol* and *Solution* contain not less than 85.0% and not more than 120% of the labeled amount of I (iodine). The *Solution* is a transparent liquid having a reddish brown color, and a pH of not more than 6.0.

Povidone-Iodine is a member of a class of compounds referred to as iodophors. Iodophors are complexes of iodine, with carrier organic molecules serving as a solubilizing agent. These complexes slowly liberate iodine in solution. Development of iodophors was begun in an effort to prepare less irritating iodine products without losing antibacterial effectiveness.

The structural nature of the complex of iodine with polyvinylpyrrolidone is not known. However, it does appear to involve chemical interaction rather than simple physical "entrapment," as demonstrated by alterations in the ultraviolet absorption spectrum of the iodine. The complexed iodine is able to be titrated against sodium thiosulfate [rx (xxxii)] according to standard assay procedures (see N.F. XIII, p. 583), and this determination provides the quantity of "available iodine." The major advantages of *Povidone-Iodine* over elemental iodine solutions include its nonirritating effects on tissue, its comparatively low oral toxicity, its water solubility, and its low iodine vapor pressure making it stable to possible iodine loss. Solutions of the complex are also nonstaining and can be washed clear from skin and clothing.

Uses. Products containing *Povidone-Iodine* have been offered for the same uses as *Iodine Solution* and *Tincture*. However, the preparations have not been shown to be as effective as aqueous or alcoholic solutions of elemental iodine. The major advantage to their use is in the lack of tissue irritation, which makes them useful for application to sensitive areas and mucous membranes. Solutions are recommended for surgical scrubs and for preoperative antisepsis of the skin.[8] *Povidone-Iodine* has been used in gargles and mouthwashes for the treatment of infections in the oral cavity (Vincent's angina). The use of this compound in the treatment of sore throats is of questionable value.

Products are available under the trade names of Betadine® and Isodine® in concentrations of 0.1 to 1% of available iodine (usually 10% of the total concentration of *Povidone-Iodine*). Preparations include an aerosol (0.5%), solutions (usually 1%), a surgical scrub (0.75%), a vaginal douche (1%), and a vaginal gel (0.1%).

Other iodophors include complexes with anionic and cationic surface active agents. One of these is a complex with the cationic agent known as undecoylium chloride (a quaternary ammonium compound). Undecoylium Chloride-Iodine is marketed (Virac®) in solutions containing up to 0.6% of available iodine as a topical skin antiseptic.

Protein Precipitant Antimicrobial Agents

Silver Nitrate, U.S.P. XVIII ($AgNO_3$; Mol. Wt. 169.87)
Silver Nitrate U.S.P. occurs as colorless or white crystals which become

gray or grayish black on exposure to light in the presence of organic matter. It is very soluble in water, sparingly soluble in alcohol, and freely soluble in boiling alcohol. The pH of a 1% solution is between 4.5 and 6.0, and solutions should be clear and colorless.

Solutions of silver nitrate in concentrations between 0.5 and 1.0% are used as antibacterial agents. The chemistry and pharmacological action of these preparations are essentially those of silver ion.

The only salts of silver are those of the monovalent cation [Ag(I)]. This ion is readily obtained from metallic silver through treatment with an oxidizing acid, e.g., cold dilute nitric acid [rx (xxxiii)].

$$3Ag + 4HNO_3 \rightarrow 3AgNO_3 + NO \uparrow + 2H_2O \qquad \text{(xxxiii)}$$

Most silver salts are insoluble (see Chapter 3), with the notable exceptions of the nitrate, chlorate, fluoride, and lactate. The sulfate and acetate are soluble to the extent of about 1%. The salts are unstable to exposure to light either in the solid state or in solution. Light will catalyze the reduction of the silver ion to free silver. The darkening produced by the metallic silver is utilized in photographic films through an emulsion containing Ag^+, which will form Ag^0 in a controlled manner when exposed to light. This same photoreduction is responsible for the production of stains when solutions of silver compounds are spilled on the skin, clothing, etc.

When solutions of soluble silver salts come into contact with a soluble source of chloride ion in neutral or acidic solution, a white, curdy precipitate of silver chloride is produced. This is the basis for using silver nitrate in the quantitative analysis of soluble chlorides. Some silver ion is produced in equilibrium with the silver chloride, allowing dissolution of the precipitate through complex ion formation with solutions containing such ligands as ammonia (NH_3) (see Chapter 3), cyanide (CN^-), and thiosulfate [$S_2O_3^{-2}$, rx (xxxiv)].

$$2AgCl + 2S_2O_3^{-2} \rightarrow [Ag_2(S_2O_3)_2]^{-2} + 2Cl^- \qquad \text{(xxxiv)}$$

Treatment of a soluble silver salt (e.g., silver nitrate) with a solution of ammonium hydroxide results in the formation of the soluble diammino-silver(I) hydroxide and a low concentration of free silver ions. In the presence of reducing agents (e.g., eugenol, formaldehyde, etc.) the free silver ions are reduced to metallic silver to form a silver mirror on available surfaces. As the silver ions are reduced, more ions are released from the diamminosilver(I) complex until all the available silver is precipitated in the metallic state. The general reaction can be represented with formaldehyde in rx (xxxv).

$$2Ag(NH_3)_2OH + HCHO \rightarrow HCOONH_4 + 3NH_3 \uparrow + \qquad \text{(xxxv)}$$
$$2Ag \downarrow + H_2O$$
$$\text{(mirror)}$$

This reaction is the basis for the not infrequent use of *Ammoniacal Silver Nitrate* (N.F. XII) in dentistry as a protective of dentin (see Chapter 10).

When soluble silver salts are treated with other alkali hydroxides, a white precipitate of silver hydroxide forms which rapidly decomposes to brown silver oxide [rx (xxxvi)]. Either the hydroxide or the oxide can be dissolved in nitric acid or ammonium hydroxide.

$$2Ag^+ + 2OH^- \rightarrow 2AgOH \downarrow \ \rightarrow Ag_2O + H_2O \qquad \text{(xxxvi)}$$

Other insoluble silver salts are formed with reagents containing phosphates and borates. This is a source of incompatibility when buffers composed of these anions are used in silver nitrate solutions. When a protein solution is treated with a solution containing a soluble silver salt, a heavy precipitate is formed involving a complex interaction between the silver ions and protein. This type of reaction is the basis for the direct antimicrobial action of silver compounds.

The protein precipitant action of silver ion, and other heavy metal ions as well, is not selective; it will precipitate both bacterial and human protein. The range of activity available includes antibacterial, astringent, irritant, and corrosive, depending upon the concentration applied. Those preparations used primarily for their effects on human protein will be discussed in the next section (see Astringents).

Silver ion precipitation of protein involves interactions between the cation and various polar groups on the protein molecule, e.g., —SH, —NH$_2$, —COOH, and heterocyclic residues, e.g., histidine. When applied to tissue in a concentration of 0.1% Ag$^+$ the activity is rapidly bactericidal. However the action is somewhat localized due to precipitation with tissue proteins and chloride ions in the tissue fluids. It is interesting to note that a bactericidal effect continues after the initial application, due to the slow production of silver ions from the silver proteinate and silver chloride. This sustained action at the tissue level has given rise to colloidal products of silver proteinate and halides as antibacterials.

Silver preparations are bacteriostatic at concentrations of silver ion below that required for protein precipitation. This concentration effect is found with many heavy metals and indicates some interaction of the metal with bacterial enzymes. In connection with this, silver presents an interesting property referred to as *oligodynamic action*, meaning it is active in small quantities. This is found in distilled water that has been in contact with metallic silver, and as a result becomes bactericidal to suspensions heavy in viable organisms. The lethal action requires a few hours, but water treated in this way, e.g., distilled through a silver condenser, will remain sterile for long periods of time. This level of activity can be found in solutions containing no more than a 1:20,000,000 concentration of silver ions.

Solutions of silver nitrate are bacteriostatic at concentration of 1:30,000 and bactericidal at 1:4,000 in the presence of organic matter. Irritation of the skin becomes a factor at concentrations above 1:1,000. Colloidal silver preparations remain bacteriostatic at concentrations of 1:20,000 but require concentrations of about 10% to be bactericidal.

Extended use of silver preparations is likely to cause a darkening of the skin due to the deposition of free silver below the epidermis. This condition is termed argyria, and is essentially irreversible (see Chapter 7).

The toxicity of silver salts when taken internally is somewhat limited by their precipitation with protein and chloride ion. The activity is quite localized, and practically no systemic effects are noted. The toxic dose of silver nitrate is about 10 g, but survival after larger doses has been observed.

Uses. Silver nitrate is the most widely used soluble silver salt. It is employed as an antibacterial in solutions ranging in concentration from 0.01 to 10%, recognizing that the higher concentrations present astringent and irritant properties to the tissues. It has been used in concentrations of about 1:10,000 on sensitive membranes, e.g., irrigation of the urethra and bladder, but this has been largely replaced by the use of colloidal forms of silver when the action of this ion is desirable. Solutions containing 10% $AgNO_3$ have been used in the treatment of infected ulcers in the mouth.

Silver Nitrate Ophthalmic Solution, U.S.P. XVIII, is a 1% solution for instillation into the eyes of newborn babies. Silver salts are quite effective against gonococcal organisms, and two drops of this solution are placed in each eye as a prophylactic measure against infections produced by these organisms (ophthalmia neonatorum). This procedure is required by law in most states, and a study has shown that it may be more effective than antibiotics, such as penicillin, which have replaced it in some areas.[9] The U.S.P. allows buffering of the *Ophthalmic Solution* with sodium acetate and states that the pH should be between 4.5 and 6.0. The solution may be rinsed from the eyes with normal saline if desired. The *Solution* is available in wax capsules containing about 0.3 ml, which should be discarded after use.

Probably the most recent important use for silver nitrate is the application of 0.5% aqueous solutions in the form of a wet dressing on burned areas of patients suffering from third-degree burns.[10] This form of therapy involves the mechanism of sustained action mentioned previously, in that the silver ions are precipitated by tissue protein and chloride ion. The antibacterial activity is then dependent upon a low but minimal concentration of ions in equilibrium with the insoluble forms.[11] This form of therapy has come to be recommended as a treatment of choice in many treatment centers, because the silver ion has seemed to be particularly effective at reducing infection due to *Staphylococcus aureus*, various species of *Proteus*, and *Pseudomonas aeruginosa*. More recent studies[12] seem to indicate that wet dressings of 0.5% silver nitrate may be recommended as a good initial treatment, due to both the antibacterial effect and the reduction in fluid evaporation and heat loss produced by the wet dressing. However, it should be alternated with other agents depending upon the types of organisms present and the status of the patient. The most common side effects associated with this treatment are those involving electrolyte imbalance due to precipitation of chloride. These include low serum sodium, metabolic acidosis (see Chapter 5), and diarrhea. In addition to these physiological considerations, the black stains produced by silver nitrate are a major housekeeping problem.

OTHER SOLUBLE SILVER SALTS. *Silver Lactate* is a salt of silver with lactic acid $(CH_3CH(OH)COO^-Ag^+)$ which is soluble in water (1:15) and provides silver ion in a manner similar to silver nitrate. It is used in solutions ranging in concentration from 0.05 to 1.0% for the same purposes as silver nitrate. A silver lactate cream containing 0.9% silver in the form of a silver-lactic acid-allantoin complex has been recommended as an alternative to silver nitrate in the treatment of third-degree burns.[13]

Silver Picrate (Silver trinitrophenolate; $2,4,6—(NO_2)_3C_6H_2O^-Ag^+$) is a yellow crystalline material which is sparingly soluble in water (1:50). This compound combines the antibacterial effects of silver ion and picrate ion. It has been used in concentrations of 1 to 2% in the form of vaginal suppositories for the treatment of *Trichomonas vaginalis* and *Monilia albicans*. Extended use of this compound has caused argyria, and the picric acid may produce nephritis.

Mild Silver Protein, N.F. XIII (Argyrol®)

Mild Silver Protein N.F. is silver rendered colloidal by the presence of, or combination with, protein. It contains not less than 19.0% and not more than 23.0% of Ag (silver). It occurs as dark brown or almost black, shining scales or granules. It is odorless, frequently hygroscopic, and is affected by light. The product is freely soluble in water, but practically insoluble in alcohol. When dissolved in water it forms a colloidal solution.

Mild Silver Protein contains very little free silver ion, thereby reducing the tissue effects seen in silver nitrate preparations. Colloidal solutions of this product cause no astringent or irritant effects, allowing them to be applied to sensitive areas and mucous membranes. This low level of tissue activity further implies that the silver is not readily precipitated by chloride or tissue proteins, indicating a greater ability to penetrate tissues than solutions containing silver ion.

Solutions employed as antibacterials may range in concentration from 0.5 to 20%. The release of ionized silver tends to increase with dilution; therefore, the more dilute solutions can be more irritating, and irritation should decrease with increasing concentration of the silver protein.

Silver protein solutions are subject to the same slow ionization of silver as mentioned earlier with tissue proteins (sustained action). For this reason, the N.F. makes the following statement: *"Caution: Solutions of Mild Silver Protein should be freshly prepared or contain a suitable stabilizer."* As solutions age, they tend to increase in irritating properties, and a precipitate forms. Calcium disodium EDTA (see Chapter 1) at a concentration of 10 mg/ml (1.0%) may be used to stabilize the solutions, e.g., Argyrol® SS. It acts by chelating the silver ions released, thereby preventing the physical breakdown of the colloidal solution, and maintaining the low irritant action of the product.

USES. The concentrations of *Mild Silver Protein* and other colloidal silver preparations usually employed provide bacteriostatic action. The solutions, due to their low irritability, are used as local antibacterials on mucous membranes in the nose, throat, and conjunctiva of the eye. They have also been used to irrigate the urethra and bladder. The major effectiveness of silver protein colloidal preparations is in the treatment of gonococcal infections, i.e., gonococcal conjunctivitis. They have been used in the treatment of infections of the respiratory tract and as a prophylactic medication against respiratory infections. This use is no longer recommended, and prophylactic use may actually be harmful. Continued use of *Mild Silver Protein* may result in argyria (see Chapter 7).

Mild Silver Protein is available as crystals under the trade name of Argyrol®. The crystals are in containers for the preparation of 5, 10, and 20% solutions. EDTA-stabilized solutions, Argyrol® SS, are also available in 10 and 20% concentrations.

OTHER COLLOIDAL SILVER PREPARATIONS. *Strong Silver Protein* (Protargin) was official in the N.F. X. It contains less silver than *Mild Silver Protein* (between 7.5 and 8.5%) but the protein is partially hydrolyzed, providing a higher concentration of silver ion. The product is more irritating than the latter preparation. Its action is more bactericidal and it was used in concentrations of 0.25 to 0.5% for irrigation of the urethra and bladder in the treatment of gonorrhea just prior to the introduction of the sulfonamides and antibiotics.

Colloidal Silver Iodide (Neo-Silvol®) contains AgI with gelatin to render it colloidally stable. It contains between 18 and 22% of AgI, and has the characteristic pale yellow color of this salt. It forms a milky or opalescent suspension in water, and solutions should be freshly prepared and protected from light. Suspensions ranging from 4 to 50% have been used in a manner similar to *Mild Silver Protein*.

Silver Chloride has also been employed in colloidal solutions with sucrose as the stabilizing agent. All the colloidal silver halides provide very little silver ion and are, therefore, nonirritating. Bactericidal activity is only available in the higher concentrations (above 20%).

SILVER STAINS. Due to their unsightly nature, some mention should be made of procedures for removing silver stains from the skin and clothing. The stain is caused by the deposition of free silver, and fresh stains on the skin may be removed by painting the area with *Tincture of Iodine* which reacts according to rx (xxxvii).

$$2Ag + I_2 \rightarrow 2AgI \qquad \text{(xxxvii)}$$

After allowing time for the reaction, the excess iodine can be removed by treating with a solution of sodium thiosulfate [rx (xxxviii)].

$$2Na_2S_2O_3 + I_2 \rightarrow 2NaI + Na_2S_4O_6 \qquad \text{(xxxviii)}$$

All products can finally be removed from the area by rinsing with water. Potassium cyanide can be used to remove stains from clothing, but caution should be observed due to the toxic nature of cyanide.

Mercury Compounds. Because of the various forms of mercury available for topical use, free metal, mercurous and mercuric salts, it is appropriate to discuss some of the general properties of this metal before examining specific compounds.

MERCURY. Physically, mercury is a bright, shiny, silvery-white metal. At ordinary temperatures, it is a liquid which is easily divisible into spherical globules. It is the heaviest metal, having a density of about 13.6. While its boiling point is quite high ($357°$ C), it has a vapor pressure at $20°$ C of 0.0013 mm. Thus, in areas where a large quantity of mercury is open to the atmosphere, appreciable amounts of mercury vapor are present, and provide a possible source of chronic toxicity (see Chapter 7). Mercury readily forms amalgams with most metals.

Some of the chemical properties of mercury relate to its ability to undergo autooxidation, and to combine with oxygen, sulfur, and halogens. Although stable in air at normal temperatures, when metallic mercury is heated in air to temperatures near its boiling point it combines with oxygen to form red mercuric oxide (HgO). Mercuric iodide (HgI_2) can be formed by simply triturating the metal with solid iodine.

Metallic mercury is insoluble in many of the common acids. However, oxidizing acids will react with mercury to form mercuric salts, e.g., rx (xxxix) with nitric acid.

$$3Hg + 8HNO_3 \rightarrow 3Hg(NO_3)_2 + 2NO \uparrow + 4H_2O \qquad \text{(xxxix)}$$

Treating the mercuric nitrate with metallic mercury demonstrates auto-oxidation-reduction to produce mercurous nitrate, [rx (xl)].

$$Hg(NO_3)_2 + Hg \rightarrow Hg_2(NO_3)_2 \qquad \text{(xl)}$$

The representation of the mercurous salt indicates the presence of two Hg^+ ions per molecule. It has been shown that the mercurous ion exists in a bimetallic state, i.e., Hg_2^{+2} (see below).

Unlike the other members of its periodic group (Group IIB—Zn, Cd, and Hg), mercury forms both mono- and divalent cations. The halides of the monovalent mercurous ion are practically insoluble in water, and are decomposed by light. The hydroxides of both the mercurous and divalent mercuric ion will spontaneously lose water to form the respective oxides, Hg_2O and HgO. The oxides are very weak bases, making it possible for salts of mercury to be hydrolyzed in neutral or alkaline solution to basic salts. Mercury will form some complex ions, e.g., $Hg(CN)_4^{-2}$ and HgI_4^{-2}.

MERCUROUS COMPOUNDS. Mercurous compounds may be considered to contain two monovalent mercury ions bonded by a covalent metal-metal bond as shown in structure III.

$$\begin{array}{c} Hg^+ \\ | \\ Hg^+ \end{array}$$

III

Heterolytic cleavage of this bond would result in the formation of metallic mercury and the mercuric ion [rx (xli)].

$$Hg^+ \overset{\frown}{\quad\quad} Hg^+ \longrightarrow Hg^{+2} + Hg \qquad \text{(xli)}$$

This may explain the ease with which mercurous salts undergo autooxidation-reduction reactions to yield the mercuric salt and free mercury. The mercuric salts formed in this manner are more stable than the initial mercurous compound.

As indicated above, mercurous salts are not as water soluble as are mercuric compounds. Those salts which are soluble (e.g., the nitrate) tend to hydrolyze in aqueous solution to form insoluble basic salts [rx (xlii)].

$$Hg_2(NO_3)_2 + H_2O \leftrightharpoons Hg_2(OH)NO_3 \downarrow + HNO_3 \qquad \text{(xlii)}$$

This is an equilibrium reaction, and an excess of nitric acid will prevent the formation of the basic unit. Hydroxides will cause the precipitation of mercurous hydroxide, which in turn forms the black-brown mercurous oxide (Hg_2O). This oxide is unstable to heat and light undergoing autooxidation-reduction to mercury and mercuric oxide (HgO).

Solutions of soluble mercurous salts and ammonium hydroxide will react to form precipitates containing mixtures of mercury and mercuric ammonolysis products, e.g., $HgNH_2NO_3$.

Soluble mercurous salts will react with soluble chlorides, including hydrochloric acid, to precipitate white mercurous chloride (*Calomel*) [rx (xliii)].

$$Hg_2(NO_3)_2 + 2HCl \rightarrow Hg_2Cl_2 \downarrow + 2HNO_3 \qquad \text{(xliii)}$$

The precipitate will slowly dissolve in hot concentrated hydrochloric acid, and will turn black in the presence of ammonia T.S. Once again, this is due to the formation of mercury and the mercuric amidochloride (*Ammoniated Mercury*) according to rx (xliv).

$$Hg_2Cl_2 + 2NH_4OH \rightarrow HgNH_2Cl \downarrow + Hg \downarrow + NH_4Cl + 2H_2O \quad \text{(xliv)}$$

This is an official test for mercurous salts.

MERCURIC COMPOUNDS. Mercuric salts also will hydrolyze to form insoluble basic salts. Using a similar example, mercuric nitrate will form basic mercuric nitrate $[Hg(OH)NO_3]$ in aqueous solution. In slightly alkaline solutions, mercuric chloride forms a brown basic mercuric chloride $[Hg(OH)Cl]$. The formation of basic mercuric salts can be inhibited by the presence of excess acid.

Treatment of solutions of mercuric salts with hydroxide bases will cause the initial precipitation of mercuric hydroxide, which spontaneously decomposes to yellow mercuric oxide and water [rx (xlv)].

$$Hg^{+2} + 2OH^- \rightarrow Hg(OH)_2 \downarrow \rightarrow HgO \downarrow + H_2O \quad \text{(xlv)}$$

Ammonium hydroxide reacts differently than the metal hydroxides in that ammonolysis occurs. For example, mercuric chloride will be precipitated as the white amidochloride (*Ammoniated Mercury*) [rx (xlvi)].

$$HgCl_2 + 2NH_3 \rightarrow Hg(NH_2)Cl \downarrow + NH_4Cl \quad \text{(xlvi)}$$

The pharmacological actions of mercury compounds are to a large extent related to the previously mentioned affinity of mercury for sulfur. Among the earliest antiseptics available, the action of mercury on microorganisms and body tissues is due primarily to the mercuric ion, which will react with many polar groups on the amino acids of proteins, but has a particular affinity for sulfhydryl (—SH) groups. This is the basis for the treatment of mercury poisoning with dimercaprol or penicillamine (see Chapter 1). The mercuric ion reacts with the cysteine sulfhydryl groups on the protein of enzymes to form mercaptides which may be represented by any of the following general structures.

Prot—S—Hg–X   Prot—S—Hg—S—Prot

X=Cl, OH, etc.

$$\text{Prot} \begin{array}{c} S \\ \diagup \diagdown \\ \diagdown \diagup \\ S \end{array} Hg$$

IV   V   VI

The antimicrobial actions of mercury compounds are best classified as bacteriostatic. This is based on the inhibition of bacterial sulfhydryl-containing enzymes by the metal ions, leading to an accompanying inhibition in metabolic function and growth. Removal of the mercuric ion from the bacterial protein results in a reactivation of the previously inhibited enzymes and a resumption of metabolism and multiplication of the organisms. This appears to be the case with all bacteria, susceptible viruses, and bacterial spores.

The protein precipitant action of mercuric ion on human tissue gives rise to a great deal of irritation. In fact, the slow release of mercuric ion from mercurous chloride (*Calomel*) is most likely the reason for the cathartic action of this compound (see Chapter 8), since the mercury(II) ion produces an irritant effect on the intestinal mucosa.

Other actions of mercury compounds include a diuretic effect, which is again related to the ability of the mercuric ion to inhibit sulfhydryl-containing enzymes. This action in the proximal kidney tubule results in an inhibition of sodium reabsorption, leading to a sodium and water diuresis (see Chapter 7).

Metallic mercury (Hg; At. Wt. 200.59) was official in the N.F. XI (hydrargyrum, quick-silver) as a standard for the metal used in the *Mild Mercurial Ointment*, N.F. XI (Blue Ointment). This ointment contained 10% mercury, and was formerly used to treat secondary lesions produced on the skin of patients infected with syphilis. This therapy has largely been replaced by antibiotics (e.g., penicillin), and to some extent by the organic arsenic compounds. The ointment has also been employed as a fungicide and parasiticide, particularly in the treatment of infestations of body and crab lice. The antimicrobial and parasiticidal uses of mercury and its salts have declined greatly during this century.

Frequent or prolonged contact of mercury with the skin should be avoided. Mercury spills in the laboratory can usually be handled by special vacuum cleaners. Household spills from broken thermometers, etc., should be handled by sweeping the droplets of mercury together with the edges of a sheet of paper, rolling the large droplet onto the paper, and pouring it into a small container that can be closed, e.g., bottle, box, envelope, etc. The mercury in the sealed container can then be discarded.

Yellow Mercuric Oxide, N.F. XIII (HgO; Mol. Wt. 216.59; Yellow Precipitate)

Yellow Mercuric Oxide N.F. is a yellow to orange-yellow, heavy, impalpable powder which is odorless and stable in air. It is practically insoluble in water and alcohol, but is readily soluble in diluted hydrochloric and nitric acids, forming colorless solutions. On exposure to light, the red form of mercuric oxide slowly develops.

The red oxide appears as heavy, orange-red crystalline scales or as a crystalline powder. When finely divided, it acquires a yellow color. Thus, the only apparent difference between yellow and red mercuric oxides is the state of subdivision, the yellow being the more finely divided material. Red mercuric oxide is no longer official. When either oxide is heated to a red heat, it will decompose to metallic mercury and oxygen.

Uses. *Yellow Mercuric Oxide*, due to its insolubility, slowly releases Hg^{+2}, providing a mild sustained antibacterial effect. This action is made use of in *Yellow Mercuric Oxide Ophthalmic Ointment*, N.F. XIII, which contains between 0.9 and 1.1% of HgO in *White Ointment*. Due to the destructive action of mercury compounds on metals, the N.F. makes the following statement: *"Caution: During its manufacture and storage, Yellow Mercuric Oxide Ophthalmic Ointment must not come into contact with metallic utensils or containers except those made of stainless steel, tin, or tin-coated material."* This caution is necessary to avoid the formation of free mercury, and the development of small metallic particles that may be damaging to the eye.

The *Ointment* is used in the treatment of mild infections and inflammations of the eye including conjunctivitis. It has also been used for non-ophthalmic treatments of fungal infections of the skin and infestations of body lice.

The ointment is available from various manufacturers under the generic (official) name.

Ammoniated Mercury U.S.P., XVIII ($Hg(NH_2)Cl$; Mol. Wt. 252.07; White Precipitate)

Ammoniated Mercury U.S.P., in white, pulverulent (easily powdered) pieces or as a white, amorphous powder, is odorless and stable in air, but darkens on exposure to light. It is insoluble in water and alcohol, but readily dissolves in warm hydrochloric, nitric, and acetic acids.

Although insoluble, prolonged contact with water will cause the formation of a yellow basic compound having the formula $H_2N(HgO)HgCl$. This necessitates the use of oil in lotions containing this compound.

Uses. This is another insoluble source of Hg^{+2} ion. It is mildly antiseptic and is used primarily in ointments. *Ammoniated Mercury Ointment*, U.S.P. XVIII contains between 4.5 and 5.5% of *Ammoniated Mercury*. This ointment has been used in the treatment of impetigo contagiosa and fungal infections of the skin (dermatomycoses). It has also found some use in the treatment of crab louse infestations. Its use in lessening the scaling of psoriasis is not well justified in light of the availability of more efficacious products. Overuse of ammoniated mercury products has produced chronic toxicities; therefore, applications to large areas for prolonged periods of time are certainly not recommended.

In addition to the 5% ointment, there is also a 3% ophthalmic ointment available.

Mercuric Chloride, ($HgCl_2$; Mol. Wt. 271.50; Mercury Bichloride, N.F. XII, Corrosive Sublimate)

Mercuric chloride occurs as heavy, white, rhombic prisms, large crystalline masses, or as a white powder. It is odorless. Although weakly ionized, it is soluble in water and alcohol. When the salt is heated, it melts at 277° C to provide a liquid which boils at 300° C, giving off dense white vapors of $HgCl_2$.

Dissolving mercuric chloride in water does not provide a simple solution. The salt forms an autocomplex as shown in rx (xlvii).

$$2HgCl_2 + 2H_2O \rightarrow (HgOH)^+ + (HgCl_4)^{-2} + H_3O^+ \qquad \text{(xlvii)}$$

As the reaction indicates, the hydrolysis of the mercuric ion provides a slightly acidic solution. The right-hand side of rx (xxxix) contains free chloride ion (not shown) due to the dissociation of the tetrachloro complex ion. The acidity of the solution can be reduced to neutral by the addition of sodium chloride because the excess chloride ion produces a common ion effect, forcing the equilibrium to the left and suppressing the ionization of the mercuric chloride.

USES. Mercuric chloride, although only slightly ionized in solution, is sufficiently reactive with certain substances (e.g., proteins) to provide appreciable amounts of mercuric ion. For this reason, mercuric chloride solutions are quite irritating when applied to the skin, and should never be used as an antiseptic for wounds or abraded skin.

Its primary use is in disinfectant solutions (1:1,000) for utensils and surgical instruments. Disadvantages associated with this use are the corrosive action of mercuric ion on metals, and the fact that rubber will deteriorate from constant exposure to mercuric chloride. As indicated above, the addition of sodium chloride to the solution suppresses the ionization, and therefore seems to diminish the corrosive action on metals.

A 1:2,000 solution has been employed infrequently as a surgical "scrub" (handwash). It has also been used in 0.1% concentrations in ointments for dermatomycoses. It is difficult to recommend a product containing mercuric chloride for anything other than rare use as a disinfectant.

Because of the extremely poisonous nature of mercuric chloride, the N.F. XII attached a statement to the monograph on *Mercury Bichloride Large Poison Tablets* (used for the preparation of solutions) to the effect that the tablets had to be of a distinctive color, irregular shape, and dispensed in distinctive containers. A "poison" label must also be attached to the container. This precaution was necessary to avoid confusion between mercuric and mercurous chlorides, which were usually side-by-side on the shelf. Tablets such as these are the most common form of mercuric chloride available.

It is interesting to note that recent evidence[14] has shown that certain strains of *E. coli* have a transferable resistance factor to mercuric chloride. The organisms are capable of metabolizing the compound and giving off mercury vapor.

Sulfur and Sulfur Compounds. Elemental sulfur exists in a variety of allotropic forms, including several solid and two liquid states (see Table

TABLE 9-1. ALLOTROPIC FORMS AND PROPERTIES OF SULFUR

Form	Designation	Preparation	Properties
Rhombic	α-Sulfur, S_α (*Sublimed Sulfur*, N.F. or "Flowers of Sulfur" is this form.)	Evaporation of CS_2 solvent from a solution containing sulfur	Pale yellow crystals M.pt. 112.8° C→$S\lambda$ (liquid) Density 2.07 Insol.: water, alcohol Sol.: carbon disulfide Faint odor and taste
Monoclinic	β-Sulfur, S_β	Needle-like crystals found on the walls of the vessel after decanting partially cooled molten sulfur	Nearly colorless crystals M.pt., 119.3° C→$S\lambda$ (liquid) Density 1.957 Insol.: water, alcohol Sol.: carbon disulfide $S_\beta \rightleftarrows S_\alpha$ at 96° C
Liquid	λ-Sulfur, $S\lambda$	Heating sulfur to 160° C	Pale yellow mobile liquid
	μ-Sulfur, S_μ	Heating λ-sulfur to 180° C	Dark brown viscous liquid Slow cooling→$S\lambda$ Rapid cooling→"plastic" sulfur
"Plastic" Sulfur	—	Supercooling μ-sulfur	An elastic substance that becomes hard and brittle upon standing for several days. The hard material contains: 70% rhombic 30% amorphous
Amorphous	—	Aging of "plastic" sulfur	A yellow amorphous solid which is nearly insoluble in any solvent

9–1). These forms and their properties are summarized in Table 9–1. Generally, sulfur in the solid and S_λ-liquid state is considered to exist as rings of eight sulfur atoms, S_8. In the liquid form S_μ, the S_8 rings are apparently broken down to form filaments which become entangled and increase the viscosity of the molten form. Since slow cooling will convert S_μ back to S_λ, these two forms are considered to be in equilibrium in molten sulfur; rapid cooling of the S_μ liquid produces the "plastic" or elastic substance that hardens on standing to yield a mixture of rhombic and amorphous sulfur (see Table 9–1). Amorphous sulfur represents another solid form, differing from the others by virtue of its insolubility in any solvent.

Chemically, sulfur is a very active element, reacting with metals and non-metals to form sulfides, e.g., Na_2S, CS_2, and other compounds containing the divalent S^{-2} ion. In the presence of oxidizing agents and water, sulfuric acid (H_2SO_4) is formed. When burned in an oxygen atmosphere, it evolves the suffocating fumes of sulfur dioxide (SO_2). This gas is found in the stack smoke from burning high sulfur-containing fuels. Attempts to remove it involve aqueous oxidation in "scrubbers" to form sulfuric acid.

Sulfur reacts with boiling solutions of metal hydroxides to form mixtures of metal sulfides and thiosulfates. For example, rx (xlviii) illustrates the reaction with calcium hydroxide.

$$3Ca(OH)_2 + 12S \rightarrow 2CaS_5 + CaS_2O_3 + 3H_2O \qquad \text{(xlviii)}$$

The products remain in solution, but the sulfur can be precipitated by the addition of acid, such as HCl. This procedure is employed in the preparation of *Precipitated Sulfur*, U.S.P. (see below).

Sulfur has been used therapeutically for centuries both internally (cathartic, see Chapter 8) and topically. Elemental sulfur is a very poor antiseptic. If there is any antibacterial action at all, it is due to oxidation-reduction products of the element. Antibacterial action has been attributed to the formation of sulfides and pentathionic acid ($H_2S_5O_6$) by bacteria on the surface of the skin. Most pathogenic bacteria are not harmed by sulfides, leaving the pentathionic acid (or related compounds) as the only other possible antiseptic agent.

The primary topical uses of products containing elemental sulfur are as a fungicide, parasiticide, and keratolytic agent (destruction of keratin in the skin). It is also used in the form of sulfur "candles" which, when burned, give off sulfur dioxide for fumigation or insecticide purposes. Sulfides have also been used as depilatories, due to the ability of sulfides in highly alkaline solution (pH 10) to reduce the disulfide linkage in the amino acid cystine in hair. The reduction and softening of the hair aid in its removal. Calcium, strontium, and barium sulfide have been used in the past for this purpose, but they have been largely displaced by calcium salts of thioglycolic acid.

Sublimed Sulfur, N.F. XIII (S; At. Wt. 32.06; Flowers of Sulfur)

Sublimed Sulfur N.F. is obtained by condensing the sulfur vapors produced by heating any form of sulfur. It is a fine, yellow, crystalline (rhombic) powder having a faint odor and taste. It is practically insoluble in water and alcohol, and sparingly soluble in olive oil. One gram of sublimed sulfur dissolves slowly and usually incompletely in about 2 ml of carbon disulfide.

Precipitated Sulfur, U.S.P. XVIII (Milk of Sulfur)

Precipitated Sulfur U.S.P. is prepared according to previously described reactions [see rx (xl)]. It is a very fine, pale yellow, amorphous or microcrystalline (rhombic) powder, without odor or taste. It is practically insoluble in water, very slightly soluble in alcohol, and slightly soluble in olive oil. Its solubility in carbon disulfide is the same as *Sublimed Sulfur*.

Uses. Both of these forms of sulfur are categorized by the U.S.P. and N.F. as scabicides. *Precipitated Sulfur* is used in *Sulfur Ointment*, U.S.P. XVIII, at a concentration between 9.5 and 10.5%. This and other sulfur ointments have been used in the treatment of scabies, in which it kills the mite *Sarcoptes scabiei*. The female mite lays eggs under the skin, causing the formation of multiform lesions and intense itching. The sulfur is effective against the live parasite, but has no effect on the eggs. Other scabicides (e.g., benzyl benzoate) have largely replaced sulfur-containing products for this use. Sulfur ointments may still be employed as fungicides in the treatment of superficial fungal infections.

Sulfur is used primarily as a dermatological agent in the treatment of seborrhea (an abnormal secretion of sebum from the sebaceous glands, giving an oily or scaly appearance to the skin), acne, psoriasis, etc. Its use in these cases is as a keratolytic agent, and it may be used alone or, usually, in combination with other keratolytics (e.g., salicylic acid). The activity of sulfur is one of loosening and softening horny elements and the scaly formation of sebum and removing them from the site. This aids in overcoming congestion around the hair follicles and sebaceous glands. Although there is no question that sulfur accomplishes this, it becomes primarily a symptomatic treatment for many dermatologic problems. It should be noted that questions have recently been raised concerning whether or not sulfur actually stimulates the formation of comedones (the collection of sebaceous material which forms the primary lesion in acne vulgaris), thereby promoting the continuation of acne rather than providing aid to its treatment.[15] Suffice it to say that sulfur is used as a keratolytic in ointments in concentrations as high as 20% as well as in lotions. Its effects are synergistic with other keratolytic agents. Some patients may show hypersensitivity to sulfur and should be observed for allergic manifestations.

Sulfur Dioxide, U.S.P. XVIII (SO_2; Mol. Wt. 64.06)

Sulfur Dioxide U.S.P. was discussed under Antioxidants (see Chapter 4) since this is its major area of use. As mentioned previously, it is evolved in the burning of sulfur candles and provides insecticidal action as a household fumigant.

Sulfurated Potash, N.F. XIII (Liver of Sulfur)

Sulfurated Potash N.F. is a mixture composed chiefly of potassium polysulfides and potassium thiosulfate. It contains not less than 12.8% of S (sulfur) in combination as sulfide. It occurs as irregular, liver-brown pieces when freshly made, changing to a greenish yellow color. It has an odor of hydrogen sulfide and a bitter, acrid, and alkaline taste. It decomposes on exposure to air. It is freely soluble in water, usually leaving a slight residue. Alcohol will dissolve only the sulfides. A 1:10 aqueous solution is light brown in color and alkaline to litmus.

This material is prepared by heating potassium carbonate and sulfur to a temperature no higher than 185° C [rx (xliv)].

$$3K_2CO_3 + 8S \xrightarrow{185°C} 2K_2S_3 + K_2S_2O_3 + 3CO_2 \uparrow \qquad \text{(xlix)}$$

When effervescence ceases the heat is increased until perfect fusion results. The melt is cooled by pouring onto a stone slab and covering to protect it from air. The resulting solid is broken into pieces and bottled immediately. The mixture actually consists of potassium polysulfides—K_2S_3, K_2S_4, K_2S_5, and $K_2S_2O_3$.

USES. *Sulfurated Potash* is official as a pharmaceutical aid as a source of sulfide. It is used in the preparation of *White Lotion*, N.F. XIII (see below). It has been used alone rather infrequently in lotions and ointments as a parasiticide (scabicide), and in the treatment of acne and psoriasis.

Selenium Sulfide, N.F. XIII (SeS_2; Mol. Wt. 143.09; Selenium Disulfide)

Selenium Sulfide N.F. contains not less than 52.0% and more than 55.5% of Se (selenium). It is a bright orange powder having no more than a faint odor. It is practically insoluble in water and organic solvents.

While selenium is toxic in large doses, it also appears to be an important nutrient in trace quantities. In large doses, selenium salts will cause irritation of the gastrointestinal mucosa, the appearance of fat deposits in the liver, and will damage the renal tubules. It may also impart damage to the blood-forming organs, i.e., spleen and bone marrow. Selenium is not well absorbed through the skin; therefore, topical application to limited areas of unbroken, unirritated skin will not usually result in selenium toxicity.

Selenium's periodic relationship to sulfur led to its suggestion as a topical agent over 20 years ago. It has since been employed in 2.5% suspensions in the treatment of seborrhea dermatitis (dandruff). When applied properly, selenium sulfide does not cause any severe problems. However, the preparation should be kept away from the conjunctiva of the eye, and prolonged contact with the skin will produce contact dermatitis. The compound is evidently not acting as a simple keratolytic. Recent data have shown[16] that selenium has a cytostatic action, reducing the cell turnover of corneocytes (epidermal cells around the base of hair shaft and around the follicle) in patients with and without dandruff.

USES. *Selenium Sulfide* is used in shampoos in concentrations of 1 to 2.5% (*Selenium Sulfide Detergent Suspension*, N.F. XIII, Selsun®) as an antiseborrheic. The effectiveness of the 2.5% concentration has been repeatedly demonstrated; however the same cannot be said for the recently introduced 1% product for nonprescription use. Insufficient studies have been performed to date to determine if this concentration is sufficient enough to be efficacious. The normal method of application is topically to the scalp, where it is allowed to remain for five minutes and then thoroughly washed off. Because of the danger of introduction into the eyes or mouth, the hands should be thoroughly washed and the fingernails meticulously cleaned after using selenium sulfide.

The product is available under the name of Selsun® Suspension in 1 and 2.5% concentrations.

Cadmium Sulfide, (CdS; Mol. Wt. 144.46)

Cadmium is a bivalent metal whose properties are most closely related to zinc (see below). The metal ion is a protein precipitant with a toxicity similar to mercury and greater than zinc.

Cadmium Sulfide is a light yellow or orange colored powder which is insoluble in water. The lack of cadmium ion contributes to its relatively non-irritating action on the skin. It is used in a manner similar to selenium sulfide in the treatment of seborrhea dermatitis of the scalp (seborrhea capitis). Although less irritating than selenium sulfide, it has the same precautions concerning its use. It may be less likely to produce excess oiliness of the hair and scalp.

It is applied in the form of a 1% suspension available under the trade name of Capsebon®.

Boric Acid and Sodium Borate

Boric Acid, N.F. XIII (H_3BO_3; Mol. Wt. 61.83)

The description and chemistry of *Boric Acid* were given in Chapter 4 and will not be repeated here. The compound is official as a buffer.

For many years, *Boric Acid* has been used in solutions, ointments, and dusting powders as an antiseptic. At best, the compound can be described as a weak bacteriostatic agent. The compound is nonirritating when applied to the intact skin and mucous membranes. It has therefore been used in ophthalmic preparations as a buffer and in a saturated solution (about 4.5%) as a bacteriostatic eye wash. The low level of activity and possible toxicity associated with boric acid have brought many requests for its deletion from clinical use.[17]

Boric acid is not absorbed through the intact skin, but it is highly toxic when ingested orally. It is absorbed when applied over large areas of broken skin. Toxic manifestations are first seen in the gastrointestinal tract regardless of the route of administration. These include nausea, vomiting, and diarrhea, followed by the formation of a severe rash (boiled lobster appearance) and a shedding of the layers of skin and mucous membranes. Other symptoms include general weakness and headache. Renal damage and circulatory collapse are the cause of death. Toxicity and death have been produced from accidental ingestion and application of boric acid-containing products (particularly ointments) to large denuded areas. This has most frequently occurred with infants where boric acid products were being used in the treatment of diaper rash, but also occurred during World War II when boric acid ointment was initially used for burn therapy.

USES. In addition to the uses involved with providing acidic media and buffered media for other drugs (see Chapter 4), *Boric Acid* will still be found in the form of solutions in concentrations from 2.5 to 4.5% for use as an eye wash. The more concentrated solution should be diluted with an equal volume of water before use. Also, the higher concentration is near saturation and boric acid tends to crystallize out with a slight drop in temperature. This represents a danger and solutions should be warmed or diluted before use in order to dissolve the crystals.

Boric Acid is also available in ointments containing about 5% of the compound (Borofax®) for use as an emollient antiseptic ointment in the treatment of diaper rash. Prolonged use and application to areas of broken skin should be avoided. Dusting powders containing boric acid are also available.

It appears that the low antimicrobial efficacy of boric acid, coupled with its high toxicity when misused (as it frequently is), is reasonable support for its elimination from topical preparations.

Sodium Borate, U.S.P. XVIII ($Na_2B_4O_7 \cdot 10H_2O$; Mol. Wt. 381.37; Sodium Tetraborate, Sodium Pyroborate, Borax)

Sodium Borate U.S.P. occurs as colorless, transparent crystals or as a white crystalline powder. The compound is odorless and effloresces in warm dry air, often leaving the crystals coated with a white powder. It is soluble in water and glycerin and insoluble in alcohol. Its aqueous solutions are alkaline to phenolphthalein T.S.

Most of the chemistry of sodium borate was discussed in Chapter 4. It is used or formed *in situ* as the alkaline member of the borate buffer system, which is the basis for its recognition by the U.S.P. Of interest in this connection is the fact that metal borates, with the exception of the alkali metal (Na, K, etc.) borates, are insoluble in water. A possible incompatibility is found when mixing solutions containing *Sodium Borate* with those containing soluble zinc salts. The zinc may be precipitated as the insoluble basic zinc borate or as zinc hydroxide due to the alkalinity of the sodium borate solution. This problem is avoided by use of boric acid because the acidic zinc borate is water soluble.

Uses. *Sodium Borate* has the same toxicity as boric acid; in fact, it is hydrolyzed to boric acid in aqueous solution [rxs (1)].

$$Na_2B_4O_7 + 3H_2O \rightarrow 2NaBO_2 + 2H_3BO_3 \qquad (1\text{-a})$$

$$NaBO_2 + 2H_2O \rightarrow NaOH + H_3BO_3 \qquad (1\text{-b})$$

It has been used externally in solutions containing 1 to 2% of the compound, as both an eye wash and as a wet dressing for wounds. The latter use is contraindicated due to the toxicity of borates.

It was present as an antibacterial in two products which are no longer official: N.F. *Mouth Wash* and *Compound Sodium Borate Solution*. The active ingredient in both these solutions was the potassium and sodium salt, respectively, of glyceroboric acid (see Chapter 4). As with *Boric Acid*, *Sodium Borate* is weakly bacteriostatic.

Arsenic Compounds. Although arsenic compounds are not used topically as are the other compounds discussed in this chapter, they do interact with the sulfhydryl groups of cysteine and other sulfhydryl-containing compounds (e.g., glutathione and thioglycolic acid) and cause disruptions in cell metabolism. These compounds may therefore be classified as protoplasmic poisons. This term distinguishes them from protein precipitants which will interact with many other polar groups on protein molecules to cause generalized precipitation of albumin and other nonsulfhydryl-containing proteins.

Elemental arsenic (As; At. Wt. 74.92) is amphoteric in terms of its metallic and nonmetallic properties; however, its action on cells is somewhat reminiscent of mercury and it is classified biologically as a heavy metal. The chemistry of arsenic is represented by two types of compounds: those containing trivalent arsenic (arsenous, oxidation state +3) and those containing pentavalent arsenic (arsenic, oxidation state +5).

TRIVALENT ARSENIC. The most common forms of arsenic exist in the trivalent state. In combination with halides, e.g., arsenic trichloride ($AsCl_3$), it is present as a cation, a fact which is in agreement with its metallic character. Treatment of the halides with water will cause the formation of orthoarsenous acid [rx (1i)].

$$AsCl_3 + 3H_2O \rightleftarrows 3HCl + H_3AsO_3 \qquad (1i)$$

Like orthophosphoric acid, this is a trihydroxy derivative of arsenic, and can be dehydrated to pyroarsenous acid [$(OH)_2As\!-\!O\!-\!As(OH)_2$ or $H_4As_2O_5$] through intermolecular dehydration, and to metarsenous acid

(HO—As=O or $HAsO_2$) through intramolecular dehydration. Arsenic trioxide (As_2O_3; white arsenic) may be considered as a dehydration product of pyroarsenous acid. The oxide has the trivalent structure shown in VII. Actually, the compound exists as a dimer (As_4O_6).

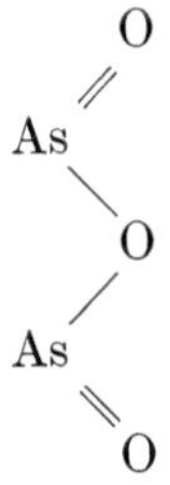

$$
\begin{array}{c}
O \\
\diagup\diagup \\
\text{As} \\
\diagdown \\
O \\
\diagup \\
\text{As} \\
\diagdown\diagdown \\
O
\end{array}
$$

VII

PENTAVALENT ARSENIC. Arsenic in this valence state can be found in the form of acids, an oxide, and a sulfide. Orthoarsenic acid (H_3AsO_4) is shown in structure VIII and arsenic pentoxide (As_2O_5) in structure IX. The latter compound, like arsenic trioxide, actually exists as a dimer (As_4O_{10}).

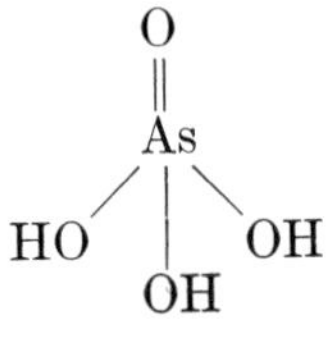

$$
\begin{array}{cc}
\begin{array}{c}
O \\
\| \\
\text{As} \\
\diagup \ | \ \diagdown \\
\text{HO} \quad | \quad \text{OH} \\
\text{OH}
\end{array}
&
\begin{array}{c}
O \\
\| \\
O{=}\text{As} \\
\diagdown \\
O \\
\diagup \\
O{=}\text{As} \\
\| \\
O
\end{array} \\
\text{VIII} & \text{IX}
\end{array}
$$

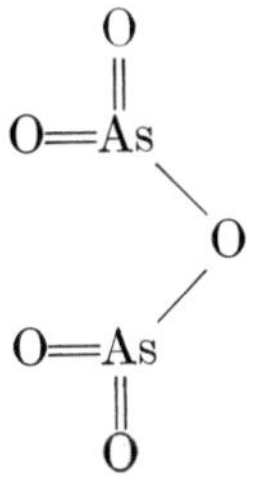

Orthoarsenic acid will undergo intermolecular dehydration to pyroarsenic acid [$(OH)_2\overset{\overset{\displaystyle O}{\|}}{As}$—O—$\overset{\overset{\displaystyle O}{\|}}{As}(OH)_2$ or $H_4As_2O_6$].

Certain aspects of the pharmacology and toxicity of arsenicals are discussed in Chapter 7. As mentioned above, arsenicals are protoplasmic poisons by virtue of their reaction with sulfhydryl-containing enzymes to block or to destroy important metabolic pathways. Structure X illustrates the type of covalent interaction possible between two sulfhydryl-containing enzymes and a compound containing trivalent arsenic.

Enz—S
\
As—R
/
Enz—S

X

Single applications of arsenic compounds to the skin produce only slight irritation. In general, there is no useful therapeutic action of topical arsenicals. However, continued exposure of the skin to these compounds can produce necrosis of the tissue and contact dermatitis.

From ancient times, arsenic in various forms has been used as a poison. In more recent history, inorganic arsenic, because of its repressive action on bone marrow, has been used in the treatment of leukemia (*Arsenic Trioxide*, N.F. XI, *Arsenous Acid, White Arsenic*) and psoriasis [*Sodium Cacodylate*, N.F. X; $(CH_3)_2AsOONa \cdot 3H_2O$]. In the case of inorganic arsenicals, mammalian cells do not distinguish significantly between trivalent and pentavalent forms. Inorganic arsenicals are no longer generally used in present-day therapy. Certain pesticides and herbicides contain arsenic, primarily in inorganic combination, e.g., arsenic trioxide, disodium orthoarsenate (*Exsiccated Sodium Arsenate*, N.F. VIII; Na_2HAsO_4), and lead arsenate ($PbHAsO_4$).

Organic arsenicals are virtually the only derivatives presently used for the therapeutic effects of arsenic. These compounds are primarily employed in the treatment of certain parasitic diseases produced by protozoa and helminths (worms or worm-like parasites). There appears to be more selectivity between host and parasite cells with regard to the valence state of arsenic in organic combination. Trivalent arsenic penetrates host (mammalian) cells to a greater extent than pentavalent organic arsenicals; therefore, the latter derivatives should be less toxic to the patient undergoing arsenical therapy.

Parasite selectivity depends upon the organism involved. Trivalent forms are more effective against the treponema organism that produces syphilis. However, organic arsenicals have been replaced by antibiotics in the treatment of this disease. Pentavalent forms are useful as veterinary anthelmintics and in the treatment of amebiasis and trypanosomiasis (protozoal infections). It has usually been thought that pentavalent arsenic had to be reduced *in vivo* to the trivalent state, which was then the active form of the organic arsenical for the parasite. Recent evidence[18] indicates that this reduction may not occur to any great extent, and that the pentavalent form may be the valence state of the active compound for certain organisms.

Although the more important therapeutic arsenicals are organic, the impact of the therapy is due to the arsenic. The organic radical is involved in distributing the compound to the site of action, reducing host cell toxicity, and improving parasite specificity. For the purpose of illustrating some of the arsenicals in use, a few of these organic derivatives are listed below.

Carbarsone, N.F. XIII (N-Carbamoylarsanilic acid)

This compound is also official in the N.F. as *Tablets* and *Capsules* containing 250 mg of the drug. It is used in doses of 250 mg two to three times a day in the treatment of intestinal amebiasis. Suppositories of *Carbarsone* are also available for vaginal use in the treatment of *Trichomonas vaginalis*.

Glycobiarsol, N.F. XIII [(Hydrogen N-glycoloylarsanilato)oxobismuth; Milibis®]

Although this compound contains bismuth in the form of the bismuthyl (BiO⁺) group, its use is based on the activity of the pentavalent arsenic. It is used in oral doses (*Glycobiarsol Tablets*, N.F. XIII) of 500 mg three times a day for seven to ten days for the treatment of intestinal amebiasis.

Tryparsamide [Sodium N-(Carbamoylmethyl)arsanilate]

Tryparsamide and *Sterile Tryparsamide* were official in the U.S.P. XVII for the treatment of trypanosomiasis, administered intravenously in doses of 1 to 3 g at weekly intervals up to a total dose of 30 to 45 g. This drug has been largely replaced by a trivalent arsenical (*Melarsoprol*) which has greater efficacy and lower toxicity.

Antimony Compounds. As is the case with arsenicals, compounds containing antimony are not employed as general protein precipitants or topical agents on the skin but are used for their specific interactions with protein.

Antimony (Sb; At. Wt. 121.75, Stibium) was known to man some 2000 years before arsenic was known. The element is below arsenic in the periodic table, indicating that its properties are metallic. Its chemical as well as biological properties are similar to those discussed above for arsenic. Like arsenic, antimony forms trivalent (Sb^{+3}, antimonous) and pentavalent (Sb^{+5}, antimonic) compounds.

Trivalent inorganic compounds include acids, the oxide, and the sulfide. Antimonous acid (H_3SbO_3) is a trihydroxy derivative which will dehydrate intramolecularly to form meta-antimonous acid ($HSbO_2$), finally yielding antimony trioxide (Sb_2O_3) by an intermolecular route. Similar reactions may be noted for pentavalent antimony. Metallic character diminishes with increased oxidation state—trivalent antimony behaves in an amphoteric manner, and pentavalent antimony acts almost exclusively as a nonmetal.

The general pharmacology and toxicity of antimony are mentioned in Chapter 7. Compounds containing antimony are less readily absorbed and produce more local irritation upon topical application to the skin. The trivalent compounds, at least, have specific interactions with sulfhydryl-containing enzymes, although this has not been established for pentavalent compounds. No inorganic antimony derivatives are used in medicine. Some pentavalent organoantimony compounds are used in therapy as antiprotozoal agents, but most of the important derivatives are trivalent.

Due to the greater irritant properties of antimony, compounds containing the metal have been used as emetics, utilizing the irritant effect on the gastric mucosa which stimulates immediate vomiting. This use, as well as uses as expectorants and nauseants, has been abandoned in modern therapeutics. The primary use of organoantimony derivatives is in the treatment of certain helminthic infections (schistosomiasis and filariasis) and infections produced by protozoa, particularly leishmaniasis. Pentavalent compounds are more active against protozoa while trivalent compounds are more useful against helminths. Some of the more important organic derivatives are given below.

Antimony Potassium Tartrate, U.S.P. XVIII [KOOC—CHOH—CHOH—COO(SbO)·$\frac{1}{2}H_2O$; Tartar Emetic]

This compound was formerly used as an emetic and an expectorant (*Brown Mixture*, N.F. XII). Its official category and only rational use is as a treatment for schistosomiasis. It is still the drug of choice in infections produced by *Schistosoma japonicum*, but better agents are available for other species. It may be given orally but intravenous doses are more effective. Doses are given as a 0.5% solution on alternate days, starting with 40 mg and increasing to 140 mg by 20-mg increments. The usual therapeutic course will provide a total dose of 2 g.

Stibophen, U.S.P. XVIII [Pentasodium bis-4,5-dihydroxy-*m*-benzene-disulfonato(4-) antimonate(5-)]

$$Na^+ {}^-O_3S \quad \text{(structure)} \quad SO_3^- Na^+ \quad \cdot 7H_2O$$

This compound is also used in the treatment of schistosomiasis. It is more effective than *Antimony Potassium Tartrate* against the Schistosomal species *haematobium* and *mansoni*, but less effective against *S. japonicum*. Several dosage schedules are employed, but the usual one involves intramuscular injection (*Stibophen Injection*, U.S.P. XVIII) of 100 mg on the first day followed by 300 mg every other day to a total dose of 2.5 to 4.6 g.

Astringents

The concept of astringency was discussed earlier in this chapter. For the purpose of understanding the compounds to follow, an appropriate definition of an astringent should be stated here. Astringents are protein precipitants of limited penetrative power. That is, they are able to coagulate protein primarily on the surface of cells, an action that does not result in the death of the cell. The general constriction of tissue, e.g., small blood vessels (smooth muscle) then occurs under the influence of the astringent, but the action is controlled by virtue of being topical, lacking deeper effects. Of course, many of the compounds considered in the previous section have astringent action, but the effect can penetrate to the point of tissue damage. The compounds to be discussed in this section also have mild antimicrobial actions, and are sometimes used for this effect. However, they are also capable of being used safely for their effects on human protein.

Some of the uses for astringent compounds include: (1) styptic, to stop bleeding from small cuts by promoting coagulation of blood and constricting small capillaries; (2) antiperspirant, to decrease secretion of perspiration by constricting pores at the surface of the skin; (3) restriction of the supply of blood to the surface of mucous membranes as a means of reducing inflammation; and (4) direct actions on skin to remove unwanted tissue. The latter use requires a higher concentration or a stronger protein precipitant, sometimes termed a corrosive. The internal use of astringents as antidiarrhetics will not be considered here (see Chapter 8). Most topical astringents are salts of aluminum, zinc, and to some extent, zirconium.

Astringent Products

Aluminum Compounds

Aluminum Chloride, N.F. XIII ($AlCl_3 \cdot 6H_2O$; Mol. Wt. 241.43)
Aluminum Chloride N.F. is a white or yellowish white, deliquescent, crystalline powder. It is nearly odorless and has a sweet, very astringent

taste. It is very soluble in water, freely soluble in alcohol, and soluble in glycerin. Its aqueous solutions are acid to litmus.

Aluminum Chloride is a Lewis acid (see Chapter 4); thus, it will accept a share in an electron pair from a base to form an adduct.

In aqueous solution, the hexahydrate may be more correctly represented as a complex ion involving the six waters of hydration in the coordination sphere around the aluminum ion: $Al(H_2O)6^{+3}3Cl^-$. Hydrolysis of this salt gives rise to an acidic solution according to rx (lii).

$$[Al(H_2O)_6]^{+3} 3Cl^- + H_2O \rightleftharpoons [Al(OH)(H_2O)_5]^{+2} 2Cl^- + H_3O^+ + Cl^- \quad \text{(lii)}$$

The hydrochloric acid formed is irritating to tissues. Adjustment of the pH to the neutral or alkaline range will reduce the irritation, but it will also reduce the concentration of aluminum ion by precipitating it as insoluble aluminum hydroxide [$Al(OH)_3$ or $Al(OH)_3(H_2O)_3$]. Thus, the astringent action will be reduced, since it is a property of the free aluminum ion.

USES. *Aluminum Chloride* is a local external astringent and mild antiseptic. It is used in aqueous solution in concentrations ranging from 10 to 25%. The solutions may be too irritating for sensitive tissues. The compound was initially employed as an antiperspirant, but was found to be too irritating and also damaged clothing due in part to the formation of HCl by hydrolysis [see rx (lii)].

Aluminum Hydroxychloride (Aluminum Chlorhydroxide, Aluminum Chlorhydrate, Aluminum Chlorhydrol)

This name actually applies to two possible compounds: the monohydroxychloride [$Al(OH)(H_2O)_5$]$^{+2}2Cl^-$, and the dihydroxychloride [$Al(OH)_2(H_2O)_4$]$^+Cl^-$. In aqueous solution, either one of these will produce a pH higher than aluminum chloride, with the dihydroxy compound existing at a higher pH than the monohydroxy compound. It should be recalled, however, that the solutions will still be acidic since a neutral pH would only be produced by the relatively nonastringent trihydroxide.

These compounds exhibit somewhat less water solubility than aluminum chloride. This might be expected considering the insolubility of aluminum hydroxide.

USES. Either one of these compounds or mixtures of the two are used as mild astringents. The primary use is as antiperspirants, where they have replaced the more acidic and more irritating *Aluminum Chloride*. Aluminum hydroxychloride (aluminum chlorhydroxide, aluminum chlorhydrate, aluminum chlorhydrol) is used in concentrations of about 20% in numerous antiperspirant-deodorant solutions, sprays and creams (Arrid®, Right-Guard®, etc.).[19]

Aluminum Sulfate, U.S.P. XVIII [$Al_2(SO_4)_3 \cdot 14H_2O$; Mol. Wt. 594.36)

Aluminum Sulfate U.S.P. may exist as a white crystalline powder, shining plates, or crystalline fragments. It is stable in air, odorless, and has a sweet taste, becoming mildly astringent. It is freely soluble in water and insoluble in alcohol. The pH of a 1:20 solution should not be less than 2.9.

This compound has the chemical properties of aluminum ion and sulfate ion.

USES. *Aluminum Sulfate* is a mild astringent by virtue of the availability of aluminum ion. Solutions containing 5 to 25% of the salt are used for their astringent and antiseptic properties. It is also found in some antiperspirant preparations.

Official recognition of *Aluminum Sulfate* is for its use in the preparation of *Aluminum Subacetate Solution*, U.S.P.

Aluminum Subacetate Solution, U.S.P. XVIII

Aluminum Subacetate Solution U.S.P. yields, from each 100 ml, not less than 2.30 g and not more than 2.60 g of aluminum oxide (Al_2O_3), and not less than 5.43 g and not more than 6.13 g of acetic acid ($C_2H_4O_2$). It may be stabilized by the addition of not more than 0.9% of boric acid.

The solution is prepared by adding *Precipitated Calcium Carbonate* to a solution of *Aluminum Sulfate* to form insoluble aluminum hydroxide and calcium sulfate [rx (liii)].

$$Al_2(SO_4)_3 + 3CaCO_3 + 3H_2O \rightarrow Al(OH)_3 \downarrow + 3CaSO_4 \downarrow \qquad \text{(liii)}$$

Acetic acid is then added and the aluminum subacetate allowed to form over 24 hours [rx (liv)].

$$2Al(OH)_3 + 4CH_3COOH \rightarrow 2Al(CH_3COO)_2OH + 4H_2O \qquad \text{(liv)}$$

The solution is filtered and the magma washed with sufficient water to make up the required volume.

The solution is a clear, colorless or faintly yellow liquid, having an acetous odor and an acid reaction to litmus. It will gradually become turbid on standing due to the formation of a more basic salt.

USES. This solution should be diluted with 20 to 40 parts of water before using. It is used as a topical astringent either as a soak or wash, or applied in the form of wet dressings.

Aluminum Acetate Solution, U.S.P. XVIII (Burow's Solution)

This solution yields, from each 100 ml, not less than 1.20 g and not more than 1.45 g of aluminum oxide (Al_2O_3), and not less than 4.24 g and not more than 5.12 g of acetic acid ($C_2H_4O_2$), corresponding to not less than 4.8 g and not more than 5.8 g of aluminum acetate ($C_6H_9AlO_6$). *Aluminum Acetate Solution* may be stabilized by the addition of not more than 0.6% of boric acid.

The solution is prepared from *Aluminum Subacetate Solution* (545 ml) by adding glacial acetic acid (15 ml), followed by addition of purified water to a total volume of 1 liter [rx (lv)].

$$Al(CH_3COO)_2OH + CH_3COOH \rightarrow Al(CH_3COO)_3 + H_2O \qquad \text{(lv)}$$

The solution should be well mixed and may be filtered if necessary. The final solution should be a clear colorless liquid having a faint acetous odor, and a sweetish, astringent taste. Its specific gravity should be about 1.02 and it should be dispensed only when clear. The pH of the solution should be between 3.6 and 4.4.

Uses. *Aluminum Acetate Solution* or Burow's solution is used as a topical antiseptic and astringent. When diluted with 10 to 40 volumes of water, the solution may be used as a soak or applied as a wet dressing. The solution seems to have good antiseptic properties even at a 1% concentration.

There are several powdered or tableted products which can be used to prepare Burow's solution by adding water alone (Domeboro®) or water and acetic acid. A colloidal suspension of aluminum acetate is available as Hydrosal®.

Alum, N.F. XIII (Aluminum Ammonium Sulfate, $(AlNH_4(SO_4)_2 \cdot 12H_2O$; Mol. Wt. 453.33; or Aluminum Potassium Sulfate, $(AlK(SO_4)_2 \cdot 12H_2O$; Mol. Wt. 474.39)

Alum N.F. can be either the ammonium or the potassium salt, and the label on the container must indicate which salt is being dispensed.

The official compound exists as large, colorless crystals, crystalline fragments, or as a white powder. It is odorless, and has a sweetish, strongly astringent taste. Its solutions are acid to litmus. The compound is freely soluble in water, freely but slowly soluble in glycerin, and insoluble in alcohol.

An alternate representation of the composition of *Alum* is seen rather frequently as $K_2SO_4 \cdot Al_2(SO_4)_3 \cdot 24H_2O$ and $(NH_4)_2SO_4 \cdot Al_2(SO_4)_3 \cdot 24H_2O$. The compounds are double salts, and the term "alum" is used to describe a double salt of a trivalent and a univalent element containing 12 molecules of water of hydration. The potassium or ammonium aluminum sulfates are the best known alums. The chemistry of *Alum* is that of aluminum ion, potassium and ammonium ions, and sulfate.

Uses. Alum serves as a source of aluminum ion, making it useful as a topical astringent. The rather high astringency of the compounds makes it possible for certain preparations to be used as irritants or caustics. It can be used in footbaths as a means of toughening the skin. Astringent solutions of *Alum* will usually contain between 0.5 and 5% of the compound. It is frequently the active ingredient in styptic pencils, where it is used to stop bleeding from small cuts.

The protein precipitant properties of *Alum* are utilized in the preparation of precipitated diphtheria and tetanus toxoids.

Soluble Zinc Compounds

Zinc Chloride, U.S.P. XVIII ($ZnCl_2$; Mol. Wt. 136.28)
The official description of *Zinc Chloride*, U.S.P., is given in Chapter 10.

Like aluminum salts, the acidity of aqueous solutions of zinc chloride is due to its hydrolysis to form hydrochloric acid and a basic zinc chloride [rx (lvi)].

$$ZnCl_2 + H_2O \leftrightharpoons Zn(OH)Cl + H^+ + Cl^- \tag{lvi}$$

Solutions of zinc chloride should be filtered through asbestos or glass wool because they will dissolve paper and cotton.

Some dental cements are formed by mixing a concentrated solution of zinc chloride with zinc oxide. The product, which sets into a hard mass, is a zinc oxychloride (see Chapter 10).

Uses. *Zinc Chloride* is used for the activity of zinc ion, which is a very strong protein precipitant. The compound is a powerful astringent in solution and a mild antiseptic. It is thought that its antiseptic properties are not totally a function of its ability to precipitate protein, but may be involved with an interaction of the metal with certain bacterial enzymes, inhibiting their function. The strong astringent properties of zinc chloride make the compound useful as an escharotic (an agent which causes the sloughing of tissue, aiding in the formation of scar tissue to improve healing), which is more of a caustic than astringent action.

The compound is applied as a solution containing from 0.5 to 2% of zinc chloride. The lower concentration may be applied to mucous membranes and is used as a nasal spray in office procedures to aid drainage from infected sinuses.

The U.S.P. also recognizes the compound as a desensitizer of dentin. For this purpose, a 10% solution is applied topically to the teeth (see Chapter 10).

Zinc Sulfate, U.S.P. XVIII ($ZnSO_4 \cdot 7H_2O$; Mol. Wt. 287.54)

Zinc Sulfate U.S.P. appears as colorless, transparent prisms, or as small needles. It may also be a granular, crystalline powder. It is odorless and effloresces in dry air. Like *Zinc Chloride*, it is very soluble in water, freely soluble in glycerin, and insoluble in alcohol. Aqueous solutions of *Zinc Sulfate* are acid to litmus due to hydrolysis of the salt [see rx (lvi)], producing a solution with a pH of about 5.

Uses. The major use of *Zinc Sulfate* is externally as an ophthalmic astringent (*Zinc Sulfate Ophthalmic Solution*, U.S.P. XVIII) in 0.25% aqueous solution. The acidic nature of zinc sulfate solutions requires some buffering. The U.S.P. *Ophthalmic Solution* should have a pH between 5.8 and 6.2, which can be achieved with an appropriate borate buffer, e.g., Gifford's (see Buffers in Chapter 4). More alkaline solutions may be prepared if required, but sodium citrate should be added to keep the zinc ion from precipitating as the hydroxide. *Zinc Sulfate Ophthalmic Solution* is usually applied in 0.1-ml doses to the conjunctiva. *Zinc Sulfate* may also be found as an ingredient in some of the astringent solutions available.[20]

Zinc Sulfate has also been used internally as an emetic. It may be given in 1- to 2-g doses in a 1% solution for this purpose. The emetic action of zinc ion is so rapid that very little, if any, local irritation to the gastric mucosa can occur.

One interesting experimental use of zinc sulfate internally is its oral administration to aid in the healing of wounds.[21,22] This topic is discussed in detail in Chapter 6.

White Lotion, N.F. XIII

This is a topical preparation prepared by adding a solution of *Sulfurated Potash* (40 g in 450 ml of water) slowly to a solution of *Zinc Sulfate* (40 g in 450 ml of water) and then adding water to a volume of 1000 ml. When completed, according to rx (lvii), the resulting product is a suspension of zinc sulfides.

$$K_2S + ZnSO_4 \rightarrow ZnS \downarrow + K_2SO_4 \qquad \text{(lvii)}$$

The order of addition of the components is very important. If the order is reversed, the hydroxide ions in the *Sulfurated Potash* cause the formation of basic zinc salts and zinc hydroxide, rather than the zinc sulfides. Also, the particle sizes of the suspended compounds are larger. When made in the prescribed manner, zinc sulfides are practically the only zinc compounds formed, and their precipitation in the alkaline medium of the suspension is essentially complete. Potassium sulfate and excess *Sulfurated Potash* is all that remains in solution.

The precipitate tends to become lumpy upon standing. This requires that the suspension be freshly prepared and shaken thoroughly before using.

USES. *White Lotion* is used topically for the effects of the sulfide ion (see above) and the astringent action of the zinc ion. It has been used in the treatment of acne vulgaris, seborrhea dermatitis, and other dermatological problems.

Toughened Silver Nitrate, U.S.P. XVIII (Lunar Caustic, Fused Silver Nitrate, Silver Nitrate Pencils)

Toughened Silver Nitrate U.S.P. contains not less than 94.5% of $AgNO_3$, the remainder consisting of silver chloride (AgCl). It is a white crystalline mass generally molded as pencils or cones. It breaks with a fibrous fracture, becoming gray or grayish black on exposure to light. It is soluble in water to the extent of its nitrate content always leaving a residue of silver chloride. It is partially soluble in alcohol and insoluble in ether. Its aqueous solutions are neutral to litmus.

A common procedure for making toughened silver nitrate involves adding hydrochloric acid in a quantity equal to 4% of the weight of the salt, melting the resulting mixture at as low a temperature as possible, and casting in silver molds. The presence of the silver chloride reduces the tendency of the molded product to crumble. Some of the commercial products contain anywhere from 20 to 55% less silver nitrate than the official preparation. Potassium nitrate is usually substituted for the missing portion of silver nitrate.

USES. The concentration of silver ion in this preparation (about 95%) places it in the caustic category. When moistened by dipping in water, toughened silver nitrate can be used to cauterize wounds (coagulate blood and tissue), remove warts and other unwanted growths of tissue on the skin, e.g., granulomatous tissue.

Zirconium Compounds. Zirconium is a transition metal (Group IVB) which has found little use in medicine. Zirconium silicate ($ZrSiO_2$) has properties similar to those of titanium dioxide, i.e., applicability (slip) and clinging power on the skin, causing it to be recommended for use in cosmetics.

Zirconium carbonate and oxide have been used in a number of topical applications, including medications for athlete's foot, antiperspirants, and the like. Zirconium oxide has also been included in some preparations used to treat poison ivy eruptions, where it was claimed to have a specific action against the urushiol toxin from the plant. It is doubtful that this is true.[23] Zirconium salts in these preparations have produced severe skin irritations and induced the formation of granulomas. For these reasons, it has been recommended that their use in topical medications be discontinued.[24]

References

1. Antopol, W. Lycopodium granuloma. Arch. Path. (Chicago), **16**:326, 1933.
2. Torosian, G., and Lemberger, M. A. O-T-C sunscreen and suntan products. J. Amer. Pharm. Ass., **NS12**:571, 1972.
3. Darlington, R. C. O-T-C topical oral antiseptics and mouthwashes. J. Amer. Pharm. Ass., **NS8**:484, 1968.
4. Meleney, F. L. Present role of zinc peroxide in treatment of surgical infections. J.A.M.A., **149**:1450, 1952.
5. A.M.A. Drug Evaluations. Chicago: American Medical Association, p. 492, 1971.
6. Gershenfeld, L. *In:* Lawrence, C. A., and Block, S. S., eds. Disinfection, Sterilization, and Preservation. Philadelphia: Lea and Febiger, p. 329, 1968.
7. A.M.A. Drug Evaluations. Chicago: American Medical Association, p. 485, 1971.
8. Povidone-iodine (Betadine) for surgical antisepsis. Med. Lett., **11**:99, 1969.
9. Barsam, P. C. Specific prophylaxis of gonorrheal ophthalmia neonatorum. New Eng. J. Med., **274**:731, 1966.
10. Moyer, C. A., Brentono, L., Grovens, D. L., Margraf, H. W., and Monafo, W. W. Treatment of large human burns with 0.5% silver nitrate solution. Arch. Surg. (Chicago), **90**:812, 1965.
11. Ricketts, C. R., Lowbury, E. J. L., Lawrence, J. C., Hall, M., and Wilkins, M. D. Mechanism of prophylaxis by silver compounds against infection of burns. Brit. Med. J., **2**:444, 1970.
12. Lowbury, E. J. L., Jackson, D. M., Lilly, H. A., Bull, J. P., Cason, J. S., Davies, J. W. L., and Ford, P. M. Alternative forms of local treatment for burns. Lancet, **2**:1105, 1971.

13. Hoopes, J. E., Butcher, H. R., Jr., Margraf, H. W., and Gravens, D. L.   Silver lactate burn cream. Surgery, **70**:29, 1971.
14. Komura, I., Funaba, T., and Izaki, K.   Mechanism of mercuric chloride resistance in microorganisms, II. J. Biochem., **70**:895, 1971.
15. Mills, O. H., Jr., and Kligman, A. M.   Is sulfur helpful or harmful in acne vulgaris? Brit. J. Derm., **86**:620, 1972.
16. Plewig, G., and Kligman, A. M.   The effect of selenium sulfide on epidermal turnover of normal and dandruff scalps. J. Soc. Cosmetic Chemists, **20**:767, 1969.
17. A.M.A. Drug Evaluations. Chicago: American Medical Association, pp. 488, 492, 1971.
18. Frost, D. V.   Arsenicals in biology. Fed. Proc., **26**:194, 1967.
19. Robinson, J. R.   *In:* Griffenhagen, G. P., and Hawkins, L. L., eds. Handbook of Non-Prescription Drugs. Washington: American Pharmaceutical Association, p. 209, 1973.
20. Walker, B. C., and Swafford, W. B.   *Ibid.,* pp. 135–137, 1973.
21. Pories, W. J., Henzel H. H., Strain, W. H., and Rob, C. G.   Zinc sulfate administered orally: Wound reported to heal faster. J.A.M.A., **196**:33, 1966.
22. Husain, S. L.   Oral zinc sulfate in leg ulcers. Lancet, **1**:1069, 1969
23. Wormser, H. C.   *In:* Griffenhagen, G. P., and Hawkins, L. L., eds. Handbook of Non-Prescription Drugs. Washington: American Pharmaceutical Association, p. 172, 1973.
24. A.M.A. Drug Evaluations. Chicago: American Medical Association, p. 498, 1971.

10
Dental Products

A wide variety of inorganic compounds used in dentistry are of interest to pharmacists. These include polishing, cleaning, and anticaries agents. The pharmacist, on occasion, may have found it difficult to obtain information about these agents and the specific products containing them. An excellent source is *Accepted Dental Therapeutics*, published every two years by the American Dental Association (A.D.A.). This publication not only contains descriptions of the active agents used, but also has recommendations concerning their therapeutic usefulness as evaluated by the A.D.A. Council on Dental Therapeutics. This chapter will be limited to those agents used in the prevention of caries (dentifrices and fluoride salts), polishing agents, and desensitizing agents.

Anticaries Agents

Fluorides. Today, when considering caries (tooth decay) prevention, one usually thinks of fluorides, since fluoridated water, fluoride drops, topical fluoride application to teeth, fluoride-containing vitamins, and fluoride dentifrices are now commonplace. The exact cause and mechanism of caries have still not been completely elucidated. It is probably associated with diet in that people consuming a diet high in fermentable carbohydrates have a higher incidence of dental caries than those on low carbohydrate diets. On the other hand, it is difficult to correlate nutrition with dental caries incidence. Also, there is no definite correlation between the inorganic calcium level in the saliva and the incidence of dental caries.[1]

The formation of caries is attributed to the action of acids, mostly lactic, obtained from oral bacterial metabolism of dietary carbohydrate. The buildup of plaque on the tooth surface usually aids the decay process by forming pockets or crevices on the tooth surface in which food particles can lodge and be degraded by the bacteria of the mouth.[2] One of the objectives of brushing the teeth is to remove material from the tooth surface before it hardens into calculus, since a smooth surface makes it more difficult for the adherence of food particles and bacteria.

Oral hygiene has evolved from the simple cleaning of teeth to attempts at actually interfering with the decay process. Ammoniated toothpastes were supposed either to neutralize lactic acid or to cause a drop in lacto-bacillus counts. There is some doubt that lactobacillus, by itself, plays a significant role in the decay process.[3] The Council on Dental Therapeutics has classified dentifrices containing urea or urea and ammonium compounds as the active compounds in Group C: "require further study by qualified investigators."[4]

Antibiotic-containing dentifrices have been tried with mixed results. There are the twin problems of: (1) sensitization of a significant portion of the population to the antibiotic; and (2) development of antibiotic-resistant microorganisms.

Dentifrices which are claimed to have antienzyme properties have been developed. These have included sodium N-lauroyl sarcosinate,

$$CH_3-NCH_2CO_2Na$$
$$|$$
$$C(CH_2)_{10}CH_3$$
$$||$$
$$O$$

(Colgate's Gardol®) which is a foaming agent. The Council on Dental Therapeutics states that "the usefulness of [these] dentifrices in caries control has not been adequately established."[5]

Currently accepted and documented approaches to caries prevention include flossing and brushing accompanied by fluoride, administered either internally or topically to the teeth. When taken internally, fluoride in solution or in rapidly soluble salts is absorbed almost completely from the gastrointestinal tract. Much of the flouride in food is present as poorly soluble calcium salts, which are poorly absorbed and to a somewhat variable degree. The absorbed fluoride is partially deposited in the bone or developing teeth, with the remainder excreted by the kidneys.

Deposition of fluoride in the bones and teeth is currently believed to occur by fluoride replacing the hydroxyl and possibly carbonate anions of hydroxyapatite (a mixed calcium salt of carbonate, phosphate, and hydroxide). Of the total fluoride retained by the body, only a small amount is found in teeth, due to their small mass relative to the total skeletal tissue of the body. Also, in contrast to the fluoride in the skeletal tissue, the fluoride deposited in teeth is apparently not subject to appreciable reabsorption.

Careful analysis of teeth has shown that the concentration of fluoride is greater in the surface layer of enamel in both erupted and unerupted teeth. This is explained by the fact that the enamel loses contact with the tissue fluids after calcification is completed. Thus, fluoride uptake from tissue fluids is limited to the external surface. The outer surface will continue to pick up fluoride until eruption, with the result that the longer the period between tooth formation and tooth eruption, the higher the degree of fluoride saturation. The fluoride content is also dependent upon the concentration of fluoride in the tissue fluids. There is some evidence that a posteruptive tooth can continue to receive fluoride from topical sources. Of course, this fluoride will definitely be confined to the external layers of the tooth. In both pre- and posteruptive teeth, most fluoride is confined to the outer 0.1-mm layer, with a limiting concentration of 1%.[6]

When too much fluoride is present in the tissue fluids, the condition known as dental fluorosis (mottled enamel) can develop. This usually occurs in areas where the fluoride concentration of drinking water exceeds 2 ppm. Dental fluorosis only occurs during excessive ingestion of fluoride during the period of tooth development and over a prolonged period. Thus, a short-term accidental increase in the fluoride concentration during the addition of fluoride to drinking water would have little or no effect. There is no evidence of a systemic toxicity associated with long-term dental fluorosis in those communities using drinking water containing high concentrations of natural fluorides. The lethal adult dose of sodium fluoride is 2 to 5 g, as compared to the average ingestion of 2.2 mg NaF per day or its equivalent in drinking water containing 1 ppm fluoride. Nevertheless, concentrates used to administer fluorides to children who are drinking nonfluoridated water should definitely be kept out of the child's reach.

The enamel of mottled teeth is chalky and soft, indicating a loosely bound structure.[6] The pigments found in foods apparently adsorb to this modified enamel, causing patches of darkening which give a mottled appearance. If severe, the teeth can become pitted and may have to be removed by early adulthood. The pitting is not universal and, even in communities with drinking water containing a high fluoride content, the severity of mottling will vary considerably among individuals. Nevertheless, it is possible to reduce the fluoride content of drinking water down to the recommended 1 ppm by using an activated alumina filtering plant.

The mechanism by which fluoride inhibits caries formation is still to be completely elucidated. There are two current hypotheses: (1) decreased acid solubility of enamel; and (2) bacterial inhibition.

It is quite easy to demonstrate that fluoride decreases the solubility of enamel in acid.[7] This lowering of solubility is more pronounced in enamel already attacked by caries. Caries lesions cause the enamel to become more permeable to fluoride which, along with the lower pH found in a caries lesion, favors fluoride uptake by hydroxyapatite. Enamel already showing carious changes may be 20% less soluble than intact enamel in the same teeth.[8] While accepting the fact that fluoride does reduce the acid solubility of enamel, it is not known precisely how this occurs. Even if one assumes that fluorapatite is less soluble than hydroxyapatite, the outer enamel will have at the most only one hydroxyl in 50 replaced by fluoride, which would not affect the overall decrease in observed solubility. Other mechanisms that have been postulated as to how fluoride might decrease enamel solubility include: (1) reduction of the number of defects in apatite crystals; and (2) competition with carbonate during apatite formation.

The bacterial inhibition hypothesis is based on the enzyme inhibitory properties of fluoride. This in itself is difficult to justify, as a fluoride concentration of 10 ppm is needed for any significant inhibition of many of the oral bacteria. Saliva contains between 0.02 to 0.1 ppm fluoride.[9] However, the layer of dental plaque has been shown to contain from 25 ppm fluoride in low fluoride areas to 47 ppm in areas where the drinking water contains 2 ppm fluoride.[10] Even though plaque fluoride is bound fluoride (as evidenced by its accumulation in plaque), it still seems to reduce acid production in the plaque.[11] In contrast, topically applied sodium fluoride loses its ability to inhibit acidogenesis within one week of application, indicating that fluoride reduces the incidence of caries by some other mechanism.[12] While it might appear that the fluoridated plaque is beneficial, its presence still promotes inflammation of the gingivae (gums), making its removal desirable.

The question still to be answered is the source of plaque fluoride. The probable source is the saliva, but food and the enamel surface with its approximate 1000 ppm fluoride concentration have been considered. The problem with the latter source is that one would expect the fluoride concentration of enamel to decrease with age if fluoride were constantly diffusing from the enamel into plaque. If anything, enamel fluoride concentration increases with age and, furthermore, the fluoride is too tightly bound to apatite to be released. Most foods have a fluoride concentration too low to have any marked caries-reducing effect. This leaves the continuous flow of saliva over the plaque as the most likely fluoride source.

It is also very likely that both the solubility and bacterial inhibition hypotheses are operative. Concomitant with both theories is the assumption that lactic acid production by the oral bacteria acting on food particles is the main cause of caries. If this theory is incorrect, then fluoride action could probably best be explained in terms of apatite structure.[13]

Fluoride is administered by two routes, orally and topically. Both are effective, but the oral route places fluoride into systemic circulation, allowing the fluoride to be laid down in unerupted teeth as they are formed. The most convenient dosage form is fluoridation of the public water supply. This is usually done by adding sodium fluoride or a fluorosilicate, yielding a fluoride concentration of 0.7 to 1 ppm. This is equivalent to an average daily intake of 2.2 mg of sodium fluoride based on a person drinking six 6-oz glasses of water. During summer months and in hot areas, it has been recommended that the fluoride concentration be reduced in order to compensate for the increased water intake by the local population. There has been a recent report that home water softeners cause a mean reduction of 23% in the fluoride concentration of the domestic water supply. This implies that communities with hard water should maintain a fluoride concentration of 1.2 ppm.[14]

The effectiveness of fluoridated water in reducing dental caries is well documented.[15-21] There has been a vocal and well organized opposition to adding fluorides to drinking water as a public health measure. To date, there has been no greater incidence of any disease or difference in mortality rates in those areas where the water is naturally fluoridated, as compared to nonfluoridated areas. This observation becomes more important when one considers that the fluoride content of the drinking water in many of these areas is greater than the recommended 0.7–1 ppm fluoride concentration. The fluoridation controversy has provided a classical example of the need for the scientist to communicate adequately with the lay public.[22]

A study on the effect of water fluoridation on dental practice and dental manpower illustrates the effect of large scale preventive medicine. When comparing dental practice in communities with and without fluoridated water, the dentists in fluoridated communities have a patient load 14.5% larger than dentists in nonfluoridated communities. Nevertheless, dentists in fluoridated communities feel less overworked, spend on the average slightly more time on each patient sitting, invest less direct (hours and days per week of their own labor) and indirect (auxiliary personnel) effort in their practices and earn as much, if not slightly more, than dentists in fluoride-deficient communities. It would appear that fluoridation helps to extend the existing pool of dental manpower to cover a substantially larger population.[23]

There have been some reports that patients using a renal dialysis unit (artificial kidney) or with severe kidney disease have a higher incidence of renal azotemic osteodystrophy (defective bone formation). However, osteodystrophy is common in patients with chronic renal failure, including those receiving long-term dialysis with nonfluoridated water.[24] Furthermore, it is recommended that tap water first be treated by reverse osmosis membranes, deionization columns, and stills, before being used in a renal dialysis unit.[25]

Keeping in mind that internal administration of fluoride works best when fluoride is available as the unerupted tooth is laying down the enamel, the question arises as to whether an infant will obtain enough fluoride. Human breast milk contains 0.05 ppm fluoride, which is about the same as plasma and saliva and is independent of the mother's fluoride ingestion.[26] In contrast, undiluted cow's milk contains 0.03 ppm while a 1:1 dilution of cow's milk with water containing 1 ppm fluoride will have a final fluoride concentration of 0.52 ppm. The average dry milk formula made from water containing 1 ppm fluoride will have a final concentration of slightly over 1 ppm, depending upon the fluoride content of the dry milk. In one study, no difference was found between breast-fed and bottle-fed children in communities using fluoridated drinking water.[27] The Council on Dental Therapeutics currently suggests that children living in areas where the drinking water contains 0.7 ppm or higher fluoride not receive any additional fluoride supplement.[28]

Because fluoridation is most effective when it can be incorporated into enamel of the unerupted tooth as it is forming there has been investigation into the feasibility of administering fluoride to the pregnant female, usually as fluoridated prenatal vitamins. It is generally accepted that fluoride is transferred to the fetus and that the fetus of a mother drinking fluoridated water will have a larger fluoride content than the fetus of a mother in a nonfluoride area.[29] The results are mixed as to whether the infant whose mother takes fluoride internally during pregnancy actually has increased caries resistance.[30–33] It does appear that prenatal fluoride has no statistically significant benefit on caries reduction for the permanent teeth.[30] Other studies indicate that there is a greater degree of protection, particularly for the deciduous ("baby") teeth, when fluoride is administered prenatally.[32,33] On the other hand, because calcification of the entire permanent tooth and a large portion of the deciduous teeth is a postnatal process, it has been concluded that there is no real need for prenatal prescribing of fluoride.[31]

Since internal fluoride intake is considered to confer better caries resistance than topical application, there have been investigations of other "dosage forms" in addition to drinking water. These have included sodium fluoride supplements, either as tablets dissolved in water or fruit juice or drops added to water or fruit juice. Either way, the dilutions should equal 2.2 mg of sodium fluoride (equivalent to 1 mg of fluoride anion) per day for children over three years of age. Half this amount (1.1 mg sodium fluoride) is recommended for children between two and three years of age. No specific daily fluoride allowance is suggested for children under two years fo age, although water containing 1 ppm fluoride (domestic or commercially bottled) is recommended. When the water already contains fluoride, the above amounts should be reduced to prevent dental fluorosis. Table 10–1 shows the adjusted sodium fluoride doses.

TABLE 10–1. ADJUSTED ALLOWANCE OF FLUORIDE[b]

Water Fluoride ppm	Sodium fluoride mg per day[a]	Provide fluoride ion mg per day
0.0	2.2	1.0
0.2	1.8	0.8
0.4	1.3	0.6
0.6	0.9	0.4

[a] For children three years of age or older. Use one half the amounts for children between two and three years of age.

[b] From Accepted Dental Therapeutics, 34th ed., Council on Dental Therapeutics, American Dental Association, Chicago, p. 241, 1973/1974. By permission.

There are fluoridated pediatric vitamin preparations on the market. These are as effective as the sodium fluoride drops and tablets, but their usefulness is limited as the dose is based on both the vitamin and fluoride content. Thus, it is more difficult to adjust the fluoride intake where the drinking water contains substantial but inadequate levels of fluoride. For this reason, the Council on Dental Therapeutics has accepted brands of sodium fluoride as dietary supplements but has not accepted combinations with vitamins.[34]

A problem with sodium fluoride drops or tablets is that the parents must be highly motivated, since the administration of internal fluoride should continue until at least 12 to 14 years of age when the crowns of all teeth except the third molars should be completed. Even when sodium fluoride tablets have been supplied by communities as a public health measure, the initial requests for the tablets have been only a fraction of the number required for an effective program, and the individual participation declines steadily with time.[35] There have been attempts to find other dietary components that could be fluoridated. Table salt has been one suggestion, since the addition of potassium iodide has been so successful in the prevention of iodine deficiency goiter. Salt containing 250 ppm fluoride (as sodium fluoride) has effected a reduction in dental caries and produced urine fluoride levels somewhat greater than those obtained from ingesting water containing 1 ppm fluoride.[36,37] The problems of proper salt intake for young children, plus the wide variation in salt intake, might not make this route as feasible. Nevertheless, salt containing 90 ppm fluoride is sold in Switzerland.[37]

Another means of conveniently administering fluoride, particularly to children, has been in milk and the school water supply. A pilot study in which children drank a half pint of milk containing 2.2 mg of sodium fluoride showed that a good reduction in dental caries can occur.[38]

A 12-year study in which the water supply at a rural school was fluoridated at a level of 5 ppm (4.5 times the optimum level for community fluoridation in the area) showed a 39% reduction in decayed, missing, and filled teeth (DMFT) as compared to a control group. The findings of this study strongly suggest that fluoridation of a school water supply is a very efficient means of supplying fluoride to large numbers of children living in areas not served by central water supplies.[39]

For a large group of people, internal administration of fluoride is not feasible or possible, and for the postadolescent individual, it probably is not useful. Topical fluoride application can be of help. Currently, there are two approaches: topical application of a fluoride solution by a dentist, or use of a fluoridated dentifrice by the lay public. The use of fluoridated mouth washes is also under investigation.

Sodium fluoride, usually as a 2% aqueous solution, has been widely used topically. The usual procedure is a series of four treatments given several days apart, beginning at age three for the deciduous teeth. Other applications are then given at ages 7, 11, and 13 as the permanent teeth erupt. The procedure requires that the teeth be thoroughly dried prior to each application. They should be cleaned by the dentist prior to the first application with no cleaning prior to the second, third, or fourth applications of sodium fluoride. It appears that a 30 to 40% reduction of dental caries can result in children. The technique does not seem to be as beneficial in young adults. This may be because teeth that have been exposed for long periods of time have a greater fluoride content derived from fluoride from dietary sources.[6] For effective prevention of caries, the parents must be motivated enough to take their children to a dentist at regular intervals. Thus, sodium fluoride mouth washes have been investigated. A controlled study involving weekly rinses with a 0.2% sodium fluoride solution in a public grade school showed a reduction of caries of 30–40%. This included no rinsing during the summer months.[40] The use of a chewable sodium fluoride tablet showed only a supplemental benefit for children receiving fluorides systemically.[41]

A modification of sodium fluoride application is the use of acidulated phosphate-sodium fluoride gels. These preparations are of various concentrations, but usually contain the equivalent of approximately 1.23% fluoride and 1% phosphoric acid (orthophosphoric acid). Several of the commercial preparations accepted in Group A (Accepted) by the Council on Dental Therapeutics contain 2 g sodium fluoride, 0.34 g hydrogen fluoride, and 0.98 g orthophosphoric acid in each 100 ml.[42]

A modification is a gel dosage form in which the patient is fitted with a tray around the teeth. A ribbon of gel is placed in the trough of the tray. In order to obtain maximum flow of the gel between the teeth, pressure is applied by squeezing the buccal (cheek) and lingual (tongue) surfaces, and the patient is requested to bite lightly, thereby molding the tray tightly around the teeth. The gel should remain in contact with the teeth for four minutes, and the patient should not eat, drink, or rinse his mouth for 30 minutes following treatment. The effectiveness of acidulated phosphate-fluoride as both a solution and gel has been reported to provide a 30–40% caries reduction.[43] The gels are currently in Group B (Provisionally Accepted) since their effectiveness as compared to the solutions is equivocal.[42]

Recent studies in Charlotte, North Carolina, which has had fluoridated water since 1949, showed that self-application of acidulated gel three times a week using a custom-made mouth-piece greatly increased caries resistance as compared to controls using only fluoridated water.[44,45]

Stannous fluoride solutions are extensively used for topical fluoride application. A simple appliction of a freshly prepared 8% solution at 6- to 12-month intervals is used. Stannous fluoride requires only one application per treatment as compared to the series of four applications per treatment of sodium fluoride. The solution is applied to cleaned, dry teeth.

Stannous fluoride solution deteriorates on standing with oxidation of the stannous cation (Sn^{+2}) to the stannic cation (Sn^{+4}), resulting in a precipitate that makes the solution ineffective. Thus, stannous fluoride solution must be freshly made. A detailed study of the stability of stannous fluoride solution shows that (1) the stability is increased with increasing stannous fluoride concentration, (2) the rate of turbidity formation is temperature-dependent, and (3) the loss of stannous cation is both pH-dependent and dependent upon the buffer or complexing agent used.[46] The reason for concern over loss of stannous ion activity is because of reports that the stannous cation is itself anticariogenic (see discussion of stannous fluoride dentifrices). In a recent study of adults, a greater fluoride uptake by enamel of teeth was found for acidulated phosphate-fluoride solution and acidulated phosphate-fluoride-silicon dioxide paste as compared with a stannous fluoride solution and a stannous fluoride-zirconium silicate paste. The test was too brief to be able to evaluate any anticariogenic effect.[47] A three-year study of children in Albany, New York, in which sodium fluoride solution, stannous fluoride solution, acidulated phosphate fluoride solution, and acidulated phosphate fluoride gel were compared, showed that the gel might have a slight advantage.[48] The authors conclude that application of any of these agents should probably be increased since loss of fluoride from enamel is high when topical agents are used.[49,50] This may explain why repeated brushing with a fluoridated toothpaste (discussed below) can be effective.

Other uses for stannous fluoride include being a component of silver amalgam restorative material (silver filling) and zinc phosphate cement.[51,52] In the former case, it was found that silver amalgam containing stannous fluoride resulted in a decrease in enamel solubility due to a possible increase of fluoride content of tooth structure adjacent to the restoration. There is the possibility that recurrent decay around the restoration will be reduced. The incorporation of stannous fluoride into zinc phosphate cement may harden the enamel around an orthodontic brace when it is cemented on to the tooth.

A commercial preparation containing 1.5% stannous fluoride mixed with the silver-tin alloy used for restorations is now available. There has been a report that stannous fluoride, when incorporated into a silver-tin-mercury amalgam, decreases the corrosion resistance of the amalgam.[53] The question that must still be answered is whether the possible reduction in decay around the restoration is more of an advantage when compared with the reported decreased resistance to corrosion of the amalgam, resulting from the incorporation of the stannous fluoride into the silver-tin amalgam restorative material. Another result of the above study showed that the addition of sodium fluoride to a test solution increased the corrosion rate of dental amalgam (not containing any additional stannous fluoride). When the above experiment was repeated using a solution of stannous fluoride, there was no effect on the corrosion rate. If these results can be confirmed, it would appear that addition of sodium fluoride to toothpaste or its use for clinical topical application may increase the corrosion rate of amalgam restorations.[53]

Official Products

Sodium Fluoride, U.S.P. XVIII (NaF; Mol. Wt. 41.99)
Sodium Fluoride U.S.P. XVIII occurs as a white odorless powder which is soluble in water and insoluble in alcohol.

USUAL DOSE: 2.2 mg (the equivalent of 1 mg of fluoride ion) once a day.

APPLICATION: 1.5 to 3 ppm (equivalent to 0.7 to 1.3 ppm of fluoride ion) in drinking water; topically, as a 2% solution to the teeth.

Stannous Fluoride, N.F. XIII (SnF_2; Mol. Wt. 156.69)
Stannous Fluoride N.F. XIII occurs as a white, crystalline powder and has a bitter, salty taste. It melts at about 213° C. Stannous fluoride is freely soluble in water and is practically insoluble in alcohol, ether, and chloroform. It is for topical use only.

Phosphates. Another approach to caries reduction has been the possible supplementation of food with a phosphate salt, usually inorganic. From 1931 through the 1950s, evidence has accumulated that phosphates will reduce the incidence of caries in rodents put on caries-producing diets.[54] Early trials on human subjects produced equivocal results. A controlled study of Swedish school children showed that those children eating school lunches in which the soft bread, hard bread, wheat flour used for cooking, and sugar contained 2% calcium monohydrogen phosphate ($CaHPO_4$) had a 40% reduction in caries of the four proximal (touching) surfaces of the maxillary (upper) incisors.[55] In another controlled study of children attending boarding schools in North and South Dakota, there was no statistical difference in caries reduction between those children eating foods made with baking flour containing 2% calcium monohydrogen phosphate and those eating foods prepared from untreated flour. A complete mouth examination was used.[56] A similar study of institutionalized children in New York also showed no statistical reduction in dental caries for those children eating food prepared from flour and sugar containing 2% calcium monohydrogen phosphate.[57]

On the assumption that a more soluble phosphate salt might be more effective, calcium sucrose phosphate was incorporated into the refined carbohydrates eaten by a group of children so that each child consumed 4.3 g per day of phosphate as compared to a control group. There was an overall reduction of approximately 25% in caries increment with a 50% reduction in the proximal (touching) surfaces of the posterior teeth.[58]

Incorporation of a 1% mixture of sodium dihydrogen phosphate (NaH_2PO_4) and sodium monohydrogen phosphate (Na_2HPO_4), adjusted so that the pH of the final product was near neutrality, into four pre-sweetened ready-to-eat breakfast cereals (sugar-coated corn flakes, sweetened puffed wheat, sugar-coated alphabet-shaped oat cereal, and sweetened crisp rice) caused a 20–40% reduction in caries as compared to a control group. Again, the reduction was greater on the proximal surfaces.[59] This may be of practical importance, since it is more difficult to detect caries and then restore the tooth in those surfaces where the teeth touch each other. To date, it appears that phosphates can cause caries reduction in man provided a soluble salt is used. Nevertheless, the results are still not conclusive, and the Council on Dental Therapeutics has not recognized any phosphate product for the purpose of reducing the incidence of caries.[60]

Dentifrices

Dentifrices Containing Fluorides. Since fluorides are effective topically, it seemed that a fluoride-containing tooth paste would be a logical way for repeated self-administration of fluoride. The compounding of a fluoride-containing tooth paste is difficult. A tooth paste not only contains the anticaries agents, but also has polishing agents, thickening agents, surfactants, and humectants, and there is a chemical incompatibility between the fluoride anion and the calcium cation of the traditional polishing agents.

Calcium carbonate has been the traditional cleaning-polishing agent for most tooth pastes and tooth powders. However, even this water insoluble salt is soluble enough to provide enough free calcium cation to cause formation of the even more insoluble calcium fluoride (CaF_2). The net result is a reduction of fluoride anion that will be applied to the teeth. A similar incompatibility occurs when calcium monohydrogen phosphate is used. Furthermore, the anions of both calcium salts form insoluble stannous salts. Thus, the traditional calcium salts could not be used as cleaning-polishing agents when stannous fluoride was to be the fluoride salt. A workable solution is the use of calcium pyrophosphate ($Ca_2P_2O_7$), which is so water insoluble that the concentration of calcium will be low enough that acceptable fluoride anion levels will be maintained. Further, the pyro-phosphate anion forms soluble complexes rather than precipitates with the stannous cation. Sodium polyphosphate ($[NaPO_3]_n$) of molecular weight high enough to have reduced aqueous solubility can also be used.

Currently, Crest® is the one stannous fluoride tooth paste accepted in Group A by the Council on Dental Therapeutics. The indicated composition is:[61]

Stannous fluoride	0.4%
Stannous pyrophosphate	1 %
Calcium pyrophosphate	39 %
Glycerin	10 %
Sorbitol (70% solution)	20 %
Water	29.6%
Miscellaneous formulating agents	

It was stated in the discussion on the topical use of stannous fluoride that the stannous cation was also anticariogenic. *In vitro* tests of enamel dissolution rate and enamel hardness have shown that stannous fluoride is superior to sodium fluoride at equivalent concentrations.[62,63] Two kinds of reaction layers on the tooth enamel surface from exposure to stannous solutions have been proposed. Initially, there is an acid-resistant layer composed of basic stannous phosphate.[64] With prolonged reaction, higher pH's, and increased concentration of stannous cation, a second type of layer forms consisting of an amorphous, hydrated stannous oxide. This layer is poorly acid-resistant. Both layers are found, but it would be the basic stannous phosphate layer that would be anticariogenic.[65-67] Whether stannous cation is an effective anticariogenic agent *in vivo* is still open for further investigation. An early comparative *in vivo* study showed that a stannous fluoride dentifrice caused a 36% reduction in caries as compared to a control, while a sodium fluoride dentifrice caused only a 10% reduction, based on an evaluation of dental surfaces. Except for the fluoride salt, the composition of both dentifrices was identical.[68]

The variation in percent reduction of caries attributed to brushing with Crest® can usually be explained by the degree of supervision of the subjects and/or treatment of the teeth prior to the beginning of the study. These studies were all controlled and usually double-blind. It must also be pointed out that for maximum benefit, the fluoride dentifrices must be used frequently and the patient must seek regular professional care. The following examples of percent caries reduction can serve as illustrations: 23% reduction using Crest® under normal home conditions; 34% reduction when brushing was supervised once daily; 57% reduction when brushing was supervised three times daily; and 58% reduction when, in addition, an 8% stannous fluoride solution was applied every six months.[69]

The only other fluoride dentifrice in Group A is Colgate with MFP fluoride®. MFP refers to sodium monofluorophosphate (Na_2PO_3F). The indicated composition of the final product is:[70]

Sodium monofluorophosphate (MFP)	0.76%
Insoluble sodium metaphosphate	41.85%
Anhydrous dicalcium phosphate ($CaHPO_4$)	5 %
Sorbitol	11.9 %
Glycerol	9.9 %
Sodium N-lauroyl sarcosinate (Gardol®)	2 %
Water	24.4 %
Miscellaneous formulating agents	4.2 %

Double-blind controlled studies indicated a 17 to 34% reduction in incidence of caries.[71] Currently, it cannot be stated whether either Crest® or Colgate with MFP® is superior as compared to each other.[70]

Dentifrices Containing Polishing Agents. Dentifrices, whether or not they contain fluorides, contain agents for cleaning tooth surfaces. Because these cleaning agents must remove stains from the teeth, they must be abrasive to some degree. The problem is to design agents whose abrasiveness is adequate to remove stains with average, but correct, brushing, and still not wear away significant enamel and dentin. The evaluation of abrasiveness is very difficult. First, the degree of abrasiveness for a specific agent is not necessarily the same in different products, as the other agents in the dentifrice can easily have an effect. Second, the design of the test protocol is important if it is to have any *in vivo* significance. *In vitro* test procedures have included: (1) brushing machines that evaluated wear on a tooth surface; (2) loss of tooth surface as measured by radiotracer techniques; and (3) enamel-polishing capability as measured by a reflectometer. The two most common *in vivo* evaluations were: (1) measuring the degree of stain buildup or removal over a period of time; and (2) the degree of abrasiveness on acrylic surfaces of veneer crowns.[72-75]

Common cleaning agents include insoluble sodium metaphosphate ($NaPO_3$), anhydrous and hydrous calcium monohydrogen phosphate (dicalcium phosphate; $CaHPO_4$ and $CaHPO_4 \cdot 2H_2O$), calcium pyrophosphate ($Ca_2P_2O_7$), and calcium carbonate ($CaCO_3$). Pumice is also used by

TABLE 10–2. ABRASIVITY OF DENTIFRICE PRODUCT‡

Product	Manufacturer	No. of Lots	Abrasivity Index
T-Lak	Laboratoires Cazé	2	20(20–21)†
Thermodent	Chas. Pfizer & Co.	2	24(23–24)
Listerine	Warner-Lambert Pharm. Co.	3	26(22–30)
Pepsodent with zirconium silicate	Lever Brothers Co.	3	26(23–29)
Amm-i-dent	Block Drug Co.	2	33(31–34)
Colgate with MFP	Colgate-Palmolive Co.	3	51(46–56)
Ultra-Brite	Colgate-Palmolive Co.	4	64(52–82)
Macleans* (regular and spearmint)	Beecham, Inc.	4	68(66–72)
Pearl Drops	Cameo Chemicals	4	72(65–83)
Close-up	Lever Brothers Co.	5	87(70–101)
Crest (mint and regular)	Procter & Gamble Co.	10	88(70–110)
Macleans (regular and spearmint)	Beecham, Inc	6	93(74–103)
Crest, regular	Procter & Gamble Co.	5	95(77–110)
Gleem II	Procter & Gamble Co.	5	106(88–136)
Plus White	Bishop Industries, Inc.	5	110(91–141)
Phillips	Sterling Drug Inc.	2	114(111–116)
Plus White Plus	Bishops Industries, Inc.	4	132(96–181)
Vote	Bristol-Myers Co.	6	134(112–162)
Sensodyne	Block Drug Co., Inc.	3	157(151–168)
Iodent #2	Iodent Co.	2	174(172–176)
Smokers Tooth Paste	Walgreen Lab., Inc.	2	202(198–205)

* New formulation
† Average and range. The higher the number, the more abrasive.
‡ Formulation available as of July 1970.
From Accepted Dental Therapeutics, 34th ed., Council on Dental Therapeutics, American Dental Association, Chicago, p. 258, 1971/72. By permission.

dentists for prophylaxis (teeth cleaning). However, it is too abrasive for daily use in a dentifrice.

Table 10–2 contains a ranking of dentifrices as evaluated by the American Dental Association's Division of Chemistry. It must be kept in mind that the data represent formulations available on the market as of July 1970.

Several of the dentifrices are promoted as whiteners of the teeth. They all contain some type of abrasive polishing agent. The Council on Dental Therapeutics does not evaluate these cosmetic dentifrices, but does include a brief description of them in *Accepted Dental Therapeutics* for informational purposes. Table 10–3 summarizes this information.

TABLE 10–3. COSMETIC DENTIFRICES*

Brand	Composition	Abrasivity
Close Up®	Silicas of controlled particle size	Moderate
Excitement®	$CaHPO_2 \cdot 2H_2O$	Low
Macleans®	$CaHPO_4 \cdot 2H_2O$	Moderate
Pearl Drops Tooth Polish®	Hydrated $AlPO_4$ and $CaHPO_4 \cdot 2H_2O$	Moderate
Plus White® Plus White Plus®	$CaHPO_4$ and $CaHPO_4 \cdot 2H_2O$	Moderate to high
Ultra Brite®	$CaHPO_4$ and $CaHPO_4 \cdot 2H_2O$	Moderate
Vote®	Silica	Moderate to high

* From Accepted Dental Therapeutics, 34th ed., Council on Dental Therapeutics, American Dental Association, Chicago, p. 298, 1971/72. By permission.

Official Product

Pumice, N.F. XIII

Pumice is a substance of volcanic origin, consisting chiefly of complex silicates of aluminum, potassium, and sodium. It occurs as very light, hard, rough, porous, grayish masses, or as a gritty, grayish powder. Pumice is odorless and tasteless, stable in air, practically insoluble in water, and is not attacked by acids. The fineness of the powder must be indicated on the label.

Dentifrices Containing Desensitizing Agents. There are two dentifrices currently promoted as products that will reduce the sensitivity of the teeth to heat and cold. Sensodyne® contains strontium chloride. The Council on Dental Therapeutics does not feel that there is enough evidence to justify the claims made for this product. Thermodent® contains formalin. This product is not described in *Accepted Dental Therapeutics* 1971/72.

Other Desensitizing Agents. *Ammoniacal Silver Nitrate Solution* N.F. XII was official as a dental protective and desensitizing agent. It was mixed with a reducing agent such as 10% formaldehyde and then applied to exposed dentin or small lesions, with a resulting deposit of metallic silver [rx (i)]. It is no longer an accepted procedure.

$$\text{HCHO} + 2\,\text{Ag(NH}_3)_2\text{OH} \rightarrow \text{HCOONH}_4 + 2\,\text{Ag} + 3\,\text{NH}_3 \uparrow + \text{H}_2\text{O} \qquad \text{(i)}$$

Official Products

Zinc Chloride, U.S.P. XVIII (ZnCl$_2$; Mol. Wt. 136.28)

Zinc Chloride U.S.P. XVIII occurs as a white or practically white, odorless, crystalline powder or white or practically white crystalline granules. It may also be found as porcelain-like masses or molded into cylinders. *Zinc Chloride* U.S.P. is very deliquescent. A 1 in 10 solution is acid to litmus. Zinc chloride is very soluble in water and freely soluble in alcohol and glycerin. Its solution in water or in alcohol is usually slightly turbid, but the turbidity disappears when a small quantity of hydrochloric acid is added.

CATEGORY: Astringent; dentin desensitizer.

FOR EXTERNAL USE: Topical to the teeth as a 10% solution; to the skin and mucous membranes, as a 0.5 to 2% solution.

Zinc-Eugenol Cement, N.F. XIII

Zinc-Eugenol Cement N.F. XIII consists of two parts: (1) the powder and (2) the liquid. Their respective formulas are:

1. The powder

Zinc acetate	0.5 g
Zinc stearate	1 g
Zinc oxide	70 g
Rosin	28.5 g

2. The liquid

Eugenol	85 ml
Cottonseed oil	15 ml

To prepare the cement, mix ten parts of the powder with one part of the liquid to form a thick paste immediately before use. The mixture will set into a hardened mass consisting of zinc oxide embedded in a matrix of long, sheathlike crystals of zinc eugenolate ([C$_{10}$H$_{11}$O$_2$]$_2$Zn).[65] The zinc acetate accelerates the setting time.[76]

Zinc oxide-eugenol preparations are widely used by dentists for their sedative effect on pulpal pain, particularly when restoring teeth with deep carious lesions.

References

1. Council on Dental Therapeutics. Accepted Dental Therapeutics, 34th ed. Chicago: American Dental Association, pp. 80–81, 1971/72.
2. Poole, D. F. G., and Neuman, H. N. Dental plaque and oral health. Nature, **234**: 329, 1971, and references cited therein.
3. Petersen, J. K. Therapeutic dentifrices. Advances Appl. Microbiol., **13**:343, 1970.
4. Council on Dental Therapeutics, American Dental Association. Council announces classification of additional products. J. Amer. Dent. Ass., **40**:625, 1950.
5. Accepted Dental Therapeutics, 1971/72, *loc. cit.*, p. 259.
6. Brudevold, F., Gardner, D. E., and Smith, F. A. The distribution of fluoride in human enamel. J. Dent. Res., **35**:420, 1956.
7. Issac, S., Brudevold, F., Smith, F. A., and Gardner, D. E. Solubility and natural fluoride content of surface and subsurface enamel. J. Dent. Res., **37**:254, 1958.
8. Dowse, C. M., and Jenkins, G. N. Fluoride uptake in vivo in enamel defects and its significance. J. Dent. Res., **36**:816, 1957.
9. Grøn, P., McCann, H. G., and Brudevold, F. The direct determination of fluoride in human saline by a fluoride electrode. Arch. Oral Biol., **13**:203, 1968.
10. Dawes, C., Jenkins, G. N., Hardwick, J. L., and Leach, S. A. The relation between the fluoride concentration in the dental plaque and in drinking water. Brit. Dent. J., **119**:164, 1965.
11. Jenkins, G. N., Edgar, W. M., and Ferguson, D. B. The distribution and metabolic effects of human plaque fluorine. Arch. Oral Biol., **14**:105, 1969.
12. Woolley, L. H., and Rickles, N. H. Inhibition of acidogenesis in human dental plaque *in situ* following the use of topical sodium fluoride. Arch. Oral Biol., **16**:1187, 1971.
13. Jenkins, G. N. Mechanism of action of fluoride in reducing dental caries. Fluoride Quart. Rep., **2**:236, 1969.
14. Full, C. A., and Parkins, F. M. Fluoride reduction by home water softeners. J. Dent. Res., **51**:666, 1972.
15. Dean, H. T. Fluorine in the control of dental caries. J. Amer. Dent. Ass., **52**:1, 1956.
16. Hayes, R. L., Littleton, N. W., and White, C. L. Posteruptive effects of fluoridation on first permanent molars of children in Grand Rapids, Michigan. Amer. J. Public Health, **47**:192, 1957.
17. Arnold, F. A., Jr., Dean, H. T., Jay, P., and Knutson, J. W. Effect of fluoridated public water supplies on dental caries prevalence. Tenth year of the Grand Rapids-Muskegon study. Public Health Rep., **71**:652, 1956.
18. Ast, D. B., Smith, D. J., Wachs, B., and Cantwell, K. T. Newburgh-Kingston caries-fluorine study. XIV. Combined clinical and roentgenographic dental findings after ten years of fluoride experience. J. Amer. Dent. Ass., **52**:314, 1956.
19. Hutton, W. L., Linscott, B. W., and Williams, D. B. Final report of local studies on water fluoridation in Brantford. Canad. J. Public Health, **47**:89, 1956.
20. Lemke, C. W., Daherty, J. M., and Arra, M. C. Controlled fluoridation: The dental effects of discontinuation in Antigo, Wisconsin. J. Amer. Dent. Ass., **80**: 782, 1970.
21. Ast, D. B., Cons, N. C., Pollard, S. T., and Garfinkel, J. Time and cost factors to provide regular, periodic dental care for children in a fluoridated and nonfluoridated area: Final report. J. Amer. Dent. Ass., **80**:770, 1970.
22. Sapolsky, H. M. Social science views of a controversy in science and politics. Amer. J. Clin. Nutr., **22**:1397, 1969.
23. Douglas, B. L., Wallace, D. A., Lerner, M., and Coppersmith, S. B. Impact of water fluoridation on dental practice and dental manpower. J. Amer. Dent. Ass., **84**:355, 1972.
24. Fluoridated water, hemodialysis, and renal disease. Med. Lett., **11**:67, 1969.
25. Rubini, M. E. Fluoridation: The unpopular controversy. Amer. J. Clin. Nutr., **22**:1343, 1969.

26. Ericsson, Y.   Fluoride excretion in human saliva and milk.   Caries Res., 3:159, 1969.
27. Ericsson, Y., and Ribelius, U.   Wide variations of fluoride supply to infants and their effect.   Caries Res., 5:78, 1971.
28. Accepted Dental Therapeutics 1971/72, *loc. cit.*, p. 205.
29. Zipkin, I., and Babeaux, W. L.   Maternal transfer of fluoride.   J. Oral Ther. Pharmacol., 1:652, 1965, and references cited therein.
30. Zipkin, I., and Babeaux, W. L.   Dental aspects of the prenatal administration of fluoride.   J. Oral Ther. Pharmacol., 3:124, 1966, and references cited therein.
31. Dale, P. B.   Prenatal fluorides: The value of fluoride during pregnancy.   J. Amer. Dent. Ass., 68:530, 1964, and references cited therein.
32. Blayney, J. R., and Hill, I. N.   Evanston dental caries study XXIV, prenatal fluorides—value of waterborne fluorides during pregnancy.   J. Amer. Dent. Ass., 69:291, 1964.
33. Tank, G., and Storvick, C. A.   Caries experience of children one to six years old in two Oregon communities (Corvallis and Albany). 1. Effect of fluorides on caries experience and eruption of teeth.   J. Amer. Dent. Ass., 69:749, 1964.
34. Accepted Dental Therapeutics 1971/72, *loc. cit.*, p. 206.
35. Greene, J. C.   Fluoridation—pill or public water supply.   J.A.M.A., 207:2292, 1969.
36. Toth, K.   Increment of dental caries over two years of fluoridation of domestic salt. A preliminary report.   Caries Res., 4:293, 1970.
37. Wespe, H. J., and Bürgi, W.   Salt-fluoridation and urinary fluoride excretion. Caries Res., 5:89, 1971.
38. Rusoff, L. L., Konikoff, B. S., Frye, J. B., Jr., Johnston, J. E., and Frye, W. W. Fluoride addition to milk and its effect on dental caries in school children.   Amer. J. Clin. Nutr., 11:94, 1962.
39. Horowitz, H. S., Heifetz, S. B., and Law, F. E.   Effect of school fluoridation on dental caries: Final results in Elk Lake, Pa., after 12 years.   J. Amer. Dent. Ass., 84:832, 1972.
40. Horowitz, H. S., Creighton, W. E., and McClendon, B. J.   The effect on human dental caries of weekly oral rinsing with a sodium fluoride mouthwash. A final report.   Arch. Oral Biol., 16:609, 1971.
41. Mellberg, J. R., Nicholson, C. R., and Law, F. E.   Fluoride concentrations in deciduous tooth enamel of children chewing sodium fluoride tablets.   J. Dent. Res., 51:551, 1972.
42. Accepted Dental Therapeutics 1971/72, *loc. cit.*, p. 209.
43. Horowitz, H., and Doyle, J.   The effect on dental caries of topically applied acidulated phosphate-fluoride: Results after three years.   J. Amer. Dent. Ass., 82:359, 1971, and references cited therein.
44. Mellberg, J. R., Nicholson C. R., Miller, B. G., and Englander, H. R.   Acquisition of fluoride in vivo by enamel from repeated topical sodium fluoride applications in a fluoridated area: Final report.   J. Dent. Res., 49:1473, 1970.
45. Englander, H. R., Sherrill, L. T., Miller, B. G., Carlos, J. P., Mellberg, J. R., and Senning, R. S.   Incremental rates of dental caries after repeated topical sodium fluoride applications in children with lifelong consumption of fluoridated water. J. Amer. Dent. Ass., 82:354, 1971.
46. Lim, J. K. J.   Precipitate-free, dilute aqueous solution of stannous fluoride for topical application: I. Simple and mixed mediums.   J. Dent. Res., 49:760, 1970.
47. Heifetz, S. B., Mellberg, J. R., Winter, S. J., and Doyle, J.   *In vivo* fluoride uptake by enamel of teeth of human adults from various topical fluoride procedures.   Arch. Oral Biol., 15:1171, 1970.
48. Cons, N. C., Janerich D. T., and Senning, R. S.   Albany topical fluoride study. J. Amer. Dent. Ass., 80:777, 1970.
49. Mellberg, J. R., Laakso, P. V., and Nicholson, C. R.   The acquisition and loss of fluoride by topically fluoridated human tooth enamel.   Arch. Oral Biol., 11:1213, 1966.
50. Aasenden, R., Brudevold, F., and Richardson, B.   Clearance of fluoride from the mouth after topical treatment or the use of a fluoride mouthrinse.   Arch. Oral Biol., 13:625, 1968.

51. Jerman, A. C.   Silver amalgam restorative material with stannous fluoride. J. Amer. Dent. Ass., **80**:787, 1970.
52. Wei, S. H. Y., and Sierk, D. L.   Fluoride uptake by enamel from zinc phosphate cement containing SnF₂. J. Amer. Dent. Ass., **83**:621, 1971.
53. Stoner, G. E., Senti, S. E., and Gileadi, E.   Effect of sodium fluoride and stannous fluoride on the rate of corrosion of dental amalgams. J. Dent. Res., **50**:1647, 1971.
54. Nizel, A. E., and Harris, R. S.   The effects of phosphates on experimental dental caries: A literature review. J. Dent. Res., **43**:1123, 1964, and references cited therein.
55. Stralfors, A.   The effect of calcium phosphate on dental caries in school children. J. Dent. Res., **43**:1137 1964.
56. Ship, I. I., and Mickelsen, O.   The effects of calcium acid phosphate on dental caries in children: A controlled clinical trial. J. Dent. Res., **43**:1144, 1964.
57. Averill, H. M., and Bibby, B. G.   A clinical test of additions of phosphate to the diet of children. J. Dent. Res., **43**:1150, 1964.
58. Harris, R., Schamschula, R. G., Beveridge, J., and Gregory, G.   The cariostatic effect of calcium sucrose phosphate in a group of children aged 5–17 years. Aust. Dent. J., **13**:32, 1968.
59. Stookey, G. K., Carroll, R. A., and Muhler, J. C.   The clinical effectiveness of phosphate-enriched breakfast cereals on the incidence of dental caries in children: Results after 2 years. J. Amer. Dent. Ass., **74**:752, 1967.
60. Accepted Dental Therapeutics 1971/72, *loc. cit.*, p. 81.
61. *Ibid.*, p. 213.
62. Gray, J. A.   Unpublished data, Miami Valley Laboratories.   The Procter and Gamble Co.   Cited in Cooley, W. E.   Applied research in the development of anti-caries dentifrices. J. Chem. Ed., **47**:177, 1970.
63. Cooley, W. E.   Reactions of tin (II) and fluoride anions with etched enamel. J. Dent. Res., **40**:1199, 1961.
64. Gray, J. A., Schweizer, H. C., Rosevear, F. B., and Broge, R. W.   Electron microscope observation of the differences in the effects of stannous fluoride and sodium fluoride on dental enamel. J. Dent. Res., **37**:638, 1958.
65. Langer, H. G., and Nebergall, W. H.   Identification of the reaction products of stannous fluoride and sodium fluoride with powdered dental enamel. J. Dent. Res., **37**:58, 1958.
66. Collins, R., Nebergall, W., and Langer, H.   A study of the reactions of various tin (II) compounds with calcium hydroxyl-apatite. J. Amer. Chem. Soc., **83**:3724, 1961.
67. Scott, D. B.   Electron-microscope evidence of fluoride-enamel reaction. J. Dent. Res., **39**:1117, 1960.
68. Muhler, J. C., Radike, A. W., Nebergall, W. H., and Day, H. G.   A comparison between the anticariogenic effects of dentifrices containing stannous fluoride and sodium fluoride. J. Amer. Dent. Ass., **51**:556, 1955.
69. Accepted Dental Therapeutics 1971/72, *loc. cit.*, p. 212.
70. *Ibid.*, p. 211.
71. Council on Dental Therapeutics.   Council classifies Colgate with MFP (sodium, monofluorophosphate) in Group A. J. Amer. Dent. Ass., **79**:937, 1969.
72. Robinson, H. B. G.   Individualizing dentifrices: The dentist's responsibility. J. Amer. Dent. Ass., **79**:633, 1969.
73. Wilkinson, J. B., and Pugh, B. R.   Toothpastes—cleaning and abrasion. J. Soc. Cosmetic Chem., **21**:595, 1970.
74. Facq, J. M., and Volpe, H. R.   In vivo actual abrasiveness of three dentifrices against acrylic surfaces of veneer crowns. J. Amer. Dent. Ass., **80**:317, 1970.
75. Stookey, G. K., and Muhler, J. C.   Laboratory studies concerning the enamel and dentin abrasion properties of common dentifrice polishing agents. J. Dent. Res. **47**:524, 1968.
76. Copeland, H. I., Jr., Brauer, G., and Forziati, A. F.   The setting mechanism of zinc oxide and eugenol mixtures. J. Dent. Res., **34**:740, 1955.

11

Radiopharmaceuticals and Contrast Media

Throughout this text discussions of chemistry have centered upon those properties and reactions of atoms and molecules involving electrons in the outermost atomic and molecular orbitals. In direct contrast, this chapter will deal primarily with nuclear reactions, that is, reactions that occur within nuclei of certain atoms or between the nucleus and the innermost electrons.

Radiopharmaceuticals

Radioisotopes. As was discussed in Chapter 1, every atom of an element is composed of a nucleus, containing protons and neutrons, surrounded by electrons. In the electrically neutral atom, the number of electrons is equal to the number of protons in the nucleus. Furthermore, the number of protons in the nucleus is equal to the atomic number of the atom, which determines its properties. The atomic number of an atom is also characteristic of the particular element of which the atom is a part. In fact, the atomic number is the determinant of elements, in that all the atoms composing an element have the same number of protons and, therefore, the same number of electrons. However, all atoms of an element are not exactly alike. Most elements contain a certain percentage of atoms which differ in atomic weight or mass from the majority of the atoms present. These different forms of an element are known as *isotopes*, and they vary in the number of neutrons contained in the nuclei of their atoms. Isotopes of a particular element, then, have the same atomic number (same number of protons) but different mass numbers (differing numbers of neutrons).

The isotopes of a particular element have the same chemical and physical properties. The only variation that is usually found is in the kinetics or rates of chemical reactions involving the isotopes, since the mass is a very important aspect of reaction rates. Other general facts about isotopes are that the natural abundance of stable forms in elements is constant regardless of the form in which the element is found. For example, carbon always contains about 1% of $^{13}_{6}C$ and 99% of $^{12}_{6}C$, and chlorine contains about 75% $^{35}_{17}Cl$ and about 25% of $^{37}_{17}Cl$. (The preceding notation will be utilized throughout this chapter. The superscript is the mass number and the subscript is the atomic number of the element presented.) There is an average of about eight isotopes per element considering all the known isotopic forms and, as a general rule, those elements with even atomic numbers have more isotopes than those with odd atomic numbers. The heavier elements tend to have more isotopes than the lighter elements. And, finally, the existence of fractional atomic weights for an element is due largely to the natural presence of isotopic forms, which contribute to the atomic weight in proportion to their natural abundance.

Two major types of isotopes are found in nature. Stable isotopes maintain their elemental integrity, and do not decompose to other isotopic or elemental forms. Unstable or radioactive isotopes, however, decompose or decay, by emission of nuclear particles, into other isotopes of the same or different elements. The decay is characteristic for a particular isotope, and continues until a stable isotopic level is achieved. Since the transition from one isotope to another, whether it is within the same element or not, involves nuclear transformations, the chemistry of radioactive isotopes differs from the chemistry of stable isotopes by the additional aspect of nuclear reactions (see below).

Some unstable isotopes exist as or pass through metastable states. Metastable isotopes are characterized by slow rates of decay, allowing an accumulation of the metastable form which often gives the appearance of a stable form of the isotope. In reality the metastable form is decaying at a rate significantly slower than the parent isotope.

Not all radioactive isotopes are found naturally in elements. In fact, a large number of unstable isotopes are produced synthetically for their usefulness in chemical, geological, and biological applications. The production of radioactive isotopes usually involves the bombardment of atomic nuclei with subatomic particles, i.e., neutrons or electrons, to produce unstable nuclei of the same or a different element. Examples of this will be illustrated later in this chapter.

Radioactive Decay Particles. When a radioactive isotope decays, it does so with the emission of certain particles or quantities of energy that are characteristic of the particular isotope involved. There are numerous species that have been identified as particles of nuclear origin. However, the major decay particles of interest here include alpha particles, beta particles, including negatrons (electrons of nuclear origin), and positrons, gamma rays, and x-rays. The properties of these particles are presented in the following discussions.

Alpha Particles (α, $^{4}_{2}He^{+2}$). These radiations are by far the heaviest and slowest of all radioactive emissions. The particle is actually a helium nucleus, containing two protons and two neutrons for an atomic mass of 4 (weight approximately 6.6×10^{-24} g) and an atomic number of 2. It is the most highly charged nuclear species, with a charge of $+2$, giving it a very high ionizing power upon interaction with air or other media. Alpha particles move at a relatively slow speed, averaging about 0.1 the speed of light (the speed of light is 3.0×10^{10} cm/sec or 186,000 miles/sec), and their penetrating power is very low. They can be stopped by a sheet of paper or a very thin sheet of aluminum foil. These particles will travel only 3 to 8 cm in air.

Alpha radiation is usually emitted only from elements having atomic numbers greater than 82. Isotopes emitting alpha particles will decay to the element having a mass number of four less and an atomic number of two less than the original isotope. However, the emission of alpha radiation is usually accompanied by other radiations. The emission of alpha radiation is illustrated below with radium-226 (the radium isotope having a mass number of 226) in rx (i).

$$^{226}_{88}\text{Ra} \rightarrow {}^{222}_{86}\text{Rn} + \alpha \text{ (or } {}^{4}_{2}\text{He}^{+2}) \tag{i}$$

The low penetrating power of alpha particles makes isotopes emitting this type of radiation rather useless for biological applications because these particles cannot penetrate tissue.

Beta Particles (β^- or β^+). This is the terminology normally applied to radiation which may be further described as an electron of nuclear origin. Beta particles are negatively charged species having the mass of an electron (approximately 9.1×10^{-28} g). Since these radiations are lighter than alpha particles by a factor of 10^{-4}, they move at faster velocities often approaching 0.9 the speed of light. Their emissions from elements do not alter the atomic mass but do alter the atomic number [see rx (iii)]. The negative charge causes beta particles to produce ionizations of molecules when they pass through various media.

Depending upon their energy, beta particles have more penetrating power than alpha particles and are often able to travel from 10 to 15 mm in water or penetrate almost 1-inch thicknesses of aluminum.

Beta particles, sometimes called negatrons, are emitted by unstable nuclei having neutrons in excess of protons. If the neutron/proton ratio exceeds stable limits, a transformation of a neutron to a proton will occur, with the expulsion of beta radiation [rx (ii)]. Elements undergoing this type of transformation will decay to the element having the next highest atomic number.

$$^{1}_{0}\text{n} \rightarrow {}^{1}_{1}\text{p} + \beta^- \tag{ii}$$

An example of beta decay is illustrated in rx (iii) with carbon-14.

$$^{14}_{6}C \rightarrow {}^{14}_{7}N + \beta^{-} \tag{iii}$$

Elements may emit beta radiation alone or in combination with other types of radiation. As mentioned above, beta particles may produce ionizations as they pass through various media; however, interactions with surrounding nuclei may cause an acceleration in the beta particle depending upon the atomic number or charge of the nucleus. As a result of this process, some of the energy of the beta particle is seen as electromagnetic radiation known as *bremstrahlung*, or "braking radiation." This radiation is similar to gamma radiation (see below). Many isotopes emitting beta radiation have useful biological applications since the radiation will penetrate tissues.

Another type of beta radiation that is not as common as the negatively charged particle discussed above is seen in the emission of *positrons* (β^{+}). This particle is identical to the electron with the exception of having a positive charge. It is emitted from nuclei having a proton/neutron ratio above stable limits. According to rx (iv), a proton can be transformed into a neutron, accompanied by the emission of the positron.

$$^{1}_{1}p \rightarrow {}^{1}_{0}n + \beta^{+} \tag{iv}$$

Elements emitting positron radiation will do so only to a small fraction of their total radioactive emissions, and will decay to the element having the next lowest atomic number. An illustration of positron beta decay is shown in rx (v) with zinc-65.

$$^{65}_{30}Zn \rightarrow {}^{65}_{29}Cu + \beta^{+} \tag{v}$$

Positrons are of little importance in biological applications. They are very short-lived species in that they undergo annihilation reactions with electrons to produce gamma radiation or annihilation photons [rx (vi)].

$$\beta^{+} + e^{-} \rightarrow 2\gamma \tag{vi}$$

Gamma Radiation (γ). The gamma ray may be best described as a photon of electromagnetic radiation. It demonstrates both wave and particle properties, as do electrons and beta particles. The rays are of short wavelength (10^{-10} to 10^{-8} cm) similar to x-rays, and travel at the speed of light. They have no mass and no charge but they are of very high energy, giving them excellent penetrating power. Very thick lead or concrete shielding is required to protect against the radiation from strong gamma-emitting sources. The fact that this radiation is uncharged indicates that its ionizing power is poor. However, gamma rays can interact with atoms and molecules in a particular medium to produce ions and free radicals secondarily by dislodging electrons from orbitals. The high penetrating power of these packets of energy increases the opportunity for secondary ionizations.

The emission of gamma rays from an element results in a lowering of the nuclear energy level of the element, but no elemental change is noted unless one of the other types of radiation is also emitted. The latter is usually the case, in that the emission of gamma rays is almost always accompanied by the emission of other forms of radiation. If gamma emission does occur alone, it will involve the transition of a metastable state of an isotope either to a stable form or to a form of the same isotope which will continue to decay by other means. For example, consider rx (vii) involving the metastable state of cobalt-60, cobalt-60m, which is produced by neutron bombardment of stable cobalt-59 with the subsequent emission of a gamma (n, γ) as indicated by the notation on the far left of the reaction.

$$\ce{^{59}_{27}Co} \; (n, \gamma) \; \ce{^{60m}_{27}Co} \rightarrow \ce{^{60}_{27}Co} + \gamma \rightarrow \ce{^{60}_{28}Ni} + \beta^- + 2\gamma \qquad \text{(vii)}$$

K-Capture. A type of radiation similar to gamma rays is seen in the emission of x-rays through a process known as K-capture. This type of radiation is produced by isotopes with an unstable proton/neutron ratio but with insufficient energy to emit a positron. Alternatively, the nucleus "captures" an electron from the so-called K shell ($1s$ orbital) (this process can also occur from the L shell or $2s$ orbital) which combines with a proton to form a neutron. The rearrangement of the orbital electrons takes place with the release of energy in the form of x-rays. Although the process involves the nucleus, the energy released comes largely from the necessary electronic rearrangements. The loss of a proton in the nucleus indicates that the isotope will decay to the element having the next lowest atomic number. Mercury-197 is an example of an isotope emitting radiation by K-capture [rx (viii)].

$$\ce{^{197}_{80}Hg} \xrightarrow{\text{K-capture}} \ce{^{197}_{79}Au} + \text{x-ray} \qquad \text{(viii)}$$

Isotopes emitting gamma radiation are used frequently in biological applications. The penetrating power of this form of radiation is sufficient to reach deep into tissues, and to be detected outside of the body.

Before leaving this topic, there are some points that should be mentioned concerning the energy characteristics of beta and gamma radiations. The energies of the radiations emitted by particular isotopes can be determined

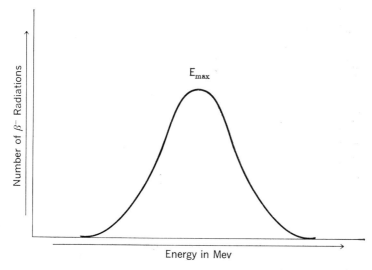

Fɪɢ. 11–1.   Plot of the number of β^- radiations from an isotope vs. the energy of the radiations, showing the E_{max} for the β emissions.

by instrumental techniques, and the energies expressed in units of million electron volts (Mev).   One electron volt (ev) is defined as the energy acquired by an electron when accelerated through a potential of one volt. For comparison, *one* Mev is equivalent to 1.6×10^{-6} ergs/molecule or 2.3×10^7 kcal/mole.

An istope may emit more than one beta and/or gamma radiations having different energies.   For example, cobalt-60 in rx (vii) above emits one beta particle and two gamma rays.   The beta radiation has an energy of about 0.31 Mev and the two gamma rays are 1.17 and 1.33 Mev, respectively. The beta emission takes place over a continuous spectrum in which the value of 0.31 Mev represents the energy of the largest number of beta particles, or the E_{max} (see Fig. 11–1).   On the other hand, gamma radiation is emitted in single energy packets, or photons.   The two gamma rays from cobalt-60 are the only two gamma energies found in that particular isotope, and there are no gamma energies in between (see Fig. 11–2).   The energy values of the various radiations are important for identification and measurement of radioactive isotopes.

Table 11–1 summarizes some properties of the common radiations produced during radioactive decay.

Kinetics of Isotope Decay.   The activity (not to be confused with the energy) of radioactive isotopes is generally expressed in terms of the *curie*, c, which is equivalent to the amount of radioactive material providing 3.7×10^{10} atomic disintegrations per second (dps).   The actual quantities of material corresponding to one curie will vary with the particular isotope

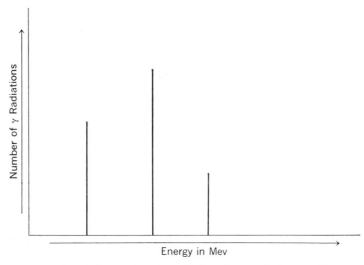

FIG. 11–2. Plot of the number of γ radiations from an isotope vs. the energy of the radiations, illustrating three separate gamma emissions.

involved. The definition does not specify a weight or concentration; neither does it specify the number of radiations nor their energies. The curie represents a large amount of radioactivity for biological purposes; therefore, the millicurie, mc, $(3.7 \times 10^7$ dps) and the microcurie, μc, $(3.7 \times 10^4$ dps) are used most frequently in pharmaceutical or medical applications.

Due to spontaneous decay, the activity of a radioactive sample diminishes with time. The rate of isotopic decay is independent of the concentration of the radioactive atoms present. This means that the decay process should follow zero order kinetics, which describe the rates of reactions proceeding in a manner independent of reactant concentrations. In radioactivity, the only changing quantities are the activity, A, with respect to time, t. The zero order expression for the disappearance in activity with respect to time can be written as the differential equation (Eq. 1), where λ is the decay constant for the particular isotope involved.

$$\frac{-dA}{dt} = \lambda A \qquad \text{(Eq. 1)}$$

Eq. 1 can be rearranged to give Eq. 2, which can then be integrated between the limits of the initial activity, A_0, and the activity at some finite time to give Eq 3.

$$\frac{-dA}{A} = \lambda dt \qquad \text{(Eq. 2)}$$

$$\ln \frac{A_0}{A} = \lambda t \qquad \text{(Eq. 3)}$$

TABLE 11-1. PROPERTIES OF RADIOACTIVE DECAY PARTICLES

Radiation	Particle Type	Mass	Charge	Velocity*	Effect on Emission		Penetration Power	Ionizing Power
					Atomic No.	Mass No.		
α	$^4_2He^{+2}$	6.6×10^{-24} g (4 atomic mass units)	+2	0.1 c	−2	−4	Low	High
β⁻	electron	9.1×10^{-28} g	−1	0.4–0.9 c	+1	0	Moderate	Moderate
β⁺	positron	9.1×10^{-28} g	+1	0.4–0.9 c	−1	0	Low due to annihilation reactions	Moderate
γ	photon	0	0	c	0	0	High	Low
K-capture x-ray	photon	0	0	c	−1	0	High	Low

* The velocity is given in terms of the speed of light, $c = 3.0 \times 10^{10}$ cm/sec.

An equivalent form of Eq. 3 is the familiar exponential equation for zero order reactions shown in Eq. 4.

$$A = A_0 e^{-\lambda t} \tag{Eq. 4}$$

The decay of every radioactive isotope may be described in terms of its half-life, $t_{\frac{1}{2}}$, which is the time during which one half of the remaining radioactive atoms in a particular sample will decay. The specific half-life is a characteristic quantity for each radionuclide. Since the passage of one half-life would leave one half the initial activity, Eq. 3 can be rewritten to give:

$$\ln \frac{A_0}{(A_0/2)} = \lambda t_{\frac{1}{2}} \tag{Eq. 5}$$

or

$$\ln 2 = \lambda t_{\frac{1}{2}} \tag{Eq. 6}$$

Solving Eq. 6 for λ and substituting the value of the natural logarithm of 2 provides Eq. 7.

$$\lambda = 0.693/t_{\frac{1}{2}} \tag{Eq. 7}$$

Substituting Eq. 7 into Eq. 3 and converting the logarithm to the base 10 provides:

$$\log \frac{A_0}{A} = \frac{0.693}{2.303} (t/t_{\frac{1}{2}}) = 0.301 \, t/t_{\frac{1}{2}} \tag{Eq. 8}$$

Eq. 8 may be rearranged to give:

$$\log A = \log A_0 - 0.301 (t/t_{\frac{1}{2}}) \tag{Eq. 9}$$

Given the initial activity of a radioisotope preparation, the date of measurement, and the half-life, Eq. 9 can be used to calculate the activity remaining in the preparation at any time. This must be done whenever a specific quantity of radioactivity is required for chemical or therapeutic purposes.

The use of Eq. 9 can be illustrated by considering a preparation containing sodium phosphate P-32 (U.S.P. XVIII) having an initial activity of 500 μc. Given that the $t_{\frac{1}{2}} = 14.2$ days for phosphorus-32, what is the activity present 30 days after the product was prepared? Substituting the appropriate quantities into Eq. 9:

$$\log A = \log 500 - 0.301 \, (30/14.2)$$
$$\log A = 2.699 - 0.635 = 2.064$$
$$A = 115.9 \; \mu c$$

Radiation Dosimetry. A complete discussion of radiation dosimetry is beyond the scope of this chapter, and further information can be found in Reference 1. The problem of dosimetry refers to both animate and inanimate objects; however, the prime interest here is the effect of radiation on man and animals. There are two areas of concern in radiological health standards that require description and units of measurement. These two areas can best be described by the terms "exposure dose" and "absorbed dose."

The exposure dose of radioactivity refers to the quantity of potentially harmful radiation in the air surrounding a particular area, or the amount of radiation available for interaction with some target material. The unit of measurement for this aspect of radiation dosage is the *roentgen*, r, named after W. C. Röntgen, the discoverer of x-rays. The roentgen was defined in 1937 in terms of x- or gamma radiation, but the original definition has been expanded to various convenient units of volume, weight, and charge. A common definition of one roentgen (1 r) is the quantity of x- or gamma radiation that will produce sufficient ion pairs to have one electrostatic unit (e.s.u.) of charge of either sign in 1 cc of dry air at standard temperature and pressure (STP). This is equivalent to 2.083×10^9 ion pairs per cc of dry air at STP, or 1.610×10^{12} ion pairs per gram of air. The energy dissipated in the formation of the ion pairs can be calculated on the basis of about 35 ev per ion pair, to give 6.8×10^4 Mev per cc or 5.2×10^7 Mev per gram of air or 83.8 ergs per gram of air. The roentgen is technically limited to x- or gamma radiation, and to the effect produced in air only. There is a rough equivalency of 1 r to about 93 ergs per gram of tissue.

The exposure dose is usually measured as a dose rate at a particular distance from the radiation source. For example, the dose may be reported in units of roentgens per hour at one meter (r/hr/m or rhm) indicating a dose rate of so many roentgens per hour measured at one meter from the source. Measurements may also be made in milliroentgens, mr, (1 mr = 10^{-3} r).

The absorbed dose of radiation is somewhat more complicated. The basic unit employed in specifying an absorbed dose is the *rad* (radiation *a*bsorbed *d*ose). The rad measures the amount of energy transferred to a particular medium, and 1 rad was defined, in 1954, to be equal to the absorption of 100 ergs per gram of absorbing medium regardless of the medium. This term does not distinguish the type of material; therefore, a modifying term has been introduced to aid in describing potential damage to biological tissue due to different types of radiation. This term is known as the relative biological effectiveness or RBE, and expresses the relative effects of each type of radioactive particle on tissue. For x- and gamma radiation and beta particles, the RBE is assigned a value of 1. Due to their greater ionizing power when absorbed, alpha particles have a range

TABLE 11-2. RADIATION DOSIMETRY UNITS

Name	Symbol	Radiation Type	Equivalent Units
Roentgen	r	x, γ	2.083×10^9 ion pairs/cc dry air at STP 83.8 ergs/g dry air at STP 6.8×10^4 Mev/cc dry air at STP 5.2×10^7 Mev/cc dry air at STP
Radiation Absorbed Dose	rad	all	100 ergs/g absorbed by any medium
Relative Biological Effectiveness	RBE	α β x, γ	10–20 1 1
Roentgen Equivalent Man	rem	all	Dose in rads $\times$ RBE

in RBE from 10 to 20, depending upon the particular tissue or organ involved. This indicates that alpha radiation, if it were absorbed, would be at least ten times more effective at producing a particular result in tissue as the same number of rads of gamma radiation.

When the dose in rads is multiplied by the appropriate RBE a more biologically useful dosage is obtained and carries the unit known as the *rem* (*r*oentgen *e*quivalent *m*an). The rem is commonly used to determine doses received by those working with radioisotopes and participating in required monitoring programs for radioactive protection. For example, an established standard dose limit of 0.3 rem per week would correspond to 300 mrads (millirads) per week of x-, gamma, or beta radiation, or 30 mrads of alpha radiation (RBE = 10). Radiation dosimetry units are summarized in Table 11-2.

The measurement of radiation dosage is accomplished through the use of instruments known as rate meters, or other devices to be described later, calibrated in units of r/hr or rads/hr. The measurements so obtained follow the inverse square law, meaning that the radioactivity from a particular source is inversely proportional to the square of the distance from the source. Consider a sample radiating 3.0×10^5 rads at 10 cm. The radiation at one meter (100 cm), x, can be calculated from the following relationship:

$$\frac{3.0 \times 10^5}{x} = \frac{100^2}{10^2}$$

$$x = 3.0 \times 10^3 \text{ rads}$$

This relationship neglects the possible absorption of the radiation in air at the greater distance.

Biological Effects of Radiation. The effect of radioactive particles impinging upon biological tissue depends upon a number of factors related to the ability of the radiation to penetrate tissue, the energy of the radiation, the particular tissue and surface area exposed, and the dose rate of the radiation. The destructive aspect of radioactivity is directly related to its interaction with molecules present in the tissue to form abnormal amounts of ions and/or free radicals. These chemical species can alter the local pH or serve to initiate free radical chain reactions, resulting in the production of peroxides or other toxic compounds. These and other events can create a hostile environment for tissue cells, leading to necrosis and, ultimately, complete destruction of the tissue or organ. Water is the most abundant molecule in most tissues and is the most probable reactive species in the path of ionizing radiation, although other biochemicals may be involved. An illustration of free radical formation and reaction to form hydrogen peroxide is shown in rx (ix). It should be stressed that the free radicals formed from water can also abstract radicals from other molecules, resulting in the production of a variety of potentially toxic species which can alter the DNA in cells and cause crosslinking between certain amino acids in proteins.

$$xH_2O \xrightarrow[\text{particles}]{\text{radioactive}} xH\cdot + xHO\cdot \longrightarrow \text{Other products} \qquad \text{(ix)}$$
$$\qquad\qquad\qquad \downarrow \quad\;\; \downarrow$$
$$\qquad\qquad\qquad yH_2 \quad yH_2O_2$$

Before radiation can produce any damage, it must first gain entry into the tissue. As has been indicated previously, the various types of radiation differ significantly in their abilities to penetrate tissue or other media. Although alpha particles have a potential to produce a tremendous amount of ionization or free radicals, isotopes emitting alpha particles must be directly applied to the tissue, in most cases, in order to observe biological effects. The range and penetration of these particles are so slight that even if an individual were close enough for the radiation to reach the skin, the particles would not penetrate the surface. The opposite characteristic is found in gamma radiation. Although the ionizing power of gamma rays is relatively low, the range and penetrating ability of this type of radiation are high enough to produce significant damage at distances of several meters from the source. The damage is produced through collision reactions with atoms comprising the tissue.

There are several approaches to the monitoring of radiation dosage for the protection of personnel working in areas where isotopes are being manufactured, stored, or utilized. Work areas are usually monitored with a Geiger-Müller (GM) counter, an instrument designed to measure the extent of ionization being produced by radiation. Other detectors are also worn or carried by individuals to aid in maintaining personal records of radiation exposure. Two common types of detectors in this category are the film badge and the pocket dosimeter. The latter measures ionizations, as does the GM counter, and is actually designed along the lines of a pocket electroscope. The film badge provides a permanent record of exposure based upon the amount of darkening of the emulsion on the film. The darkening is proportional to the amount of radiation and can be calibrated for different types of films to provide an actual dosage level.

Shielding is another required protection in areas where radioactive materials are stored or handled. Isotopes that present the greatest problem in terms of shielding are those emitting gamma radiation with greater penetrating power. For this reason, shielding material should be of rather high density and be composed of atoms having high atomic numbers. The high electron densities in materials having these properties offer a higher probability that gamma rays will be absorbed than lighter materials. The usual shielding materials are lead and/or concrete, since these are most efficient in absorbing gamma rays. Because there is always some residual radiation that will pass through shielding, the thicknesses of various types of material are usually expressed in terms of half-thickness for particular energies of gamma radiation. The half-thickness is the thickness of material which will reduce the amount of measured radiation from a particular source by one half. If a particular source is emitting 100 mr/hr, and the half-thickness of concrete is 2.2 inches, then the placement of that thickness between the source and the meter will reduce the radiation to 50 mr/hr, and 4.4 inches will reduce it to 25 mr/hr. In this fashion, the amount of radiation being released into a work area can be reduced to an acceptable level by using appropriate thicknesses of lead and concrete. If the isotope is a pure alpha or beta emitter, shielding requirements are less stringent. Similar considerations must be applied to shipping containers.

Internal Administration of Radioisotopes. There are a number of preparations containing radioisotopes which are used internally for therapeutic and diagnostic purposes. These preparations are referred to collectively as *radiopharmaceuticals* and the important products are discussed in the following sections. Rather extensive reviews of radiopharmaceuticals have appeared by Christian[2] and Wolf and Tubis,[3] and are recommended to those wishing information on other applications and references beyond the discussion presented here.

Isotopes important as radiopharmaceuticals are, first of all, those emitting beta or gamma radiation. The beta radiation is usually of the negatron variety, although, positron emitters are sometimes used. Table 11–3 lists some of the important isotopes used in medicine and biology. Secondly, many of these isotopes are concentrated in a specific manner in certain organs or cells, e.g., ^{131}I in thyroid tissue. Areas of heavy concentration are known as "hot spots" while areas where concentration is light are known as "cold spots" (see Fig. 11–3). Sometimes the isotope must be incorporated

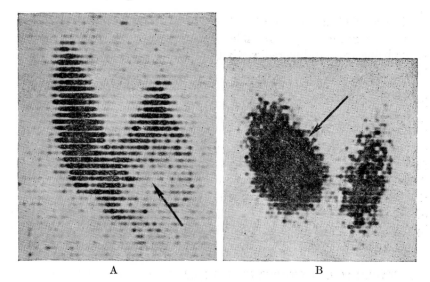

into or "tagged" on to a molecule, which aids in directing it into a particular tissue with some degree of specificity. Due to the potential hazard of radioisotopes, selective absorption and distribution are important factors to be considered in their use. Thirdly, the isotopes should be able to be eliminated from the body easily and, aside from the associated radioactivity, they and the decay products should be of low toxicity.

Isotopes employed in diagnostic procedures must be of sufficient energy to allow measurement of radioactivity outside the body. Common means of measuring radioactivity in diagnostic procedures involve autoradiography, scintillation scanning, and detection with GM tubes. The former is accomplished by exposing a photographic plate over the appropriate area of the body. The autoradiographic picture produced may show an outline of the tissue with heavy concentrations of radioactivity ("hot spots"), indicating possible tumor sites or other areas of abnormal activity. Scintillation scanning utilizes the ability of gamma radiation to excite certain molecules to higher electronic states. When these molecules return to their ground state, light is emitted (phosphorescence) at low intensity which is multiplied through several stages of a photomultiplier tube. The number of phosphorescent events is proportional to the intensity of radioactivity. When the detector scans a particular area of the body, it can map the organs where an isotope has been concentrated. GM tubes are also used to detect and to follow radiation from diagnostic isotopes by measuring the ionizations produced in the gas within the tube.

Therapeutic isotopes are utilized for their destructive effects on tissue. It is necessary that they have sufficient energy to penetrate throughout the tissue being treated, but radioactivity spreading to surrounding tissues is undesirable, however, and difficult to control.

The time element involved in the medical use of radioisotopes is an important consideration. Diagnostic procedures are usually rather short, and it is desirable to have the radioactive compound eliminated from the body within a matter of hours. On the other hand, therapeutic procedures will usually require the presence of the isotope for a longer period of time (days or weeks). The duration of activity of the isotope in the body is determined by two time factors, the physical half-life of the isotope and the biological half-life of the preparation. The mathematical definition of the physical half-life or the half-life of isotopic decay ($t_{\frac{1}{2}}$) can be found by solving Eq. 7 above for $t_{\frac{1}{2}}$. An identical expression given in Eq. 10

TABLE 11–3.   ISOTOPES USED IN RADIOPHARMACEUTICALS
AND BIOLOGICAL RESEARCH

| Isotope | $t_{\frac{1}{2}}$ | Energy (Mev) and Type of Radiation | | | | Application |
		β^-	β^+	γ, x^a	α	
^{198}Au	2.7 d	0.959		0.412		Therapeutic, Diagnostic
^{14}C	5700 y	0.16				Research
^{45}Ca	165 d	0.26				Diagnostic
^{47}Ca	4.5 d	1.1		1.3		Diagnostic
^{57}Co	270 d			0.137 0.123		Diagnostic
^{58}Co	71 d		0.47	0.81, K		Diagnostic
^{60}Co	5.27 y	0.312		1.172 1.332		Therapeutic, Diagnostic
^{51}Cr	27.8 d			0.321, K		Diagnostic
^{131}Cs	9.7 d			0.029, K		Diagnostic
^{137}Cs	30 y	0.518 1.17		0.66		Research
^{18}F	1.7 h		0.6			Diagnostic
^{59}Fe	45 d	0.462 0.271		1.30 1.10		Diagnostic
^{3}H	12.3 y	0.018				Diagnostic, Research
^{197}Hg	2.7 d			0.077		Diagnostic
^{203}Hg	46.9 d	0.21		0.279		Diagnostic

TABLE 11–3. ISOTOPES USED IN RADIOPHARMACEUTICALS
AND BIOLOGICAL RESEARCH (*continued*)

Isotope	$t_{\frac{1}{2}}$	β^-	β^+	γ, x^a	α	Application
^{125}I	60 d			0.027, K 0.035		Diagnostic, Therapeutic
^{131}I	8.08 d	0.608 0.335 0.250		0.722 0.637 0.364 0.284 0.080		Diagnostic, Therapeutic, Research
^{113m}In	1.66 h			0.390		Diagnostic
^{192}Ir	74.4 d	0.67		0.32 0.47		Therapeutic
^{42}K	12.4 h	2.04 3.58		1.5		Research
^{99}Mo	2.8 d	1.23		0.14		Source of ^{99m}Tc
^{22}Na	2.6 y		0.54	1.28, K		Diagnostic
^{24}Na	15 h	1.39		1.38 2.75		Diagnostic
^{32}P	14.3 d	1.71				Therapeutic, Diagnostic, Research
^{226}Ra	1620 y			0.19	4.77	Therapeutic
^{86}Rb	18.8 d	1.77		1.08		Diagnostic
^{222}Rn	3.8 d				5.5	Therapeutic
^{35}S	88 d	0.167				Research
^{75}Se	120 d			0.136 0.265 0.401, K		Diagnostic
^{85}Sr	64 d			0.513, K		Diagnostic
^{90}Sr	28 y	0.54				Therapeutic
^{182}Ta	115 d	0.36 0.44 0.51		0.068 1.12 1.22		Therapeutic
^{99m}Tc	6.0 h			0.140		Diagnostic
^{90}Y	2.6 d	2.26				Diagnostic, Therapeutic
^{169}Yb	32 d			0.063 0.198		Diagnostic
^{65}Zn	245 d		0.32	1.11, K		Research

a "K" indicates that x-rays are emitted by K-capture.

defines the biological half-life, t_b, where k_b is the rate constant for the elimination of the radiopharmaceutical (or any drug) from the body.

$$t_b = \frac{0.693}{k_b} \qquad \text{(Eq. 10)}$$

These two half-lives are combined to provide an expression for the effective half-life,[1,3] t_{eff}, shown in Eq. 11. This equation can be used, when the biological half-life is known, to calculate the time required to reduce the radioactivity in the body to one half the administered activity.

$$t_{eff} = \frac{(t_{\frac{1}{2}})\,(t_b)}{(t_{\frac{1}{2}}) + (t_b)} \qquad \text{(Eq. 11)}$$

Wolf and Tubis[3] have suggested a classification of radiopharmaceuticals on the basis of their effective half-lives. Essentially they propose the terminology of "very short-lived," "short-lived," and "long-lived" to describe preparations having t_{eff} values of less than, up to three times, and greater than five times the particular clinical procedure, respectively. After an isotope has gone through ten half-lives, it essentially has lost all its radioactivity. The physical half-lives of some representative isotopes are listed in Table 11–3, along with the types of radiation emitted and the usual area of application.

This discussion has been centered on the internal administration of radio-isotopes, orally or intravenously, for diagnostic or therapeutic purposes. There are other means of utilizing isotopes for therapeutic benefit. A procedure known as *teletherapy* employing gamma-emitting isotopes, e.g., ^{60}Co, with activity as high as 2000 c, focuses radiation directly on the area under treatment. The patient is exposed to the radiation for a prescribed period of time through a remote-controlled shutter. *Implantation therapy* or interstitial irradiation describes various procedures involving direct introduction of sealed radioactive sources into tumor tissue. Isotopes normally utilized in this way include ^{60}Co, ^{192}Ir, ^{198}Au, and ^{182}Ta. The sources are available as small encapsulated "seeds," needles, or wires. *Contact therapy* through the use of applicators containing beta-emitting isotopes, e.g., ^{90}Sr and ^{32}P, is a means of applying radioactivity to superficial sites such as dermatological areas or ophthalmic tumors. This dosage form can be removed as necessary and placed where desired. The specific isotopes and instances where these and other techniques are employed are discussed in the following section.

Radiopharmaceutical Preparations
Chromium-51

Sodium Chromate Cr 51 Injection, U.S.P. XVIII (Na$_2$ ^{51}CrO$_4$; Chromitope® Sodium, Rachromate-51®)

As is true of many of the isotopes employed in medicine, chromium-51 is artificially produced by neutron bombardment of chromium-50. The isotope is produced with the emission of a gamma ray, and the reaction is represented by $^{50}_{24}Cr$ (n,γ) $^{51}_{24}Cr$. Chromium-51 decays by emitting a 0.320 Mev gamma ray by K-capture, and decaying to vanadium-51. The half-life is 27.8 days.

$$^{51}_{24}Cr \xrightarrow[\text{K-capture}]{27.8\ d} {}^{51}_{23}V + 0.320 \text{ Mev } \gamma$$

Sodium Chromate Cr 51 injection is a clear to slightly yellow solution having a pH between 7.5 and 8.5. The solution should be prepared so that at the expiration date (no later than three months after standardization) the specific activity (activity per unit weight) should be not less than 10 mc per mg of sodium chromate.

Sodium Chromate Cr 51 is used diagnostically to determine red blood cell mass, volume, and survival time, and for scanning the spleen. Chromium in the +6 oxidation state [Cr(VI)] is readily taken up by erythrocytes and becomes fixed to the globin portion of hemoglobin as chromium (III). This process is usually done *in vitro* and the cells replaced in the blood to measure blood cell volume and mass. Survival time is determined as the cells are destroyed, releasing ^{51}Cr which is excreted in the urine. The rate of excretion of the isotope is then directly related to the time of survival of the erythrocytes.[5] Radioactive iron will not work in this type of study because the iron is stored and recycled in the synthesis of new erythrocytes (see Chapter 6).

The use of *Sodium Chromate Cr 51* in scanning the spleen involves damaging the red blood cells with heat after incubation with the isotope. The damaged cells are then reinjected intravenously where they are rapidly taken up by the spleen. The concentration of radioactivity by this organ is an indication of its ability to function properly. Figure 11–4 illustrates a normal and an abnormal spleen scan.

Preparations and Doses. *Sodium Chromate Cr 51* is available as Chromitope® Sodium and Rachromate-51®. Intravenous doses for red blood cell volume range from 25 to 100 μc. Red blood cell survival times require doses of 150 to 250 μc. These doses are usually administered as previously labeled erythrocytes.

Radio-Chromated Serum Albumin Cr 51 (Chromalbin®)

The ^{51}Cr-labeled sodium chromate discussed above is not capable of chemically labeling plasma proteins with the isotope. The problem is associated with the Cr(VI) oxidation state (the oxidation state of Cr in chromate ion) and the involvement of the metal in the chromate anion. When chromium is in the +3 oxidation state [Cr(III)] and available as

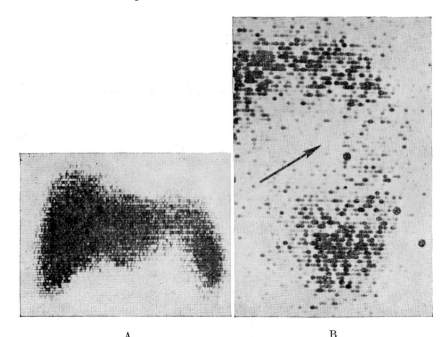

A   B

FIG. 11–4. (A) A normal spleen scan. (B) Spleen scan showing incomplete filling due to a malignant lesion (^{51}Cr-tagged erythrocytes). From Clinical Nuclear Medicine, by C. Douglas Maynard, Lea & Febiger, Philadelphia, pp. 73, 75, 1969. By permission.

a cation, i.e., $CrCl_3$, the metal readily interacts with plasma proteins, but has virtually no interaction with erythrocytes. Therefore, radioactive $^{51}CrCl_3$ can be used to label serum albumin, a plasma protein, to determine plasma volume.

A product containing ^{51}Cr-labeled serum albumin is utilized in placental localization procedures,[6] that is, as a means of visualizing the position, size, etc., of the placenta. This preparation has the advantage over the commonly employed iodinated I-131 serum albumin (see below) in that procedures to reduce uptake of the iodine isotope by the thyroid gland are unnecessary. Radiation exposure to the mother and fetus is also reduced.

Preparations and Dose. Radio-Chromated Serum Albumin Cr 51 is available as Chromalbin®. The usual dose is 30 to 35 μc given I.V. with the radioactivity determined ten minutes after injection by placing scanning probes over several predetermined areas of the abdomen. The areas of heavy localization of radioactivity serve to outline the placenta.

Cobalt-57 and -60

Cyanocobalamin Co 57 Capsules and Solution, U.S.P. XVIII (Racobalamin®-57, Rubratope®-57)

Cyanocobalamin Co 60 Capsules and Solution, N.F. XIII (Racobalamin®-60, Rubratope®-60)

These two isotopes are discussed together because they are both used in the same diagnostic procedure. Cobalt-57 can be produced by several methods. One of these methods involves gamma irradiation of ^{58}Ni: $^{58}_{28}$Ni (γ,p) $^{57}_{27}$Co, and another is accomplished through proton bombardment of ^{56}Fe: $^{56}_{26}$Fe (p,γ) $^{57}_{27}$Co. Cobalt-57 decays by K-capture and emits a 0.123 Mev gamma ray. The half-life of the isotope is 270 days.

$$^{57}_{27}\text{Co} \xrightarrow[\text{K-capture}]{270\ \text{d}} {}^{57}_{26}\text{Fe} + 0.123\ \text{Mev } \gamma$$

On the other hand, cobalt-60 is produced by bombardment of the stable cobalt-59 in a neutron reactor, $^{59}_{27}$Co (n,γ) $^{60}_{27}$Co. The isotope has a half-life of 5.27 years, emitting both beta and gamma radiation. For the purposes of diagnostic measurement, the 1.17 and 1.33 Mev gamma rays are the most important.

$$^{60}_{27}\text{Co} \xrightarrow{5.27\ \text{y}} {}^{60}_{28}\text{Ni} + 0.312\ \text{Mev } \beta^- + 1.17\ \text{Mev } \gamma + 1.33\ \text{Mev } \gamma$$

Cobalt-60 is present in the fallout from nuclear bomb explosions. Its half-life and emissions are responsible for the public health hazard associated with fallout contamination.

The official capsules of *Cyanocobalamin Cobalt 57 and 60* may contain a small amount of solid material or may actually appear to be empty. The solutions are clear, colorless to pink, having a pH between 4.0 and 5.5, and are preserved with a suitable bacteriostatic agent. The specific activity of the capsules or solutions of either isotope should be not less than 0.5 μc per μg.

Cyanocobalamin Co 57 or Co 60 is vitamin B_{12} in which a portion of the molecules contains radioactive cobalt in place of the stable isotope of the metal. The radioactive forms of the vitamin are used in diagnostic procedures for pernicious anemia. The basis of the test was developed by Schilling[7] on the premise that if vitamin B_{12} is absorbed from the gastrointestinal tract, it will be excreted in the urine. Therefore, the radioactivity from an oral dose of ^{60}Co-labeled vitamin B_{12} should be detectable in the urine of the normal patient, and absent or at significantly lower levels in the urine of the patient with pernicious anemia, since these patients lack intrinsic factor which is necessary for the proper intestinal absorption of vitamin B_{12}. Both oral doses of the capsules and injected doses of the solutions may be used alternately to study the effect of the liver on the intestinal absorption of the vitamin.

Recently, cobalt-57 has become preferred over cobalt-60 for various reasons.[8] Primary among these is the fact that cobalt-57 offers greater radiation counting efficiency in that the scintillation crystal detects only the single gamma at 0.123 Mev. In the case of cobalt-60 or other isotopes of cobalt, interactions of the crystal with the gamma rays produce a "scattering" of the radiation (Compton effect) which reduces the efficiency of the detector. Another aspect of the preference for cobalt-57 is the shorter half-life, no beta radiation, and the lower energy of the gamma emission. All of these mean a lower radiation exposure to the patient and particularly the patient's liver, the organ which stores most of the unexcreted vitamin B_{12} and receives the largest dose of radioactivity.

Preparations and Doses. Cyanocobalamin *Co 57* and *Co 60* are available as both capsules and solutions under the trade names of Racobalamin®-57 and Racobalamin®-60. Concentrates are available as Rubratope®-57 and Rubratope®-60. The usual dose for oral and intramuscular injection is 0.5 μc, and may range to 1.0 μc. This dose corresponds to 0.5 to 2 μg of cyanocobalamin.

Cobalt-60 Metallic Sources

Cobalt-60 is sometimes used in metallic form where the metal has been encased in a polymer or alloyed with a corrosion-resistant metal. These exist in the form of "seeds," needles, or wires which are implanted into certain body cavities (intracavitary implantation) or directly into tumor tissue (interstitial implantation). The major use for this type of isotope therapy is in the treatment of the advanced stages of cancer involving the cervix, vagina, uterus, and bladder. It has also been used in carcinoma of the mouth, tongue, and lip.

Highly active cobalt-60 sources (up to 2000 c) are used in external radiation therapy in a manner similar to x-rays. This requires the focusing of the gamma radiation from the cobalt on a particular area, e.g., the lymph nodes in the neck region, for a prescribed period of time. This type of therapy makes use of the penetrating power of the gamma rays from the isotope. Although the treatment produces cell damage, the patient does not carry any radioactivity after the treatment period is completed.

Iron-59

Ferrous Citrate Fe 59 and Ferric Chloride Fe 59 (Ferrutope®).

Iron-59 is a beta- and gamma-emitting isotope prepared by neutron activation of iron-58, $^{58}_{26}$Fe (n,γ) $^{59}_{26}$Fe, which is a stable isotope in iron metal occurring in 0.33% abundance. The half-life of iron-59 is 45 days.

$$^{59}_{26}\text{Fe} \xrightarrow{\ 45\ d\ } {}^{59}_{27}\text{Co} + 0.462\ \text{Mev}\ \beta^- + 0.271\ \text{Mev}\ \beta^- + 1.30\ \text{Mev}\ \gamma + 1.10\ \text{Mev}\ \gamma$$

Ferrous Citrate Fe 59 is normally provided in a sterile solution containing about 30 μc per ml. Specific activities may range from 5 to 200 $\mu c/\mu g$. The solutions are preserved (benzyl alcohol), and contain an antioxidant (ascorbic acid).

The isotope is employed in diagnostic procedures relating to various aspects of iron metabolism and red blood cell formation. The preparation can be administered orally to study the absorption of iron from the G.I. tract,[9] and injected intravenously for determinations of plasma iron clearance and turnover,[10] and the incorporation of iron into erythrocytes.[11] Iron-59 has sufficient gamma radiation energy to allow scintillation counting of the radioactivity in various tissues associated with erythrocyte formation and destruction, i.e., spleen, sacrum, and liver, from outside the body.

Preparations and Doses. Ferrous Citrate Fe 59 is available under this generic name and under the trade name of Ferrutope®. The usual oral and intravenous dose is 2 to 5 μc, and may range as high as 10 μc.

Another preparation which is used for the same purposes and at the same doses as the above is *Ferric Chloride Fe 59* in a sterile solution.

Gold-198

Gold Au 198 Injection, U.S.P. XVIII (Aurcoloid®-198, Aureotope®, Aurcoscan®-198).

Radioactive gold-198 is a short half-life isotope ($t_{\frac{1}{2}} = 2.7$ days) emitting both beta and gamma radiation. It is produced by neutron bombardment of stable gold-197 according to the reaction: $^{197}_{79}Au$ (n,γ) $^{198}_{79}Au$.

$$^{198}_{79}Au \xrightarrow{\quad 2.7\,d \quad} {}^{198}_{80}Hg + 0.959 \text{ Mev } \beta^- + 0.412 \text{ Mev } \gamma$$

Gold Au 198 Injection is a sterile, distinctly red, colloidal solution of radioactive gold. It is stabilized by the addition of gelatin and reducing agents. The colloidal particle size ranges from 2 to 60 mμ (nm). The pH of the solution is between 4.3 and 7.5, and the radiation may cause both the solution and the glass container to darken over a period of time. The solution is categorized as both a diagnostic preparation for scintillation scanning of the liver, and a therapeutic preparation in the treatment of disorders secondary to neoplastic diseases.

Gold-198 solution is most frequently used therapeutically. The solution is administered by intracavitary injection into the pleural and peritoneal cavities (the potential space within the membrane enveloping the lung and viscera, respectively) as an aid in the management of pleural effusion (the accumulation of serous fluid in the pleural cavity) and ascites (the accumulation of serous fluid in the peritoneal cavity).[12] These fluid accumulations, when secondary to neoplastic disease in the area, can be inhibited by the effect of the beta radiation on the cancerous tissue cells. The short range of this radiation (about 4 mm in tissue) requires that, for a beneficial effect, the radioactive gold particles must be placed directly on tissue in the cavity. Another use for this preparation is based on a possible prophylactic benefit against the growth of more tumors after surgical removal of tumors from a major cavity. Because the intracavitary administration of gold-198 solution is contraindicated in the presence of unhealed surgical wounds, exposed cavities, or ulcerative tumors, the solution should not be used until healing of the surgical wound is well under way.

Radioactive colloidal gold solutions are also used at a lower dose of radioactivity to perform diagnostic scanning of the liver.[13] This colloidal solution, as well as many other colloidal solutions of dyes, e.g., rose bengal, is taken up by and stored in the reticuloendothelial cells of the liver, frequently called Kupffer's cells (these are phagocytic cells sometimes called macrophages). Except in cases where this cell system is functionally impaired, gold-198 will usually not be localized in other tissues. The bone marrow will demonstrate some radioactivity, and the spleen will concentrate detectable amounts when the reticuloendothelial system is impaired. The liver will store the isotope for long periods of time, and in most instances, the radioactivity will be lost while the isotope is still present in the body. Liver scanning with gold-198 aids in determining the position, shape, and size of the organ, and the distribution of the isotope provides information concerning the functioning of the Kupffer's cells. The isotope will not enter tumor tissue, abscesses, or cysts; therefore, these will appear as light areas ("cold" spots) in a scan of the liver (see Fig. 11–5). However, gold-198 cannot be used to distinguish between various growths in terms of type or origin.

FIG. 11–5.   Anterior view of a liver with metastatic carcinoma. ([198]Au colloid, 3-inch scanner.)   From Clinical Nuclear Medicine by C. Douglas Maynard, Lea & Febiger, Philadelphia, p. 123, 1969.   By permission.

Preparations and Doses. Gold Au 198 Injection is available for therapeutic purposes as Aurcoloid®-198 (40 to 90 mc/ml) and Aureotope® (25 to 250 mc). The usual dose for pleural effusions is 35 to 75 mc, and for ascites 100 to 125 mc. It may take three to four weeks to notice clinical benefit. It is important to note that in no instance should dosage be repeated at intervals of less than four weeks, and then only if necessitated by the rate of fluid formation.

Gold Au 198 Injection is available for diagnostic purposes as Aurcoscan®-198 at a concentration of about 100 μc/ml. Intravenous doses have been variable, averaging around 300 μc with a range of 100 to 500 μc.

Gold-198 metal is also available as seeds or needles for implantation therapy.

Iodine-125 and -131

Sodium Iodide I 125 Solution, U.S.P. XVIII (Iodotope® I 125).
Sodium Iodide I 131 Capsules and Solution, U.S.P. XVIII (Iodotope® I 131, Radiocaps®-131).

Iodine-131, and to some extent iodine-125, occurs as an isotope in numerous radiopharmaceuticals for diagnostic and therapeutic purposes. The radiochemical properties of these two isotopes will be discussed here, and the various preparations and uses will be discussed below.

Iodine-131, presently the most frequently employed of the two isotopes, emits both beta and gamma radiation to produce a rather complex emission spectrum. The isotope is present in the products of uranium fission or it may be produced through neutron bombardment of tellurium 130, yielding the isotope along with gamma and beta emission: $^{130}_{52}\text{Te}\ (n,\gamma)\beta^-\ ^{131}_{53}\text{I}$. The important emissions from iodine-131 for medical purposes are the 0.608 Mev beta and the 0.364 Mev gamma (from metastable xenon-131) and the half-life is 8.08 days.

$$^{131}_{53}\text{I} \xrightarrow{\ 8.08\ \text{d}\ } {}^{131}_{54}\text{Xe} + 0.608\ \text{Mev}\ \beta^- + 0.364\ \text{Mev}\ \gamma$$

Iodine-125 emits significantly lower energy radiation than iodine-131. Produced in a neutron reactor, the isotope is formed from the conversion of xenon, with the emission of gamma and K-capture x-radiation: $^{124}_{54}\text{Xe}\ (n,\gamma)\text{K}\ ^{125}_{53}\text{I}$. The iodine-125 then decays with a half-life of 60 days, first emitting 0.027 Mev x-rays produced by K-capture and yielding $^{125\text{m}}\text{Te}$. The metastable tellurium decays to the ground state, ^{125}Te, with the emission of a 0.035-Mev gamma ray. In contrast to iodine-131, there is no beta radiation from this isotope. Both iodine-125 and -131 can be produced in the reactor to yield essentially carrier-free isotopes; that is, they are free of, or contain only trace amounts of, nonradioactive isotopes of the same element (iodine-127) in the same chemical form.

$$^{125}_{53}\text{I} \xrightarrow[\text{K-capture}]{60 \text{ d}} {}^{125\text{m}}_{52}\text{Te} + 0.027 \text{ Mev x}$$
$$\quad\quad\quad\quad\quad\quad\quad\quad \longrightarrow {}^{125}_{52}\text{Te} + 0.035 \text{ Mev } \gamma$$

Sodium Iodide I 131 Capsules may contain a small amount of solid material or they may appear empty. *Sodium Iodide I 131* and *I 125 Solutions* are suitable for either oral or I.V. administration. The solutions are clear and colorless, but over a period of time both the solution and glass may darken due to the effects of radiation. The pH of both solutions is between 7.5 and 9.0.

Sodium Iodide I-131 is the most common isotope and chemical form in use as a diagnostic aid in the study of the functioning of the thyroid gland, and in scanning the thyroid to determine size, position, and possible tumor location. The usual procedure in the study of thyroid function is to measure the uptake of radioactive iodine in a 24-hour period.[14] The isotope is administered orally or by intravenous injection and the activity measured over the thyroid at various time intervals up to and including 24 hours after the dose was given. The euthyroid (normal) patient will take up from 15 to 45% of the administered dose in 24 hours. If the uptake is less than 10%, the patient is hypothyroid, and an uptake of over 50% is an indication of hyperthyroidism. Other measurements can be made to confirm the results of uptake procedures, i.e., plasma clearance and urinary excretion of the isotope. Modifications of the uptake study using thyroid-stimulating hormone or blocking agents must be done to discover the etiology of any abnormality conclusively.

Thyroid scanning procedures require about two to three times the radioactive dose used in uptake studies. Scanning of the gland in the neck area will provide a picture of its size, shape, and location, as well as areas of high and low iodine-concentrating ability.[15] This procedure will show thyroid tumors as areas of low concentration ("cold" spots; see Fig. 11–3A); however, this occurrence is not sufficient proof of malignancy. Scanning procedures can be done with iodine-125 which has the advantage of lower radiation exposure to the patient due to the lower energy.[16] Most of the tissue effect of I-131 is due to the beta radiation which will penetrate 2 to 3 mm into tissue. I-125 has no beta radiation; therefore, its short-term damage potential is minimized. The longer half-life of iodine-125 preparations gives them a longer shelf life in the laboratory, and also contributes to a longer effective half-life.

Sodium Iodide I-131 is also employed therapeutically to destroy thyroid tissue or at least to alter the function of the tissue cells. The particular disease states in which this isotope comes into use are hyperthyroidism[17] (thyrotoxicosis), thyroid carcinoma,[18] and severe cardiac disease.[19] The activity of the isotope in these areas of treatment is dependent upon the ability of the thyroid to concentrate the iodine. As mentioned above, the tissue effect is primarily due to the beta radiation which is of relatively short range. Therefore, the isotope must become concentrated in the iodine storage areas (colloid) of the gland in order to have any effect on cells synthesizing thyroid hormone (thyroxine) or on any adjacent tumor cells.

The use of Sodium Iodide I-131 in the treatment of hyperthyroidism is done with the intention of impairing the hormone-synthesizing capability of the apex of the thyroid cells. Doses must be carefully calculated since doses that are too high will destroy cell division capabilities in the nucleus leading to eventual hypothyroidism, and doses that are too low will require eventual retreatment. Recent studies have demonstrated that the properties of iodine-125 may make it more desirable than iodine-131 in this type of therapy.[20] The lower energy, shorter path-length radiation from iodine-125 serves to limit its deleterious effects on cell nuclei while maintaining its effect on hormone synthesis areas. The concept is to avoid overtreatment resulting in hypothyroidism. Iodine-125 is experimental in this particular application.

Radioactive iodine in thyroid carcinoma is at best only palliative, offering no direct cure. The isotope is used most frequently after surgical removal of a cancerous thyroid as a means of treating any residual tumor tissue.

Severe cardiac diseases may be eased through the use of sodium iodide I-131, which is used to induce a hypothyroid state as a means of reducing the work load on the heart. This treatment is used only when conventional therapy has not been successful, and in the conditions known as angina pectoris and congestive heart diseases.

Preparations and Doses. Diagnostic preparations containing sodium iodide I-131 or I-125 are available in capsule or solution form under the name Iodotope®. Other names include Radiocaps®-131 (capsules) and Tracervial®-131 (solution).

Oral or intravenous doses for uptake or general thyroid scanning range from 5 to 50 μc. Scanning for metastatic thyroid cancer requires doses around 300 μc.

Therapeutic preparations of sodium iodide I-131 are available under the names Iodotope® Therapeutic (capsules and solution), Theriodide®-131 (capsules), and Oriodide®-131 (oral solution).

Oral or intravenous doses for therapeutic purposes are quite variable. In hyperthyroidism, doses are calculated on the estimated weight of the gland giving 80 to 120 μc/g of tissue. Cancer of the thyroid will usually be treated with doses ranging from 100 to 200 mc. Cardiac disease usually requires 10 to 25 mc, which may be repeated in two to six months.

Iodinated I 125 Serum Albumin, U.S.P. XVIII (Radio-iodinated [125I] Serum Albumin [Human], RISA®-125).

Iodinated I 131 Serum Albumin, U.S.P. XVIII (Radio-iodinated [131I] Serum Albumin [Human], RISA®-131, RISA®-131-H, Albumatope® I-131, Albumatope-LS®, Macroscan®-131).

Both of these products are sterile, buffered, isotonic solutions prepared to contain at least 10 mg of radioiodinated normal human serum albumin per ml. The iodination procedure is done under mild conditions so as to introduce not more than 1 gram-atom of iodine for each gram-molecule (60,000 g) of albumin. The solutions are clear and colorless to slightly yellow, but they and the glass container may darken upon standing due to the radiation. The pH of each solution is between 7.0 and 8.5.

When injected I.V., the radioiodinated serum albumin will mix homogeneously with the plasma proteins in 10 to 15 minutes. The radioactivity of a withdrawn blood sample can then be determined in counts per minute (cpm) and compared with a standard to obtain the circulating blood volume. It can also be used to determine the plasma volume, and simultaneous use with sodium chromate Cr-51 and ferrous citrate Fe-59 to determine total blood volumes is also possible. Radioiodinated serum albumin is also used to study circulation time and cardiac output. Brain tumors have an affinity for this form of the isotope, making it a useful diagnostic aid for localizing neoplasms in this area.[21] It has also been used to evaluate the circulation of cerebrospinal fluid.[22] These two isotopes of iodine (131I and 125I) when found in this chemical form are not used in any techniques related to the thyroid gland. The radioactive iodine serves only as a source of radiation, and the serum albumin serves as a carrier molecule.

Due to low energy gamma radiation, the preparation containing iodine-125 is not suitable for brain scans, but can be employed in the determination of blood and plasma volumes and determinations of cardiac output. It offers the advantages of emitting low energy radiation with no beta particles; therefore, radiation exposure to the patient is minimized. The shelf life of the product is also longer, due to the longer half-life of I-125 as compared with I-131.

Radioiodinated I-131 serum albumin is available in a macroaggregated form having particles averaging from 10 to 40 microns in size. Due to the particle size, these aggregates are filtered by the blood vessels in the lungs and allow specific scanning of lung tissue.[23] This preparation is used in the diagnosis of problems relating to pulmonary blood flow, emphysema, and lung and bronchial tumors. A microaggregated form is also available. This form has particles averaging around 0.4 microns in size, and is used in scanning the liver.

In all of the above uses of radioiodinated serum albumin I-125 and I-131, there is always a certain amount of metabolism of the albumin, allowing the release of the iodine which can in turn be taken up by the thyroid gland. It is, therefore, normal procedure to give the patient Lugol's solution (i.e. *Strong Iodine Solution*, U.S.P.; see Chapters 6 and 9) at least 24 hours before the administration of the radioactive compound in order to saturate the thyroid gland with iodine and thus prevent the uptake of radioactivity.

Preparations and Doses. Iodinated I 125 Serum Albumin is available as RISA®-125. Iodinated I-131 serum albumin is available as RISA®-131, RISA®-131-H, and Albumatope® I-131. Macroaggregates of iodinated I-131 serum albumin are available as Albumatope-LS® and Macroscan®-131.

Doses are administered intravenously. For blood and plasma volume studies, doses range from 3 to 20 μc for both I-125 and I-131. For scanning procedures including brain, lung, and liver, doses range between 200 and 500 μc.

Sodium Iodohippurate I 131 Injection, U.S.P. XVIII (Hippuran®-131, Hipputope®; Sodium o-Iodohippurate-^{131}I, $C_9H_7{}^{131}INNaO_3$)

This preparation is a clear, colorless, sterile solution of sodium o-iodohippurate in which a portion of the iodohippuric acid molecules are labeled with ^{131}I. The pH of the solution is between 7.0 and 8.5.

Sodium o-iodohippurate is cleared from the blood, collected, and excreted only by the kidneys in the same manner as p-aminohippuric acid. In addition, 75 to 80% of the administered dose will be excreted within the first 30 to 90 minutes. Sodium iodohippurate I-131 is, therefore, a useful diagnostic agent for determining renal (kidney) function.[24] The general procedure involves measuring the rate of collection and elimination of the isotope in both kidneys to provide a graphic display known as a renogram (see Fig. 11–6). Small doses of Hg-197- or Hg-203-labeled chlormerodrin (see below) can be used to position the scintillation detectors over the kidneys properly. The appearance of the renograms along with other diagnostic data aids in the diagnosis of various diseases of renal origin, i.e., renal hypertension and pyelonephritis (inflammation of the kidney).[25] Radioiodinated iodohippurate is not satisfactory for renal scanning since its passage through the kidney is too rapid to give accurate pictures of isotope localization.

Preparations and Doses. Sodium Iodohippurate I 131 Injection is available as Hippuran®-131 and Hipputope®.

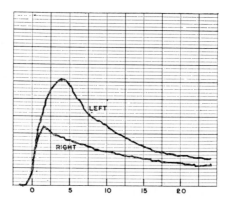

FIG. 11–6.   A renogram of right and left kidneys illustrating a nonfunctioning right kidney.   The renogram gives the time-course of accumulation and elimination of radioactivity as measured by scintillation detectors.   From Clinical Nuclear Medicine, by C. Douglas Maynard, Lea & Febiger, Philadelphia, p. 228, 1969.   By permission.

The dose is administered intravenously and ranges between 5 and 30 μc, depending upon the weight of the patient and the type of counting equipment.

Sodium Rose Bengal I 131 Injection, U.S.P. XVIII (Robengatope® I-131; 4,5,6,7-Tetrachloro-2′,4′,5′,7′-tetraiodoflourescein Disodium Salt-131I; C_{20}-$H_2Cl_4{}^{131}I_4Na_2O_5$)

This preparation is a clear, deep red, sterile solution of sodium rose bengal in which a portion of the molecules is labeled with I-131.   The pH of the solution is between 7.0 and 8.5, and may contain a suitable buffer.

Rose bengal is a dye which was used for many years as a colorimetric diagnostic aid in liver function determinations.   When injected intravenously, the dye is rapidly and selectively taken up by the polygonal cells of the normally functioning liver.   Maximum concentrations are usually achieved within 30 minutes.   After this time, the dye begins to appear in the bile, and is later excreted into the intestine.   Sodium rose bengal labeled with I-131 is then a useful radioactive tracer in the determination of liver function.   Through the use of properly placed scintillation probes the disappearance of an injected dose of the radioactive dye from the blood

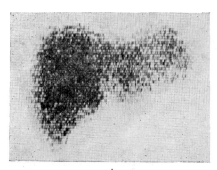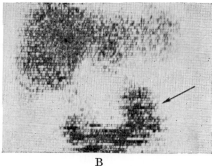

A   B

Fig. 11-7. (A) Anterior view of the liver. (B) Same anterior view performed 30 minutes later demonstrates the presence of labeled rose bengal in the small intestine ([131]I Rose Bengal, 3-inch scanner). From Clinical Nuclear Medicine, by C. Douglas Maynard, Lea & Febiger, Philadelphia, p. 128, 1969. By permission.

can be measured,[26] as well as the appearance of radioactivity in the intestine[27] (see Fig. 11-7). These procedures provide information concerning blood flow through the liver and the presence of possible obstructions. This preparation also remains in the liver long enough to provide radioactive photoscans of the organ. Liver scans can serve as a means of determining size and location as well as the presence of abscesses, cysts, and tumors. These latter growths are usually unable to concentrate the labeled dye, and are observed as "cold" spots or "holes" in the scintigram.[28] Since metabolism of the dye and its uptake by the thyroid are possibilities, the patient should be given Lugol's solution at least 24 hours before the diagnostic procedure is begun to prevent the concentration of radioactive iodine in the thyroid.

Preparations and Doses. Sodium Rose Bengal I 131 Injection is available under the trade name of Robengatope® I-131, and also by generic name.

Counting doses for determining blood clearance and bile clearance, and for liver counting will usually be in the range of 10 to 25 μc diluted with about 5 ml of normal saline. For making liver scans, doses will range from 100 to 200 μc in adults, and around 50 μc in children.

Mercury-197 and -203

Chlormerodrin Hg 197 Injection, U.S.P. XVIII
Chlormerodrin Hg 203 Injection, U.S.P. XVIII ([3-(Chloromercuri)-2-methoxypropyl] urea, $C_5H_{11}ClHgN_2O_2$; Mol. Wt. 367.20)

$$H_2N-\overset{\overset{\textstyle O}{\|}}{C}-NH-CH_2-\underset{\underset{\textstyle OCH_3}{|}}{CH}-CH_2-HgCl$$

Mercury-197 is produced by neutron bombardment of naturally occurring mercury-196, $^{196}_{80}Hg$ (n,γ) $^{197}_{80}Hg$. It has a half-life of 2.7 days (65 hours), and emits gamma radiation by electron capture with the major peak having an energy of 0.077 Mev.

$$^{197}_{80}Hg \xrightarrow[\text{Electron capture}]{2.7\ d} {}^{197}_{79}Au + 0.077\ \text{Mev}\ \gamma + 0.268\ \text{Mev}\ \gamma$$

Mercury-203 is also produced by neutron bombardment with the target isotope being the naturally occurring mercury-202, $^{202}_{80}Hg$ (n,γ) $^{203}_{80}Hg$. This isotope has a longer half-life of 46.9 days, and emits both beta and gamma radiation having energies of 0.21 Mev and 0.279 Mev, respectively.

$$^{203}_{80}Hg \xrightarrow{46.9\ d} {}^{203}_{81}Tl + 0.21\ \text{Mev}\ \beta^- + 0.279\ \text{Mev}\ \gamma$$

Chlormerodrin Hg 197 and *Hg 203 Injections* are clear, colorless, sterile solutions of chlormerodrin in which a portion of the molecules contains radioactive mercury. The solutions have a pH between 5.5 and 8.5.

Chlormerodrin labeled with either ^{197}Hg or ^{203}Hg is a special radioactive tracer for making scintillation scans of the kidneys or the brain. This compound is in the chemical class of mercurial diuretics, which are generally taken up by the cells of the proximal kidney tubules in the renal cortex. The excretion from these cells is slow enough to allow scanning procedures of the kidneys (see Fig. 11–8) to determine the presence and location of cysts, tumors, or other abnormalities.[29] Results must usually be corroborated with other diagnostic procedures, i.e., pyelography (radiography of the ureter and kidney).

Neoplastic and many nonneoplastic lesions in the brain will concentrate radioactive chlormerodrin in less time than radioiodinated serum albumin. Therefore, chlormerodrin labeled with ^{197}Hg or ^{203}Hg is useful in scanning procedures to locate brain tumors and other lesions in the brain.[30] Other

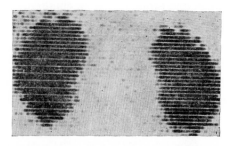

FIG. 11–8. A normal renal scan with *Chlormerodrin Hg 197*. From Clinical Nuclear Medicine, by C. Douglas Maynard, Lea & Febiger, Philadelphia, p. 236, 1969. By permission.

advantages of this preparation over I-131 serum albumin include the simpler decay pattern, which simplifies the instrumentation, and a lower radiation exposure to the patient. Radiation exposure to the kidneys when labeled chlormerodrin is used in brain scanning may be minimized by injecting a dose of a mercurial diuretic I.M. on the day before the procedure. The drug usually used for this is meralluride, which theoretically saturates the sites in the kidney which would normally bind the chlormerodrin.

Chlormerodrin Hg 197 and *Hg 203* appear to give comparable results in both kidney and brain scans. Mercury-197, however, has an advantage over mercury-203 in that the half-life is shorter, the gamma radiation is of lower energy, and there is no beta emission. All of these signify a lower radiation exposure to the patient.

Preparations and Doses. Chlormerodrin Hg 197 and Hg 203 are available under the trade names of Neohydrin®-197 and Neohydrin®-203.

Doses for kidney scanning range from 50 to 200 μc, while brain scanning usually employs a dose calculated on the basis of 10 μc/kg of body weight. An upper limit of the latter dose of 700 μc is generally observed.

Phosphorus-32

Sodium Phosphate P 32 Solution, U.S.P. XVIII (Phosphotope®).

Phosphorus-32 is prepared by neutron bombardment of elemental sulfur, yielding the radioactive isotope with the emission of a proton, $^{32}_{16}S$ (n,p) $^{32}_{15}P$. The isotope decays by beta emission with a maximum energy of 1.71 Mev, and a half-life of 14.3 days.

$$^{32}_{15}P \xrightarrow{\text{14.3 d}} {}^{32}_{16}S + 1.71 \text{ Mev } \beta^-$$

Sodium Phosphate P 32 Solution is suitable for either oral or intravenous administration. The solution is clear and colorless; however, the solution and the glass container may darken upon standing due to the radiation. The pH of the solution is between 5.0 and 6.0, indicating that the chemical form of the preparation is a mixture of disodium hydrogen phosphate (Na_2HPO_4) and sodium dihydrogen phosphate (NaH_2PO_4) (see Chapter 4).

Sodium Phosphate P 32 Solution is used for both diagnosis and treatment of various neoplastic diseases. Phosphate is utilized in cell metabolism. Those cells which are rapidly proliferating have the highest turnover of phosphate. Tumor cells are characterized as being rapidly proliferating, and will, therefore, accumulate phosphate labeled with ^{32}P to a greater extent than noncancerous cells.

The primary diagnostic use for sodium phosphate P-32 is in the localization of intraocular tumors.[31] It also finds infrequent use in the localization of cerebral tumors. The beta radiation from ^{32}P is of sufficient energy to give it a maximum path length in tissue of about 8 mm. Since eye tumors are relatively near the surface, they can be located through the beta radiation by use of specially designed Geiger counter probes. The difference in radioactivity measured over a suspected tumor site and a normal site should be 20 to 30%.

The therapeutic uses of sodium phosphate P-32 are found in two diseases associated with both red and white blood cells. The principal use is in the treatment of polycythemia vera,[32] a disease characterized by the increase in the number and absolute mass of the red blood cells. The effect of the radioactivity is primarily to reduce the formation of erythrocytes; other measures must be taken to reduce the number of erythrocytes already circulating. Phlebotomy, removing blood through an incision in the vein, is usually done prior to or in conjunction with P-32 therapy.

Sodium phosphate P-32 is used less frequently in the palliative treatment of chronic granulocytic or myelocytic leukemia.[33] This is the form of leukemia in which there is an increase in the number of white blood cells from the granulocytic series, including a number of immature forms. These leukocytes are quite sensitive to the effects of radiation, giving rise to the efficacy of [32]P. The preparation may also be used in chronic lymphocytic leukemia, but this form of the disease usually responds better to chemotherapy. Radiation therapy is never used in acute forms of the disease.

Preparations and Doses. *Sodium Phosphate P 32 Solution* is available either as a sterile solution for injection or as an oral solution under the trade name of Phosphotope® and under the generic names of *Sodium Phosphate P 32 Sterile Solution* and *Sodium Phosphate P 32 Oral Solution.*

Because the isotope emits high energy ionizing radiation, dosage is carefully monitored. Intravenous administration usually requires about 75% of the oral dose, and both routes of administration will accomplish the same therapeutic results. The effective half-life in blood cells is about eight days; in other tissues the half-life is eight to ten days, except in bone and brain where labeled phosphate does not accumulate in adults unless neoplastic disorders are present. The clinical and product literature should be consulted concerning the dosage requirements in sodium phosphate P-32 therapy.

For diagnostic purposes, doses will range from 250 to 500 μc with counts being taken at 1 hour, 24 hours, and 48 hours over both eyes. The maximum diagnostic dose is 1 mc.

Therapeutic doses in polycythemia vera range from 2 to 10 mc orally (I.V. dose is 75% of this) with an average of 6 mc. The dosage is dependent upon body weight, and erythrocyte, leukocyte, and platelet counts.

In chronic granulocytic leukemia, the dosage is based on the leukocyte count. Initial doses will generally follow the schedule:

3 mc if count is below 40,000
4 mc if count is between 40,000 and 100,000
5 mc if count is over 100,000

Subsequent doses are dependent upon the response of the patient, and will generally follow weekly intervals.

Chromic Phosphate P 32 Suspension ($Cr^{32}PO_4$; Chromphosphotope®)

This nonofficial preparation is a grayish-green to brown colloidal suspension of radioactive chromic phosphate in various types of suspending vehicles.

The suspension is used therapeutically by intracavitary injection in the treatment of pleural effusions and ascites in much the same manner as colloidal preparations of gold-198 (see *Gold Au 198 Injection*).[34] This isotope has certain advantages over gold-198 in that it has no gamma radiation, thus, it has a lessened hazard potential to personnel. Its beta radiation is more penetrating, which may improve its efficacy, and it is more economical than gold preparations. Chromic phosphate P-32 has also been injected interstitially into tumors (i.e., directly into the tumor tissue). The insoluble nature of the compound renders it physiologically inactive as a source of phosphate. The radioactivity remains localized since the compound is not absorbed or transported to sites of high phosphate turnover.

Preparations and Doses. Chromic Phosphate P 32 Suspension is available as Chromphosphotope® and under the generic designation.

Doses for intracavitary injection range from 6 to 9 mc for pleural effusions, and 9 to 12 mc for ascites. Interstitial administration uses variable doses ranging between 6 and 40 mc.

The same cautions and contraindications apply to chromic phosphate P-32 as were mentioned under *Gold Au 198 Injection*.

Technetium-99m

Technetium Tc 99m Injections

Technetium is an artificial element (not present in nature); all of its isotopes are radioactive. Technetium-99m is produced by the decay of molybdenum-99, which is produced by neutron bombardment of molybdenum-98. The preparation and decay scheme of the isotope are shown in rx (x).

$$\text{$^{98}_{42}$Mo (n,γ) $^{99}_{42}$Mo} \xrightarrow{\text{67 h}} \text{$^{99m}_{43}$Tc} + 3\beta^- + 6\gamma$$

$$\downarrow 6.0 \text{ h}$$

$$\text{$^{99}_{43}$Tc} \xrightarrow{2.1 \times 10^5 \text{ y}} \text{$^{99}_{44}$Ru} + \beta^- \qquad (x)$$

$$+ \\ 2\gamma$$

14

The half-life of Tc-99m is 6.0 hours and over 98% of the radiation occurs as 0.140-Mev gamma rays. This radiation accompanies the isomeric transition to Tc-99, which decays by 0.3-Mev beta emission to stable ruthenium-99. The slow decay of the last step in rx (x) contributes very little to the radioactivity of Tc-99m.

The radioisotope is obtained as sodium pertechnetate Tc-99m ($Na^{99m}TcO_4$) by elution with sodium chloride injection from an alumina column which has been loaded with the parent Mo-99. The entire borosilicate glass column with the elution and collection systems, packed with Mo-99, sterilized, and shielded, is available as a technetium-99m generator. Fig. 11–9 is an illustration of a prototype generator set up for manual (nonvacuum) use. The sterile eluent is usually an isotonic sodium chloride solution. This form of preparation provides Tc-99m in an absolutely carrier-free state. The Mo-99, when prepared by neutron bombardment as shown in rx (x), is free of the side products found when the isotope is obtained from uranium fission, i.e., Ru-103, I-131, and Tc-132. Therefore, this represents a convenient way to get very pure Tc-99m for radiopharmaceutical and medicinal use.

Technetium-99m in several different chemical forms is employed in an ever-widening number of diagnostic applications. The advantages of this isotope over those discussed heretofore are its short half-life, its single gamma photon which simplifies instrumentation, the absence of beta radiation and its accompanying damaging effects on tissue, and the penetrability of the radiation, making it easy to detect outside the body. The carrier-free nature of the isotope eliminates the possibility of chemical toxicity. All of these contribute to the fact that large doses of radioactivity may be administered to the patient, thereby improving resolution and scanning rate with decreased radiation burden to the patient.

The common chemical forms of Tc-99m are the sodium pertechnetate salt and a colloidal preparation of technetium sulfide. Other forms coming into more frequent use include ^{99m}Tc-labeled serum albumin macroaggregates and ^{99m}Tc-iron-ascorbic acid complex. The applications of the various forms of ^{99m}Tc have been reviewed by Gottschalk.[35] Most of the derivatives can be prepared from the pertechnetate salt obtained directly from the generator.

Sodium pertechnetate Tc-99m is used diagnostically for obtaining brain scans to aid in determining the presence and location of neoplastic, and some nonneoplastic, lesions[36] (see Fig. 11–10). The pertechnetate anion, TcO_4^-, has an initial physiological distribution similar to iodide. When administered either orally or intravenously, a small percentage of the ion will become entrapped in the normal thyroid gland, and larger amounts in hyperthyroid states. Since perchlorates, ClO_4^-, are similarly entrapped by the thyroid and by other organs which concentrate iodides, a dose of 200 to 250 mg of potassium perchlorate is usually administered prior to

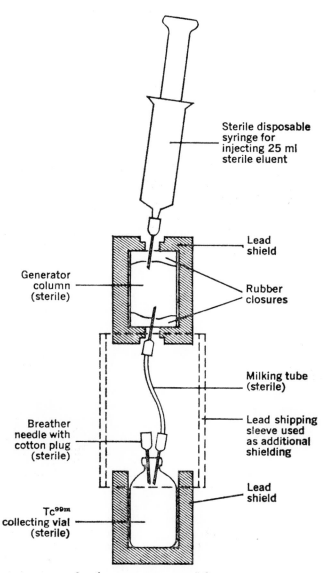

Sterile disposable
syringe for
injecting 25 ml
sterile eluent

Lead
shield

Generator
column
(sterile)

Rubber
closures

Milking tube
(sterile)

Breather
needle with
cotton plug
(sterile)

Lead shipping
sleeve used
as additional
shielding

Lead
shield

Tc^{99m}
collecting vial
(sterile)

Sterile generator set up for use.

Fig. 11–9. Prototype of a Tc-99m sterile generator illustrating the column and shielding. From Roger's Inorganic Pharmaceutical Chemistry, 8th ed., by T. O. Soine and C. O. Wilson, Lea & Febiger, Philadelphia, p. 663, 1967. By permission.

14a

using pertechnetate Tc-99m in order to block uptake by the thyroid, stomach, etc., diminishing radiation exposure to these organs during brain scanning. The advantages of this isotope have led to the replacement of older preparations used for brain scanning, i.e., Hg-197, Hg-203, and radio-iodinated serum albumin.

The uptake of pertechnetate Tc-99m by the thyroid gland renders this form of the isotope useful for thyroid function studies.[37] Since the ion is entrapped but not organically bound, it is not useful for determining all aspects of thyroid activity. However, it will serve in basic diagnosis of hyperthyroidism, and in initial diagnostic steps with lower exposure to the patient than with ^{131}I preparations.

Technetium sulfide Tc-99m colloidal solution can be prepared extemporaneously by reducing an acidic solution of pertechnetate Tc-99m from the generator with sodium thiosulfate.[38] It is also available in a manufacturer-prepared form ready for use. Like gold-198 colloid, technetium sulfide Tc-99m is a colloidal solution of $^{99m}Tc_2S_7$ which, when injected intravenously, is taken up by the reticuloendothelial cells in the liver, spleen,

A   B

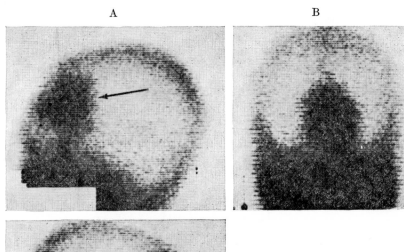

Fig. 11–10. A large meningioma visualized in the frontal region. (A) Left lateral view. (B) Anterior view. (C) Right lateral view. From Clinical Nuclear Medicine, by C. Douglas Maynard, Lea & Febiger, Philadelphia, p. 166, 1969. By permission.

C

and bone marrow, in that order. Depending upon dosage and procedure, this preparation can be used to obtain scintillation scans of all three organs.[35] The quality of the scans is improved over what is obtainable with gold-198, and the patient is exposed to less radiation.

Preparations and Doses. Sodium pertechnetate Tc-99m is available under the trade names of Pertgen®-99m and Technetope® II as sterile generator kits. It is also available in a sterile solution for injection as Pertscan®-99m.

For brain scanning, sodium pertechnetate Tc-99m may be administered orally, after fasting, or intravenously in doses of 200 μc/kg of body weight. The average dose for adults is 10 mc. Oral doses are poorly absorbed in about 10% of the patients.

Thyroid function tests can be performed with doses of around 500 μc.

Technetium sulfide Tc-99m colloid is available in a kit for extemporaneous preparation under the trade name of Tesuloid® Kit. It is also available as a sterile colloidal solution for injection as Colloscan®-99m.

Liver and spleen scanning can be carried out with doses of 1 to 3 mc. Bone marrow studies require a higher dosage, ranging from 4 to 10 mc. The colloidal solution is injected intravenously in all instances.

Several other chemical forms of ^{99m}Tc are being prepared and studied for specific diagnostic procedures.[35] Macroaggregates of ^{99m}Tc-labeled human serum albumin are used to obtain lung scans. The preparation of these aggregates has been described by Bowen and Wood.[39] Renal function and scanning have been done with Tc-99m-iron-ascorbic acid complex and Tc-99m-citrate complex.

Indium-113m

Another generator-produced isotope being employed in diagnosis is indium-113m (Indikow®). This isotope has a half-life of 1.66 hours and is produced by the decay of neutron reactor-produced Sn-113 ($t_{\frac{1}{2}} = 115$ days). It has been utilized in various chelates and other compounds for renal and brain scanning and for the determination of plasma volume.[40,41]

Miscellaneous Isotopes (See Table 11-3)

Strontium-85 as $^{85}SrCl_2$ (Stronscan®-85) and fluorine-18 as $Na^{18}F$ have been used for their ability to localize in bone to provide information concerning bone metabolism and for bone scanning.[42] Calcium-45 and -47 may also be used to study calcium metabolism in diagnosed bone cancer and other bone lesions.

Selenium-75 is incorporated into the amino acid methionine in place of the sulfur to make selenomethionine Se-75 (Sethotope®). This radiopharmaceutical is concentrated by the pancreas, where it is used for the synthesis of digestive enzymes and other proteins as a replacement for methionine. The compound is used diagnostically in scintillation scanning of the pancreas where tumors and other growths or lesions are imaged as "cold" spots.[43] Selenomethionine Se-75 has also been used to scan the parathyroid gland.[44]

Ytterbium-169, a gamma emitter with a half-life of 32 days, has been shown to be a useful isotope for brain scanning and for determining glomerular filtration rates in kidneys.[45] In order to be used in these procedures, the isotope must be chelated with an agent like diethylenetriaminepenta-acetate to form the ^{169}Yb-DTPA complex.

Two isotopes which are implanted into tumors or areas surrounding tumors for their therapeutic effects are iridium-192 and radium-226. Iridium-192 is used in a form known as iridium seeds, which contain the isotope encapsulated in a nylon ribbon. When the seeds are implanted in tumor tissue, the beta radiation as well as the gamma radiation produces local destructive effects on cells. The gamma radiation does not penetrate as far into tissue as the rays from many of the isotopes mentioned heretofore. However, they are energetic enough to avoid absorption in bone, which is an important consideration for any gamma-emitting isotope used in a local manner at therapeutic levels for extended periods of time. Salts of radium-226, i.e., $RaBr_2$, are available in capsules which can be implanted into tumor tissue in much the same manner as ^{192}Ir seeds. The radiation level from this source is about one third more energetic than iridium-192, thus exposing the patient and personnel to a greater hazard.

Other isotopes being used in medicine and research along with appropriate references may be found in Refs. 2 and 3.

Radiopaque Contrast Media

Radiopaque media are chemical compounds containing elements of high atomic number which will stop the passage of x-rays. These types of compounds are used as diagnostic aids in radiology or roentgenology. Roentgenology involves the use of x-rays (roentgen-rays), which are short wavelength electromagnetic radiation, in the imaging or shadowing of various internal organ structures. X-rays are capable of passing through most soft tissue so that when special photographic film or a photosensitive plate is placed on the side of the patient opposite to the x-ray source, the film or plate will become darkened in an amount proportional to the number of x-rays that are able to pass. Bone and teeth are the only types of tissue capable of significantly arresting the passage of x-rays. These structures will appear light on exposed x-ray film, allowing their visualization for the diagnosis of fractures, malformations, and the like. The chemical constituents of bone and teeth which give them the ability to stop this type of radiation are the large concentrations of calcium and phosphorus.

Although these elements do not have tremendously high atomic numbers, they represent the highest available in biological systems in any significant concentration. Furthermore, they occur in close-packed structures providing large localizations of electron density. As a general rule, the more electrons in an atom or molecule the greater the chance of stopping the passage of x-rays. Soft tissues, being less dense and composed primarily of carbon, hydrogen, and oxygen, which are relatively low in atomic number, do not present a dense enough electron "screen" or barrier. For this reason skin and soft organs appear only as shadows, if at all, on x-ray film.

The most common radiopaques contain barium and iodine. The iodine compounds are covalently bonded organic iodides. These compounds will not be discussed to any great extent here except to mention that they are generally used by I.V. injection or retrograde (mechanical instillation into the organ) administration. They are used for x-ray examinations of the kidney, liver, blood vessels, heart, and brain. Covalent organic iodides are nonionic, and the organic molecule usually has some affinity for the organ system to be studied.

Only one compound of barium is useful as a radiopaque, barium sulfate, which is discussed below. Other salts of barium exhibit some solubility and thus provide toxic barium ion (see Chapter 7). Therefore, barium sulfate is essentially the only inorganic compound in this class of agents. Although both barium and iodine do not have the highest atomic numbers, they are the most easily incorporated into molecules exhibiting relatively low toxicity. Their opacity to x-rays is dependent upon their being highly concentrated in the organ to be studied, and they have served quite well as contrast media in the x-ray examination of soft tissues.

Barium Sulfate

Barium Sulfate, U.S.P. XVIII ($BaSO_4$; Mol. Wt. 233.40; Esophotrast®)
Barium sulfate is a fine, white, odorless, tasteless, bulky powder, free from grittiness. It is practically insoluble in water, organic solvents, and solutions of acids and alkalies.

The insolubility of barium sulfate limits the ability of the compound to enter into chemical reactions. It will react with concentrated sulfuric acid to form the soluble bisulfate salt.

Barium sulfate is the agent of choice in roentgenographic studies of the gastrointestinal tract. Its insolubility in acidic gastric juice is a major criterion for this use, since soluble salts would produce toxic barium ion. Other insoluble compounds, i.e., the oxide, carbonate, sulfide, and phosphate, will exhibit some solubility in the acidic medium of the stomach. In order to avoid confusion in product names, the U.S.P. makes the following statement:

Caution—When Barium Sulfate is prescribed, the title always should be written out in full to avoid confusion with the poisonous barium sulfide or barium sulfite.

Although barium ion generally is not available in this product, certain aspects of its pharmacology should be mentioned with respect to its toxicity. Barium ion will produce a stimulation of all muscles. In the gastro-intestinal tract, this is seen as a stimulation of the smooth muscle resulting in vomiting, severe cramps, diarrhea, and possible hemorrhage. Stimulation of the heart muscle can produce cardiac arrest as the cause of death. In fact, caution has been suggested in the use of $BaSO_4$ in cardiac patients.[46] Hypertension can result from constriction of the smooth muscle of the arteries. The effect on skeletal muscle is similar, producing tremors and spasms. The systemic absorption of 800 mg of a soluble barium salt is sufficient to produce death.

Barium sulfate is employed in suspensions of various concentrations for use in the G.I. tract. A paste of the compound will remain in the esophagus long enough for roentgenographic or fluoroscopic study. The suspensions are administered orally or by enema after fasting. The major side effect associated with the use of barium sulfate suspensions is constipation. Accidental entry of barium sulfate into the peritoneal and other cavities through perforations and the like has produced little in the way of severe reactions or inflammation.[47]

Preparations and Doses. Barium Sulfate, U.S.P. XVIII is available as the powder for use in making oral suspensions. It may contain suspending and flavoring agents, and may be found under certain trade names, i.e., Esophotrast®, and by its generic name.

The dosage for most procedures will utilize a suspension of 200 to 300 g orally, or 400 to 750 g for rectal administration.

Organoiodine Radiopaque Compounds

There are a number of official iodine-containing organic compounds used in diagnosis by roentgenography. The following is a partial listing of these agents and their primary uses. More complete information may be found in Ref. 48.

Meglumine Diatrizoate, U.S.P. XVIII
Sodium Diatrizoate, U.S.P. XVIII

Depending upon concentrations of the particular salt and mixtures of the two salts, these agents are used in cerebral angiography (visualization of cerebral blood vessels), urography, pyelography, and gastrointestinal studies. The meglumine salt is used in coronary angiography.

Meglumine Iodipamide, U.S.P. XVIII
Sodium Iodipamide, U.S.P. XVIII

Both of these are used in cholangiography (roentgenography of the bile duct).

Iodized Oil, N.F. XIII

This is a preparation of iodized poppy seed oil used in hysterosalpingography (roentgenography of the uterus and oviducts).

Iopanoic Acid, U.S.P. XVIII

This is a compound of low toxicity used in cholecystography (visualization of the gall bladder).

Iophendylate, U.S.P. XVIII

This is the ethyl ester of iophenylundecanoic acid. It is used in myelography (visualization of the spinal subarachnoid space).

Calcium Ipodate, N.F. XIII

Sodium Ipodate, N.F. XIII

These are cholecystographic agents for oral use.

REFERENCES

1. Overman, R. T., and Clark, H. M.   Radioisotope Techniques.   New York: McGraw-Hill, Chapter 4, 1960.
2. Christian, J. E.   Radioisotopes in the pharmaceutical sciences and industry. J. Pharm. Sci., **50**:1, 1961.
3. Wolf, W., and Tubis, M.   Radiopharmaceuticals.   J. Pharm. Sci., **56**:1, 1967.
4. Quimby, E. H., and Feitelberg, S.   Radioactive Isotopes in Medicine and Biology. Basic Physics and Instrumentation, 2nd ed.   Philadelphia: Lea and Febiger, 1963.
5. Shih, S. C., Tauxe, W. N., Fairbanks, V. F., and Taswell, H. F.   Urinary excretion of ^{51}Cr from labeled erythrocytes.   An index of erythrocyte survival.   J.A.M.A., **220**:814, 1972.
6. Johnson, P. M., Sciarra, J. J., and Bragg, D. G.   Placental localization with radioisotopes.   Amer. J. Roentgen., **96**:677, 1966.
7. Schilling, R. F.   Intrinsic factor studies.   II.   The effect of gastric juice on the urinary excretion of radioactivity after the oral administration of radioactive vitamin B_{12}.   J. Lab. Clin. Med., **42**:860, 1953.
8. Rosenblum, C.   Radiation comparison of cobalt isotopes.   Amer. J. Clin. Nutr., **8**: 276, 1960.
9. Bonnet, J. D., Hagedom, A. B., and Owen, C. A., Jr.   A quantitative method for measuring the gastrointestinal absorption of iron.   Blood, **15**:36, 1960.
10. Teng, C. T., Collins, V. P., and West, W. D.   A critical appraisal of the diagnostic and prognostic value of radioiron study in hematologic disorders.   Amer. J. Roentgen., **84**:687, 1960.
11. Silver, S.   Radioactive Isotopes in Medicine and Biology, 2nd ed.   Philadelphia: Lea and Febiger, p. 186, 1962.
12. Rose, R. G.   Intracavitary radioactive colloidal gold, results of 257 cancer patients. J. Nucl. Med., **3**:323, 1962.
13. Czerniak, P., Lubin, E., Djaldetti, M., and de Vries, A.   Scintillographic follow-up of amoebic abscesses and hydatid cysts of the liver.   J. Nucl. Med., **4**:35, 1963.
14. Dyrbe, M. O., Peitersen, E., and Friis, Th.   Diagnostic value of ^{131}I in thyroid disorders.   Acta Med. Scand., **176**:91, 1964.
15. Croll, M. N., and Brady, L. W.   Thyroid scintillation scanning: Methodology and interpretation.   New York J. Med., **63**:211, 1963.
16. Charkes, N. D.   Utilization of iodine-125 for thyroid scanning.   J. Nucl. Med., **5**: 312, 1964.
17. Nofal, M. M., Beierwaltes, W. H., and Patno M. E.   Treatment of hyperthyroidism with sodium iodide I-131.   J.A.M.A., **197**:605, 1966.
18. Haynie, T. P., Nofal, M. M., and Beierwaltes, W. H.   Treatment of thyroid carcinoma with ^{131}I.   J.A.M.A., **183**:303, 1963.
19. Delit, C., Silver, S., Yohalem, S. B., and Segal, R. L.   Thyrocardiac disease and its management with radioactive iodine ^{131}I.   J.A.M.A., **176**:262, 1961.

20. McDougall, I. R., Greig, W. R., and Gillespie, F. C. Radioactive iodine (^{125}I) therapy for thyrotoxicosis. New Engl. J. Med., **285**:1099, 1971.
21. Schlesinger, E. B., De Bores, S., and Taveras, J. Localization of brain tumors using radioiodinated human serum albumin. Amer. J. Roentgen., **87**:449, 1962.
22. Alker, J. G., Jr., Glasauer, F. E., and Leslie, E. V. Long-term experience with isotope cisternography. J.A.M.A., **219**:1005, 1972.
23. Wagner, H. N. Diagnosis of massive pulmonary embolism in man by radioisotope scanning. New Engl. J. Med., **271**:377, 1964.
24. Nordyke, R. A. Use of radioiodinated hippuran for individual kidney function tests. J. Lab. Clin. Med., **56**:438 1960.
25. Meier, D. A., and Beierwaltes, W. H. Radioisotope renal studies on renal hypertension. J.A.M.A., **198**:119, 1966.
26. Nordyke, R. A., and Blahd, W. H. Blood disappearance of radioactive rose bengal. J.A.M.A., **170**:1159, 1959.
27. Nordyke, R. A. Biliary tract obstruction and its localization with radioiodinated rose bengal. Amer. J. Gastroent., **33**:563, 1960.
28. Boute, F. J. Scintillation scanning of the liver. Amer. J. Roentgen., **88**:275, 1962.
29. Woodruff, M. W., Kibler, R. S., Bender, M. A., and Blau, M. Hg-203 neohydrin kidney photoscan: An adjuvant to diagnosis of renal disease. J. Urol., **89**:746, 1963
30. Blau, M., and Bender, M. A. Radiomercury (Hg^{203}) labeled neohydrin: A new agent for brain tumor localization. J. Nucl. Med., **3**:83, 1962.
31. Goldberg, B. The use of phosphorus 32 in the diagnosis of ocular tumors. Arch. Ophthal. (Chicago), **65**:196, 1961.
32. Calabresi, P., and Meyer, O. O. Polycythmeia vera. II. Course and therapy. Ann. Intern. Med., **50**:1203, 1959.
33. Osgood, E. E. Treatment of chronic leukemia. J. Nucl. Med., **5**:139, 1964.
34. Jacobs, M. L. Radioactive colloidal chromic phosphate to control pleural effusions and ascites. J.A.M.A., **166**:597, 1958.
35. Gottschalk, A. Technetium-99m in clinical nuclear medicine. Ann. Rev. Med., **20**:131, 1969.
36. Quinn, J. L., Ciric, I., and Hauser, W. N. Analysis of 96 abnormal brain scans using technetium 99m. J.A.M.A., **194**:157, 1965.
37. Golden, A. W. G., Glass, H. I., and Williams, E. D. Use of Tc 99m for the routine assessment of thyroid function. Brit. Med. J., **4**:396, 1971.
38. Stern, H. S., McAlee, J. G., and Subramanian, G. Preparation, distribution, and utilization of technetium 99m sulfur colloid. J. Nucl. Med., **7**:665, 1966.
39. Bowen, B. M., and Wood, D. E. Preparation of technetium 99m, human serum albumin macroaggregates. Amer. J. Hosp. Pharm., **26**:529, 1969.
40. Burdine, J. A., Jr., Waltz, T. A., Matsen, F. A., and Rapp, F. Localization of In 113m chelates compared with Tc 99m sodium pertechnetate in experimental cerebral lesions. J. Nucl. Med., **10**:290, 1969.
41. Hosain, P., Hosain, F., Iqbal, Q. M., Carulli, N., and Wagner, H. N. Measurement of plasma volume using Tc 99m and In 113m labeled proteins. Brit. J. Radiol., **42**:627, 1969.
42. Fleming, W. H., McIlraith J. D., and King, E. R. Photoscanning of bone lesions utilizing strontium 85. Radiology, **77**:635, 1961.
43. Collela, A. C., and Pigorini, F. Experiences in pancreas scanning using Se-75-selenomethionine. Brit. J. Radiol., **40**:662, 1967.
44. DiGiulio, W., and Morales, J. D. The value of the selenomethionine Se-75 scan in the preoperative localization of parathyroid adenomas. J.A.M.A., **209**:1873, 1969.
45. Wagner, H. N., Jr., Hosain, F., and Rhodes, B. A. Recently developed radiopharmaceuticals: Ytterbium-169 DTPA and technetium-99m microspheres. Radiol. Clin. N. Amer., **7**:233, 1969.
46. Eastwood, G. L. ECG abnormalities associated with the barium enema. J.A.M.A., **219**:719, 1972.
47. Gaston, E. A. Barium granuloma of the rectum. Dis. Colon Rectum, **12**:241, 1969.
48. Doerge, R. F. In: Wilson, C. O., Gisvold, O., and Doerge, R. F. (eds.) Textbook of Organic, Medicinal and Pharmaceutical Chemistry, 6th ed. Philadelphia: J. B. Lippincott, p. 999, 1971.

12
Miscellaneous Inorganic Pharmaceutical Agents

This chapter will discuss inhalants, expectorants and emetics, antidotes, tableting aids, and suspending agents. Some of these agents (expectorants, suspending agents) will be components of finished products. Others (gases, antidotes) will be found in emergency rooms and operating rooms, and the pharmacist may be responsible for their inventory.

Inhalants

Currently there are five gases that are in the U.S.P.: oxygen, carbon dioxide, helium, nitrogen, and nitrous oxide. They will be discussed in that order.

Oxygen. Oxygen is necessary in normal oxidative metabolism for the production of useful energy. Located in the mitochondria of each cell is a tightly integrated enzyme system that receives electrons from the reduced coenzymes nicotinamide adenine dinucleotide (NADH) and flavin adenine dinucleotide ($FADH_2$), and transports them in a series of oxidation-reduction steps ending with the reduction of atomic oxygen and forming the oxide anion (Fig. 12–1). The oxide anion quickly combines with protons forming water which, along with carbon dioxide, is the end product of combustion. Energy is released during each of the oxidation-reduction steps in the mitochondrial enzyme system. This energy, by a process not yet fully understood, is used by the cell to synthesize adenosine triphosphate (ATP). When ATP is hydrolyzed, energy is released to drive otherwise energetically unfavorable processes to completion. The formation of ATP by release of energy from the electron transport system is known as oxidative phosphorylation.

Because land mammals live submerged in an "ocean" of air, they do not have any apparatus for extensive storage of oxygen during periods of deprivation. It is for this reason that severe oxygen deprivation causes anatomic and physiologic disorders that can lead to permanent local damage. The utilization of atmospheric oxygen is illustrated in Fig. 12–2.

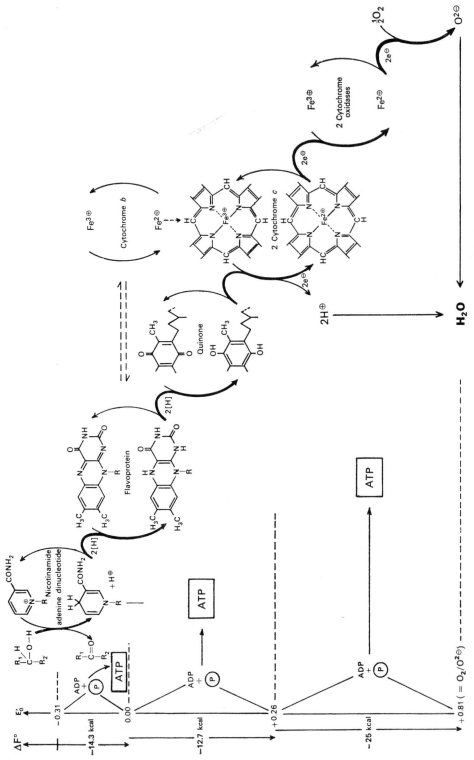

FIG. 12-1. Sequence of redox systems in the respiratory chain. From Kurzes Lehrbuch der Biochemie, 7, by P. Karlson, Georg Thieme Verlag, Stuttgart, 1970.

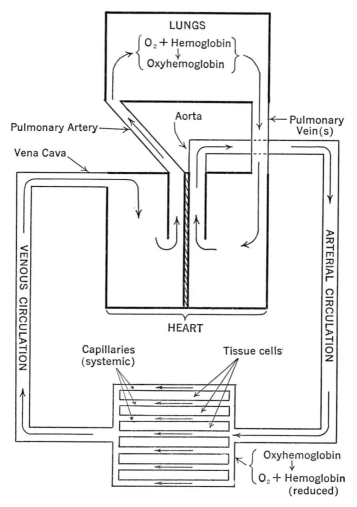

FIG. 12–2. Diagram of oxygen circulation. From Rogers' Inorganic Pharmaceutical Chemistry, 8th ed., by T. O. Soine and C. O. Wilson, Lea & Febiger, Philadelphia, p. 24, 1967. By permission.

The venous blood enters the heart from where it is pumped to the lungs for aeration and then travels back to the heart for pumping into the arterial system. The arterial system carries the blood to the capillaries, where it is in intimate contact with tissue cells and from which it is discharged into the venous system for conveyance back to the heart and lungs. The entire process takes about one minute in the normal state and as little as 10 to 15 seconds when the body is being exerted. The oxygenation of the blood takes place from the alveoli of the lungs where the alveolar air and venous blood are separated by a pulmonary membrane (0.004 mm thick) consisting of the alveolar and capillary wall. The surface area of the membrane is estimated at from 70 to 90 square meters. All the red blood cells, which contain hemoglobin, must pass through the lung capillaries in single file, thus offering a tremendously large surface for oxygenation. The process is continuous because there is always residual air in the alveoli.

For the successful transport of oxygen from air to the tissues it is necessary that there be a progressive lessening of oxygen tension in each of the steps, i.e., lower tension in plasma than in alveolar air, lower in red blood cells than in the plasma, and lower in the tissues than in the red cells.

As shown in Fig. 12–3, human plasma would carry very little oxygen if it were not for the hemoglobin in the red blood cells. It has been estimated that it would require 300 quarts of plasma to carry all the oxygen necessary for body functions if the body were dependent upon only the amount of oxygen that would diffuse into the plasma and be held in solution. The body contains an average of 40 ml plasma/kg body weight. Each gram of hemoglobin can hold 1.34 ml of oxygen. Blood contains 15 g of hemoglobin per 100 ml and 100 ml of blood can hold 20 ml of oxygen. Under normal conditions, the arterial blood is approximately saturated with oxygen.

Oxygen requirements in the body can be conveniently classified into four major divisions: (1) anoxic, (2) anemic, (3) stagnant, and (4) histotoxic.

(1) In the anoxic type the oxygen supply to the tissues is inadequate because the blood arrives with its oxygen at a lowered tension. The cause for this may be lowered oxygen tension in the inspired air as a result of high altitudes, increase in the inert gases normally present, or abnormal presence of other inert gases. Lowered oxygen tension in the plasma may

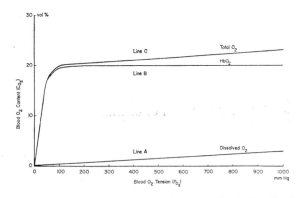

FIG. 12–3. *Line A* describes the amount of oxygen physically *dissolved* in plasma. *Line B* is the oxyhemoglobin dissociation curve for normal adults at pH 7.4 drawn under the assumption of an oxygen capacity of 20 ml of oxygen per 100 ml of blood; *line B*, therefore, describes the amount of oxygen combined with hemoglobin at any oxygen tension up to 1000 mm Hg. *Line C* represents the total amount of oxygen contained in blood at a given oxygen tension. From Hyperbaric oxygen in patients with venoarterial shunts: Theoretical implications, by N. M. Nelson and E. O. R. Reynolds, New Eng. J. Med., **271**:497, 1964. By permission.

be another cause, resulting from interference with diffusion of alveolar air into the plasma because of disturbed pulmonary function or a defect in the cardiac septum (wall between the auricle and ventricle), allowing mixing of arterial and venous blood. Studies showing that the inhalation of oxygen-enriched atmospheres containing 40 to 60% oxygen raised the oxygen saturation of arterial blood to or near normal value in patients with pneumonia and cardiac insufficiency placed this therapy on a sound basis. Oxygen therapy has since found use in such conditions as asthma, massive collapse of the lungs, atelectasis of the newborn (incomplete expansion of the lungs at birth), bronchopneumonia, congestive heart failure, coronary thrombosis, cerebral thrombosis, etc. In these cases the oxygen is administered by nasal tubes, mask, or in tents. Apparently its use in high concentrations is quite safe although uninterrupted inhalation of pure oxygen for one or two days could cause harmful effects (edema of the lungs, etc.) if the dose is not regulated. (See later discussion of oxygen toxicity.) In premature infants a condition known as retrolental fibroplasia (RLF) may occur due to the administration of high concentrations of oxygen at birth. This is a vascular proliferative disease of the retina and may lead to blindness due to retinal detachment. Limitation of the concentration of inhaled oxygen to 35 to 40% when possible minimizes the danger.

(2) In the anemic type the oxygen tension is normal, but the amount of hemoglobin is inadequate to supply enough oxygen to the tissues. This condition may result from hemorrhage, decreased or defective red blood cell formation, or carbon monoxide poisoning. In the latter, carbon monoxide has a much greater affinity for hemoglobin than does oxygen, and results in the formation of carboxyhemoglobin from hemoglobin. Oxygen administration for many of these anemic anoxias is not effective, but, in the case of carbon monoxide poisoning, a carbon dioxide-oxygen mixture or hyperbaric oxygen administration (administration of oxygen at greater than atmospheric pressure) is said to be of value. Carbon dioxide has a specific stimulating effect on the respiratory center and aids in the swifter elimination of carbon monoxide. Hyperbaric oxygen facilitates the dissociation of carbon monoxide from carboxyhemoglobin and sufficiently increases the amount of oxygen dissolved in the plasma for the patient to survive until the carbon monoxide is removed and the red cells can carry oxygen again.

(3) The stagnant type of anoxia occurs when the general circulation is inadequate or when circulation is locally retarded. Therapy for this type with oxygen is usually not indicated. It is best treated with cardiotonic drugs to speed up the circulation.

(4) In the histotoxic type, tissue cell oxidation (Fig. 12–1) may be interfered with in several ways. In some cases it is due to an "uncoupling" of oxidative phosphorylation. While the transfer of electrons continues, with the resulting reduction of atomic oxygen to the oxide anion and

formation of water, the released energy cannot be used by the cell to form ATP. The end result is that the cell cannot obtain useful energy to carry on essential metabolic processes. Other toxic substances, such as cyanide, actually block electron transport, thereby stopping the cell's capacity to carry out metabolism requiring oxidation and reduction. Thus glycolysis, the tricarboxylic acid cycle, the hexose monophosphate shunt, β-oxidation of fatty acids, and other cycles which are essential for life to continue are stopped. Again, oxygen administration is theoretically without value and treatment has traditionally been directed toward neutralization of the toxic materials, but some clinicians suggest that hyperbaric oxygen administration, particularly at about 2.5 atmospheres pressure, is dramatic in countering the effects of lethal doses of cyanide. Many suggest that it be a routine procedure, together with sodium nitrite and sodium thiosulfate, in the antidotal treatment of cyanide poisoning. (See later discussion on cyanide poisoning.)

Oxygen is commonly administered by nasal tubes, mask, or in tents at atmospheric pressure in concentrations ranging from 30 to 80%. Occasionally 100% oxygen is administered for short periods of time.

Another method of administration is hyperbaric oxygen therapy in which 100 percent oxygen is given at 1.5 to 3 atmospheres absolute pressure. The technique has been used for air embolism (air bubbles in the blood), carbon monoxide poisoning, gas gangrene, crush injury, myocardial infarction, and peripheral and cerebral vascular disease. A pressure chamber is required.

Hyperbaric oxygen therapy is based on increasing the content and partial pressure of oxygen in arterial blood, an increase that occurs mostly in the plasma since the hemoglobin in normal arterial blood is already 97% saturated. In patients whose oxygenation is limited by defective pulmonary gas exchange, such as found in atelectasis (lung collapse) and pulmonary edema, an increase in plasma oxygen may be difficult to achieve. Further hyperbaric oxygen may not be able to compensate for impaired oxygen delivery due to circulatory impairment, shock, or ischemic (localized blood deficiency) disorders. While used for a wide variety of conditions that supposedly should benefit from oxygenation, doubt has been expressed whether hyperbaric therapy is any better than alternate procedures.[1,2] In addition to the conditions already mentioned, hyperbaric therapy has been used in the healing of bedsores and other skin lesions, including a device for localized administration of high pressure oxygen, and as an adjunct to radiation therapy in cancer treatment.[3–5]

The latter is based on the observation that a cell deprived of oxygen is less sensitive to damage by x-rays. Since a great number of cells in a large malignant tumor mass appear to be anoxic due to an abnormal blood supply, there will be a group of cells that could survive a large dose of x-rays. These surviving cells can become the nucleus for regrowth of the tumor. Use of hyperbaric therapy, which would increase the oxygen supply to the anoxic cells and thereby make them more susceptible to radiation damage, has increased the survival time in patients with advanced carcinoma.[6]

Administration of oxygen at atmospheric pressure at a high concentration can be toxic to the adult. As contrasted with the previously described retrolental fibroplasia that can occur when oxygen is administered to premature infants, oxygen toxicity in adults is usually associated with pulmonary distress leading to edema in the lungs. The mechanism is not well understood. Part of the problem may be because oxygen drawn from a cylinder is anhydrous and tends to dry the mucous membranes. This causes discomfort, thickening of natural or pathologic secretions, and diminished activity of the cilia, thereby impeding the clearing of fluids out of the lungs. The drying effect can be reduced by humidification of the oxygen by use of bubble-bottles, spray nebulizers, ultrasonic aerosol generators, and other devices. Dry oxygen, however, does not seem to be the sole cause of pulmonary oxygen toxicity. Recent work has shown that the killing of *Staphylococcus aureus* by alveolar macrophages is inhibited in mice by oxygen.[7] This inhibition is dose-dependent. Pure oxygen also sharply reduces the amount of surfactant in dog's lungs after 48 hours exposure.[8] This causes alveolar collapse, thus reducing the absorption of oxygen. The collapse may be delayed by 5% nitrogen.[9]

Most of the reports in the literature are from observations of oxygen toxicity occurring when oxygen was administered for supportive treatment in very ill patients. Just recently, there have been reported carefully designed studies of oxygen toxicity in man. Using normal, healthy young men, no statistically significant changes at the pulmonary vasculature level were reported in subjects breathing 100% oxygen for 6 to 12 hours as compared with controls breathing 21% oxygen at the same flow rate.[10] In another study, 40 cardiac patients were divided into two groups. One group received pure oxygen, and the other received the minimal inspired oxygen concentration (maximum of 42%) required to maintain an arterial oxygen tension between 80 and 120 mm of mercury. The average exposure for both groups was 21 to 24 hours. No evidence of pulmonary oxygen toxicity could be found in either group.[11]

Most studies of this type are terminated at the first sign of any toxic symptoms or are of such short duration that symptoms do not have time to develop. In a unique type of experiment, patients who were victims of massive cerebral trauma that had produced irreversible and ultimately fatal brain damage were divided into two groups. One group received ventilation with air until death and the other pure oxygen until death. The oxygen group demonstrated significantly greater impairment of lung function than the air group. While x-ray examination and total lung weight supported the physiologic findings, microscopic examination of lung tissue failed to reveal noteworthy differences between the two groups.[12] It appears that time is the important factor. Short-term ventilation with 100% oxygen is a fairly safe procedure. Nevertheless, it is probably better to titrate the patient in order to maintain an arterial oxygen tension of 70 to 100 mm of mercury, assuming that the patient's cardiac output is adequate.

Official Product

Oxygen, U.S.P. XVIII (O_2; Mol. Wt. 32.00)

Oxygen U.S.P. occurs as a colorless, odorless, tasteless gas, which supports combustion more energetically than does air. One liter at 0° C and at a pressure of 760 mm mercury (Hg) weighs about 1.429 g. One volume dissolves in about 32 volumes of water and in about 7 volumes of alcohol at 20° C and at a pressure of 760 mm of mercury. Oxygen is stored in cylinders which are usually green-colored (World Health Organization: white) or carry a green label. Since oxygen supports rapid combustion, there should be no smoking or open flames nearby, and precautions should be taken against ignition by sparks from electrical appliances, clothing, or bedding.

Carbon Dioxide

Carbon dioxide, along with water, is the normal end product of combustion, whether rapid as found in a burning flame or very slow as found in the metabolism of all aerobic organisms. In man, the enzyme carbonic anhydrase catalyzes reaction (i). The bicarbonate/carbonic acid ratio is an

$$CO_2 + H_2O \rightarrow H_2CO_3 \leftrightharpoons H^+ + HCO_3^- \qquad \text{(i)}$$

important determinant of body fluid pH (see Chapter 5). A decreased removal of carbon dioxide can lead to respiratory acidosis while an increased removal can cause respiratory alkalosis. The dizziness accompanying the latter condition can be experienced by hyperventilation (abnormally prolonged, rapid, and deep breathing).

The administration of carbon dioxide for medical purposes is limited to a few conditions. Carbon dioxide has been used as a respiratory stimulant since it is a normal respiratory stimulant. However, it is ineffective if the respiratory center is depressed. In conditions where the respiratory center is depressed (drug toxicity, disease, injury), there will be a buildup of carbon dioxide. Administration of additional carbon dioxide may further depress the respiratory center and worsen the respiratory acidosis resulting from the buildup of endogenously produced carbon dioxide. Intermittent carbon dioxide inhalation has been used postoperatively to increase breathing, but other techniques are available.

Carbon dioxide, 5 to 7% in oxygen, has been used in the treatment of carbon monoxide poisoning as carbon dioxide increases both the ventilatory exchange rate and the rate of dissociation of carbon monoxide from carboxyhemoglobin. The resulting respiratory acidosis further facilitates the dissociation of carboxyhemoglobin.

Carbon dioxide had been suggested for detoxifying heroin addicts without the painful withdrawal syndrome. The patient was given 75% carbon dioxide until he passed out. He was then revived with air or 100% oxygen.[13] The study was recently suspended due to the death of a patient, and the entire procedure is being reevaluated.[14]

The frozen form of carbon dioxide, dry ice, has been used in the treatment of such skin conditions as acne, angiomas (a tumor made up of blood or lymph vessels), corns and calluses, eczema, moles, psoriasis, and warts. Here it is used to destroy tissue by freezing it.

The soft drink industry uses solutions of carbon dioxide to make carbonated beverages. Pharmaceutically, most effervescent preparations contain sodium bicarbonate and an acid, usually citric, which react when the patient mixes the preparation with water [rx (ii)].

$$NaHCO_3 + HA \rightarrow Na^+ + A^- + H_2CO_3 \quad\quad\quad\quad\quad \atop \hookrightarrow CO_2\uparrow + H_2O \quad\quad\quad\quad (ii)$$

The U.S.P. permits carbon dioxide to be used to displace air in parenteral and topical preparations that are easily oxidized, but it must be so stated on the label.

Official Product

Carbon Dioxide, U.S.P. XVIII (CO_2; Mol. Wt. 44.01)

Carbon Dioxide U.S.P. occurs as an odorless, colorless gas. Its aqueous solutions are acid to litmus. One liter at $0°$ C and at a pressure of 760 mm Hg weighs 1.977 g. One volume dissolves in about one volume of water. Carbon dioxide is supplied in cylinders which are usually of grey metallic color. It is administered in concentrations up to 7% in oxygen. It is also available as a 5% mixture with oxygen in gray/green cylinders (W.H.O.: gray/white).

Helium

Helium, the second lightest element, is used as a diluent in oxygen administered during respiratory obstruction (laryngeal edema, mucous plug obstructions) which cause turbulent air flow. When flow is turbulent, the work required for drawing 80% helium with 20% oxygen into the lungs is one half that necessary for air.[15] This mixture has only one third the density of air and, since the ease with which gas can diffuse through a small orifice is related to the density of the gas, the 80% helium-20% oxygen mixture penetrates the restricted respiratory passages with greater facility than does air. Because helium has a much lower water and fat solubility than nitrogen, helium-oxygen mixtures are used in high pressure, underwater diving. This has reduced the risk of bends and the time of decompression when bringing the diver up (see nitrogen discussion). Helium, being lighter than air, causes the pitch of sounds uttered by the vocal cords to be increased, producing unintelligible speech with a "Donald Duck" sound.

Official Product

Helium, U.S.P. XVIII (He; Mol. Wt. 4.003)

Helium U.S.P. occurs as a colorless, odorless, tasteless gas, which is not combustible and does not support combustion. One liter at a pressure of 760 mm Hg and at 0° C weighs not less than 178 mg and not more than 189 mg, indicating not less than 99.0% of helium. It is very slightly soluble in water. Helium is usually supplied in brown cylinders and also as a mixture containing 20% or 40% oxygen in brown/green containers (W.H.O.: brown/white).

Nitrogen

Nitrogen has little therapeutic use. Rather, it is used as a diluent for pure oxygen and is official as a pharmaceutic aid to displace air, usually to increase the shelf life of an easily oxidized product. Under the high pressure conditions associated with deep sea diving, it will dissolve in the blood. If the diver is brought up too suddenly, the nitrogen comes out of solution, forming bubbles in the blood causing the condition known as the *bends* to develop. The normal procedure is to bring the diver up slowly and allow the dissolved nitrogen to be exhaled as it gradually comes out of solution. When conditions require bringing a diver up quickly, he is put into a decompression chamber which simulates the pressure changes that occur when a diver is brought up slowly. The increased dissolved nitrogen in the blood also results in a narcosis which is quite similar to the effects of alcohol intoxication. Helium, which has a very low water solubility and therefore a much lower potential for narcosis and the bends, has largely replaced nitrogen for high pressure dives.

Nitrogen is an important constituent of plant and animal tissues, largely as amino acids and as protein. The determination of a patient's nitrogen balance is an important diagnostic parameter. Essential as nitrogen is and widespread as it is in the environment, animals and higher plants cannot fix atmospheric nitrogen the way they can fix oxygen. Instead, nitrogen-fixing bacteria are required to form nitrogenous compounds, or fertilizers containing nitrates must be added to the soil. Either way the plants use these compounds to form plant protein. The plants are eaten by higher animals which, along with vegetable crops eaten by man, provide the source of man's required nitrogen. The plant and animal protein is broken down to the individual amino acids and then used to make up the necessary human protein.

The U.S.P. permits the use of nitrogen to displace air in parenteral and topical preparations providing it is so indicated on the label.

Official Product

Nitrogen, U.S.P. XVIII (N_2; Mol. Wt. 28.01)

Nitrogen U.S.P. occurs as a colorless, odorless, tasteless gas. It is nonflammable and does not support combustion. One liter at 0° C and at a pressure of 760 mm Hg weighs about 1.251 g. One volume dissolves in about 65 volumes of water and in about 9 volumes of alcohol at 20° C and at a pressure of 760 mm Hg. *Nitrogen* U.S.P. is usually sold in black cylinders.

Nitrous Oxide

Nitrous oxide is the only inorganic gas used as an anesthetic. It has been called "laughing gas" because of the delirium associated with its use and has been abused because of this. There have been deaths by suffocation and brain damage from hypoxia because most lay users are not aware that usual administration requires the concomitant administration of 20 to 25% oxygen.

The common procedure calls for a ratio of 85 parts nitrous oxide:15 parts oxygen for two to three minutes maximum for induction followed by a 80:20 ratio for maintenance. Usually barbiturates or other suitable sedatives must also be given in order to obtain a deep enough anesthesia for surgery. Also, muscle relaxants may be necessary since nitrous oxide does not have the muscle relaxant properties of other anesthetic gases. It does have a very rapid recovery time and has been used in the physician's or dentist's office.

Nitrous oxide has analgesic properties in concentrations at which the patient remains conscious. The use of pure nitrous oxide for a few minutes, together with alternate use of pure oxygen, is condoned in the second stage of labor where it is of considerable value and does not impair uterine contraction nor interfere with oxygen saturation of the patient's blood. A British study concluded that a 50% oxygen:50% nitrous oxide mixture can safely be used for the relief of pain in childbirth by unsupervised midwives.[16] The same ratio has been found to give rapid relief of pain from myocardial infarction, as has a 35% nitrous oxide:65% oxygen mixture.[17,18] A preliminary report indicates that nitrous oxide administered to the conscious dental patient would be beneficial in the treatment of the apprehensive patient.[19,20]

Official Product

Nitrous Oxide, U.S.P. XVIII (N_2O; Mol. Wt. 44.01; dinitrogen monoxide, laughing gas)

Nitrous Oxide U.S.P. occurs as a colorless gas, without appreciable odor or taste. (Note: It is also described as having a slightly sweet odor and taste.) One liter at 0° C and at a pressure of 760 mm Hg weighs about 1.97 g. One volume dissolves in about 1.4 volumes of water at 20° C and at a pressure of 760 mm Hg. *Nitrous Oxide* U.S.P. is freely soluble in ether and oils. It is available in blue metal cylinders.

Respiratory Stimulants

Currently there is one inorganic agent still official as a respiratory stimulant, *Aromatic Ammonia Spirit* N.F. The two active ingredients are *Ammonium Carbonate* N.F. and *Strong Ammonia Solution* N.F., both of which cause a reflex action of the patient taking a sudden deep breath. They have been used to revive an unconscious person who may have fainted or "had the wind knocked out of him." Ammonium carbonate has been used in the formulating of the so-called "smelling salts." *Strong Ammonia Solution* is described in Chapter 4, but it must be remembered that this is really the concentrated ammonium hydroxide solution of commerce and should only be opened in well ventilated areas. Under no circumstances should the concentrated solution be used as a respiratory stimulant.

Official Products

Ammonium Carbonate, N.F. XIII ($[NH_4HCO_3]_m[NH_2CO_2NH_4]_n$; ammonium sesquicarbonate, sal volatile, Preston's salt, hartshorn)

Ammonium Carbonate N.F. consists of varying proportions of ammonium bicarbonate (NH_4HCO_3) and ammonium carbamate ($NH_2CO_2NH_4$) such that it yields between 30 to 34% ammonia (NH_3). *Ammonium Carbonate* occurs as a white powder or as hard, white, or translucent masses, having a strong odor of ammonia, without empyreuma (the peculiar odor of animal or vegetable matter when charred in a closed vessel), and having a sharp, ammoniacal taste. Its solutions are alkaline to litmus. On exposure to air, it loses ammonia and carbon dioxide, becoming opaque, and is finally converted into friable porous lumps or a white powder of ammonium bicarbonate [rx (v)]. *Ammonium Carbonate* is freely soluble in water, but is decomposed by hot water [rxs (vi) and (vii)].

Although this salt actually contains little, if any, normal ammonium carbonate, ($NH_4)_2CO_3$, it is readily converted into the normal carbonate by dissolving it in dilute ammonia water [rxs (iii) and (iv)]. Reactions (iii) and (iv) occur during the manufacture of *Aromatic Ammonia Spirit* N.F.

$$NH_4HCO_3 + NH_3 \rightarrow (NH_4)_2CO_3 \qquad \text{(iii)}$$

$$NH_2CO_2NH_4 + H_2O \rightarrow (NH_4)_2CO_3 \qquad \text{(iv)}$$

The decomposition of the salt from its original hard, translucent state to that of a white powder is caused entirely by loss of ammonia and carbon dioxide from the ammonium carbamate, leaving the white powder, ammonium bicarbonate [rx (v)].

$$(NH_4HCO_3)_m(NH_2CO_2NH_4)_n \rightarrow nCO_2 \uparrow + 2nNH_3 \uparrow + mNH_4HCO_3 \quad (v)$$

Both ammonium bicarbonate and normal ammonium carbonate decompose to ammonia and carbon dioxide at 60° C [rxs (vi) and (vii)].

$$NH_4HCO_3 \underset{\Delta}{\rightarrow} NH_3 \uparrow + CO_2 \uparrow + H_2O \qquad \textbf{(vi)}$$

$$(NH_4)_2CO_3 \underset{\Delta}{\rightarrow} 2NH_3 \uparrow + CO_2 \uparrow + H_2O \qquad \textbf{(vii)}$$

CATEGORY: Source of ammonia.

OCCURRENCE: **Aromatic Ammonia Spirit**, N.F. XIII

Aromatic Ammonia Spirit, N.F. XIII

Aromatic Ammonia Spirit N.F. contains, in each 100 ml, 1.7 to 2.1 g of total ammonia and ammonium carbonate corresponding to 3.5 to 4.5 g of $(NH_4)_2CO_3$. In addition there are volatile oils and, as a solvent, ethyl alcohol.

Tradition called for the use of translucent (undecomposed) pieces of ammonium carbonate to be dissolved in a diluted ammonia solution, generating normal ammonium carbonate [rxs (iii) and (iv)]. However, it has been shown that by using additional ammonia solution and decomposed *Ammonium Carbonate* N.F. (essentially ammonium bicarbonate) a preparation meeting the analytical standards of the official spirit can be made.[21] The finished product is light sensitive and should be stored in light resistant bottles.[22]

CATEGORY: Respiratory stimulant.

APPLICATION: By inhalation of vapor as required.

Expectorants and Emetics

Expectorants. Expectorants are used orally to stimulate the flow of respiratory tract secretions. The rationale is that this facilitated flow will allow ciliary motion and coughing to move the loosened material toward the pharynx more easily. Expectorants are used in the treatment of respiratory disorders in which secretions are purulent, viscid, or excessive. It is difficult to design a precise evaluation of these agents since adequate methods for measuring sputum volume and viscosity have not yet been devised. Agents commonly used are terpin hydrate, which may have a direct effect on the bronchial secretory cells, and ammonium chloride, glyceryl guaiacolate, potassium guaiacol-sulfonate, syrup of ipecac, iodinated glycerin, potassium iodide, and hydroiodic acid syrup, which are all believed to act with a reflex action by irritating the gastric mucosa and thereby stimulating respiratory tract secretion. It must be emphasized that use of most of the above agents is based on tradition rather than properly designed clinical trials.[23] This section will limit detailed discussion to the official inorganic agents.

Emetics. Sometimes emetics in low doses have been used in cough preparations. The rationale is that a mild emetic response stimulates flow of respiratory tract secretions. Antimony potassium tartrate has found some use to a limited extent in cough preparations. Sometimes emetic agents have been added to cough syrups containing addicting (opiate) antitussives. The rationale is that the patient would become nauseated if an excessive dose were consumed.

Official Products

Ammonium Chloride, U.S.P. XVIII

This product is described in Chapter 5 and is official as a systemic acidifier. It is found in many cough preparations and administered as 300 mg per dose.

Potassium Iodide, U.S.P. XVIII

This product is described in Chapter 6. Although official as an expectorant, it is widely used as a source of iodine. When used as an expectorant, it is usually in combination rather than alone. The usual expectorant dose is 300 mg four times a day with a dose range of 300 mg to 2 g daily.

Antimony Potassium Tartrate, U.S.P. XVIII

This product is described in Chapter 9. Although official as an antischistosomal agent, it was used as an emetic in the formerly official cough preparation, *Brown Mixture* N.F. XII. The emetic action has a slow onset and marked depression can follow.

Antidotes

An antidote is an agent that counteracts a poison. The mechanism of antidotal action usually occurs in one of three ways: (1) by counteracting the effects of a poison by producing other effects (a physiological antidote); (2) by changing the chemical nature of the poison (a chemical antidote); and (3) by preventing the absorption of the poison into the body (a mechanical antidote). Examples of those inorganic agents used as antidotes include sodium nitrite, which converts hemoglobin into methemoglobin in order to bind cyanide (physiological antidote), sodium thiosulfate, which causes the conversion of the systemically toxic cyanide to the nontoxic thiocyanate (chemical antidote), and activated charcoal, which adsorbs the poison prior to absorption across the intestinal wall (mechanical antidote). Other examples of mechanical antidotes are cupric sulfate, magnesium sulfate, and sodium monohydrogen phosphate (Na_2HPO_4), which inactivate and precipitate the toxic material as insoluble salts.

Cyanide Poisoning. Cyanide (CN^-) readily combines with the ferric (Fe^{+3}) ion. Thus, cyanide poisons by combining with the ferric ion of cytochrome oxidase which stops electron transfer and thereby stops cellular respiration or oxidation-reduction reactions (Fig. 12–1). Cyanide poisoning is treated by a combination of sodium nitrite ($NaNO_2$) and sodium thiosulfate ($Na_2S_2O_3$). Injection of sodium nitrite causes the oxidation of the ferrous (Fe^{+2}) ion of hemoglobin to the ferric ion of methemoglobin, which then combines with the serum cyanide that has not yet entered the cell.

Following the sodium nitrite, a slow intravenous infusion of sodium thiosulfate is given. The thiosulfate anion, catalyzed by the enzyme rhodanese, reacts with cyanide to form the relatively nontoxic thiocyanate ion which is excreted in the urine [rx (viii)].

$$Na_2S_2O_3 + CN^- \underset{SCN^- \text{ oxidase}}{\overset{\text{rhodanese}}{\rightleftharpoons}} SCN^- + Na_2SO_3 \qquad \text{(viii)}$$

Because thiocyanate oxidase can reverse the reaction, although slowly toxic symptoms have returned following initial antidotal therapy.

Hyperbaric oxygen administration has been reported to be effective in cyanide poisoning. In theory this should not be true since cyanide blocks the reduction of oxygen, preventing the cell from using the oxygen already present. One explanation is that a large excess of oxygen displaces cyanide from the oxidized cytochrome.

Official Products

Sodium Nitrite, U.S.P. XVIII ($NaNO_2$; Mol. Wt. 69.00)

Sodium Nitrite U.S.P. occurs as a white to slightly yellow, granular powder, or white or nearly white, opaque, fused masses or sticks. It has a mild, saline taste and is deliquescent in air. Its solutions are alkaline to litmus. *Sodium Nitrite* U.S.P. is freely soluble in water and sparingly soluble in alcohol.

Although official as an antidote in cyanide poisoning, sodium nitrite is also used as an antioxidant (see Chapter 4). It is widely used in meat packing as it helps maintain the desired red color and restricts the growth of *Clostridium botulinum* spores. There has been some concern that sodium nitrite might facilitate methemoglobin formation in infants and, more important, may be reduced to nitroso derivatives in the intestinal tract which would react with the amines found in meat, forming nitrosamines. Some nitrosamines are known to be carcinogenic.[24] Norway has banned the use of nitrite as a flavoring and coloring agent and will only permit the level necessary as an antibotulinogenic agent.[25]

CATEGORY: Antidote to cyanide poisoning.

USUAL DOSE: Intravenous, 10 to 15 ml of a 2% solution.

Sodium Thiosulfate, U.S.P. XVIII ($Na_2S_2O_3 \cdot 5H_2O$; Mol. Wt. 248.18; incorrectly called sodium hyposulfite; "hypo")

Sodium Thiosulfate U.S.P. occurs as large, colorless crystals or as a coarse, crystalline powder. It is deliquescent in moist air and effloresces in dry air at temperatures above 33° C. Its solutions are neutral or faintly alkaline to litmus. *Sodium Thiosulfate* U.S.P. is very soluble in water and insoluble in alcohol.

The synonym "sodium hyposulfite" is a misnomer. Sodium hyposulfite is really $Na_2S_2O_4$ (also sodium hydrosulfite). Nevertheless, the term "hypo" is still used in the photographic industry.

Although official as an antidote in cyanide poisoning, sodium thiosulfate is used as a reducing agent in iodide solutions (see Chapter 4). The injection is administered for antidotal use. It must be buffered near pH 8.5 and covered with nitrogen during autoclaving to prevent degradation by acid [rx (ix)] or oxidation [rx (x)].[26]

$$Na_2S_2O_3 \xrightarrow[\Delta]{H^+} H_2SO_3 + S \text{ or } HSO_3^- + S \qquad \text{(ix)}$$

$$Na_2S_2O_3 \xrightarrow{O_2} Na_2SO_4 + S \text{ or } Na_2S_4O_6 \qquad \text{(x)}$$

CATEGORY: Antidote to cyanide poisoning.

USUAL DOSE: Intravenous 1 g in a 5 to 10% solution.

USUAL DOSE RANGE: 500 mg to 2 g.

OCCURRENCE: **Sodium Thiosulfate Injection**, U.S.P. XVIII (Each ml contains the equivalent of 100 mg $Na_2S_2O_3 \cdot 5H_2O$.)

Adsorbents

Adsorbents are used to remove a toxic substance before it can be adsorbed into the body. Two are currently official, *Kaolin* N.F. and *Activated Charcoal* U.S.P. Kaolin and the nonofficial attapulgite are both clays and are used in antidiarrheal preparations. They supposedly adsorb bacterial toxins that cause the diarrhea (see Chapter 7).

Activated charcoal is recommended as a component of first aid kits and is used to treat emergency poisoning prior to emesis. Properly prepared, it will adsorb alkaloids and other amines, gases such as ammonia, nitrous oxide, carbon monoxide, carbon dioxide, oxygen, nitrogen, and hydrogen, and certain salts of heavy metals. Activated charcoal is used in filters of gas masks because of its ability to adsorb gases. When wet, however, it will not adsorb gases and therefore its use for "gas on the stomach" is irrational.

It is found in some first aid kits as a mixture of two parts activated charcoal, one part magnesium oxide and one part tannic acid, known as "universal antidote." This mixture is considered inferior to the use of activated charcoal alone. Studies indicate that the presence of magnesium oxide and tannic acid actually interferes with activated charcoal as an adsorbent.[27]

It should also be noted that "burned toast" is useless as a substitute for activated charcoal. This fact becomes obvious when one realizes that activated charcoal is manufactured by the destructive distillation of plant substances and then must meet certain performance specifications.

Activated charcoal has a broad spectrum of activity, but it is not a universal adsorbent. It is imperative that its limitations be kept in mind when administering first aid treatment to a poisoning victim. Compounds effectively bound by activated charcoal in *in vivo* experiments include acetaminophen, (+)-amphetamine sulfate, aspirin, barbital, chlordane, chloroquine, chlorpheniramine, chlorpromazine, ethchlorvynol, glutethimide, kerosine, mefenamic acid, methyl salicylate, mercuric chloride, phenobarbital, pentobarbital, propoxyphene hydrochloride, secobarbital, strychnine, and sodium salicylate. Activated charcoal has a poor antidotal action for hexachlorophene and potassium cyanide and virtually none for imipramine hydrochloride and thallium acetate.[28]

There has been some concern that the poison could be eluted from activated charcoal as the mass moves through the intestinal tract and experiences pH changes, digestive juices, bile, and food particles. However recent time course studies with aspirin, methyl salicylate, chloroquine, pentobarbital, and ethchlorvynol suggest that the charcoal-poison complex remains stable throughout its passage along the gastrointestinal tract.[28–30] However, oral administration to adult humans of an aspirin-activated charcoal mixture equilibrated such that either 50% or more than 99% of the drug was adsorbed, resulting in 87.4% and 99% of the total aspirin dose, respectively, being recovered in the urine.[31]

Activated charcoal in a charcoal-to-poison ratio of 5:1 to 10:1 is usually administered as the pure powder dispersed in water. Higher doses are necessary if the patient ate before poisoning occurred. Charcoal tablets are not as effective, probably because the compaction of particles resulting from tablet compression and the addition of other constituents required for producing tablets diminish the adsorption efficacy of activated charcoal. Nevertheless, charcoal tablets can still be useful, but powdered charcoal is preferred.[32]

Repeated administration of activated charcoal after an adequate initial dose of the antidote appears to exert no additional inhibition of gastrointestinal absorption of chemical compounds. Activated charcoal must be administered as soon as possible following ingestion of the poison. When its administration is delayed, a substantial amount of the poison may be absorbed from the gastrointestinal tract, and that portion which remains unabsorbed may have advanced along the intestine to such an extent that there will not be any contact between the poison and charcoal.[29] On the other hand, once the poison passes out of the stomach, emesis and gastric lavage are of little benefit, and activated charcoal may be useful providing it can "catch up" to the poison in the intestinal tract.

The antidotal spectrum of charcoal must be kept in mind, and it should not be relied on unless *in vivo* evidence for a given poison is available. Further, activated charcoal should not be administered in combination with other substances that may interfere with its adsorptive capacity. Activated charcoal also nullifies the emetic effect of ipecac syrup.[29]

Other adsorbents have been tried as potential antidotes. A comparative study of activated charcoal, Arizona montmorillonite (a form of bentonite), and evaporated milk showed that activated charcoal had the greatest adsorptive spectrum, although the other two adsorbents had selective potential.[33]

Official Products

Activated Charcoal, U.S.P. XVIII

Activated Charcoal U.S.P. is the residue from the destructive distillation of various organic materials, treated to increase its adsorptive power. It occurs as a fine, black, odorless, tasteless powder, free from gritty matter.

Category: General purpose antidote.

Usual Dose: 10 g.

Usual Dose Range: 5 to 50 g.

Kaolin, N.F. XIII

Kaolin is more fully described in Chapter 8.

Precipitants

There are a group of inorganic salts used as antidotes because they form insoluble precipitates with the cations of toxic heavy metals. This prevents their absorption, since the cations must be in solution in order to be systemically absorbed. After precipitating the heavy metal, the gastric contents are removed by lavage or inducing emesis. The antidotal action of cupric sulfate is more complex than forming a simple precipitate. In treating phosphorus poisoning, the cupric ion oxidizes the phosphorus, thereby inactivating the unabsorbed particles by forming a coating of metallic copper over them [rx (xi)]. Cupric sulfate is used for both topical and gastric exposure to phosphorus.

$$2P + 6\ CuSO_4 + 8H_2O \rightarrow Cu(H_2PO_4)_2 + 5Cu \downarrow + 6H_2SO_4 \qquad (xi)$$

Official Products

Cupric Sulfate, N.F. XIII ($CuSO_4 \cdot 5H_2O$; Mol. Wt. 249.68; copper sulfate, blue vitriol, blue stone)

Cupric Sulfate N.F. occurs as deep blue, triclinic crystals, or as blue crystalline granules or powder. It has a nauseous, metallic taste, and it effloresces slowly in dry air. Its solutions are acid to litmus. *Cupric Sulfate* N.F. is freely soluble in water and in glycerin. It is very soluble in boiling water and slightly soluble in alcohol.

CATEGORY: Antidote (phosphorus).

Sodium Phosphate, N.F. XIII

This compound is more fully described in Chapter 8. Although official as a cathartic it has been recommended in the treatment of iron poisoning through formation of the insoluble iron phosphate salts.

Magnesium Sulfate, U.S.P. XVIII

This compound is more fully described in Chapter 8. Although official as a cathartic, it, along with sodium sulfate, has been used in barium poisoning to form insoluble barium sulfate and in lead poisoning to form insoluble lead sulfate.

Tableting Aids

The compounding of a tablet is a complex operation. In addition to the active ingredient, the tablet contains binders, filler material (or diluent) which brings the tablet up to acceptable size, sometimes a disintegrating agent, and lubricants. The tablet must be of such compactness that it does not break during packaging and subsequent handling, but still can disintegrate when properly located in the gastrointestinal tract. Furthermore, the inactive ingredients in the tablet must not bind the active ingredient, as this could prevent its systemic absorption. Detailed discussions on tableting can be found in several reference works on pharmaceutical manufacturing.

Diluents

These agents are physiologically inert. Without them many tablets would be too small for convenience since most doses of drugs are in the mg range. Lactose (milk sugar) is a common diluent. However, the inorganic agents calcium sulfate and colloidal silicon dioxide are also used.

Official Products

Calcium Sulfate, N.F. XIII ($CaSO_4$; Mol. Wt. 136.14)

Calcium Sulfate N.F. is anhydrous or contains two molecules of water of hydration ($CaSO_4 \cdot 2H_2O$). It occurs as a fine, white to slightly yellow-white odorless powder. *Calcium Sulfate* N.F. dissolves in diluted hydrochloric acid [rx (xii)]. It is slightly soluble in water.

$$CaSO_4 + 2HCl \rightarrow Ca^{+2} + 2Cl^- + H_2SO_4 \qquad (xii)$$

The anhydrous form is marketed as Drierite® and used as a rechargeable laboratory and industrial desiccant. The dihydrate is commonly known as gypsum, alabaster, satin spar, and light spar. Both forms are included in the N.F. in recognition of the need to provide standards for a chemical extensively used as a tablet diluent. It has little toxicity and is sufficiently inert chemically to prevent undesirable reactions with other medications when incorporated into tablets as a diluent.

Although not official, the hemihydrate of calcium sulfate ($CaSO_4 \cdot \frac{1}{2}H_2O$), commonly known as *plaster of paris*, is widely used for making supportive casts by physicians and dental impressions by dentists. The chemistry of the setting of the plaster is not well understood. Plaster of paris is prepared by heating gypsum ($CaSO_4 \cdot 2H_2O$) at 125° to 150° C. But if gypsum is heated above 200°, it loses this property of setting into a hard mass and is known as deadburnt or dead-burned gypsum. Similarly, if the plaster is exposed to moist air for any length of time, it slowly re-hydrates and will not exhibit the setting-up characteristic. The setting-up reaction is some type of rehydration process such that a portion crystallizes into long thin microscopic needles which mesh through the material, imparting the observed rigidity. A number of substances may be added to slow down the setting-up process; among these are dextrin, acacia, glue, alcohol, etc. The setting-up process is accelerated by addition of gypsum, sodium chloride, alum, and potassium sulfate. During the setting-up process, the plaster expands slightly so that it fills all minute spaces. This latter property is the reason it is used extensively in dentistry for making plaster cast impressions.

Colloidal Silicon Dioxide, N.F. XIII (SiO_2; Mol. Wt. 60.08; Cab-O-Sil®)

Colloidal Silicon Dioxide N.F. is a submicroscopic fused silica prepared by the vapor-phase hydrolysis of a silicon tetrachloride ($SiCl_4$) at 1100° C. It occurs as a light, white nongritty powder of extremely fine particle size (about 15 μ). *Colloidal Silicon Dioxide* N.F. is insoluble in water and in acids (except hydrofluoric). It is dissolved by hot solutions of alkali hydroxide.

Although it is nongritty, colloidal silicon dioxide exists in chain-like formation. These chains are branched and have surface areas of 400 m^2/g to 50 m^2/g. It is this chain-like formation which gives colloidal silicon dioxide its thickening and thixotropic properties. It also enhances the flow characteristics of dry systems where caking is a problem.

This largely chemically and physiologically inert material is new to the N.F. and is admitted as a tablet diluent and suspending and thickening agent.

Lubricants

During the compression of the granulated mixture into a tablet heat is generated and the tablet can be tightly held in the mold of the tableting machine making ejection very difficult, especially with the high speed tablet press. For this reason lubricants are added to the mixture, making it possible to eject the tablet cleanly leaving no residue behind in the mold. Most of the lubricants are insoluble soaps—heavy metal salts of fatty acids.

Official Compounds

Calcium Stearate, N.F. XIII

Calcium Stearate N.F. is a compound of calcium with a mixture of solid organic acids from fats and consists chiefly of variable proportions of calcium stearate, $Ca[O_2C(CH_2)_{16}CH_3]_2$ and calcium palmitate $Ca[O_2C-(CH_2)_{14}CH_3]_2$. It is assayed in terms of calcium oxide (CaO). *Calcium Stearate* N.F. occurs as a fine, white to yellowish white, bulky powder having a slight, characteristic odor. It is unctuous, free from grittiness, and insoluble in water, alcohol, and ether. The sole purpose for official recognition of this salt is in keeping with the need to set up standards for chemicals that are taken internally, not as medications *per se* but as pharmaceutical adjuncts. In this capacity, *Calcium Stearate* N.F. is found in many tableted preparations by virtue of its use as a lubricant in the tableting process. Its virtually nontoxic nature and unctuous properties make it ideal for this purpose.

Magnesium Stearate, U.S.P. XVIII

Magnesium Stearate U.S.P. is a compound of magnesium with a mixture of solid organic acids obtained from fats, and consists chiefly of variable proportions of magnesium stearate, $Mg[O_2C(CH_2)_{16}CH_3]_2$ and magnesium palmitate, $Mg[O_2C(CH_2)_{14}CH_3]_2$. It is assayed in terms of magnesium oxide (MgO). This salt is recognized only as a pharmaceutic aid for its use as a tablet lubricant. Its relatively nontoxic nature and unctuous character make it quite suitable as a lubricant useful in the tableting procedure. *Magnesium Stearate* U.S.P. has also been used as a baby dusting powder because of its ability to cling to the skin, but there is a potential toxicity due to inhalation by the infant.

Suspending Agents

There are a number of pharmaceutical aids used as suspending agents. Some act by altering the surface character of the solvent (surfactants) and others are thickening agents. Found in the latter category are the vegetable gums, silicates, and clays.

Official Compounds

Bentonite, U.S.P. XVIII

Bentonite U.S.P. is a native, colloidal, hydrated aluminum silicate. It occurs as a very fine, odorless, pale buff or cream-colored powder, free from grit. *Bentonite* U.S.P. has a slightly earthy taste and is hygroscopic. It is insoluble in water, but swells to approximately 12 times its volume when added to water. *Bentonite* U.S.P. is insoluble in, and does not swell in, organic solvents. A water suspension (1 in 50) has a pH between 9 and 10.

The most common type is Volclay bentonite, composed of about 90%

montmorillonite [$Al_2Si_4O_{10}(OH)_2 \cdot nH_2O$] which is clay mineral of unique characteristics. The remainder is mostly feldspar ($K_2O \cdot Al_2O_3 \cdot 6SiO_2$). Analysis shows it to be an aluminosilicate containing SiO_2, Al_2O_3, Fe_2O_3, CaO, MgO and some sodium and potassium. Particle size of bentonite is about 44 μ, or what will pass through a 325-mesh sieve. One cubic inch of bentonite is estimated to contain 9,500 billion particles having a total surface area of more than an acre.

In contact with water, each flake of bentonite attempts to surround itself with a layer or shell of water. This creates a particle several times larger than the original bentonite flake. Swelling of mass results, and bentonite can be expected to absorb up to five times its weight of water and increase from 12 to 15 times its bulk. The more or less round "balls" of bentonite-water particles give the product a smooth slippery feeling. Repeated wetting and drying of bentonite does not alter its properties. Dilutions as great as 1 to 5000 do not settle out upon standing.

OCCURRENCE: **Bentonite Magma**, U.S.P. XVIII

REFERENCES

1. Hyperbaric therapy can't cure doubt. Med. World News, **11**:16, Feb. 20, 1970.
2. Hyperbaric oxygen therapy. Med. Lett., **13**:29, April, 16, 1971.
3. Hyperbaric healing of bedsores. Med. World News, **11**:20, Sept. 18, 1970.
4. Hyperbaric oxygen goes portable. Med. World News, **12**:36D, Feb. 26, 1971.
5. van den Brenk, H. A. S. Hyperbaric oxygen breathing and radiation therapy. J.A.M.A., **217**:948, 1971.
6. Hyperbaric oxygen and radiotherapy. Brit. Med. J., **2**:368, 1972, and references cited therein.
7. Herber, G., La Force, M., and Mason, R.: Impairment and recovery of pulmonary antibacterial defense mechanisms after oxygen administration. J. Clin. Invest., **49**: 47a, 1970.
8. Morgan, T. E., Finley, T. N., Huber, G. L., and Fialkow, H. Alteration in pulmonary surface active lipids during exposure to increased oxygen tension. J. Clin. Invest., **44**:1737, 1965.
9. Lung damage by oxygen. Lancet, **2**:1292, 1970.
10. Van De Water, J. M., Kazey, K. S., Miller, T., Parker, D. A., O'Connor, N. E., Sheh, J., MacArthur, J. D., Zollinger, Jr., R. M., and Moore, F. D. Response of the lung to six to 12 hours of 100 per cent oxygen inhalation in normal man. New Eng. J. Med., **283**:621, 1970.
11. Singer, M. M., Wright, F., Stanley, L. K., Roe, B. B., and Hamilton, W. K. Oxygen toxicity in man: A prospective study in patients after open-heart surgery. New Eng. J. Med., **283**:1473, 1970.
12. Barber, R. E., Lee, J., and Hamilton W. K. Oxygen toxicity in man: A prospective study in patients with irreversible brain damage. New Eng. J. Med., **283**:1478, 1970.
13. Can CO_2 smother heroin addiction? Med. World News, **13**:5, Feb. 4, 1972.
14. Death puts a sudden halt to CO_2 therapy for drug addicts. Med. World News, **13**:36, April, 7, 1972.
15. Wollman, H., and Dripps, R. D. The therapeutic bases. *In:* Goodman, L. S., and Gilman, A., eds. The Pharmacological Bases of Therapeutics, 4th ed. New York: Macmillan, p. 926, 1970.
16. Report to the Medical Research Council. Clinical trials of different concentrations of oxygen and nitrous oxide for obstetric analgesia. Brit. Med. J., **1**:709, 1970.

17. Kerr, F., Ewing, D. J., Irving, J. B., and Kirby, B. J.   Nitrous oxide in myocardial infarction. Lancet, **1**:63, 1972.
18. Breathing cardiac pain away.   Med. World News, **13**:44E, March 24, 1972.
19. Trieger, N., Loskota, W. J., Jacobs, A. W., and Newman, M. G.   Nitrous oxide—a study of physiological and psychomotor effects. J. Amer. Dent. Ass., **82**:142, 1971.
20. Council on Dental Therapeutics.   Nitrous oxide-oxygen psychosedation. J. Amer. Dent. Ass., **84**:393, 1972.
21. Moore, W. E., and Abraham, D.   A new approach to the chemistry of aromatic ammonia spirit. J. Amer. Pharm. Ass., **45**:257, 1956.
22. Poe, C. F., and Hultquist, M. E.   A chemical study of aromatic spirit of ammonia. J. Amer. Pharm. Ass., **34**:216, 1945.
23. A.M.A. Drug Evaluations, 1st ed. Chicago: American Medical Association, p. 353, 1971.
24, Wolff, I. A., and Wasserman, A. E.   Nitrates, nitrites and nitrosamines.   Science, **177**:15, 1972.
25. Hassle over cured meat and cancer.   Med. World News, **13**:4, Feb. 25, 1972.
26. Cox, M. J.   Sterilization of sodium thiosulfate solutions. Pharm J., **171**:268, 1953.
27. Picchioni, A. L., Chin, L., Verhulst, H. L., and Dieterle, B.   Activated charcoal versus "universal antidote" as an antidote for poisons. Toxic. Appl. Pharmacol., **8**:447, 1966.
28. Picchioni, A. L.   Antidotal spectrum of activated charcoal.   Presented at the annual meeting of the Academy of Pharmaceutical Sciences, April 24, 1972, Houston, Texas, Abstracts, p. 117.
29. Picchioni, A. L.   Activated charcoal, a neglected antidote. Pediat. Clin. N. Amer., **17**:535, 1970.
30. Decker, W. J., Shpall, R. A., Corby D. G., Combs, H. F., and Payne, C. E. Inhibition of aspirin absorption by activitated charcoal and apomorphine. Clin. Pharmacol. Ther., **10**:710, 1969.
31. Levy, G., and Tsuchiya, T.   Effect of activated charcoal on aspirin absorption in man, Part I. Clin. Pharmacol. Ther., **13**:317, 1972.
32. Tsuchiya, T., and Levy, G.   Drug adsorption efficacy of commercial activated charcoal tablets *in vitro* and in man. J. Pharm. Sci., **61**:624, 1972.
33. Chin, L., Picchioni, A. L., and Duplisse, B. R.   Comparative antidotal effectiveness of activated charcoal, Arizona montmorillonite, and evaporated milk. J. Pharm. Sci., **58**:1353, 1969.

Appendix A

Expressing the Concentration of Solutions

Three means of expressing the concentration of a component in a solution include: (1) the molar (M) concentration; (2) the molal (m) concentration; and (3) the mole fraction (X_2). The relationships between these three items can be demonstrated by calculating the composition of a solution of disodium hydrogen phosphate (Na_2HPO_4) containing 52.48 g/l and having a density of 1.0496 g/cc at 20° C. The molecular weight is 141.98 g/mole.

Molar Concentration

$$\frac{52.48 \text{ g/l}}{141.98 \text{ g/mole}} = 0.3696 \text{ moles/l or Molar or M}$$

Molal Concentration

$$\frac{0.3696 \text{ moles/l}}{1049.6 \text{ g/l} - 52.48 \text{ g/l}} \times 1000 = 0.300 \text{ moles/1000 g } H_2O \text{ or } m$$

Mole Fraction

The mole fraction of solute (Na_2HPO_4) is X_2.

$$X_2 = \frac{0.3696 \text{ moles/l}}{0.3696 \text{ moles/l} + \dfrac{(1049.6 - 52.48) \text{ g/l}}{18 \text{ g/mole}}} = 0.007$$

The mole fraction of solvent (water) is X_1.

$$X_1 = 1 - X_2 = 1.000 - 0.007 = 0.993$$

Appendix B[a]

Method of Calculation for Isotonic Solutions

To cite an example, consider a 1000-ml solution containing 0.15% w/v chlorobutanol.

0.15% = 0.0015
0.0015 × 1000 = 1.5 g chlorobutanol

By referring to the table it is found that the chlorobutanol factor is 0.18. Therefore 0.18 × 1.5 = 0.27 g sodium chloride. This quantity of sodium chloride (0.27 g) when dissolved in sufficient water to produce 1000 ml will then have a tonicity equal to that of the 0.15% chlorobutanol solution. It is important to note that the quantity of sodium chloride derived in these calculations will produce a solution of equal tonicity only if it is dissolved in sufficient water to produce a volume equal to that in which the medicinal agent is dissolved. Up to this point the calculations have produced the sodium chloride equivalent of the medicinal agent. It must be borne in mind that an isotonic solution of sodium chloride contains 0.9% sodium chloride or 0.9 g per 100 ml of solution. Medicinal agents present in a solution also contribute toward tonicity and, in finding the weight of sodium chloride that is equivalent to the weight of medicinal agent present, an amount has been determined that must be subtracted from the weight of sodium chloride that would have been used if no medicinal agent were added.

Previously it was found that the sodium chloride equivalent of chlorobutanol (1.5 g) was 0.27 g of sodium chloride. In 1000 ml of solution the

[a] From Roger's Inorganic Pharmaceutical Chemistry, 8th ed., by T. O. Soine and C. O. Wilson, Lea & Febiger, Philadelphia, pp. 675–679, 1967. By permission.

15
J

amount of sodium chloride required for an isotonic solution in the absence
of chlorobutanol could be determined in the following manner.

    0.9% = 0.009
    0.009 × 1000 = 9 g sodium chloride

Therefore 9 g of sodium chloride in 1000 ml of solution would be isotonic,
but chlorobutanol (1.5 g) also contributes a tonicity equal to 0.27 g of
sodium chloride. Subtracting this quantity from 9 g gives a difference
equal to the amount of sodium chloride to be added

    9.00
  −0.27
  ―――――
   8.73 g = amt. of sodium chloride to be added to produce an isotonic
solution.

If the same quantity of chlorobutanol were dissolved in a smaller
quantity, the preliminary calculations would be the same, but the determination of the amount of sodium chloride necessary to produce isotonicity
in the absence of chlorobutanol would give a different result. Thus, if a
90-ml solution containing 1.5 g of chlorobutanol is to be made isotonic the
following calculations would be necessary.

    Wt. of chlorobutanol = 1.5 g
    Chlorobutanol factor = 0.18
        1.5 × 0.18 = 0.27 g NaCl

    Vol. of solution = 90 ml
    % NaCl required = 0.9%
        0.009 × 90 = 0.81 g sodium chloride

Weight of sodium chloride required for
isotonicity in absence of medicinal agents = 0.81 g

Sodium chloride isotonicity weight equivalent
of chlorobutanol (1.5 g)   = 0.27 g

Difference (0.81 − 0.27)   = 0.54 g

Weight of sodium chloride to be added to 90 ml = 0.54 g

Where more than one medicinal agent is present, the procedure is
virtually the same. The sodium chloride weight equivalents of all agents
are calculated, added together, and subtracted from the weight of sodium
chloride required for isotonicity in the absence of medicinal agents.

In those instances where a compound to be included in a preparation
is not listed in the accompanying table, an approximation can be relied
upon that will, in most instances, suffice. As previously stated, a molar
quantity of a relatively nonionized compound such as glucose produces

only about half as many particles in solution as does a molar quantity of one that is practically 100% ionized. Therefore 18 g (0.1 molar) of glucose would be tonically equal to 2.9 g (0.05 molar) of sodium chloride.

18 g glucose ⇌ 2.9 g sodium chloride

Divide both sides by 18 to get tonicity equivalent of 1 g of glucose.

1 g glucose ⇌ 0.16 g sodium chloride

Any quantity of glucose could then be multiplied by this factor (0.16) to obtain the tonicity weight equivalent of sodium chloride.

If the compound is highly ionized, as is sodium chloride, their equal molecular (molar) quantities will produce the same tonicity effect. Using potassium nitrate as an example of a highly ionized substance it can be seen that one gram-mole (101 g) should be equivalent tonically to one gram-mole of sodium chloride (58.5 g).

101 g potassium nitrate ⇌ 58.5 g sodium chloride

Divide both sides by 101 to get sodium chloride tonicity weight equivalent of one gram of potassium nitrate.

1 g potassium nitrate ⇌ 0.58 g sodium chloride

This approximation is very close to that given in the table (0.60) which is based on experimental results.

Occasionally a prescription may be encountered wherein directions are given to make a solution isotonic with some agent other than sodium chloride. This can readily be accomplished by proceeding in exactly the same manner as before. That is, determine the weight of sodium chloride that would be required to make the solution isotonic, and divide this weight by the factor corresponding to the desired tonicity agent. In the problem previously calculated the amount of sodium chloride required for 90 ml of solution containing 1.5 g chlorobutanol was 0.54 g. If it is desired to make the solution isotonic with boric acid instead of sodium chloride the table should be consulted to find the factor for boric acid.

Weight of sodium chloride required = 0.54 g

Boric acid weight equivalent   = 0.55 g
    0.54 ÷ 0.55 = 0.98 g

Dividing this factor into the weight of sodium chloride gives a quotient of 0.98 g of boric acid that must be added to produce the same tonicity that would result if 0.54 g of sodium chloride had been used.

SODIUM CHLORIDE EQUIVALENTS FOR 1 TO 3%
ISOTONIC SOLUTIONS*

	$NaCl$ equiv.
Alcohol, dehydrated	0.70
Ammonium chloride	1.08
Antipyrine	0.17
Antistine hydrochloride (Antazoline hydrochloride)	0.18
Ascorbic acid	0.18
Atropine sulfate	0.13
Aureomycin hydrochloride	0.11
Benadryl hydrochloride (Diphenhydramine hydrochloride)	0.20
Boric acid	0.50
Butacaine sulfate (Butyn sulfate)	0.20
Calcium chloride·$2H_2O$	0.51
Calcium gluconate	0.16
Calcium lactate	0.23
Chloramphenicol (Chloromycetin)	0.10
Chlorobutanol (Chloretone)	0.24
Cocaine hydrochloride	0.16
Cupric sulfate·$5H_2O$	0.18
Dextrose·H_2O	0.16
Dibucaine hydrochloride (Nupercaine hydrochloride)	0.13
Ephedrine hydrochloride	0.30
Ephedrine sulfate	0.23
Epinephrine bitartrate	0.18
Epinephrine hydrochloride	0.29
Glycerin	0.34
Homatropine hydrobromide	0.17
Lactose	0.07
Magnesium sulfate·$7H_2O$	0.17
Menthol	0.20
Meperidine hydrochloride (Demerol hydrochloride)	0.22
Mercuric succinimide	0.14
Methacholine chloride (Mecholyl chloride)	0.32
Metycaine hydrochloride	0.20
Mild silver protein	0.18
Naphazoline hydrochloride (Privine hydrochloride)	0.23
Neomycin sulfate	0.11
Neostigmine bromide (Prostigmine bromide)	0.18
Phenacaine hydrochloride (Holocaine hydrochloride)	0.16
Phenol	0.35
Phenylephrine hydrochloride (Neo-Synephrine hydrochloride)	0.29
Physostigmine salicylate	0.16

SODIUM CHLORIDE EQUIVALENTS FOR 1 TO 3%
ISOTONIC SOLUTIONS* (*Continued*)

	$NaCl$ equiv.
Physostigmine sulfate	0.13
Pilocarpine nitrate	0.23
Potassium acid phosphate (KH_2PO_4)	0.43
Potassium chloride	0.76
Potassium iodide	0.34
Procaine hydrochloride	0.21
Quinine hydrochloride	0.14
Scopolamine hydrobromide (Hyoscine hydrobromide)	0.12
Silver nitrate	0.33
Sodium acid phosphate ($NaH_2PO_4 \cdot H_2O$)	0.40
Sodium bicarbonate	0.65
Sodium bisulfite	0.61
Sodium borate $\cdot 10H_2O$	0.42
Sodium chloride	1.00
Sodium iodide	0.39
Sodium nitrate	0.68
Sodium phosphate, anhydrous	0.53
Sodium phosphate $\cdot 2H_2O$	0.42
Sodium phosphate $\cdot 7H_2O$	0.29
Sodium phosphate $\cdot 12H_2O$	0.22
Sodium sulfite, exsiccated	0.65
Streptomycin sulfate	0.07
Strong silver protein	0.08
Sucrose	0.08
Sulfacetamide sodium	0.23
Sulfadiazine sodium	0.24
Sulfamerazine sodium	0.23
Sulfanilamide	0.22
Sulfathiazole sodium	0.22
Tetracaine hydrochloride (Pontocaine hydrochloride)	0.18
Tetracycline hydrochloride	0.14
Tripelennamine hydrochloride (Pyribenzamine hydrochloride)	0.22
Urea	0.59
Zinc chloride	0.62
Zinc phenolsulfonate	0.18
Zinc sulfate $\cdot 7H_2O$	0.15

* The above data are taken in part from E. R. Hammarlund and K. Pedersen-Bjergaard, *J. Amer. Pharm. Assoc., Pract. Ed.*, 19:38 (1958).

Appendix C[a]

Effect of pH on Selected Drugs

I. Drugs which are unfavorably affected by alkali.

    A. Drugs which are precipitated by alkali:

> Alkaloidal salts (e.g., strychnine sulfate, codeine salts).
> Salts of local anesthetics (e.g., procaine salts, "Butyn sulfate").
> Soluble salts of aluminum, calcium, copper, iron, lead, mercury, silver, zinc and certain salts of other metals (e.g., silver nitrate, copper sulfate, mercury bichloride, ferrous sulfate, etc.).

    B. Drugs which are decomposed or inactivated by alkali:

Acetanilid	Mild mercurous chloride (Calomel)
Acetylsalicylic acid (Aspirin)	Pectin
Ammonium salts	Penicillin
Apomorphine HCl	Pepsin
Ascorbic acid (Vitamin C)	Phenacetin
Barbiturates	Phenyl salicylate (Salol)
Chloral hydrate	Pyrogallol
Chlorobutanol	Resorcinol
Cocaine (when sterilized)	Resorcinol monoacetate
Creosote carbonate	Riboflavin
Epinephrine ("Adrenalin," "Suprarenin")	Santonin
	Strophanthin
Guaiacol carbonate	Tannic acid
Hydrogen peroxide	Thiamine HCl (Vitamin B_1)
Iodine	Vanillin

[a] From Roger's Inorganic Pharmaceutical Chemistry, 8th ed., by T. O. Soine and C. O. Wilson, Lea & Febiger, Philadelphia, pp. 665–668, 1967. By permission.

449

II. Drugs which may impart an alkaline reaction to preparations—and which may, therefore, precipitate, decompose, or inactivate drugs in Part I:

Aminopyrine	Rose water ointment
Ammonia water	Sodium acetate
Ammonium carbonate	Sodium benzoate
Barbiturates, sodium salts	Sodium bicarbonate
Bentonite	Sodium borate
Calcium hydroxide	Sodium carbonate
Chloramine-T	Sodium citrate
Dilantin sodium	Sodium hydroxide
Ephedrine (base)	Sodium perborate
Fluorescein sodium	Sodium peroxide
Lead acetate	Sodium phosphate
Milk of Magnesia	Sodium sulfite
Magnesium carbonate	Sulfadiazine sodium
Magnesium oxide	Sulfapyridine sodium
Methenamine	Sulfathiazole sodium
Pentobarbital sodium (Nembutal)	Sulfurated potash
Phenobarbital sodium	Theobromine and sodium acetate
Potassium acetate	Theobromine and sodium salicylate
Potassium bicarbonate	Theophylline ethylenediamine
Potassium carbonate	(Aminophylline)
Potassium citrate	Theophylline and sodium acetate
Potassium hydroxide	

III. Drugs which may be unfavorably affected by acid:

A. Drugs which may be precipitated by acid:

Barbiturates, sodium salts	Sodium salicylate
Caffeine and sodium benzoate	Sulfadiazine sodium
Dilantin sodium	Sulfapyridine sodium
Iodophthalein sodium	Sulfathiazole sodium
Mercurochrome	Theobromine and sodium acetate
Pentobarbital sodium	Theobromine and sodium salicylate
(Nembutal)	Theophylline ethylenediamine
Phenobarbital sodium	Theophylline and sodium acetate
Sodium benzoate	

B. Drugs which may be decomposed or rendered ineffective by acid:

Bicarbonates	Penicillin
Carbonates	Sodium nitrite
Hydroxides	Sodium sulfite
Iodides	Sodium thiosulfate
Methenamine	Sulfurated lime solution
Oxides	Sulfurated potash
Paraldehyde	Salts of slightly soluble organic
Pancreatin	acids, generally

IV. Drugs which may impart an acid reaction to preparations:

Acacia
Acetylsalicylic acid (Aspirin)
Alkaloidal salts
Alum
Aluminum chloride
Aluminum sulfate
Ammonium chloride
Ammonium nitrate
Arsenic triiodide
Arsphenamine
Ascorbic acid (Vitamin C)
Bismuth subnitrate
Boric acid
Calcium chloride
Calcium lactate
Citrated caffeine
Cocaine HCl

Copper sulfate
Ephedrine HCl
Sodium borate (or boric acid) and
    glycerin
Magnesium salts (soluble)
Mercury bichloride
Potassium bitartrate
Silver nitrate
Silver picrate
Sodium biphosphate
Tannic acid (other acids)
Trinitrophenol (Picric acid)
Phenols
Zinc chloride
Zinc iodide
Zinc sulfate

V. Drugs which undergo color changes with change in pH:

Drug	Acid Solution	Alkaline Solution
Bromsulphonphthalein	Nearly colorless	Deep bluish-purple
Carmine	Red-orange	Reddish-purple
Cochineal	Orange	Reddish-purple
Cudbear	Light red	Dark red to purplish-red
Fluorescein	Yellowish-orange	Green fluorescence
Gentian violet	Blue	Purple
Methylene blue	Light blue	Blue to slightly violet
Phenolphthalein	Colorless	Red
Phenolsulfonphthalein	Orange or yellow	Red to purplish-red
Syrup cherry	Red	Nearly colorless
Syrup raspberry	Red	Nearly colorless

VI. Drugs which depend upon a suitable pH for their activity:

Drug	pH or Reaction Required or Preferable
Acriflavine	Alkaline
Benzoic Acid	Acid
Gentian violet	Alkaline
p-Hydroxybenzoic acid esters	Acid
Mandelic Acid	pH 5.5 or less
Methenamine	Acid
Pepsin	Acid
Salicylic Acid	Acid
Sulfonamides	Alkaline

VII. Recommended pH for solutions of a few drugs. With some drugs it is possible to obtain greater therapeutic efficacy, greater stability or lessened irritation or discomfort to the patient by means of adjustment of the pH of solutions of these drugs to suitable values. The pH values recommended for a few drugs are provided below:

Drugs	Recommended pH
Atropine salts	6.8
Alkaloidal salts generally	6.8
Alum	6.0
Butyn sulfate	5.0
Cocaine HCl and Procaine HCl	6.0
Ephedrine and its salts	6.8
Epinephrine ("Adrenalin," "Suprarenin")	6.0
Fluorescein sodium	9.0
Homatropine salts	6.8
Metycaine	5.0
Penicillin	6.3–6.8
Phenacaine HCl (Holocaine)	5.0
Physostigmine salts	7.6 or 6.0
Pilocarpine salts	6.8
Thiamine HCl (Vitamin B_1)	3.5–4.5 (not greater than 5)
Zinc salts	6.0

Index

Page numbers followed by f indicate figures, those followed by t tables.

Natural Numbers	0	1	2	3	4	5	6	7	8	9	1	2	3	4	5	6	7	8	9
											PROPORTIONAL PARTS								
10	0000	0043	0086	0128	0170	0212	0253	0294	0334	0374	4	8	12	17	21	25	29	33	37
11	0414	0453	0492	0531	0569	0607	0645	0682	0719	0755	4	8	11	15	19	23	26	30	34
12	0792	0828	0864	0899	0934	0969	1004	1038	1072	1106	3	7	10	14	17	21	24	28	31
13	1139	1173	1206	1239	1271	1303	1335	1367	1399	1430	3	6	10	13	16	19	23	26	29
14	1461	1492	1523	1553	1584	1614	1644	1673	1703	1732	3	6	9	12	15	18	21	24	27
15	1761	1790	1818	1847	1875	1903	1931	1959	1987	2014	3	6	8	11	14	17	20	22	25
16	2041	2068	2095	2122	2148	2175	2201	2227	2253	2279	3	5	8	11	13	16	18	21	24
17	2304	2330	2355	2380	2405	2430	2455	2480	2504	2529	2	5	7	10	12	15	17	20	22
18	2553	2577	2601	2625	2648	2672	2695	2718	2742	2765	2	5	7	9	12	14	16	19	21
19	2788	2810	2833	2856	2878	2900	2923	2945	2967	2989	2	4	7	9	11	13	16	18	20
20	3010	3032	3054	3075	3096	3118	3139	3160	3181	3201	2	4	6	8	11	13	15	17	19
21	3222	3243	3263	3284	3304	3324	3345	3365	3385	3404	2	4	6	8	10	12	14	16	18
22	3424	3444	3464	3483	3502	3522	3541	3560	3579	3598	2	4	6	8	10	12	14	15	17
23	3617	3636	3655	3674	3692	3711	3729	3747	3766	3784	2	4	6	7	9	11	13	15	17
24	3802	3820	3838	3856	3874	3892	3909	3927	3945	3962	2	4	5	7	9	11	12	14	16
25	3979	3997	4014	4031	4048	4065	4082	4099	4116	4133	2	3	5	7	9	10	12	14	15
26	4150	4166	4183	4200	4216	4232	4249	4265	4281	4298	2	3	5	7	8	10	11	13	15
27	4314	4330	4346	4362	4378	4393	4409	4425	4440	4456	2	3	5	6	8	9	11	13	14
28	4472	4487	4502	4518	4533	4548	4564	4579	4594	4609	2	3	5	6	8	9	11	12	14
29	4624	4639	4654	4669	4683	4698	4713	4728	4742	4757	1	3	4	6	7	9	10	12	13
30	4771	4786	4800	4814	4829	4843	4857	4871	4886	4900	1	3	4	6	7	9	10	11	13
31	4914	4928	4942	4955	4969	4983	4997	5011	5024	5038	1	3	4	6	7	8	10	11	12
32	5051	5065	5079	5092	5105	5119	5132	5145	5159	5172	1	3	4	5	7	8	9	11	12
33	5185	5198	5211	5224	5237	5250	5263	5276	5289	5302	1	3	4	5	6	8	9	10	12
34	5315	5328	5340	5353	5366	5378	5391	5403	5416	5428	1	3	4	5	6	8	9	10	11
35	5441	5453	5465	5478	5490	5502	5514	5527	5539	5551	1	2	4	5	6	7	9	10	11
36	5563	5575	5587	5599	5611	5623	5635	5647	5658	5670	1	2	4	5	6	7	8	10	11
37	5682	5694	5705	5717	5729	5740	5752	5763	5775	5786	1	2	3	5	6	7	8	9	10
38	5798	5809	5821	5832	5843	5855	5866	5877	5888	5899	1	2	3	5	6	7	8	9	10
39	5911	5922	5933	5944	5955	5966	5977	5988	5999	6010	1	2	3	4	5	7	8	9	10
40	6021	6031	6042	6053	6064	6075	6085	6096	6107	6117	1	2	3	4	5	6	8	9	10
41	6128	6138	6149	6160	6170	6180	6191	6201	6212	6222	1	2	3	4	5	6	7	8	9
42	6232	6243	6253	6263	6274	6284	6294	6304	6314	6325	1	2	3	4	5	6	7	8	9
43	6335	6345	6355	6365	6375	6385	6395	6405	6415	6425	1	2	3	4	5	6	7	8	9
44	6435	6444	6454	6464	6474	6484	6493	6503	6513	6522	1	2	3	4	5	6	7	8	9
45	6532	6542	6551	6561	6571	6580	6590	6599	6609	6618	1	2	3	4	5	6	7	8	9
46	6628	6637	6646	6656	6665	6675	6684	6693	6702	6712	1	2	3	4	5	6	7	7	8
47	6721	6730	6739	6749	6758	6767	6776	6785	6794	6803	1	2	3	4	5	5	6	7	8
48	6812	6821	6830	6839	6848	6857	6866	6875	6884	6893	1	2	3	4	4	5	6	7	8
49	6902	6911	6920	6928	6937	6946	6955	6964	6972	6981	1	2	3	4	4	5	6	7	8
50	6990	6998	7007	7016	7024	7033	7042	7050	7059	7067	1	2	3	3	4	5	6	7	8
51	7076	7084	7093	7101	7110	7118	7126	7135	7143	7152	1	2	3	3	4	5	6	7	8
52	7160	7168	7177	7185	7193	7202	7210	7218	7226	7235	1	2	2	3	4	5	6	7	7
53	7243	7251	7259	7267	7275	7284	7292	7300	7308	7316	1	2	2	3	4	5	6	6	7
54	7324	7332	7340	7348	7356	7364	7372	7380	7388	7396	1	2	2	3	4	5	6	6	7